UTOPIA(S) – WORLDS AND FRONTIERS OF THE IMAGINARY

PROCEEDINGS OF THE 2ND INTERNATIONAL MULTIDISCIPLINARY CONGRESS PHI 2016 – UTOPIA(S) – WORLDS AND FRONTIERS OF THE IMAGINARY, 20–22 OCTOBER, 2016, LISBON, PORTUGAL

Utopia(s) – Worlds and Frontiers of the Imaginary

Chief-Editors

Maria do Rosário Monteiro
Centro de História d'Aquém e d'Além-Mar, Faculdade de Ciências Sociais e Humanas, Universidade Nova de Lisboa, Lisbon, Portugal

Mário S. Ming Kong
Centro de Investigação Arquitetura, Urbanismo e Design, Faculdade de Arquitetura, Universidade de Lisboa, Lisbon, Portugal

Co-Editor

Maria João Pereira Neto
Centro de Investigação Arquitetura, Urbanismo e Design, Faculdade de Arquitetura, Universidade de Lisboa, Lisbon, Portugal

 CRC Press
Taylor & Francis Group
Boca Raton   London   New York   Leiden

CRC Press is an imprint of the
Taylor & Francis Group, an **informa** business

A BALKEMA BOOK

CRC Press/Balkema is an imprint of the Taylor & Francis Group, an informa business

© 2017 Taylor & Francis Group, London, UK

Typeset by V Publishing Solutions Pvt Ltd., Chennai, India
Printed and bound in Great Britain by CPI Group (UK) Ltd, Croydon, CR0 4YY

Published by: CRC Press/Balkema
      P.O. Box 11320, 2301 EH Leiden, The Netherlands
      e-mail: Pub.NL@taylorandfrancis.com
      www.crcpress.com – www.taylorandfrancis.com

ISBN: 978-1-138-19748-0 (Hbk + CD-Rom)
ISBN: 978-1-315-26532-2 (eBook PDF)

Table of contents

Table of contents

Table of contents

Editorial foreword

It is our pleasure to present the second volume of Proportion, Harmony, and Identities (PHI). It is the result of the 2ND INTERNATIONAL MULTIDISCIPLINARY CONGRESS: PHI 2016 – UTOPIA(S) – WORLDS AND FRONTIERS OF THE IMAGINARY, held in Lisbon, at Museu Oriente, October 20–22, 2016.

The congress was designed as a platform for researchers, academics, and students to present, share and exchange ideas, visions of the past and the future and research results applicable to arts, science and humanities on the importance of harmony and proportion as subjects that define, differentiate, and unite identities. This year's Congress was inspired by the 500-year anniversary of the publication of Thomas More's *Utopia*.

We received papers from seventeen countries and after an intense process of scrutiny through the double blind peer review method sixty-two papers were selected for presentation and publication as chapters of this volume. In this sense this book represents the combined effort of scholars from Brazil, Greece, Italy, Poland, Portugal, The Netherlands, Romania, Serbia, Switzerland, Sweden, Ukraine, United Kingdom, United States of America and Turkey.

We decided to organize the publication in four major sections. The first, bearing the same title as the book itself, is formed by our guest speakers' articles. They serve, in our opinion, as important introductions to both the diversity and complexity involving Utopia and Utopianism in Western culture in the past five hundred years, but also to the articles of the following sections. The second section approaches Utopia from the point of view of Architecture, Urbanism and Design, or putting it in a different formulation, this section deals with the city and its evolution. The third section assembles texts dealing with Utopia and the arts, from the 16th century to the present. The fourth and final section assembles the articles one can unite under the vast umbrella of Humanities: literature, philosophy, political sciences, and culture. All four section are organized chronologically from the myths at the base of Utopia to present day perspectives. Apart from our intervention in the organization of the volume, all individual chapters are the sole responsibility of their respective authors.

We thank all members of the review committee, partner universities, and their research centres, organizing committee members, sponsors, and especially all the congress participants for making this congress and this book of proceedings possible.

Chief-Editors, PHI 2016 Congress Organizing Chairpersons
Maria do Rosário Monteiro
Professor/Researcher, CHAM, FCSH,
Universidade NOVA de Lisboa, Universidade dos Açores

Mário S. Ming Kong
Professor/Researcher, CIAUD- FAU Lisboa/CHAM, FCSH,
Universidade NOVA de Lisboa, Universidade dos Açores

Co-Editor, PHI 2016 Congress Co-Organizing Chairperson
Maria João Pereira Neto
Professor/Researcher, CIAUD- FAU Lisboa/CHAM, FCSH,
Universidade NOVA de Lisboa, Universidade dos Açores

Preface

We are pleased to welcome you to the book Proportion, Harmony and Identities—UTOPIA(S) – WORLDS AND FRONTIERS OF THE IMAGINARY, the result of the 2ND INTERNATIONAL MULTIDISCIPLINARY CONGRESS: PHI 2016 – UTOPIA(S) – WORLDS AND FRONTIERS OF THE IMAGINARY held in Lisbon, at Museum Orient, 20–22 October.

The congresses Proportion, Harmony and Identities are international annual events for the presentation, interaction and dissemination of research related to the topic of Harmony and Proportion relevant to arts, sciences and humanities and aims to foster the awareness and discussion on the importance of this subject and its benefits for the community at large.

It is my pleasure to announce the participation of expert speakers from fourteen countries in this three day event. We have received research papers from distinguished participating academics and researchers from countries hailing from three continents. Thus, this event revealed itself as being a platform for researchers in a vast variety of fields to discuss, share, and exchange experiences. The comprehensive content of the congress has attracted immense attention and the wealth of information spread out over all the papers is, in our point of view, extremely useful for professionals working in the related fields.

This book, divided in four parts, contains a collection of research papers, organized as chapters, presented at PHI 2016, Lisbon, Portugal. This publication containing the conference proceedings documents and the presentations made at PHI 2016 is the result of the creative work of their authors and of a highly selective review process.

I would like to express my sincere thanks to all those who have contributed to the success of PHI 2016.

The second INTERNATIONAL MULTIDISCIPLINARY CONGRESS: PHI 2016 – UTOPIA(S) – WORLDS AND FRONTIERS OF THE IMAGINARY would not have been possible without the help of a group of people from the Faculty of Architecture of the University of Lisbon, and the Faculty of Human and Social Sciences (FCSH) of Universidade Nova de Lisboa, Lisbon. Researchers from two research centers of these two universities, CIAUD and CHAM, selflessly and enthusiastically supported and helped me to overcome the many obstacles that arose during the preparation and organization of this event.

I want to thank all authors of submitted papers for their participation. All contributed a great deal of effort and creativity to produce this work, and I, in my quality of organizing chair, am especially happy that they chose PHI 2016 as the place to present it. Credit also goes to all collaborators, in particular, the Scientific Committee members and reviewers, who donated substantial time from their busy schedules to carefully read and conscientiously evaluate the submissions.

In addition, in the name of the Organizing Committee, I would like to take this opportunity to extend our sincere gratitude to all supporting Organizations, namely Fundação Orient, especially Dr. João Amorim and Dr. Patrícia Moita, for their support and encouragement and for making the event a success.

Special thanks go to all our speakers, authors, and delegates for making PHI 2016 a fruitful platform for sharing, learning, networking, and inspiration.

We sincerely hope you find this publication enriching and thought provoking.

Lisbon, October 2016

Mário S. Ming Kong
PHI 2016 Congress Organizing Chairperson
Professor/Researcher, CIAUD- FA-UL/CHAM- FCSH,
Universidade NOVA de Lisboa, Universidade dos Açores, Portugal

Secção de Desenho, Geometria e Computação, Departamento de Artes,
Humanidades e Ciências Sociais, Faculdade de Arquitectura,
Universidade de Lisboa, Portugal

Introduction

Maria do Rosário Monteiro

CHAM, FCSH, Universidade NOVA de Lisboa, Universidade dos Açores, Portugal

'You mean in the city we have just been founding and describing, our hypothetical city, since I don't think it exists anywhere on earth.'

'No, though there may perhaps be a pattern or model laid up in heaven somewhere, for anyone who chooses to see it—and seeing it, chooses to found a city within himself. *It makes no difference whether it exists anywhere, or ever will. It, and no other, is the only city whose politics he would engage in.*'

Plato (2000: Book 9, 592ab)[1]

A complete community constituted out of several villages, once it reaches the limit of total SELF-SUFFICIENCY, practically speaking, is a city-state. It comes to be for the sake of living, but it remains in existence for the sake of living well. That is why every city-state exists by NATURE, since the first communities do. For the city-state is their end, and nature is an end. [...] It is evident from these considerations, then, that a city-state is among the things that exist by nature, *that a human being is by nature a political animal*, and that anyone who is without a city-state, not by luck but by nature, is either a poor specimen or else superhuman.[2]

Aristotle (Book 1, 1252b–1253a 1998: 3–4)

And it will happen that after he has brought down everything which is in the world, and has sat down in eternal peace on the throne of the kingdom, then joy will be revealed and rest will appear. And then health will descend in dew, and illness will vanish, and fear and tribulation will pass away from among men, and joy will encompass the earth. And nobody will again die untimely, nor will any adversity take place suddenly. Judgement, condemnations, contentions, revenges, blood passions, zeal, hate, and all such things will go into condemnation since they will be uprooted. For these are the things that have filled this earth with evils, and because of them life of men came in yet greater confusion. And the wild beasts will come from the wood and serve men, and the asps and dragons will come out of their holes to subject themselves to a child. [...] And it will happen in those days that the reapers will not become tired, and the farmers will not wear themselves out, because the product of themselves will shoot out speedily during the time they work on them in full tranquillity. For that time is the end of that which is corruptible and the beginning of that which is incorruptible.

Baruch (1983: 645–646)

Meanwhile, tough in other respects [Raphael] is a man of the most undoubted learning as well as of the greatest knowledge of human affairs, I cannot agree with all that he said. But I readily admit that there are very many features in the Utopian commonwealth which it is easier for me to wish for in our countries than to have any hope of seeing realized.

Thomas More (1518, 1965: 245–247)

A map of the world that does not include Utopia is not worth even glancing at, for it leaves out the one country at which Humanity is always landing. And when Humanity lands there, it looks out, and, seeing a better country, sets sail.

Progress is the realisation of Utopias.

Oscar Wilde (1st ed. 1891, 1912: 43)

Even the Catholic Church of the Middle Ages was tolerant by modern standards. Part of the reason for this was that in the past no government had the power to keep its citizens under constant surveillance. The invention of print, however, made it easier to manipulate public opinion, and the film and the radio carried the process further. With the development of television, and the technical advance which made it possible to receive and transmit simultaneously on the same instrument, private life came to an end.

George Orwell (1989: 214)

1   THE ROOTS OF UTOPIA

The idea of Utopia springs from a natural desire of transformation, of evolution pertaining to humankind and, therefore, one can find expressions of "utopian" desire in every civilisation.[3] Having to do explicitly with human condition, Utopia accompanies closely cultural evolution, almost as a symbiotic organism. Maintaining its roots deeply attached to ancient myths, utopian expression followed, and sometimes preceded cultural transformation.

In the Western culture, this mythic desire for a better life gave origin to a long tradition of narratives, philosophical conceptions, political ideologies, artistic works, architectural plans, etc.

2   MYTH BECOMES EARTHBOUND

This year we celebrate 500 years of *Utopia*, or expressing it in another form; we celebrate the author who brought Utopia from the dominions of myth, religion, and metaphysics to the concrete, actual world while keeping it in the realm of fiction.

Utopia as a mode gained form in a genre. This transformation happened, probably, in the most fortunate way. Thomas More condensed in a short narrative almost all possible ways of assuming definitely the permanent, sometimes desperate, need for transformation felt by the "political animal". At the same time, he somehow entrusted on each human the responsibility to find ways to live a better life, to assume the responsibility of seeking new paradigms, to improve existing ones.

More's ambiguous narrative has, in latent, all the forms literary utopia has been developing through the last 500 years. It also challenges architects, scientists, philosophers, politicians to explore beyond the known reality, using imagination, to discover other "better" worlds.

More's challenge is not for perfection. Being a Christian humanist, he knew better. His challenge points to the route of improvement, to a process of constant revaluation of different options, of imagined alternatives, all faulty (because they are product of human endeavour) nevertheless better solutions.

The island of Utopia is not a perfect world, but always bear in mind that it is not intended to. For us, living in the 21st century, it would definitely be a dystopia, but then, as says Lyman Sargent, if we were poor peasants in 1516 we would find life in Utopia to be extremely appealing. Or maybe not, because as peasants:

> [we] would not have heard of it at all, and even in the unlikely event that [we] could read, it was only available in Latin. [...] Secondly, if [we] did come to hear of it, [we] would have filtered it through [our] understanding of the world, an understanding in which reason played a very small part. (SARGENT 2004: 2–3)

3   UTOPIA, REASON AND IMAGINATION

More ends his narrative expressing both doubts and expectations. And he even leaves his narrative opened to further development. Many did try to develop it thought few had the ability to understand the meanings of either the doubts or the openness.

Some corrected what they felt was missing in the Island of Utopia. Just to mention two examples: Campanella and Bacon understood that reason was calling for science to emancipate itself from philosophy and assume a central role in social organisation. Centuries later, under the dominium of positivism, Oscar Wilde made a call for the importance of a central element that, since Enlightenment, was being progressively discarded: desire.

To move forward, to improve, to find Utopia, one needs reason and imagination. These two drives, so often understood as antagonists, must become complementary, balanced. If one prevails, hell gets loose. So warns Orwell, turning his dystopian *1984* into a serious warning. Dystopia results from either reason without imagination, or from imagination without the restrains of reason. But also from the lack of ethic values, of respect for the Other, or simply due to pure stupidity.

4   500 YEARS LATER

Why do we, Westerners, still celebrate Utopia?

We live better than the vast majority of world population and yet we are not pleased. Too many things just do not seem right, not to say some are preposterously wrong.

We have better technology, but its use is also responsible for, at least, a critical climatic crisis.

We have better and faster forms of communication, but this is not synonym of better communal life. Isolation, solitude, loneliness prevail in the world of instantaneous communication.

Democracy triumphed in most western cultures, but politics and some (too many) politicians seem to have stopped after reading Machiavelli's *Prince*, retaining only the ideas of how to remain in power regardless of any ethical or social values, aiming at economic individual wealth.

Knowledge has never before been so easily accessible to so many people, but bombastic headlines, scandalous *faits divers*, corruption, and lack of critical reasoning seem to threaten of extinction our millenarian culture.

Thomas More, returning to his beloved London today probably would yield due to the absence of humanist values in Western society. The vast majority of the people are no longer poor, illiterate peasants, but neither do they work six hours a day and enjoy or search for knowledge during the rest of the time.

To tell the truth, More's beloved London was hell too in the 16th century: unethical political elites, unjust judiciary system, and poverty and ignorance menacing constantly the vast majority of the people.

5 CONCLUSION

What have we learnt in 500 years?

That Utopia is definitely the country we strive for, not a perfect land, not a perfect society. However, one can surely imagine and build a better one. A society that might lay its foundations on the wealthy western culture, with the benefits of science and technology, with the respect of difference, seen as an enrichment. Human diversity is like natural diversity: it must be protected, cherished, developed, and enhanced. In addition, hubris, that ever-menacing feeling must be restrained. Utopia has to be reinvented every day, an unfinished and unfinishable task laid on the shoulders of every human being. A shared responsibility but also an individual one.

Through the next almost five hundred pages (virtually close to one for each year since *Utopia* was published) researchers on the fields of Architecture and Urbanism, Arts and Humanities present the result of their studies on the different areas of expertise under the umbrella of Utopia. Past, present, and future come together in one book. They do not offer their readers any golden key. Many will leave questions unanswered, as they should.

When we stop asking: "What if...?" we stop evolving, and then human society is as good as dead.

NOTES

[1] Emphasis added.
[2] Emphasis added.
[3] See, Gregory Clayes (2011: Chapter 1), Fátima Vieira (2010), Raymond Trousson (2000), and Frank Manuel (1973; 1979). Fortunately the list does not, end here. Therefore, consider these references as a challenge for further investigation.

BIBLIOGRAPHY

Aristotle. *Politics.* Translated, with introduction and notes, by C.D.C. Reeve. Indianapolis: Hackett Pub, 1998. lxxix, 288 p. ISBN: 978-0-87220-388-4.

Baruch. "2 Syriac Apocalypse of Baruch." In *The Old Testament Pseudepigrapha.* Ed. Charlesworth, James H. Vol. 1. Garden City: Doubleday, 1983. p. 615–652. ISBN: 0-385-09630-5.

Claeys, Gregory. *Searching for Utopia: The History of An Idea.* London: Thames & Hudson, 2011. 224 p. ISBN: 978-0-500-25174-4.

Fortunati, Vita, and Trousson, Raymond, eds. *Dictionary of Literary Utopias* Paris: Honoré Champion, 2000. 732 p. ISBN: 2-7453-0218-3.

Manuel, Frank E., ed. *Utopias and Utopian Thought.* Londres: Souvenir Press, 1973. pp. xxvi, 321. 322 cm. 0285647229.

Manuel, Frank E., and Manuel, Fritzie P. *Utopian Thought in the Western World.* Cambridge, Mass.: Belknap Press, 1979. vi, 896 p. : ill. ; 826 cm. ISBN: 978-0674931862.

More, Thomas. *Utopia.* The Complete Works of St. Thomas More. Eds. Surtz, Edward and J.H. Hexter. Vol. 4. New Haven and London: Yale University Press, 1965. cxciv, 629 p. ISBN: 9780300009828.

Orwell, George. *Nineteen Eighty-four.* Londres: Penguin, 1989. 326 p. ISBN: 9780140126716.

Plato. *The Republic.* Edited by G.R.F. Ferrari; translated by Tom Griffith. Cambridge: Cambridge University Press, 2000. 382 p. ISBN: 9780521484435.

Sargent, Lyman Tower. "The Necessity of Utopian Thinking: A Cross-National Perspective." In *Thinking Utopia: Steps into Other Worlds.* Eds. Rüsen, Jörn, Michael Fehr and Thomas Rieger. Nova Iorque: Berghahn Books, 2004. 1–16. ISBN: 978-1-57181-440-X.

Vieira, Fátima. "The concept of Utopia." In *The Cambridge Companion to Utopian Literature.* Ed. Claeys, Gregory. Cambridge: Cambridge University Press, 2010. 3–27. ISBN: 9781139798839.

Wilde, Oscar. *The Soul of man under socialism.* Londres: Arthur L. Humphreys, 1912. 99 p.

Part I *Utopia(s) – Worlds and Frontiers of the Imaginary*

Utopia: A world to be, through denegation and affirmation of collective consciousness

Aires Augusto do Nascimento
Academia das Ciências de Lisboa, Lisboa, Portugal

ABSTRACT: Utopia has a long history, from Antiquity to the present. Therefore, it has been subjected to several transformations, incorporating different cultural elements, according to the *Weltanschauung* of each moment. This paper traces some of the most influential elements and their transformations with particular emphasis on the cultural and literary foundations from Antiquity to the present. It should be read as an introduction to Utopia as a cultural and literary phenomenon and to its influence on occidental social and cultural evolution.

Keywords: Utopia, Thomas More, desire, quest

1 UTOPIA: THE NON-PLACE AS HUMAN RESOURCE

Etymologically, utopia means "non-place". If so, why does it establish itself as target of a quest? Basically, because the "place" people inhabit, due to its corporeity, is incapable of containing them, and the consciousness of that fact compels them to acknowledge their responsibility regarding all the places inhabited. Being an extension of whoever inhabits it, the place we were randomly endowed with does not possess enough space to shelter us in our whole dimension. Being space it is but a mere reference spot that gives us a location and tensionally launches us at the others (Belugue 1999).

Had we by any chance sought refuge on an Island, the meditation we would carry on out there would not fail to put forward, with more intensity than any other pledge, the fragile that seclusion would turn out to be and how crucial it would appear to depend on the sense of human involvement as to what might have been left aside. While revealing man to himself, the place means to him nothing but the starting point of a building process that is never finished.

Although inhabited, the place is never definitively concluded: before being inhabited, it is an object of desire, but once inhabited, it becomes the object of never-ending transformation. Due to depredatory invasion, the new history is driven to consider the era of the Anthropocene must fight against the destructive habits in order to find a responsible *ecotropia* for the planet, able to embrace an environment friendly culture[1].

Taken as an anthropological symbol, place is man, by "culture". The Latin verb *colo* clarifies this process: originally, it meant, "move within a given space"; then it acquired the meaning of "inhabit"; subsequently it acquired the sense of "cultivate"; finally, it favoured the attitude of "worship". Each pattern discloses an outward movement towards something new; the last step points to transcendence. The verb *colo* originated the term *culture*, being quite significant the suffix that translates both a process and an outcome (or just an outcome because it is born from a process—never finished), in itself, culture is a project rather than an outcome.

Utopia, being formally a negative denomination, adduces the expectation that provides human being with the necessary drive to give the inhabited place the human dimension it originally lacks: in that place we do not simply are for we transform it by inhabiting it.

In fact, place limits us, but if it contains us, it does not satisfy us; rather it throws us forward and becomes launching platform for new ways of being and living. In a now emblematic statement (more than a century has gone by since), Oscar Wilde wrote that a map of the world that does not include Utopia is not worthy of attention. However, even if it did, people will not focus on it, because once arrived at Utopia, as soon as one looks out of the window and sights another place, sails up and departs immediately in search of it[2].

However, it would be paradoxical that once established the need to hail the "non-place" one would simply acknowledge it, without granting it content or integration. However, futile and ridiculous would it also be if, while paying attention to the "non-place", we were to run deceitfully after a mirage.

In other words, by pursuing the difference (a transfer from what we inhabit) we are committed to turning the place we are into something it is not yet, but which, filled with hope, we quest as pilgrims.

Somewhere, in a place it took me a while to recall, I had left Oscar Wilde's little book and I felt tempted to borrow his formulation; such a contingency immediately made me feel the importance of a dwelling-place, which makes a place the right spot to dwell in. However, I also realised that such a place is not worthy in itself. A construed relationship is required to turn it into a place of return and to turn it into a dwelling worthy of lodging whatever is in it and whatever we take in. As a meeting place, it must be open to things we have not taken into yet. And as a place of return it will also become an abode of identification—so steady that I expect the postman will ring the door-bell to deliver the mail addressed to me[3] without fear of being wrong about the addressee or being informed that, after his previous passing by, I have left in quest of another destination[4].

Defined by marks, the place that is mine will have to be inhabited without restriction of space and interaction, so that it will not become a prison; extended in size, it can only be filled by the desire to make it ever larger in order to accommodate every human being and cover the arc of the universe. Being object of desire, it is not object of conquest; allowing being visited, it is not opened to the wanderings of a hasty tourist because it only reveals itself to the one who, during the quest, takes upon oneself the pilgrim's condition.

Questioned by the "non-place", I feel myself entangled, without further justification, in the construction of the place that only I can and must invent as an extension of myself. Knowing, moreover, as it happens with any architect worthy of being known as such, that only through invention is it possible to create human spaces and turn them into places of identity and interaction. In the same way, it happens with any. I am thus challenged to enhance the "non-place", which is not a place yet, but will become one. Moreover, I will not let myself be fooled by some mirage empowered only to instil courage in whoever wanders lost amidst the sands of the desert. A mirage that cannot provide more than a mere reflection of a remote reality, and will not even offer the possibility of a friendly shade. In fact, if the abode is a dimension I must build, a transitory shelter will only be meaningful to measure the capacity of personal investment.

2 BETWEEN DESIRE AND UTOPIA

In a restrained explanation, Oscar Wilde ensures that Utopia is the only country one always yearns to reach, because there resides the one place worth of such desire. However, being a desire does not imply it immediately stands for fullness—it can even derive into frustration. . . If it is a desire it is not a place to live in, rather an ambition that impels us to reach yonder, where perfection waves and happiness awaits.

With a premonitory attention, Wilde warns that (contradiction of contradictions), not even by being imperative to demand Utopia (in order to acquire human dimension), will there be someone who, inhabited by the dynamics of Desire, will tolerate to live there for a long time (let alone forever). Because, he explains, when someone casts anchor there and tries to settle in, there arises the urgent feeling to peep at the horizon, and perceiving a country that looks better, immediately sails are set for another destination.

After all, utopia always demands alterity, in a permanent quest for something that is not what is actual. It is more than a simple alternation- to change from an inhabited place to one recognised as outside. It may eventually be a strife, because it feeds on the opposition between the acquired and the desired[5]; but it only performs its function when it integrates what is further in what is nearer. If I may use a neologism, the otherness (a word we find in Fernando Pessoa) should also mean *ultridade* ["ultrotherness"], that is, as exceeding in what we demand; it is not just a state, rather a dynamic: therefore, the poet also allows himself the use of the verb "outrar" ["to other"].

Utopia will be shadow; it remains in the half-light waiting for revelation: that is why Amaurotum, the concealed city, is the capital of More's Utopia[6]; this is also the reason why in such emblematic texts as *Navegação de S. Brandão* [Voyage of Saint Brendan] (Nascimento 1998), the Promised Land lets itself be seen through the clouds[7].

However, more than shadow of others, Utopia is our own shadow; it precedes us and casts light effects on who stares at it; or it acts like the sun that, for fear of dazzling, stays in the background. When we crave for better, Utopia is the one that suggests the way we may pursue relatively confident through the distance that opens before us. At some moment, we will need that shadow to be sure that we exist. All the time we will need that a divine presence "may lengthen through the shadow what I, limited and weak, am prone to reduce to my exiguity", as José Régio taught to say as a "morning prayer"[8] (2001). However (to be platonic at least once), will we, our back turned, be able to distract ourselves from what is superfluous and focus our mind on the true reality which only from within is able to reach us?[9]

In a no lesser emblematic work than that of Wilde, named *Le città invisibili*, by Italo Calvino (1997; 2008; 2012), there is a simulated dialogue between Genghis Khan and Marco Polo (who brings news about the world to the city where Khan is confined). Here desire and memory intersect in such a way that in its centre we come across human condition. A human condition carrier of accumulated information and willing to engage in the construction of what is still unformed and will never be finished. In this way, he avoids repetitive schemes and innovates in his purposes based on what he has to offer[10].

Calvino does not innovate, because Thomas More, the inventor of the name of Utopia, had previously pointed out that in that "Land-of-Nowhere" the cities (all alike) show no more than a possibility and a glimpse of what we are permanently seeking for, by contrast or by desire of what we keep but is never enough to fulfil us.

The theme Marco Polo introduces in his conversation with the Tartars' emperor is subtle. So much, that the latter cannot restrain himself from asking Marco if, by any chance, he walks head turned backwards and if, this way (urgent and vital question) he travels to revisit his past or to meet his future. Marco's response comes promptly:

> Somewhere there is a mirror in negative. The traveller recognises the little part that is his, while discovering how much he did not have, neither will. (Calvino 2012: 27)

Utopia is that mirror of a Dystopia, land of quest, where we incessantly move.

The "Land of Nowhere" opposes and is equivalent to the "Land of Somewhere" because Desire expands as it fills itself, but never quenches because the plenitude is the outcome of a process that has to mature in a constant dynamism towards a remote point we will never be able to determine.

Utopia is Land of Desire. Daniel praised in the Bible precisely because he is a man of desires (Dan. 9, 23). Who expresses this praise is Angel Gabriel, the one who announced the Incarnation of the Word in Mary. She will receive Him, or rather conceive Him, allowing a new life to the One who has no place He can fit in, who is the supreme plenitude and can satisfy completely all desires.

To the Land of Wishes one does carry all the lands ever discovered and treasured. However, Man neither shuts himself in the Land of Wishes, nor does he make a stand there if it will not help him achieve a new dimension and so it gets indefinitely closer to the Infinite. As Saint-Exupéry

recalled in *Lettre à un otage* (2004; Saint-Exupéry 2015):

> It is always in the caves of oppression that the novelty of new times is prepared. [...] that is why, dear friend, I feel the need of your friendship: I thirst for a companion who, outside the disputes of reason, respects me as the pilgrim I have been; I need to enjoy, sometimes by anticipation, the promised warmness and rest from my own self, a little beyond myself, in that meeting place which will be ours"[11]

3 THE HORIZON OF HOPE

The ability to dare reach the "non-place" allows one to open himself to Hope. It is not even worthwhile to let unrestrained desires growing within. Once again, Italo Calvino forestalls:

> This melancholic emperor realised his extreme power has little value because the world is heading for ruin. Moreover, a visionary traveller tells him about impossible cities—for example, a microscopic city that expands, extends, and ends up being formed by numerous concentric cities in expansion, a spider's web suspended over an abyss, or by a bi-dimensional city like Moriana. (2012: VIII)[12].

In More's *Utopia*, Hythlodaeus, the new land sailor, is the one who tears horizons. He just needs his listeners to accept being confronted with the reality that surrounds them in order to avoid being buried in it and postulate another one. We do know, because we live in them, about tentacular cities branching into satellites, without known mentor—whether in London, in Paris or in Lisbon. . .

The identification between place and resident leads us to perceive in that web the image of the human being entangling himself in what one builds until ending suspended on the abyss of desires, waiting for a moment of redemption. As for Moriana, it carries the epitome of contradictions in the lucidness it promotes. It is a transparent city—tells us Marco: the doors of Moriana are made of alabaster and the sun freely shines through them; it has coral columns crowned by serpentine encrusted pediments; all the houses are made of glass, like aquariums, where the shadows of dancers swim, covered with silver spangles, under jellyfish shaped chandeliers.

Is it a dazzling city or does it create aimless illusions? There is always the danger of the image being taken for what it represents, or of the messenger by the recipient. It is up to the Angel of Revelation to avoid humans from prostrating themselves before the light that points the way. In Apoc. 22, 8–9, John kneels before the Angel who must refuse the gesture and point to the Eternal. The desired City is not always univocal in its suggestions; if it illuminates, it merely retrofits our steps, by divergence in what it allows us to perceive.

In the image created by Italo Calvino, Moriana, like any other city, has a reverse: heaps of rusty plates, burlaps scattered on the floor, wooden planks bristling with nails, black pipes soot clogged, piles of cans, walls covered in graffiti poorly cleared, chairs with torn seats, and ropes with knots that would only be useful for someone to hang himself from a rotten beam.

In Thomas More's representation, it is the kingdom of East Anglia furrowed by everyday petty and degrading interests. Both city and kingdom have no thickness, just front and back side, like a sheet of paper (Calvino 2008: 107). Only critical awareness can lend some life to its design, which remains on each side, and can give some human dimension to the place inhabited, by showing humans their ephemeris. What we see from the outside creates a form: but it does not reveal humanity.

4   THE PLACE TO BE BUILT—FROM INSIDE, IN THE LIGHT

Transparency was, in due time and for political reasons, a category brought under discussion: *glasnost*[13]; not even that succeed in changing the course of events for the better. Yet, Dostoyevsky had already attributed to one of his characters "the shine of crystal"[14]. Is this a sign of openness and sharing? A search for clarity, there is no doubt.

Synonym of purity, it is a human exercise, though not a spontaneous one. Pointing to a hint of the new being, it cannot be but the outcome of a purification achieved by oneself from inside. It is not enough to expose oneself to the sun; clarity has to pass through the pleats created inside. Thanks to clarity, it is possible to give radiance to people, agent of the circumstances (which Moriana brings under judgment).

Being, first, matter for insight, it qualifies the one who faces an ever-unfinished reality. In this case, the contrast brings the observer before the dreamlike vision of something he cannot set specific dimensions to, but through what one imagines and intuits, releases possibilities of altering the things one touches and feeling their real weight... To the indefinable nature of the city, there corresponds naturally the subtlety of wit, which overlays the sense of things, trying and acquiring them from the inside.

Astonished, Kublai Khan asks Marco how can he attest to what he describes since he did not move from the garden where he was welcomed. The Venetian replies:

> Everything I see and do assumes meaning in a mental space where the same calm reigns as here, the same penumbra, the same silence streaked by the rustling of leaves. [...] Perhaps this garden exists only in the shadow of our lowered eyelids, and we have never stopped: you, from raising dust on the fields of battle; and I, from bargaining for sacks of pepper in distant bazaars. But each time we half-close our eyes, in the midst of the din and the throng, we are allowed to withdraw here, dressed in silk kimonos, to ponder what we are seeing and living, to draw conclusions, to contemplate from the distance.(Calvino 2012: 103–104)

Therefore, the dimension of the gap between Kublai Khan's gardens and the filthy neighbourhoods of cities scattered everywhere is no larger than the blink of an eye. But, Marco warns, "... we cannot know which is inside and which outside" (2012: 104).

This inner dimension gives light and plenitude to the places we go through during the journey and it alone can transform them into gardens of identity and engagement.

In the utopias that have been created, there should be no place for shadows, at least theoretically, of either nature or the inhabitants. In The Glass Man[15], everything is a reflection of his intelligence, everything obeys the rules of square and compasses he wields like a demiurge[16]. Everything should spring from a program where nothing was left to chance...

Then, what would be left for humanity to carry out in the process of experimenting oneself? In Utopias, everything is devised as if ensuing from a demiurge who orders everything. The truth is that these works only relate to real beings in order to confront them with patterns of happiness. However, not even so is the project they establish less necessary and indelibly solidary with humanity, because in them converges the desire to overcome all human.

5   UNFINISHED AND CURSED PLACES

There are places one must restore in their dimension and density to grant them dignity.

In the Bible, there is an unfinished place, and quite understandably, because God condemned it for having had its origin in excessive human hubris: the Tower of Babel. Its construction is not brought to its fullness because, according to *Genesis* 11, the rising tower became suspicious in the eyes of the Immortals and that was reason enough for them to come down and put an end to a boldness only comparable to that of Giants climbing the skies to challenge divinity. Suddenly, humanity was divided by languages, and in the subsequent confusion, ended up dispersed in nations.

Must human constructions remain unfinished? The usual reading of this etiological myth (regarding the diversity of languages) ends in condemnation and punishment. Are we, therefore, facing the myth of a negative place, denouncer of accumulated illusions?

The exegetical scholarship today offers another reading, recovering readings from the Alexandrian Fathers, particularly Origen. According to a literary analysis that takes into account the narrative sequence and does not isolate the episode, it is a negative project by the very absurdity of leading into slavery passive people subject to Nimrod, an arrogant and violent warrior, who seeks to impose a totalitarian regime where there would be only one language and one people. This is conflicting to the diversifying project of primitive creation. Therefore, the God of Creation counteracts the Tower of Babel[17].

Opting for the mythological deconstruction, George Steiner comes to similar conclusions: in a restrained analysis (not explicitly put forward), he points out that the biblical account of Babel hides a previous one (of Babylonian origin) and does not mean a damnation. Instead it adduces the gift of languages as a divine gift and a blessing, Pentecost being its most obvious expression, overcoming differences by means of a mutual understanding among all those who open themselves to the insertion of the Holy Spirit (Steiner 1992; 1998: 108-ff.)[18].

Pertinently, the Cambridge Scholar recovers a forgotten lesson. Namely that Babel had a distinct dimension from the one the formulation of the tale pointed to and as Origen, one of the greatest geniuses of early Christianity, had already insinuated (Origen 1980)[19]. For the Fathers of the Church, Babel prefigures Pentecost as an antitype (Acts 2: 1–11.). Assuming that humankind tends to dominate each other (a message from the Biblical Wisdom), accepting also that excess brings along nothing but punishment (a message they received from the Greek culture), they contrastively celebrated the fullness of Pentecost, in which the multiplicity of languages congregates in a unit where everyone hears the others in one own language. Celebrating diversity, the Fathers of the Church achieve the gathering of all in celebration of the reencounter of God with humanity.

It is a fact: the Adamic language was tautological, stresses Steiner; Babel is the place where human language acquires dimension and so comes closer to the divine plenitude. This, being untranslatable

into human language, postulates the diversity of languages in order to make itself known. It is the plenitude humanity must wait for, but he has also to become involved, for it would be useless to consider it only possible as a gift from Heaven without human commitment in receiving and assuming it through an intense personal participation. The new Tower of Babel is Jerusalem: transfigured, it welcomes the diversity of people; without annulling, it sends back again to the Diaspora so that the envoy will transform it and give it a meaning within a composite unit where differences are complementary.

It is not easy to recover the sense of a place, once its function is misconstrued. The biblical Paradise truly was a finished place. Humanity turned away from that primitive and original place, but ever seeks to return—sometimes angry, other times hopeful.

By chance, I laid hands on a *Sermo* (exposition) which in the past had been attributed to St. Bernard and the Benedictines still make available, although they claim that it belongs to another author[20]. It is an exposition on the well-known theme of *Orto do Esposo*[21] that is part of Portuguese medieval tradition: "És um jardim concluso, ó minha esposa, minha irmã, minha amada" [You are a concluded garden, O my wife, my sister, my beloved], (Cant. 4, 12). The commentator remarks (I translate):

> Adam kept poorly the garden in which he had been placed; he did not prevent the creeping snake from treacherously penetrating there. That garden, covered with trees, was pleasant: but I give preference to another one I see implanted in his soul. The coming of the snake into the first garden would not have been meaningful had it not penetrated this second one too. It was allowed in, and immediately did Adam feel its poison. Whoever agitates the hedge, will be bitten by the serpent (Eccl. 10: 8). The hedge is a good protection; it separates farmland from what is not, and protects the first. Put that hedge round your garden and watch so that it will not be withdrawn and no one will destroy it, lest you may be handed over to enemies who will tear you, and step over you.(Bernardo 2003).

The *Sermo* further warns:

> It is not easy for Adam to return to Paradise, which he had left with such an inconsiderate lightness.

It is not easy to get back, but it is possible. This thought underlies the comment and is the fundamental message of the Bible. Despite the exclusion, man is not hopelessly excluded from Paradise; the return is promised and the entire human wandering described in the Bible represents the process towards a new Paradise—which will be eschatological, but not necessarily in the end of time. It will take place in a New Town (the heavenly Jerusalem that reveals the immolated Lamb at its centre).Above all, it will be the place of a Gathering of love (because the final, decisive words of the Apocalypse/Revelation are, *Maranatha* (Come), exchanged between the Bridegroom and the Bride).

6 TRAVEL OR PILGRIMAGE?

Placed in the distance, Utopia is the remote, rebuilt, specular, and rationalised face of Paradise humanity aspires to go back to or aims at reaching. For that it has to reverse its aimless drift and become a *viator*, in the ancient sense of the word (builder of paths who goes ahead, even not knowing immediately the way nor the term of his effort) (Nascimento 2004: 173–188). Thus transformed into *peregrinus* (coming from outside and away from his land), he knows his only expectation concerning the place is that it has a distant end and so all lodgings are no more than temporary places to recover one's strengths. A lodging is, in fact, a fortunate place, because there the dream is fed which comforts and restores the image of human life[22].

In its representations, Utopia appears out of the immediate reach: separated by sea, the object of hope and expectation continually promised and always poorly spotted. The foresights of the *Apocalypse* (21.1) give it a specifically eschatological character and cannot but assign it a divine origin. The Fortunate Islands of Antiquity also belong to the world inhabited by the Immortals and destined to the heroes the gods want to admit to their conviviality. Transferred to a human geography they will remain on the horizon as the indefinite fate of the blessed ones. In *Navegações de S. Brandão* [*Voyage of Saint Brendan the Abbot*],which has a prestigious history among us (Nascimento 2006), that island of happiness shows itself at a distance from the travellers who repeatedly follow the same courses and progress steadily, but out of reach, because travellers are forbidden to enter it. In a short narrative from the 10th century, an enigmatic character named Trezenzónio is blessed with a short stay in this Promised Land, but at the end of his journey, he cannot stay there forever. The Lost Island the Portuguese king promises to all those who will find it remains undiscovered, although it figures in the maps of the Discoveries.

In the figuration of Thomas Morus, Rafael Hythlodaeus is more fortunate because not only is he admitted to admire the organisation that reigns in that Elsewhere, where he arrives, but he also seems to want to make it a definitive place, when

he disappears in the horizon of the text (remaining only in our memory)[23].

Returning just for a while to a world that is not even his homeland, and in the context of an ambiguous and complex world (Flanders, where peace treaties are negotiated and exotic products smuggled); Hythlodaeus described that remote Land/ Island. It is clear that utopian value judgments are different from those chosen in the ancient world. No pacts are necessary, because they are as easy to violate, as it is to tear the documents ratifying them. There is no appetite for rare metals, because only vanity could give them value. There is a rotational and delegated participation in the administrative agencies. Communitarianism solves everything and everyone renounces ambition.

The new land is *Utopia*, a land of Nowhere, but in the fiction of contrast, it is the projection of Desire and in the passion of the quest, it sounds like *Eutopia*. It did not initially occur to his inventor (a true *euretês* by name and content) to choose this name as main title of the work he was about to make public. Others (1519 edition, away from the eyes of the author) brought the title to the main entry. More chose to keep at all times in the foreground a formulation that would take into account the governmental function (The best form of government)[24]. This works as a premonition, after all, that human space is dignified by what we dare to invest in it, concerning the strategic management of the existing resources and the pursuing of goals that go beyond mere conservation.

It is worthwhile to warn about the genesis of Thomas More's fiction, at least as far as we can identify it. Initially, More intended to formulate a theory about an ideal republic (most probably after reading Plato). Soon reality set the narrative framework and gave him the value of experience as shaper of human aspirations. He was judicious enough to understand that the historical moment offered him conditions to measure human limitations and possibilities. The testimonies about a New World that came to him through the narratives of the Portuguese, gave him the hope that it would be possible to find a world different from the one he lived in and where he felt the effects of injustice, of arbitrariness, of opportunism, of flattery, of unlimited ambition. On the other hand, the lesson from the past (by means of a new and intensive reading of texts of the tradition) allowed him the audacity to suggest the amendment of the present through a new form to exercise power, in which new agents would not deprive their subjects of the hope.

Utopia first readers, More's closest friends, quickly understood the message. In the Morian structure, Utopia is a place of Nowhere; it is also, in a positive way of thinking, not an "anti-place", but a "fore-place": the one that is in front and functions as a mirror. Its image is so real that, if denied, the opposite would be lost, but once affirmed, it enhances the value of its own meaning[25].

Due to its strangeness (almost absurdity), in More novelty prevails in the distance: Utopia, as the humanists did not fail to point out, rather than "no-place" was a "strange land"[26] and thereby provocative, as much as the old *mirabilia*. Enjoying it will be the result of the audacious attitude of crossing the ocean into the unknown, a journey of discovery. The description will be the extension of the finding.

By admitting that description, More (through the voice of Hythlodaeus) is faced with the difficulty of naming, which, following the biblical style should reveal what it points to. Thus, he reinvents a double-faced language with a multivalence of meanings, which will enable the association of two worlds, the old and the new, the real and the desired, the measurable and the dreamed one[27]. The exercise is about denial and overcoming; here the philological lesson is mandatory and necessary. Using mainly the prefixion a- the outcome is an ambivalent denomination, which on the one hand denies by contrast but on the other opens glades for a human plenitude. In fact, the prefix can be either denial—*a* in Greek—or accumulation and motion—Latin *ad*. The authority of Dionysius the (Pseudo) Areopagite had validated this type of language in the denomination of divine realities[28]. The new language becomes now also epiphanous of human resilience.

It is the appropriate language for the pilgrim's exercise: on Monte do Gozo [the Hill of Joy] and later in the sanctuary (whether in Compostela or in Jerusalem—sources point that way), the pilgrim experiences the highest stage of happiness. Accomplished the visit to the targeted place, he can now return home and tell what he had behold. It is not a feeling or a state of happiness that one can endorse, but an experience that will challenge others. As example none would better ignore, rather assume.

7 UTOPIA AND HAPPINESS

In a six line stanza added by More at the beginning of his work, and which he claims to have translated from a Utopian original set by an enigmatic Anemolius, "nephew of Hythlodaeus by his sister" (MORE 1965: 20–21; 2006: 364–365; 2009: 207), and consequently Portuguese and as much fictional as his uncle (using a name that can either mean "subtle", "witty" or "weathercock"), he celebrates *Utopia* (the "non-place") as *Eutopia* (the "place of happiness"). The transition from one meaning to the other was possible by the use of a pronunciation that would make equal the reading of the initial elements of *Outopia* and *Eutopia*, "Country of Happiness", "Land of the Blessed". By denying what his place does not allow him to have, he postulates what he wishes for.

Under its new form, the word takes over the long tradition that goes back to Hesiod, Plato, and Lucian, and extends to the *Fortunate Islands*. According to the Budé's letter to Lupset, published at the beginning of Mores' *Utopia* (1965: 5–15; 2006: 346–361; 2009: 197–206), the island inhabitants live in innocence and happiness; therefore, their capital would be named Hagnopolis (City of Innocence). Such a designation cannot to be found in More's text. He was careful in predicting something that would go beyond the possible. If the collective aspirations remain cumulated, they are subject to a rationality of restraint, appeasing individual and collective ambitions.

Hythlodaeus does not state (at least not clearly enough to let More retain it) where this desired land is. Despite all, he attempts to describe it. More than simply describing it, he names it and places it in the limit of perfection. It is an island out of reach of human contacts; before it was called Abraxa, city full of limitations (because it lacks a symbol that would allow it to reach the perfection of Abraxas, Kabbalistic name of gnostic tradition that embraces the full circle of a year of 365 days). Now its name is Utopia, derived from the name of its founder. As for the capital, it has the designation of Amaurotum—city where the citizens gather to decide about collective life—but that means "hidden city". The name implies it is a shrouded city, but it is not a mirage. That is the reason why, in the last edition revised by the author, Amaurotum replaces the name "Mentirano", used in the 1516 first edition.

The author faltered in the game of the names correlations (*nomina sunt numina*, as Max Müller uttered), but he bet on imagining Utopia at the limit of human perfection, without illusions and with subtle irony.

There are differences to the biblical Paradise. One of them touches us closely. The biblical Paradise was given to humanity as an open space to be cultivated. In it, the only point of reference was one Tree, planted in the centre (source of Wisdom that would be destined for a greater experience and could not be seized otherwise). The new Paradise, according to a built model (and meant to be continuously invented), is presented as a concluded space[29], where everything is programmed by the power of an absolute Organizer (Utopus, the philosopher-king) that arranges and foresees all[30]. It must certainly be a concluded place, as final step of a construction. However, not so closed that it will not accept being subject to judgment. Predicted as such, it forestalls itself only as perspective of a horizon. There is the risk of the wanderer, losing the reasons for enduring his effort until the end and focusing only in a single isolated part, missing the sense of the purpose, failing to assume his commitment with the construction as his own[31]. The apocalyptic Angel will have to remind and free the wanderer from mistakes. There is also the risk that in the effort of constructing the pilgrim may believe that the rationality he admires is the one that may be imposed (the tyranny of intelligence or effectiveness are no less restrictive than weapons; any of them leads to death).

More's speech acquires more acceptability as fiction approaches and differentiates the experience lived in the historical context (as the Anglian kingdom where Thomas More was living). The alternative is suggested with the smile of benevolent irony. It is legitimised by the lesson brought by an actor of an adventure validated by experience, Raphael Hythlodaeus[32]. It does not lack the lesson of the past, obtained predominantly in Greek sources, then retrieved through erudition. We may do without either and will benefit from it only those of good will, inwardly free to be ready for permanent wandering. The prophet will eventually have to pay with his own life for daring to proclaim and be consistent in his convictions (as happened to More).

8 UTOPIA AS AN APPEAL

Anyway, it is inevitable the perplexity generated by utopias: Christian utopia itself, living from hope, is only rewarding for "who is capable of losing everything to have life" (Math 16, 25)[33]. Thomas More's *Utopia*, which aims to oppose an ideal against experienced reality, is located in *Nusquama*, in Nowhere, a non-place, not as a vanishing point of every perspective (without having the arch supported in any known point) but as an instance of judgment.

Each follows a personal path during one's pilgrimage. More sets himself in his own time and in the place where he fully undertakes his duties in the city of men. The Carthusian life seduces him, but also allows him to feel the attraction of the numinous parting from the kingdom of men. He does not even restrain himself from visiting the homeland of learning in his days: Italy. He deviates himself from the ancient Platonic utopia that still dependents on the dream of the golden age, lost in primordial times or in a lost Atlantis[34]. Neither does he remain entangled in the images of a biblical Paradise—for reasons easily understood (doctrinal orthodoxy and respect for a sacredness the author

does not discuss and gives him the assurance that the return will only be real through eschatology).

Set in his country and in his time, he judges behaviour with detachment and irony, and engages himself in the renewal of his City, focusing on the reform of habits and denouncing the manoeuvres of the law to curb excesses.

The ending of *Utopia* is quite meaningful:

… também me é fácil confessar que muitíssimas coisas há na terra da Utopia que gostaria de ver implantadas nas nossas cidades, em toda a verdade e não apenas em expectativa. (More 2006: 673; More 2009: 415)
[There are very many features in the Utopian commonwealth which it is easier for me to wish for in our country than to have any hope of seeing realised.] (More 1965: 247)

Through dream, More intends to provoke a *catharsis*. Through fiction, he aims to determine the form[35]. Through irony he leads humanity to plunge inside itself, to reform it, and then leading it outside, bring it back to a new dimension[36]. He does not appeal to the supernatural as a way to solve human problems, neither those resulting from evil nor expression of limitations of human nature. He does not appeal to any theology either, although it would be easy to mention St. Augustine, as the thought of this son of Hippo was familiar to him, especially in the *City of God*, where the opposition of the "two cities" is obvious[37]. More remains within the rationality of the means that the "cultural" man (based on the polyvalence of *colo*) has at his disposal and which allow him to pass from "moving" freely to "inhabit" responsibly in order to achieve the "cultivate/worship" in plenitude, surpassing in every action what had been achieved in previous stage. His *Utopia* is a perennial appeal always open…

NOTES

Translation of the original paper from Portuguese to English by Anabela Monteiro Nunes; Revision Maria do Rosário Monteiro.

[1] We use the term "Anthropocene" which is a neologism (άνθρωπος, human, καινός, new): it aims at expressing the harmful effects that man provokes on the earth system, particularly on his environment, by endangering the existing balance ("ecotropia"). The data we know of, reveal egregious distortions: Britain and the United States account for about 55% of the effects of accumulated gas emissions with greenhouse effects in a fraction of time between 1750 and 1950, but today the rest of the world already exceeds the share of these two countries.
[2] Oscar Wilde, *The Soul of Man under Socialism*: "A map of the world that does not include Utopia is not worth even glancing at, for it leaves out the one country at which

Humanity is always landing. And when Humanity lands there, it looks out, and, seeing a better country, sets sail. Progress is the realisation of Utopias". It is on p. 36 of the recent Portuguese translation under the title (somehow distorted) *O espírito humano no socialismo* (Wilde 1912: 43; Wilde 2005: 36). This quote is also used by P. Versins (Versins 1984: 917).

[3] Here the association is made to Antonio Skármeta's *El Cartero De Neruda* (2002), thinking of metaphors that transform the world, after having transfigured the places.

[4] The challenge is also to be found in Oscar Wilde, *op. cit., loc. cit.*, as he perceives the human dissatisfaction when arriving in Utopia: once there, after rising to the window and becoming aware of the place he found, man immediately sets out in search of another one.

[5] We employ here the categories formulated by Henri Desroche, in *Encyclopedia Universalis*, s. v. "Utopie": the alternation is but the replacement at the same level, in which reality admits or tolerates its opposite within a dominant paradigm (the feast of fools will be the accurate example, because it lasts one day and functions as an escape or source of amusement). Strife implies enthusiasm and the defence of a credible and sustainable opposition (although there is no certainty the sustained paradigm may survive verification). The alternative lies on the level of credibility and of suitable strategies adequate to come forward as better one.

[6] For all references to the work of Thomas More we follow our edition: Thomas Morus, *Vtopia*, critical edition, translation and commentary by Aires A. Nascimento; introductory study by Joseph V. Pina Martins, Lisboa, Fundação Calouste Gulbenkian, 2006 (More 2006). [T.N.: in the translation from Portuguese to English, quotes from *Utopia* will follow Yale edition].

[7] See A *Navegação de S. Brandão nas fontes medievais (Navigatio Brendani—I; Benedeit, Navigatio Brendani—II; Trezenzonii In Solistionis magna Insula, Conto de Amaro)*—critical edition, translation, introduction, and commentary notes, by Aires A. Nascimento, Lisboa, Colibri 1998 (Nascimento 1998).

[8] To be found in "Cancioneiro João Bensaúde".

[9] We are obviously referring to the "Myth of the Cave"; cf. Plato, *Republic*, Book VII (Plato 2000: Book VII, 514a-517c, p. 220–223).

[10] This is not the Great Emperor, but his grandson, Kublai Khan, shut in his palace from where he should never leave.

[11] The author evokes the quietness of the city of Lisbon, where he had just attended the celebrations of the Centennials and is enchanted by a city that offered him the smile he missed in other European cities, overwhelmed by the war that had fallen on them.

[12] The quotation belongs to the introduction to the Italian edition, which cannot be found in the Portuguese translation.

[13] During the XXVII Congress of the SUCP, in February 1986, Mikhail Gorbachev, who had only recently become Secretary-General, launched his slogan of "glasnost": freed Soviet minds so that they could fight the administrative bureaucracy, and bring into daylight collective attacks committed in Stalin's times.

[14] Italo Calvino knows utopias are encircled by transparency. As remarked in the comment to an exhibition of the National Library of France, in 2000, Dostoyevsky speaks of "crystal cities". Through one of his characters, Zamyatin foresees that tomorrow all shadows will disappear. Not a shadow from any person or anything! The sun will be shining through everything" (Zamiatin 1924: 170).

[15] *The Transparent Man* had its public appearance at the 1930 Universal Exhibition in Dresden. The model maker Franz Tschakert created it using transparent materials and revealing the entire anatomy in three dimensions, in a room plunged into gloom: thus man was celebrated as a masterpiece and model of unitary and hierarchical organization.

[16] Medieval men left wonderful miniatures of the *Pantocrator* who, compass in hand and leaning over the natural elements, builds everything in due measure—as read in the Bible: cf. *Biblia Moralizada* (Viena, Ms. 2554), Madrid, Casariego, 1992; Biblia de S. Luís (Toledo), theoretically. Visit http://www.moleiro.com/en/biblical-books/the-bible-of-st-louis/miniatura/581.

[17] This is how André Wénin explains it: "Un refus du totalitarisme (Wénin 2002); François Marty, *Bénédiction de Babel: Vérité et communication* (Marty 1990); Yeshayahou Leibowitz, *Brèves leçons bibliques* (Leibowitz 1995).

[18] Pertinently, the author seeks to restore consistency to the biblical speech—which, as at other times, only retains a fragment to form the basic message.

[19] In *Against Celsus:* "As far as the form lets you understand, the narrative does not contain only a certain historical truth, but also reveals a hidden meaning. It is important to know that some have retained their native language, because they did not move towards Levant, they remained in the Levant and kept the language of the Levant. It is important to know that only they have become part of the Lord and His people under the name of Jacob, thus becoming "part of his inheritance, Israel"; only these were subjected to a sovereign who did not accept, like the others, his subjects to punish them" (Book V, Chap. 30–31). Trad. Aires do Nascimento.

[20] The editors explain that the author is Gilberto Holandia. The epithet comes from the name of an island or zone located on the border between Anglos and Scots, which Velando and Beautiful surround with their waters; on this island, which belongs to the bishopric of Lincoln, there stood the monastery of Swinhetense, that Gilbert administrated by chairing the two communities, one female and one male. He died in 1172.

[21] Digital copy in http://purl.pt/24118.

[22] Inspired by Dante and *Roman de la Rose*, the Cistercian Guillaume Digulleville (c. 1330), opens the doors to the dream as a literary process in his *Pilgrimage of human life*, an allegory. Granted with a vision of Jerusalem, this appears to him as a fortress guarded by the Fathers of the Church (especially by Augustine), where only as a pilgrim will he be allowed to enter, wrapped up by Faith, and supported by the rod of Hope. He walks towards it, swaying between vices and virtues and accompanied by maiden Reason. For editions of the trilogy of his text see Bibliography: (Camille 1985; Deguileville 1893; Deguileville 1895; Digulleville 1897; Faral 1952; Langlois 1928; Maupeu 2005).

[23] See the texts in question in *A Navegação de S. Brandão nas fontes portuguesas medievais (Navigatio Brendani—I; Benedeit, Navigatio Brendani—II; Trezenzonii In Solistionis Insula magna; Conto de Amaro)* (Nascimento 1998).

[24] Some questions which are not developed here can be read in the introduction and in the notes of comment we affix to the edition of *Utopiana Moreana* we were asked to collaborate in by Prof. José V. Pina Martins (More 2006).

[25] As much as this specular function is distorted in the term "image" (resulting from a Platonism where only the shadows can be perceived), the connotations that one can collect from its Latin origin is of exaltation and potentiation of something it stands for. See Aires A. Nascimento, "Pictura tacitum poem Texto e imagem no livro medieval" (Nascimento 2002).

[26] The acceptance of the name had opponents: Germain de Brie writes to a colleague of his in Paris: "*Vdepotiam non-Vtopiam, si quid volebat Graece recte formare, appellare debuit*, in *Antimorus*", Paris, 15 19, fol. G 2. J. Scaliger also writes: "*Verbum Vtopia, quamquam ab ornato & docto viro Anglo confictum, tamen Graecum non est; et qui Vtopiam Graecum putant ii quid sit Graeca componere nesciunt*". cf. Paul A. Sawada (Sawada 1971).

[27] In a letter dated September 3, 1516 to Erasmus, he names his island *Nusquama*, "Nowhere". Erasmus takes this meaning (Epist. 467; 474, 477; 481). However, soon Guillaume Budé, who knows of the dialogue between the two humanists, his friends, quickly turns into Greek the Latin word *Nusquama* and forges the word Oudepotía, the country of nowhere, or the untraceable country. This is the meaning More already emphasizes in the letter addressed to Peter Gilles. There he states no one would be able to locate the Island Utopia, for the simple reason that the sailor who had visited it had not bothered to reveal it and the others, for lack of knowledge in cartography, forgot to ask. The functionality of the name (as indicator of a Land we can idealize) is the most important factor. However, its truth is legitimized by the consistency of its underlying idealization; credibility is acquired in the invention/discovery witnessed by a scholar who is also a skilled sailor (the Portuguese Raphael Hythlodaeus). In another moment, a different semantic pole is created again by Guillaume Budé (Letter to Lupset): to the first designation he adds a Greek adverb of the same family: oudépote, "never" As well as Land with unknown place, Oudepotí will be Land that cannot be located in time, but is beyond Time, although it is not eschatological. A game played by scholarly minds? Budé passes the concept of Utopia from the category of place to the category of time. Becoming timeless in order to be for all ages, its mutation implies it being of an internal order.

[28] This is what the Pseudo-Dionysius considers "negative theology" to assert a supreme reality by starting from below. See the treaty "Mystical Theology", and particularly the cap. III (Pseudo Dionisio 1990). In 1501, Thomas More takes a course on the Celestial Hierarchies of the Pseudo-Dionysius taught at S. Paul by his peer Grocyn.

[29] Urbanism theorizers wish for a city in a rational and geometric style: "Medieval cities had grown at random, under the anarchic impulse of individual initiatives; but behold, the centrifugal communal structures give room to princes concerned with centralization and order", says Raymond Trousson (1999: 41). Cf. Georg Braun et Frans Hogenberg, *Civitates Orbis Terrarum: Nova Palmae civitas in patria Foroiuliensi ad maris Adriatici ostium con-*

tra Barbarorum incursum a Venetis ædificata (Braun and Hogenberg 1598) (Palmanova, near Venice: spider web structure).

[30] The notion of civilizations older than ours is an ancient one. Atlantis described by Plato in *Critias* is present in the spirit of all authors of Utopias, particularly Campanella when he writes *La Città del Sole* or Francis Bacon when he writes *New Atlantis*. Both tried to recreate the mysterious atmosphere in which Plato surrounds his imaginary island. In a very different way, More concerns himself primarily in describing the people, a society with qualities and skills. About its history, he decides to emphasise a turning moment, the contact with another civilization, in 315, precisely when Utopus was structuring the new society. This "historical fact" gave More ground to put forward another moment, the arrival of the Portuguese, and Christianity.

[31] "Marco Polo descrive un ponte, pietra per pietra./ — Ma qual è la pietra che sostiene il ponte?—chiede Kublai Khan./ — Il ponte non è sostenuto da questa o quella pietra,—risponde Marco—ma dalla linea dell'arco che esse formano./Kublai Khan rimane silenzioso, riflettendo. Poi soggiunge:—Perché mi parli delle pietre? È solo dell'arco che m'importa./Polo risponde: —Senza pietre non c'è arco" (Calvino 2012: 80).
[Marco Polo describes a bridge, stone by stone./"But which is the stone that supports the bridge?" Kublai Khan asks./ "The bridge is not supported by one stone or another," Marco answers, "but by the line of the arch that they form." /Kublai Khan remains silent, reflecting. Then he adds: "Why do you speak to me of the stones? It is only the arch that matters to me." /Polo answers: "Without stones there is no arch"].

[32] Primarily reconstituted from *Itinerarium Portugallensium and Lusitania in Indiam, et inde in occidentem et demum ad aquilonem*. See Luís de Matos (Montalboddo 1992).

[33] One can recognize them, at least in prophetic utterances, in medieval times; cf. Janet Coleman (Coleman 1982).

[34] Plato creates the myth of Atlantis tracing it back to distant ancestors as a counterpart to an Athens where democracy leads nowhere; but not even so does he give up dreaming of a Republic in which the regime will lead to a complete communitarianism and a planning that would be demographic, economic, educational, political, religious, and legal.

[35] The passage is from Varro: "Ut fictor cum dicit fingo, figuram imponit, quom dicit formo, formam, sic cum dicit facio, faciem imponit; a qua facie discernitur, ut possit aliud".

[36] He can rightly take from the stanza in Utopian alphabet:
"Única de todas as terras sem filosofia, /A cidade da filosofia descrevia eu aos mortais. / De bom grado partilho o que é meu; sem pesar, recebo o que for melhor" (More 2006: 363; 2009: 209).
[Sole among all lands without philosophy, / The town of philosophy I described to mortals. /Willingly do I share what is mine; without regret, I accept what is best].

[37] "Two loves have made two cities: the love of oneself which led to the contempt of God, the earthly city; the love of God as far as the underestimation of himself, the heavenly city". Augustine, *City of God*, 14, 28.

BIBLIOGRAPHY

Belugue, Genevieve. Le lieu dans l'oeuvre romanesque d'Edouard Glissant. Paris: Universite de Paris IV-Sorbonne, 1999. 568 p. ISBN: 9782284014898.

Bernardo, S. "SERMON XXXV. Vous êtes un jardin fermé, ô mon épouse, ma soeur, etc. (Cant. IV, 12.)." Traduction Nouvelle par M. L'Abbé Charpentier. Librairie Louis De Vivès, Éditeur. http://www.abbaye-saint-benoit.ch/saints/bernard/tome05/gillebert/cantique/cantique035.htm. (accessed 22nd April 2016).

Braun, Georg, and Hogenberg, Frans. Civitates Orbis Terrarum. Vol. 5. Cologne:1598.

Calvino, Italo. Le città invisibili. 2nd ed. Milan: Mondadori, 2012. 176 p. ISBN: 978-8804425540.

——. As Cidades Invisíveis. Tradução de José Colaço Barreiros. Lisboa: Teorema, 2008.

——. Invisible Cities. 2nd ed. Translator William Weaver. London: Vintage Classics, 1997. 160 p. ISBN: 978-0099429838.

Camille, Michael. "The Illustrated Manuscripts of Guillaume de Deguileville's, "Pèlerinages", 1330–1420." PhD. Cambridge University Press, 1985.

Coleman, Janet. "The continuity of utopian thought in the Middle Ages: a reassessment." Vivarium 20.1 (1982): 1–23.

Deguileville, Guillaume de. Le pelerinage de l'ame de Guillaume de Deguileville. Ed. Stürzinger, Johann Jakob. London: Club Roxburghe, 1895. 307 p.

——. Le Pelerinage de Vie humaine de Guillaume de Deguileville. Ed. Stürzinger, Johann Jakob. London: Club Roxburghe, 1893.

Digulleville, Guillaume de. Le pèlerinage Jhésu Crist. Ed. Stürzinger, Johann Jakob. London: Nichols and sons, 1897. 372 p.

Faral, Edmond. "Guillaume de Digulleville, moine de Chaalis." In Histoire littéraire de la France. Vol. 39. Paris: Imprimerie Nationale, 1952. 1–132 p.

Langlois, Charles-Victor, ed. Pèlerinages par Guillaume de Digulleville. Paris: Hachette, 1928.

Leibowitz, Yeshayahou. Brèves leçons bibliques. Paris: Desclée de Brouwer, 1995. 290 p.

Marty, François. Bénédiction de Babel: Vérité et communication. Paris: Editions du Cerf, 1990. 274 p. ISBN: 978-2204040372.

Maupeu, Philippe. "Pèlerins de vie humaine. Autobiographie et allégorie de Guillaume de Digulleville à Octovien de Saint-Gelais." PhD. Univ. Toulouse-Le Mirail, 2005.

Montalboddo, Fracanzano da. Itinerarium Portugallensium. Facsimile da edição de Milão, 1508. Estudo Introdutório por Luís de Matos. Lisboa: Fundação Calouste Gulbenkian, 1992. LXXVII, 197, [192] p. ISBN: 9789723105728.

More, Thomas. Utopia ou a Melhor Forma de Governo. 2ª ed. Tradução, prefácio e notas de comentário de Aires do Nascimento. Estudo Introdutório de José V de Pina Martins. Lisboa: Fundação Calouste Gulbenkian, 2009. 425. ISBN: 978-972-31-1309-9.

——. Utopia. 1ª ed. Estudo Introdutório de José v de Pina Martins; Inclui edição facsimilada Basileia, Joannes Froben, Novembro, 1518. Edição crítica, tradução e notas de comentário de Aires do Nascimento. Lisboa: Fundação Calouste Gulbenkian, 2006. 715. ISBN: 972-31-1169-1.

——. Utopia. The Complete Works of St. Thomas More. Eds. Surtz, Edward and J.H. Hexter. Vol. 4. New Haven and London: Yale University Press, 1965: cxciv, 629 p. ISBN: 9780300009828.

Nascimento, Aires Augusto do. "The Hispanic Version of the Navigatio Sancti Brendani: Tradition or form of reception of the text?" In The Brendan Legend—Texts and Versions. Eds. Burgess, Glyn S. and Clara Strijbosch. Boston: Brill, 2006. 193–220.

——. "Viator e Peregrinus: Registos da construção da viagem." In Homo Viator; Estudos em homenagem a Fernando Cristóvão. Eds. Amorim, Maria Adelina, Maria José Craveiro and Maria Lúcia Gracia Marques. Lisboa: Colibri, 2004. 173–188 p. ISBN: 972-772-510-4.

——. "Pictura tacitum poem; Texto e imagem no livro medieval." In Actas del III Congreso Hispánico de Latin Medieval. Ed. González, Maurilio Perez. Vol. 1. León: Universidad de León, 2002. 31–52 p.

——, ed. Navegação de S. Brandão nas Fontes Portuguesas Medievais; Navigatio Brendani I; Benedeit, Navigatio Brendani II; Trezenzonii In Solistionis magna Insula, Conto de Amaro. Edição crítica de textos latinos, tradução, estudo introdutório e notas de comentário de Aires do Nascimento. Lisboa: Colibri, 1998. 281 p. ISBN: 972-772-041-2.

Origen. Contra Celsum. rev. ed. Trad. Henry Chadwick. Cambridge: Cambridge University Press, 1980. 572 p. ISBN: 978-0521295765.

Plato. The republic. Edited by G.R.F. Ferrari; translated by Tom Griffith. Cambridge: Cambridge University Press, 2000. 382 p. ISBN: 9780521484435.

Pseudo Dionisio, Areopagita. Obras completas. Biblioteca de autores cristianos. Edición preparada por, Teodoro H. Martín; presentación por Olegario González de Cardenal. Madrid: Biblioteca de Autores Cristianos, 1990. 418 p. ISBN: 84-7914-014-3.

Régio, José. Poesia II. Obra Completa de José Régio. Introdução de José Augusto Seabra; revisão de Luís Amaro. Vol. 2. Lisboa: Imprensa Nacional-Casa da Moeda, 2001. 467 p. ISBN: 972-27-1059-1.

Saint-Exupéry, Antoine de. Carta a um Refém. Trad. Júlia Ferreira, José Cláudio; Nota Introd. e Pósf. Francisco Vale. Lisboa: Relógio d'Água, 2015. 64 p. ISBN: 9789896414962.

——. Lettre à un otage. Paris: Gallimard, 2004. ISBN: 78-20703170359.

Sawada, P.A. "Towards the Definition of Utopia." Moreana. 31–32 (1971): 135–146.

Skármeta, Antonio. El cartero de Neruda. 1ª ed. 1985. La Habana: Casa de las Américas, 2002. ISBN: 9592600422.

Steiner, George. Errata—El examen de una vida. Madrid: Siruela, 1998.

——. After Babel. Aspects of Language and Translation. 2nd ed. revised and enlarged. Oxford: Oxford University Press, 1992.

Trousson, Raymond. Voyages aux Pays de Nulle Part. 3ª ed revista e aumentada. Bruxelas: Éd. de l'Université de Bruxelles, 1999. 318 p. ISBN: 2-8004-1220-8.

Versins, Pierre. Encyclopédie de l'utopie, des voyages extraordinaires et de la science-fiction. 2ª ed.
Lausanne: l'Âge d'homme, 1984. 1037 p.

Wénin, André. "Un refus du totalitarisme." Biblia. 12
(2002): 18–25.

Wilde, Oscar. O espírito humano no socialismo.
Comentário de George Orwell; trad. de Ana Barra-
das. Lisboa: Dinossauro, 2005. 63 p. ISBN: 972-8165-
41-2.

——. The Soul of man under socialism. Londres: Arthur
L Humphreys, 1912. x, 99, p.

Zamiatin, Evgenii Ivanovich. We. New York: Dutton,
1924. xiv, 286 p.

The four modes of thinking framed by utopian discursivity. Or why we need Utopia

Fátima Vieira

CETAPS—Centre for English, Translation and Anglo-Portuguese Studies, University of Porto, Porto, Portugal

ABSTRACT: This article stands for the idea that Thomas More was a "founder of discursivity". Based on this notion (which Michel Foucault applied to Karl Marx and Sigmund Freud), the article examines the four modes of thinking—*prospective thinking, critical thinking, holistic thinking and creative thinking*—that are framed by "utopian discursivity". Ultimately, the description of these modes of thinking aim to ground the author's conviction that, because of the way they organise our reflection on the possibilities for the development of our society, they are the tools we need to construct a better future.

Keywords: Thomas More, utopian thinking, discursivity, future

1 UTOPIAN DISCURSIVITY

In his book on curiosity (2015), Alberto Manguel explains what happens when we read a piece of "great literature", the sort of book that is characterised by a "multi-layered complexity": in spite of all our efforts, we never manage to capture its essence, and this is why we are to return to the book, over and over again, hoping, if not to reach its depths, at least to go a step further in its understanding. This, Manguel suggests, is a never-ending task: "generations of readers cannot exhaust these books", but have instead contributed to the construction of a "palimpsest of readings that continuously re-establishes the book's authority, always under a different guise" (7). In the end, the book is richer as a result of what Manguel calls "the art of reading":

> Reading is a craft that enriches the text conceived by the author, deepening it and rendering it more complex, concentrating it to reflect the reader's personal experience and expanding it to reach the farthest confines of the reader's universe and beyond. (9)

Further in his book, Manguel describes how he has learned to find in books "clues to [his own] identity" (49), how the words of others, being "valid instruments for inquiry", have helped him think (83).

Although Manguel does not mention *Utopia* in his book, Thomas More's masterpiece illustrates rather well the case of a complex and multi-layered book. The history of utopian literature is in fact based on different interpretations of it by generations of readers who have tried to update the book's message to their own age, reflecting on different ways of constructing the future. But I believe that *Utopia* goes beyond the idea of a mere "palimpsest of readings": I find in Thomas More the qualities that Michel Foucault acknowledged in Karl Marx and Sigmund Freud when he depicted them, in "What is an author?", as "founders of discursivity",[1] claiming that they have produced the "possibilities and the rules for the formation of other texts" (Foucault 1984: 114).[2] I am not thinking here of the fact that Thomas More invented a new literary genre, with a set of narrative conventions; I am thinking, instead, of the way More offered a totally different perspective on the world, no doubt framed by the age he lived in (and for the advancement of which he contributed), namely by the way Humanism valued the agency of human beings. This new perspective, which in rigour corresponded to a revolution in thought, had as its foundation act a practice of thinking where the discourse on the Other is centred on oneself. In true fact, as Andrew Hadfield has noted, the inhabitants of the island of Utopia are "dislocated Europeans", dealing with the same problems that afflicted the Europeans of the 16th century (Hadfield 2007: 7).

Describing the characteristics of the texts by the "founders of discursivity", Foucault further clarifies that these are texts we are always to go back to; he suggests, however, that

> [t]he return is not a historical supplement which would be added to the discursivity, or merely an

ornament; on the contrary, it constitutes an effective and necessary task of transforming the discursive practice itself." (Foucault 1984: 116)

This, according to Foucault, would be the main difference between the revolution in thought orchestrated by Freud or Marx, and the revolution motivated by scientific discoveries:

a re-examination of Freud and Marx will inevitably" result in changes in Freudianism or Marxism, whereas to re-examine Galileo's texts would not bring a change in mechanics. (116)

This acknowledgement that "the initiation of a discursive practice is heterogeneous to its subsequent transformations" (115) is instrumental to the understanding of utopian discursivity. In fact, I believe that the expansion of *utopian discursivity* is crucial for the understanding not only of contemporary utopian thinking, but also of Thomas More's founding text as well.

Foucault suggests that:

to limit psychoanalysis as a type of discursivity is (…) to try to isolate in the founding act an eventually restricted number of propositions or statements to which, alone, one grants a founding value, and in relation to which certain concepts or theories accepted by Freud might be considered as derived, secondary, and accessory." (116)

This is what I am trying to examine in this article. As I said above, I believe the founding act of utopian discursivity to be found in the way the discourse on the Other is transformed into a discourse on oneself. I further believe that it is possible to identify in utopian discursivity a number of propositions or, to be more precise, modes of thinking, approaches to the world and to the role that we are to play on it, which I will label as *prospective thinking, critical thinking, holistic thinking,* and *creative thinking*. My main aim here is to build an argument in defence of utopian thought. I will argue that we need utopia because, through the four modes of thinking it entails, it provides us with the tools we need to change society.

2 THE FOUR MODES OF UTOPIAN THINKING

2.1 *Prospective thinking*

We need utopia, first of all, because it helps us to set goals. The best definition of utopia, in this sense, I have ever come across has been offered by the Argentinian film director Fernando Bírri, who famously said that utopia is something that we set

on your horizon: we know that we will never reach it, that every time we take ten steps forward, it will walk ten steps away; but we need it to proceed, as it forces us to walk.[3]

Utopian thinking inspires us to be ambitious while asking the inaugural utopian question "what do we want for our society?" Furthermore, it impels us to inflate the possibilities of our future with what Ernst Bloch called a "surplus of desire". For Bloch, it is the subsistence of desire (something he recognises to exist in man as an ontological category), even after our dreams have been fulfilled, that ensures the permanence and dynamics of utopia (Bloch 1976: 216–217). The notion of incompleteness is thus a vital propeller for the development of societies.

It is important to note that this prospective attitude, even though it is inflated by a surplus of desire, is not disconnected from reality. It is quite the opposite: in all truth utopian thinking always moves from the real, which it rejects; as Paul Ricoeur put it, the utopian element implies a capacity for denial and refusal of the prevailing ideology (Ricoeur 1986: 313). But just as the utopian encounters with the other entail a search for a better solution for our society, the imagination of different futures for humanity implies a search for recipes to be applied in the here and now, as Pierre Furter has contended. Furter has further argued that this search is carried on with the awareness that utopia will never offer final truths: the truths it offers are provisional, as they result from the dialectic transforming movement they establish with the real world (apud.Greis 1996: 32). The truths put forward in the framework of utopian thinking are thus transitory; to resort to Fernando Bírri's metaphor of the ten steps towards the horizon, utopian discursivity provides us with time to, after every ten steps, critically observe reality and reset the horizon for the inaugural question "what do we want for our society?"

There are two more things that are specific to prospective thinking. The first has to do with the idea of future it deals with; the second, with its intentionality. In utopian thinking, the future is not to be seen as something that is bound to happen, but as a network of possibilities, a distinction which is conveniently conveyed in French by the existence of two parallel concepts, those of *futur* and *avenir*.[4] As Gérard Klein explains, the concept of *futur* translates the perspective which dominated from Antiquity to the 17th century, the idea that the future is the continuity of the past and of the present, and that it is certain that it will happen; for that reason people tried to predict it, in a prophetic attitude. But the birth of euchronia, in the late 18th century, signalled a new way of thinking about the future, adequately represented by the concept of *avenir*. As Klein clarifies, the *avenir* is more complex and larger than the *futur*, as it includes everything that may happen, everything we may invent (255–7). It is this notion of *avenir*, of a multiple, plural, unpredictable future offering a myriad of very interesting possibilities that utopian prospective thinking takes on.

But the idea of prospective thinking also implies a particular intentionality when we are thinking about the future. The etymological Latin roots of *prospective* indicate that it is the result of the juxtaposition of the prefix *pro* (forward) and of the suffix *tivus* (intensity of action) to the verb *specere* (*pro+specere+tivus*). Thus, what is important about prospective thinking is the intensity of the action of looking into the future which results in an intentionality: the future is researched and explored so that we may anticipate what will happen. It is important to note here that when the word prospective was imported into English it lost part of its original meaning, which has been retained, however, by Romance languages: while in English prospective means "likely to be or become something specified in the future" or "likely to happen",[5] in Romance languages such as Portuguese, Spanish, French and Italian the word conveys the notion of exploring ahead, an idea which is still to be found in the English noun prospection (as in *gold* or *oil prospection*).

There is another reason I am calling this kind of thinking prospective. I believe it resonates the principles of the system of thinking which Gaston Berger described in the mid-20th century (Berger 1964; Berger 1967). As Michel Godet has clarified, la prospective, as it is called in French, is a state of mind which entails imagination and anticipation that inspires hope and leads to will (1999: 8); it is a methodological tool and a system of analysis that integrates a set of techniques that promotes imagination and is based on the idea that anticipation encourages action (1999: 5). The prospective, as Godet calls it in English,[6] goes beyond the imagination of future scenarios. It starts by looking at the future as the object of desire, as the utopia which provides the present with a direction and a meaning (2006: 334).[7] The prospective differs from extrapolative thought, which is a common method in social sciences. Instead of trying to devise which tendencies may shape the future, the prospective gives priority to desire and only afterwards reflects on the strategy, i.e., on the path to reach the goals that have been set; in the late 1980s, this came to be known as strategic prospective (Godet 1999: 4). It is thus a methodology that moves from the future to the present, as the reflection on the present is promoted in light of what the future may become.

Utopian discursivity is also deeply focused on images of a desirable future. However, its logic differs from that of strategic prospective in that it relies on a hypothesis approach. Although it is guided by the inaugural question "what do we want for our society?" that provides us with a horizon towards which we are to walk, next utopian prospective thinking tests the exploratory question "what if?", in order to assess the different possibilities of the paths available. There is, besides, the very basic difference derived from the fact that strategic prospective is, above all, a tool for the strategic management of companies and other institutions. Nevertheless, I believe that Gaston Berger's famous assertion that "looking at the future disturbs not only the future but also the present" (Apud. Godet 2006: 5) may well be applied to prospective utopian thinking as well.

2.2 Critical thinking

The second reason why we need utopia has to do with the critical thinking logic incorporated into the utopian discursivity. In fact, how could we ever responsibly decide which path to take without being sure of the accuracy, credibility, impartiality, relevance, and substantiality of the data that we will have to use in order to validate our choice? As it happens with critical thinking, our conclusions (that will validate the choice of the path to take) must be consistent and reliable, and be the logical, sequential, and progressive outcome of a careful consideration of data that can be easily verified and validated. Thus, the choice of the utopian road will imply the mastery of the six core skills of critical thinking: interpretation, analysis, evaluation, inference, explanation, and self-regulation. Utopian thinkers must have the characteristics of critical thinkers, as well as a strong sense of consequence that will provide them with the capacity to readily reconsider and revise their views when necessary. This is a quality that utopian thinkers should never lack, as it is the instrument that will allow for the continuous verification and redefinition of the horizons they are aiming at.

There are three more things that utopian thinkers have in common with critical thinkers. First, the capability to decentre the issue they are dealing with from their own interests and to include, in the redefinition of their interests, the interests of other people (that is why the inaugural question is inclusive in its formulation: "what do we want for our society?"); second, their ability to resist the influence of preconceived ideas and prejudices; and third, their capacity for thinking differently, for trying to devise new ideas, new solutions, and for being creative.

2.3 Holistic thinking

The third reason why we need utopia is because it provides us with holistic thinking, the only mode of thinking that, according to Edgar Morin, will eventually lead us to understanding. In his book *La Voie pour l'avenir de l'humanité* [The Path to the future of Humanity] (2011) Morin stands for the idea that the crisis we are living in—which is also a crisis of imagination—has a cognitive nature. As knowledge has been cut into many small bits, being taught in non-communicating subjects that are offered to students at university, we are led to misunderstand the information we are given: because we lack a larger view, we will never be able to contextualise, organise and understand it (2011: 239). As Morin says, if we observe separately each thread that composes a tapestry, we will never be able to perceive and enjoy the beauty of its pattern (253).

Morin contended in his *Introduction à la Pensée Complexe* [Introduction to the Complex Thought] (1990) that ideas and thoughts are complex; the modern tendency to clarify them by reducing them to small units and describing them partially prevents us from understanding their complexity, the way they relate to each other (thus establishing a relationship of complementarity) and the way they contradict each other (thus establishing a relationship of antagonism). It is only when we come to understand ideas in their complexity, when we accept that they are multidimensional and that they can be at the same time complementary and contradictory, that we will have access to knowledge (Morin 1990: 10)—and, I would add, that we will be able to practise utopian thinking.

Utopian holistic thinking is capable of giving an adequate answer to the problems we are facing today because the crisis we live in is systemic: it has an economic, social, political, and ecological nature, and each part is interconnected with the other. It has, furthermore, a global expression: although some of the problems have clear local origins, they have given rise to other problems on a universal scale. As Ruth Levitas has explained, utopian models rely on a holistic view, as they reveal an acute awareness of the fact that societies work as systems (2007: 69). By incorporating the principles of multidimensional analysis, utopian thinking provides utopian thinkers with the qualities they need to address systemic problems. Moreover, this awareness of the systemic nature of the problems leads to decisions entailed by attentiveness to the possible consequences of actions and paths we may have opted to take. In a world that has local problems that have global roots or repercussions (and vice-versa), only a holistic approach will provide us with reasonable, practical answers.

2.4 Creative thinking

Finally, we need utopian thinking because, by inviting us to think about alternatives, it fosters creativity. Utopian accounts always involve estrangement as they confront us with a diversity of new possibilities, with the description of multiple worlds that test a myriad of "what if?" hypotheses. This exposure to difference is of paramount importance: with utopian thinking, we escape the mere replication of existing knowledge and walk towards the creation of new knowledge.

This creation of new knowledge happens when someone dares to look out of the corner of his or her eye, as the Portuguese writer Gonçalo M. Tavares explains:

> What is observed from the centre of one's eye is the obvious, what is shared by the multitude.
> In Science, as in the world of inventions, observing out of the corner of one's eye is seeing the detail, that thing which is different and which may be the start of something meaningful.
> Observing the reality out of the corner of one's eye, i.e.: thinking slightly to the side. From here all the important scientific theories were born. (2006: 75)

Thinking out of the corner of one's eye is then having the capacity to see things that other people are not able to see; it is having a wider angle of vision. When talking about utopia, Eduardo Galeano insists on this idea: "There are many realities wishing to be born"—he says. "They are just waiting for us to imagine them so that they can be born."[8] The problem is that we are not trained to see things that are not familiar to our eyes. The famous example of the Portuguese vessels and the Indians indicate that this is not a new problem. I am referring to the fact that, according to historical accounts, when the Portuguese vessels arrived in Brazil, the natives did not understand what they were. They had never seen a large ship before, so they perceived them as gods watching them from the horizon, as they believed that gods lived in the water. The lack of a familiar reference prevented them from seeing what they really were—large ships. In *Tools for Conviviality*, Ivan Illich explained what the problem is:

> Our brain has been so deformed by the society we live in and by our habits of thought that we do not even dare to think about other possible forms of organisation. The question is that we are used to accepting that there is just one way of using things, but there is always not one way but at least two ways of using, for instance, a scientific discovery, two ways that are fundamentally contradictory. (1975: 12)

According to Illich, we have been trained to accept instead of to imagine; we have almost completely lost the power to dream of a world where all people would have the right to speak and to be heard, where there are no limits to creativity, where everyone and every single person may change his or her life (34).

And still, the sort of creativity required by utopian thinking does not imply the creation of anything radically new. Boaventura de Sousa Santos explains that what utopian thinking requires us to do is to combine what already exists in new ways and new scales. As the Portuguese sociologist says, it is a matter of taking to the centre what used to be on the margins, and to devise the consequences of that change (1995: 479).

3 CONCLUSION

Utopian discursivity has no doubt changed since the publication of More's *Utopia*. To start with, the prospective attitude related to the concept of future as an *avenir* is not to be found in the founding text, as the society Raphael Hythloday describes is not set on the future but on a distant island. The exploration of the "what if?" hypotheses is at the basis of More's text, though. The prospective attitude also permeates Utopia in the sense that, as has so often been remarked, the description of the utopian island is not to be seen as More's ideal, as the final truth, but as the illustration of an alternative and an incentive for the reader to search for other alternatives. Right from the beginning, the utopian discourse was thus set as a movement of denial of the prevailing ideology; the idea of incompleteness that ensures the dynamics of utopia is made clear when Thomas More, the narrator, dismisses Raphael Hythloday "praising both [the Utopian's] institutions and his communication" and takes him by the hand saying that they would choose another time to weigh and examine the same matters" (1988: 135).

Critical thinking is also present in More's *Utopia* and is rendered evident by the analysis of the European society of the early 16th century, which Hythloday rejects and against which the island of Utopia is described. Holistic thinking too prevails in More's text, where the fact that societies work as systems is evinced by the careful description of the consequences of the abolition of private property at the level of political, economic, and social life.

Creative thinking is at the very basis of the Morean text. In fact, the "what if?" hypothesis that inspires utopian discursivity is founded upon a combination of what already existed, but on a new scale. The founding text of the utopian tradition, though,

was restricted by a negative perspective on the capability of human beings to overcome the limitations cast on them because of the original sin. Creative thinking was thus conditioned by the coeval conviction that, as the human being was sinful, the only way of ensuring social peace would be by creating a set of laws destined to restrain his or her actions. A more optimistic view, which would be typical of the Enlightenment, would eventually lead to the imagination of societies based on an unlimited exploration of human capabilities. Nowadays, since Modernity has put the human being right at the centre of the universe, utopias are based on the presupposition that change will be brought about by human action.

Utopian discursivity has changed over the centuries; the four modes of thinking I have described certainly testify to those changes. But it is because of these changes, this adaptation to new times and new challenges, that utopian discursivity is so meaningful to our days—and this is also why More's text proves to be so exhaustingly re-readable.

NOTES

[1] In many theoretical texts written in English, the word Foucault has coined (*discursivity*) is spelt differently (*discoursivity*). In this article, I am using the word as it was proposed by Foucault to evince that I am subscribing to his definition of the concept.
[2] According to Foucault, this discursivity only emerges in the 19th century, Marx and Freud being the most obvious examples. However, I believe that More falls into this category as well, as he offered a way of critically reflecting about the world that has been transformed by subsequent authors.
[3] It was Eduardo Galeano who attributed this definition of utopia to Fernando Birri in a YouTube video clip that became viral on the Internet (available at https://www.youtube.com/watch?v=m-pgHlB8QdQ).
[4] This is in fact the case with all the Romance languages.
[5] http://www.merriam-webster.com/dictionary/prospective
[6] La prospective is often translated into English as future studies or just foresight, but the translation limits the scope of the concept to little more than the imagination of future scenarios, often within the logic of scenario entertainment. I do subscribe to Godet's option of anglicizing the French word in order to keep its original meaning (Godet 2006: 329, 333).
[7] Godet explains how the *prospective,* as a system of thinking, regains the idea of utopia as a powerful tool for the construction of the future: "By considering desire as a productive force for the future, we rehabilitate the concept of subjective utopia, thus reuniting imagination and scientific logic. Like many terms, utopia is too often used as synonymous with impossible. Etymologically *ou-topos*, coming from Greek, means a non-place; i.e., a place that does not exist. However, this does not exclude an eventual future existence. Utopia, as a virtual object of desire, is the source from which action takes its meaning and direction (Godet, 2006: 8).

[8] Cf. TV show *O Tempo e o Modo*, broadcast by Portuguese National Television. Available at https://www.youtube.com/watch?v=LmwiYp-kbKM.

BIBLIOGRAPHY

Berger, G. *Étapes de la prospective*. Paris: Presses Universitaires de France, 1967.
——. *Phénoménologie du temps et prospective*. Paris: Presses Universitaires de France, 1964.
Bloch, Ernst. *Le Principe Espérance*. Transl. Françoise Wuilmart. Paris: Gallimard, 1976.
Foucault, Michel. *Foucault Reader*. Ed. Rabinow, Paul London: Penguin Books, 1984.
Godet, Michel. *Creating Futures: Scenario Planning as a Strategic Management Tool*. Preface by Joseph F. Coates. Transl. Adam Gerber and Kathryn Radford. London: Economica, 2006.
——. *Manuel de prospective stratégique. Paris*: Dunod, 1999.
Greis, Yvonne dos Soares dos Santos. "O Elemento Utópico no Pensamento de Pierre Furter." MA thesis. Universidade Estadual de Campinas, 1996.
Hadfield, Andrew. *Literature, travel, and colonial writing in the English Renaissance, 1545–1625*. Oxford: Clarendon Press, 2007. xiv, 305 p. ISBN: 978-0-19-923365-6.
Illich, Ivan. *Tools for Conviviality*. Glasgow: Fontana & Collins, 1975.
Levitas, Ruth. "The Imaginary Reconstruction of Society: Utopia as Method." In *Utopia Method Vision: The Use Value of Social Dreaming*. Eds. Moylan, Tom and Raffaella Baccolini. Ralahine utopian studies v. 1. Berne: Peter Lang, 2007. 47–68. ISBN: 978-3-03910-912-8.
Manguel, Alberto. *Curiosity*. New Haven: Yale, 2015.
More, Thomas. *Utopia*. Transl. Ralph Robinson. London: Everyman's Library, 1988.
Morin, Edgar. *La voie: pour l'avenir de l'humanité*. Paris: Fayard, 2011. ISBN: 9782213655604.
——. *Introduction à la pensée complexe*. Paris: Esf, 1990. 158 p. ISBN: 2-7101-0800-3.
Ricoeur, Paul. *Lectures on Ideology and Utopia*. Ed. Taylor, George H. New York: Columbia University Pres, 1986.
Santos, Boaventura de Sousa. Toward a New Common Sense: Law, Science and Politics in the Paradigmatic Tradition. New York: Routledge, 1995.
Tavares, Gonçalo M. *Breves Notas sobre Ciência*. Lisboa: Relógio D'Água, 2006.

Miguel Real's *O Último Europeu 2284*, or a utopian questioning of our individual and collective freedom

Anna Kalewska

Institute of Iberian and Ibero-American Studies, Warsaw University. Warsaw, Poland
CHAM, FCSH, Universidade NOVA de Lisboa, Universidade dos Açores, Lisboa, Portugal

ABSTRACT: In *O Último Europeu 2284* [*The Last European 2284*] (2015), Miguel Real inserts into the fictional narrative both literary and philosophical theory. Thus, he abandons the historiographical metafiction and adopts the utopian novel structure. The rereading of the past gave way to a vision of the future—a Utopia (with dystopic touches) that turns out as an invention of the sensible (post)modern humanism. After a brief incursion into the history of literary Utopias beginning with Plato (2001; 2000), Thomas More (1965; 1978; 2009), Francis Bacon (2002; 2008), Jonathan Swift (2005), Aldous Huxley (2004; 2006), George Orwell (1998; 2004), J. V. Pina Martins (1989; 2005), the narrative reveals an (im) perfect society: Nova Europa [New Europe]. This New Europe is defined by an unrestrained technological progress, democratic crises, alienation, and the erasure of basic human features. The Headmaster, as a new Raphael Hythlodaeus, experiences a Utopia. After having suffered the attack from the Barbarians of the Westlands, The Headmaster sails to Azores to recreate the old European democracy. The last surviving European citizen and his work (a synecdoche of both Author and the novel *O Último Europeu*) by sharing with the islanders the tasks of repopulating the islands and restructuring its civilization, giving a testimony of an ethical responsibility for individual and collective freedom. The utopian "happy ending", however, is not possible, because the American Technological Democracy reclaims the rights for the creation of another "brave new world". Miguel Real's utopian metafiction is marked by the reiterative historicism in the image of the repopulation of the Azorean Archipelago. This is a multifaceted novel, with a strong character, the brave Headmaster, possible avatar of a new King Sebastian and the "last European".

Keywords: Historiographic Metafiction, Utopia vs dystopia, (post) modernity; democratic crises, ideas, beliefs and religions

A novel continuously assumes new forms and expresses new contents in a singular manifestation of the perennial aesthetical and spiritual human restlessness.

(Aguiar e Silva 2007: 684)

We took pride in the vast progress performed in ten years, endowed with an excellent moral plan and a quite solid historical optimism. [...] To expand and continue our ancestors' civilization. Noble the responsibility, huge the will, grand the road travelled.] [1]

(Real 2015: 204)

Portuguese literature does not have many examples of successful or renowned utopias, though having many readers of utopias, judging by the number of published translations. There are several reasons that may explain this phenomenon (...), one, and probably the most self-evident, being the almost continuous strong exercise of religious and/or political censorship imposed in Portugal from the late sixteenth century to the last quarter of the twentieth century.

(Monteiro 2015: 280)

1   PRESENTING THE AUTHOR (MIGUEL REAL'S BRIEF BIO-BIBLIOGRAPHIC SKETCH)

Miguel Real (Luís Martins' literary pseudonym) was born in Lisbon in 1953. Graduated in Philosophy by Universidade de Lisboa, and later Master of Portuguese Studies by Universidade Aberta, where he presented a dissertation on Eduardo Lourenço. This dissertation was afterwards divided in two bio-bibliographic books (Real 2003b; Real 2007)[2]. His first novel, *O Outro e o Mesmo*, earned him the Prize Revelation in Fiction, awarded by APE/IPLB. From then onward he is known as writer of fiction and essays, and as philosophy teacher.

In 1998 Miguel Real won another award by APE/IPBL, this time for his literary essay *Portugal; Ser e Representação* (1998). Another important award was won in 2000, Prémio LER/Círculo de Leitores, to the book *A visão de Túndalo por Eça de Queirós* (Real 2000).

In 2001, Miguel Real received a scholarship given by Centro Nacional de Cultura (Centre for National Culture), which allowed him to follow Father António Vieira itinerary through Brazil. The writer kept a diary he named *Atlântico: a viagem e os escravos* (Real 2005b), as well as a "tribute", recently published with Filomena Vieira (2015).

From 2003 onward, with the novel *Memórias de Branca Dias* (Real 2003a)[3], Miguel Real started writing separately novels and essays thus avoiding mixing philosophical and literary theory inside the fiction. In *O Último Europeu,* the author abandons this practice.

Miguel Real's novels fit into the concept of historiographic metafiction defined by Linda Hutcheon (1988). Therefore one can base critical analysis of most of Miguel Real's novels using the concept that views modern historical novels as critical reinterpretations of the past.

One can consider that historiographic metafiction started, in Portugal, with Saramago's *O Levantado do Chão* (1980) where fictional action takes place between "history and fiction" developing itself on a vast scenery displaying the life peasants in Alentejo, exhibiting a clear antifascist connotation.

After the publication of Saramago's book, Portuguese writers had a new point of view for figures until then either marginalized or deleted from historical narratives. In *O Último Europeu 2284*, Miguel Real instead of focusing directly his fiction on present times or history, uses them as the foundations, focusing on the future, on ethical options, on political tendencies, on human divergences, on ethnic differences, on war and on cultural contradictions.

In 2005, Miguel Real published *A Voz da Terra* [The Earth's Voice], a historical novel on Marquis of Pombal and the 1755 Earthquake. This novel was well received by both readers and critics. There he fictionalizes the historical transformation of Lisbon from a superstitious, naval, and imperial capital, that of the Discoveries—the city of Saint Anthony governed by the Voice from Heaven—to a bourgeois Lisbon, rational and geometrically rebuilt according to Marquis of Pombal's planning. In the old Lisbon, image of Portugal, pullulating with black slaves, Moors, Galician, *calhandreiras* [washerwomen], *colarejas* [Lisbon market-women], artisans with their shops always opened, the workers of the dockyards in Ribeira das Naus, nobles, friars and priests surrounding the King, buying English flour, iron and coal, French fabrics and paying with golden doubloon mined in Brazil. In the New Lisbon reborn from the earthquake, under the control of Marquis of Pombal, pullulate the civil servants, the magistrates, the purveyors,

the new public teachers, politicians and intellectuals educated under the Enlightenment philosophy.

The focus of Real's narrative is the earthquake as it was lived by each of the characters in this remarkable historical novel: Júlio Telles Fernandes (known as Julinho), a wealthy Brazilian widower, secret messenger for the *mascates* [Brazilian wandering seller of jewels and fabrics], herald of Pernambuco's independence, Father Malagrida, (burnt to death in an auto-de-fé in Rossio in 1761), Violante Dias (a Brazilian jewess who mysteriously disappears among the ruins), Miss Smith, daughter of the purveyor of the English trading-station in Lisbon. There is also the widow Passarinho who suffers from night ailment and the greedy canon, always seeking money and power.

After the earthquake, Lisbon's social organization changed, as did the politics and the scientific academies, and the same happened to Julinho and to the younger Portuguese generations.

… learn French and laws, Descartes' analytical geometry, Leibniz and Newton's infinitesimal calculus would be taught for the first time in Portugal. Teachers were allowed to teach that the earth is a sphere, but planets' orbits are elliptical, just as Kepler had proved two hundred years before […]. Everything changed in less than ten years everything kept on changing. Lisbon spread itself, became geometrical, rational, industrial, learned how to read and write. Lisbon put on another skin, bourgeois, rich, European. But regardless of what skin she put on, she would never stop being faithful, rude, superstitious, the city of St. Anthony, homeland of jealousy, always sighing for an enlightened elite to guide her. (Real 2005a: 364)[4]

A Voz da Terra earned Miguel Real the Prémio Literário Fernando Namora 2006 [Literary Award Fernando Namora], one of the most prestigious Portuguese literary awards. The solid historical knowledge behind this historical novel is based on four excellent books: *Lisboa Seiscentista* (Branco 1990), *Lisboa Pombalina e o Iluminismo* (França 1983), *O Mal sobre a Terra; Uma História do Terramoto de Lisboa* (Priore 2003) e *A Vida Quotidiana em Portugal no Tempo do Terramoto* (Chantal 1986). Equally important were paintings and images that inspired the author to describe Lisbon in the 17th century: the characters, their expressions, the personality, and behaviour of Marquis of Pombal, as well as the bullfights, the attempted assassination of king D. José I, the sermons of Father Gabriel Malagrida in the streets of Lisbon after the earthquake, the auto-de-fé where Father Malagrida was executed and Violante Dias and the Knight of Oliveira burnt in effigies, the execution of the Marquis and Marquise of Távora, etc. All this information reveals a thorough research, uniting the historian's labour and literary imagination of one of the best contemporary Portuguese writers.

Simultaneously with the edition of *A Voz da Terra*, Miguel Real published the essay *O Marquês de Pombal e a Cultura Portuguesa* [Marquis of Pombal and Portuguese Culture] (2006b). By the end of the same year is published *O Último Negreiro* [The Last Slave Trader] (Real 2006b) a novel focusing on the slave trader Francisco de Félix de Sousa, who lived in São Salvador da Bahia and in Ajudá, in Benin.

Miguel Real writing activities expand, regularly, from the areas of fiction and essays to writing textbooks on history, philosophy, cultural studies, as well as theatre adaptation, in collaboration with Filomena Oliveira[5]. Since the year 2000, he has been writing regularly in the literary journal *JL-Jornal de Letras, artes e ideias*. He also took part in a radio program ("Um certo Olhar" in Antena 2), with other authors and critics.

Real's *opus magnum* is the novel *O Último Europeu 2284*, simultaneously a novel and a philosophical essay "dedicated to Thomas More celebrating the 500 years of *Utopia*, 1516" (2015: 7), as can be read in one of the three dedications. In it, the reader finds a harsh portrait of modern societies and the dangers they engender. The historical metafictionality is achieved through the critic imagination of the future. The intellectual and biographical evolution of the main character and fictional author of a *Chronic of New Europe* is the narrative central focus. He may even be considered an alter ego of Miguel Real, who visited the Warsaw University during the celebration of the fortieth anniversary of Portuguese revolution, 25th April 2014.

2 UTOPIA, THE ESSENTIAL MOTIF OF POSTMODERN HUMANISM

Written in Latin, More's *Utopia*, according to José V. Pina Martins, is a fundamental book produced by the humanist movement and the text that instituted Thomas More as an undisputed Renaissance figure, in the sense of a promoter of the *studia humanitatis* and as a defender of civic rights[6].

More's utopian society or new republic was inspired in Plato's *Republic*, imagining that only a different place or island in the New World would be able to ensure a life different from that of the unfair *polis*. A place where citizens would cooperate, where intolerance and fanatics would be punished with exile and slavery, and where "owning nothing all were rich". The human being is placed at the centre of the world, holding his/her destiny in his/her hands. No one would be oppressed in the

name of religion and the people was free to choose different beliefs and cults, all living in ecumenical harmony.

According to several historians, More was fascinated with the narratives of Amerigo Vespucci about the recent discovery of Fernando de Noronha Island, in 1503.

More's *Utopia* is divided in two parts. In the first there is the criticism to his contemporary way of life in the city of London; the second part, influenced by Plato's *Republic*, presents and an alternative society.

However, More's aim was not to defend the view of the Portuguese sailor Raphael Hythlodaeus and the utopian society as models, but to encourage the reader to exercise a critical reflection on both the positive and the negative aspects of the alternative society. The word Utopia, was created using two Greek radicals, one being a suffix of negation and *topos* meaning place: "no" *topos*, a place that does not exist, a nowhere. Moreover, this is an example of how More played with irony. From 1516 onward the word "utopia" became synonym for ideal society, though of impossible existence, or a generous idea, though unrealizable.

In *O Último Europeu 2284* Portuguese readers may find echoes of Agostinho da Silva's philosophy, or a mythic V Empire Europe is looking for so that its citizens will not need to work, but might always be active.

Not to work but always be busy—the lemma of another philosopher, born in Portugal, when the nations and the languages were considered first priorities in the education of any citizen—this became one of our essential mottos for our collective living. (33)[7]

In 2016, we celebrate the 500 years of More's Utopia. This small book continues to be read, discussed, and interpreted. More's imaginative and criticism model as well as the book's message(s) still makes sense today. There are alternatives to the social-political world order being imposed. Just as the island of Utopia offered an antithetical society to the sixteenth century Europe. According to Pina Martins, More's work is the product of "renaissance humanism". I believe that *O Último Europeu 2284* is a novel-essay of contemporary humanism, being inspired in the same ideas of alterity in what concerns the concepts of citizenship, nationality, social organization, and philosophical message. The New Utopia is a project of social transformation and it represents crucial aspects of contemporary humanism. There is still the two-party division between the description of Old Europe, in which the first person narrator lived, and the New Europe that faces its definite decline in 2284. However, there is also the vision of a possible future Europe, an alternative society surviving in Azores, after the attack of Great Asia, which is destroyed by the cynical American Empire. There is food and houses for all, schools for the children, work, or activities that fulfil everybody's needs, hospitals for the elderly and a kind teacher, former Headmaster:

> I volunteer myself as teacher, Ancient word that designates the one who has the prior task of educating the new generations from the moment they are weaned. There are no fractions among us; we are not a fragmented society... (180)[8]

Miguel Real's contemporary humanism has its foundation on Thomas More's Renaissance humanism. To have a better understanding of the ideas, structure and problems raised by the novel, one must make use of the basic concepts of utopias and utopianism.

Utopia has as common meaning the idea of an ideal, imaginary, and fantastic civilization. It may be centred in a city or in a whole world. It may be placed in the future or in the present. Either wat in Real's novel utopia is achieved through the succession of past memories that took place in the future. However, the time movement in the novel is not linear. Instead, there are several variations of present/past and future/present information.

Utopia may also be defined as a fantasy, a strong hope, a dream cherished for a very long time. In Real's novel it is a dream that came true in the future (2294), with the building of a Base, a school, a community of new settlers and, at last, the conclusion and improvement of the *Crónica da Criação e Extinção da Nova Europa* [Chronicle of the creation and extinction of New Europe], concluded at Pico island, in Azores.

> We performed a new colonization of the Island; a scientific colonization [...] after the Portuguese had done, in Ancient times, a human colonization, a geographical one, aiming only to people it. (197)[9]

These narrative threads tie historical, utopian, and narrative time to the Island of Newest Europe [Ilha da Novíssima Europa]. It is a new utopia, based on democratic principles, with the alternatives, freedom of expression, of reunion, and even the peculiar election of three new companions to replace the old Headmaster.

> Each of the three [companions] does not represent a third of the whole Assembly, as if we were back to the Ancient proportional representation in use in the 20th century democracies, as if each were to represent a fraction of the whole, or a "part" of the whole... (180)[10]

More's *Utopia* has marks left by contemporary reality, but it is a fiction. More draws a very critical picture of his contemporary England [and Europe] in what concerns its society, politic practises, economic structure, and religious acting. These are the true causes for promoting poverty, theft, and hunger. Book II is dedicated to describing the island of Utopia, its organization, and principles. The hero and traveller to Utopia, Raphael Hythlodaeus, as almost every word used throughout the narrative, is full of connotative meanings: while the name "Raphael" points to the Archangel who carried the divine cure for blindness, the name Hythlodaeus means he is an "expert in trifles" or "well-learned in nonsense" (More 1965: 301). The structure, the contradictions, the products of the imagination, they all place *Utopia* in the realm of ideas.

Several basic principles of the island of Utopia are part of our common heritage and goals western societies strive to maintain. In spite of the closeness and dictatorial practices, it was a democracy, with free access to education, respect for the law and the restrictions on private property did not deprive people of whatever they needed and avoided social inequalities.

Miguel Real breaks from this paradigm, as well as from the idea of an Old or a New Europe founded on the historical idea of creating a society completely communitarian, equalitarian and fair (2015: 41).

In the island of Utopia there is religious freedom, peace is privileged over war, this being used as a last resource but considered as contrary to human, social and common good. In *The Last European*, the Headmaster explains that:

> Everything in Newest Europe flowed normally, according to the resources available, the basic needs already supplied and the desire to build the most advanced society one could from the technological point of view. (207)[11]

It took Europeans two hundred years to build a new society, but it collapsed because the pre-historical barbarians, inhabitants of the Waste Lands, proved to be resistant and untameable (2015: 21)

Though not being the first text on the utopian ideal, More's *Utopia* is the first of a long tradition of texts with a common origin: the proposal of alternative social projects. The first person narrator in *The Last European*, assuming the quality of *vox populi*, presents a somewhat naïve proposal of a new alternative social order unquestionably utopian:

> We favour leisure, the pleasure of waking and being available, with nothing to do except what one decides to do, a delightful feeling of the existence of

a time permanently empty, filled in a singular way each day, outcome of the wise and selected choice of our actions, which correspond intrinsically to our desires, even though being accomplished only mentally. (32)[12]

Literary utopia, due to its narrative strategies (shared in Portugal by J. V. Pina Martins (1989) and Miguel Real) made possible, in the last five hundred years, the criticism of the authors' contemporary society and the transmission of subversive messages, either in times of oppression, under totalitarian governments, or in the time of crises in parliamentary democracies. The latter situation made possible, in *The Last European*, the fantasy of the destruction of the EU, threatened by terrorist attacks and migration crises. The "old Europeans", according to the narrator are those that rule Europe nowadays and govern it as if it were a grocery. The "Neo Europeans", according to the Headmaster, led a parasitical existence, surely as happy as useless and delightful.

> To favour pleasure instead of work, that is one of our distinctive marks, and one Confucian Mandarins despise because either they ignore its sublime virtue, or due to their technological incapacity, their scientific backlog compared with New Europe. [...] We, the neo-Europeans, fulfil ourselves in the entertainment, free, without schedules nor any other end than the delight and relish achieved by the action itself, as making collections, as I do, since I am the main organizer and director of museums. (33)[13]

The inhabitants of the Wastelands, regions out of the control of the Great Electronic Brain, kept outside the Clusters[14] by the Security Green Thread (2015: 31), did not wait for the extermination of the neo-Europeans. Once the Security Green Thread was disabled, they invaded the New Europe and wildly attacked the neo-Europeans, as reported by the Headmaster:

> Screams, the blasting of mechanical fire weapons, shrills of couples fornicating, the groans of the dying people, the sound of barbarian music, mainly produced by percussion, every sound arriving stiffly to my ears. (89)[15]

The Headmaster's mission is extraordinary: to rebuild a utopian world, two hundred years after the foundation of New Europe, now destroyed by the attacks performed by the Barbarians, after the destruction of the security mechanisms led by the totalitarian Mandarins, in the year 2284. The Newest Europe will be built in Pico Island, in Azores. However, this new utopia will be destroyed in 2299, exterminated this time by the "Democratic American Technocracy".

In the end of *Utopia*, the fictional More tries to persuade Raphael Hythlodaeus to work as counsellor, but the wise mariner refuses, saying his opinion is too radical and he would not be listened. At the end of *O Último Europeu*, the Headmaster is one hundred and thirty years old. He had put an end to his mission of being Counsellor of the Pantocrates and recreating the Newest Europe. He is about to die after resurrecting the humanist Europe in a small island in Azores. His only friend is Jorge Tomás Evangelista, a descendant of the first Portuguese to live in the island, in the 16th century: Fernão Álvares Evangelista. What the Headmaster has to leave as legacy is the manuscript he wrote on paper and with a pen, telling the history of humanity and the remembrance of a European utopian democracy. The *Chronicle of New Europe*, entrusted to the daughter-in-law of the islander Jorge Tomás, will be a living legacy of "the last European citizen", the only one who remained always faithful to the civilizational, cultural, and ethical principles of a humanist Europe.

However, *O Último Europeu 2284* is also a narrative on the process of writing a literary text. Therefore, it belongs to the long tradition of discursive dialogical novels, inviting the reader to take part in the discussion. This way, the narrative becomes an evidence that modernity is not over yet, idea defended by Jürgen Habermas, Zygmut Bauman, Onésimo Teutónio Almeida and Miguel Real (Almeida, Ribeiro and Real 2015).

3   THE FALL OF THE MYTH OF "A NEW EUROPE INVINCIBLE BY SEA, EARTH OR AIR", OR THE NOT SO ADMIRABLE NEW WORLD

Maria de Fátima Marinho presents a study concerning the evolution of this contemporary epic subgenre, from Romanticism to the present, clarifying concepts, themes, and categories (Marinho 1999). The Portuguese researcher starts her analysis with Georg Lukacs' *The Historical Novel* (Lukács 1983), and then proceeds to the present Portuguese writers in whose novels one can find a new point of view that distinguishes them in what regards the debate, the re-reading, the commentary (sometimes ironical) of traditional historiographic discourse. This was considered only as a starting point for the elaboration of a world simultaneously real and fictional: the novel. Following this reasoning, authors as Miguel Real, Fernando Campos, Pedro Almeida Vieira, Mário de Carvalho, Agustina Bessa-Luís, António Cândido Franco, and Carlos Alberto Machado[16] all consider the past as a kind of legitimator of the present[17] for they frequently imply the reconstruction of characters' lives that have their counterpart in history. That is what happens, for instance, in Real's novels depicting characters as King Sebastian, Queen Amélia, and Marquis of Pombal.

While contemporary historical novels display, each on its own way, an original form of reworking the past, integrating it in the present and playing with the need to contest it, the utopian novel invents a future and plays with the the need of a feeling of belonging felt by the reader, the main character, or the first-person narrator. The imponderability of past time gives way to the philosophical invention of the future, in an attempt to denounce (or to exorcise) the present through a future that is perceived as a technological tyranny where all social systems and invented ideologies have already exhaust all their potentials.

O Último Europeu 2284 is a long and descriptive fictional essay (with soliloquies and personal commentaries produced by the main character) that questions the possibility of building a Utopia in the third millennium. Nowadays, after two world wars, the publication of *Brave New World* (Huxley 2004; 2006), exploring the theme of individual freedom versus State authoritarianism, and *Nineteen Eighty Four* (Orwell 1998), every utopia must come accompanied by the description of a dystopia, and express the need for demanding a third route by inventing future universes. Maria do Rosário Monteiro expressed this idea.

> For some time, dystopia and contemporary culture seemed to have decreed a death penalty on utopia, the eutopia, the promise of happiness. Fortunately, several writers found a middle way, a third route, or a compromising position. (2015: 281)

Miguel Real follows this third route.

The long tradition of literary utopia still endures both the memory and practice of More's *Utopia*, and of Bacon's *New Atlantis*, the latter being the first scientific utopia[18]. However, in *O Último Europeu*, there is also the construction of a phalanstery, an international community, harmonious and liberal as the one imagined by the French philosopher Charles Fourier[19]. Miguel Real's narrative echoes all these past literary and philosophical proposals, but there also echo some other contemporary literary utopias.

Using the long utopian tradition, Miguel Real mixed, shaped and recreated it to fictionalize the "Old" and the New" Europe in a narrative that echoes More and Orwell. But the fictional Headmaster reminds readers (particularly Portuguese ones) of another first-person narrator, Miguel Marcos Hitlodeu, in *Utopia III* (Pina Martins 1989; 2005). The outcome of Real's fictional and intertextual work is the daily portrait of a totalitar-

ian and repressive regime, cherishing the memories of a still possible happiness, all this in the fateful year 2284, beginning of the narrative, ruled by futuristic prolepses and analepses.

Maybe tired of historical novel and readers and writers historiographic ignorance, Miguel Real took a temporary leave on his historical research and decided to jump to the future. Inspired by More's *Utopia*, now celebrating 500 years. Therefore, sharing Maria do Rosário Monteiro opinion on More's text defending it is "an open narrative *avant la lettre*" (Monteiro 2015), one may expand this to *O Último Europeu*. The last comments and the way the novel ends leave many questions unanswered, awaiting to continue the dialogue with the new Hythlodaeus and/or the new reader.

Every utopian narrative needs to be contextualized.

> Every utopian text gains its full meaning when its reading is integrated in the political and cultural milieu that triggered it, for utopia presents implicit and explicit political differences meant as responses to the actual society the author lives in. Therefore, they become obvious targets for censorship. However, since the last quarter of the twentieth century, Portugal has become a democratic political system enjoying freedom of speech. (2015: 280)

Moreover, in the Portuguese context, utopian fictions are still very rare[20].

> If censorship was the only reason for the scarcity of Portuguese utopias, the regained freedom of speech should have allowed for the development of Portuguese utopian literature, but unfortunately, it did not. (Monteiro 2015: 280)

Therefore, this paper has the intention of contributing to the study of the second utopian text written after 25 April 1974, the one that followed *Utopia III* (1989). Hence, its focus on the first-person narrator, the Headmaster, on his actions and ideas, in their implementation on an imagined future society, the endless search for a paradise on earth or a perfect democracy in Azores, pleading strongly for a humanist removal and the defence or European citizenship.

In *O Último Europeu*, he leaves the past that had been his field for many years, and creates a sinister vision of Europe divided in wastelands governed by martial clans that enslave starving populations, a "medieval" future. Among the wastelands, there is an isolated region that looks like a paradise amid hell, isolated due to highly advanced technology, a small enclave with only one hundred inhabitants called New Europe. Work was abolished and replaced by the activities each inhabitant chose according to his free will, aiming at the full development of the human being. Amusements, radical experiences, everything was allowed in a sort of virtual experience that all can accede and that leaves implanted in their brains the memories felt as being real.

The inhabitants of New Europe, connected to the Great Electronic Brain, give up their individuality to become a collective entity: there are no names, no birthdays. All citizens celebrate their birthday on the same day, a festivity that is meant to develop the notion of collective common good, to achieve kind of ethereal perfection in which all take part, without vanity or loftiness, enjoying stability, personal fulfilment, maintained through constant evolution and research.

Focusing on the egalitarian and utilitarian division of resources, as well as on the freedom of association that gives origin to different familiar models, life in New Europe elapses long and easily, in complete isolation from the other civilizations. Until Great Asia strikes.

In spite of all the evolution, this civilization is naïve in what concerns personal and collective military protection, placing their confidence on the protective shields, supported by the inexhaustible power of earth's core. When Great Asia manages to disconnect that energy, New Europe is utterly defenceless.

Having eliminated their natural instincts of survival and self-preservation, the inhabitants cannot organize themselves to fight the Barbarians. They can but run, hide or be slaughtered.

Utopia or dystopia? This society is presented from a single point of view, that of the Headmaster. Therefore, it would be sufficient that one inhabitant might feel himself oppressed and this utopia would be a dystopia. However, for the inhabitants, New Europe is apparently a utopia. From the point of view of the reader, it might not be so. One can understand the way how instinctive behaviour are controlled, how conflict is minimized, but in this society there is no place for shared emotions or passions, the fuel that causes conflicts but also the one that promises intellectual and artistic evolution.

The New Europeans are peaceful, intelligent and rational in such a complete and superior way that they seem strange, maybe even plainly handicapped. This fictional future offers several choices, some possible within the concept of utopia, but some announcing dystopia, a dark warning about the future we are walking into, a tyrannical technocracy.

O Último Europeu portrays a civilization despising the self, a society based solely on the collective. Their main goal is the fulfilment of all possible individual desires, if and only if they never become a threat to common good. Nevertheless, to achieve such a rational perfection it was necessary to add an artificial rationality (by introducing in each brain an electronic hyper-cortex), thus removing individual personality. Ironically, the inhabitants of this brave new world consider themselves as humanists. In fact, in this society that values freedom that freedom is under the control of a Great Brain, commanding, shaping, deciding.

The narrator of *O Último Europeu* is a Headmaster living in a small enclave surrounded by an enormous wasteland. In that small stronghold of peace, protected from the rest of the world, a group of scholars managed to build a perfect society, a bit cold, but whose inhabitants are provided for, free to desire and living a technological happiness. There are no problems, until the Great Asia, dealing with serious demographic problems decided to reclaim European territory to accommodate the exceeding Asians.

Founded in 2184[21], New Europe was a paradise governed by advanced technology, a space without "religious or metaphysical beliefs, different visions of the world, without systems of ideas or ideologies" (Real 2015: 54). There is no end for leisure, controlled reproduction, a Great Brain with its electronic and neurological nets controlling "every buildings, streets, shops, houses and individual minds" (Real 2015: 45), along with the Protective Hyper-atomic Bubble, the "safeguard shield"(Real 2015: 31).

In 2284 utopia ends and becomes a *locus horrendus*, with pillaging, poverty, slavery and the erasing of basic human features.

Armed and pompous barbarian hordes from Old Europe, incited by primitive drums, flags hoisted and the roar of scowling captains, encircle our citizens, orderly grouped in the squares and invade our Depositories of Nutrition, searching for food, trying to dissolve the nutritive powder in water, exploding the hermetically pressurized boxes where we preserve our vegetables, greens and natural fruits intact. (2015: 45)[22].

Mournful and powerless, the Headmaster observed the result of "the Old European Barbarian's attacks, due to the prevalence of their reptilian and mammal brains impulses, controlling their behaviours" (2015: 38): men being tortured, women raped, New European corpses rotting in the streets, forming anarchic piles.

Without the connection to the Great Electronic Brain, [Neo-European citizens] can only register an existential memory, the remembrance of actions, behaviours, and reasoning that constitute the unit of our personal identity, our biography not seldom fictionalized from the virtual realization of our desires. (46)[23]

This means that it becomes impossible for them the only capacity of dreaming that the New European had, based on technology and nanoscience. Their identities as well as their individual and collective freedom fall down. The utopian vision of a future Europe disintegrates itself. The Headmaster is "as all the Neo-Europeans, absolutely normal, born, integrated and grown up within the rectorial institutions of New Europe [...] later [became] Headmaster [one of the] two superior technical categories before the level of Pantocrate, this only accessible due to old age" (49) and unlike the inhabitants of Old Europe wastelands, he has neither mother nor father.

After the invasion of the Barbarians, the Asian Mandarins blocked the mega power centrals installed in Earth's core, and reintroduced polygamy. The builders of New Europe "considered the monogamic family the most perverted institution of European civilization, spring of all educational, formative and existential perversions" (65), as well as the old "baseless myth of the existence of a permanent motherly instinct justification for the traditional family" (66). It is important to make note of the Headmaster's complicity regarding the bureaucratic and technological mechanisms of New Europe.

Waiting for the eminent Asian attack, the Barbarians loot the shops, load their primitive trucks with their sack, sell the electronic equipment they find, for the Old Europeans live without electricity, starve, and pigeons, dogs and cats become the inhabitants food. Meanwhile in South America breaks the war between Brazil and Venezuela and three hundred million South-Americans are killed.

The Headmaster decides to dedicate all his energy to writing the *Chronicle of New Europe*, working night and day, sleepless, "because the writing of the book is crucial and its future survival even more" (78). It is as if the "last European" had a mission to accomplish, as a new King Sebastian, or a Marquis of Pombal, herald of science and progress in the midst of vilified crowds.

Relating the present and recording the past, this is the life of the Headmaster, simultaneously an apparatchik and a "golden citizen" of the technocratic system of New Europe. This imbues in *O Último Europeu*, a slight biographic nature, the new Raphael Hythlodaeus acting as former abettor, then as dissident and refugee of the new democratic technological utopia.

The narrative unveiling itself for the reader is an insurance for the future, always more utopian than dystopian, because it will be entrusted to a young girl, guardian of the History of New Europe, mother of neo-Europeans, married with a descendant of the first settlers in Pico Island.

Apparently, utopia and dystopia, past and future join in a continuous fight against the American Empire. In the last scene, the dying Headmaster closes his eyes and "opened his heart to the truth", imagining another utopia, post-future, post everything, sure there was no possible peaceful relationship between Americans and Europeans, for the latter would surely lose.

The chief technicians had concluded that the majority of the adolescents and adults Neo-Europeans would never be integrated in the American Empire, obeying scrupulously to every regulation.

Due to their upbringing and personality, the Neo-Europeans become outcast citizens, contestants, disturbers of regulated normality, foundation of the Empire's happiness.

Their privileged past in New Europe, and as fighters for survival in Newest Europe indicated that the majority would rebel, continuously demanding new conditions of life, better scientific instruments, eventually better equipped laboratories, contact with universities and even, probably, demands of freedom and social equality. These demands would never be accepted by the Empire, which decided for them not a life as chief technicians, but a life as regular citizens.

All evidence, assured by the best technicians in psychology, pointed to a scenario were the majority would become, eventually, a social problem; some, possibly stimulated by the difficulties, might become leaders of future regional insurrections [...].

Soldiers were ordered to kill—painlessly, assured me Jorge Tomás with an inadequate passion, raising the tone of his words—all the grown-ups and adolescents that boarded the military spacecraft.

How they were killed, I do not know.

More than five hundred children would be delivered by the three militaries to an agency that would sell them to American families wishing to become parents without the sufferings of pregnancy and the worries of taking care of infants.

They are white people families, said Jorge Tomás, comforting me.

The children will be all right, he said, softening my pain, scared by me raising my hand placing it over my heart.

In the new family, the younger will easily forget the Newest Europe and they will become happy adults, according to regulations.

There was no point in keeping on living.

I asked to be left alone. I was harsh.

Leave me, I said.

I decided I would die that same day, at sunset, I, the last true European.

Tomorrow I would not see the sun rising. (275–277)[24]

This new (Portuguese, Azorean, someway dystopic) utopia was terminated by the American invasion. The time of discourse matching the time of the story, the year 2299, only in appearance resembles science fiction. *O Último Europeu 2284* is an essay, with long narrative sections and many Headmaster's ramblings, and this turns the novel into a philosophical reflection extremely present about what we can do now concerning our collective future. A novel/essay and an interesting utopian/dystopian philosophical dissertation, even if one does not agree with some cause/effect sequences or with the naïveté of the organization of some scenes. It is, for instance the case of The Flight, Azores, Moral Principles, The State of Nature, Jorge Tomás, Individual Identity, Formation of Families, Vegetarianism, etc. There are also disquieting diachronies in the working of the discourse in the future that become historical with intermittent lapses.

A scientific and philosophical novel, *tout court*, the Headmaster as philosopher and leader of the surviving Europeans in Azores, forces us to ask some questions and look for some answers. Sometimes the plot is a bit weak, and this does not allow us to make a quick reading of *O Último Europeu*, because the exposition of the several aspects of European, Asian, American and Azorean civilizations is developed slowly, in detail, methodical, but easily understood.

Concluding, taking into account the values and the ancient historical references, the process of imagining the future on different layers seems much more important than the description of the past. In this novel, the future is extremely stimulating, since the references to the present (i. e. the time of the discourse) are vague, always overlapped by the biographical, scientific and technological innovation. The future promises what is new and has a romantic appeal, since it is a utopia deeply grounded on a social, political, and cultural analysis of the future.

As Vítor Aguiar e Silva states:

According to some critics, the novel today, after so many and profound metamorphoses and adventures, suffers an undisputable crises, approaching its own decline and exhaustion. Regardless of the value of such prophecies, one fact remains undisputable: the novel endures the most important literary form, due to the expressive possibilities it offers the author and to the dissemination, and influence it has in the public. (684).

Therefore, novels and essays based on utopianism had quite an important role during the 20th century, beginning with H. G. Wells *A Modern Utopia* (1905), one pioneer of science fiction, Lewis Mumford's (1923; 2007), the already mentioned Huxley and Orwell's novels, and more recently Herbert Marcuse's *The End of Utopia*, Leopoldo

Brizuela *Inglaterra, una fábula* (1999), and *L'ile des gauchers* de Alexandre Jardin (1995). More recently, Joe Oliver published *Uniorder: Faça você mesmo o paraíso* (2014), based on More's Utopia and modern technology.

These works are meta-utopians, for they are rational mental exercises on utopian thinking, and they try to validate utopia as a strategy for building a future society. Without being necessarily negative dystopias, each work mentioned above results from a re-evaluation of previous utopias and their actualization. The outcome is the increased value of utopia as strategy, still as valid, when imagening alternative forms of social organization[25].

O Último Europeu 2284, being simultaneously utopian and dystopian, addresses Europe inciting it to build a society where technology might allow a harmonious relationship between human beings and nature. May it ever be possible, with present technology to enlighten and free society, put an end to injustices and inequalities and the investment in future wars? Miguel Real does not seem pessimist, and his "old/new/ Europe" still has a future and some chances of survival as "classical Europe", democratically governed and without the political interference from The Unites States of America.

4 THE FICTIONAL AUTHOR PRESENTS HIMSELF, OR THE THEORY OF UTOPIAN NOVEL

The critical assessment of *O Último Europeu* is guided by the utopian vision one has of the future, in actual Europe, and which gives birth to a new "construction" of historical novel (Medeiros 2015: 10). I believe this is a necessity in what concerns utopian novels, future vision of the future based on the same rules Linda Hutcheon established for the historiographic metafiction, but focusing in another direction, "for not all historical novels are historiographic metafiction" (Medeiros 2015: 11). In a utopian novel, the future is invented and commented on *a posteriori*. Meanwhile, contemporary historical novel presents a new point of view in what concerns either the winners' history or the losers'. Therefore, there is clear difference between contemporary discourse and the narrative time—which is historicized in the Headmaster's point of view and enunciation—and the previous futuristic discourse in which the utopian historiographer relates events after the year 2284, until the moment he dies, in 2299.

The Headmaster's metanarrative comments, his omnipresence, essence the future's present and the future that becomes history, drawing nearer his writing to utopian metafiction, and this happens both in his narrative as in his *Chronic of New*

Europe, which ends his mission as kind of new King Sebastian.

In his typology of the novel, Vítor Manuel Aguiar e Silva, after having discussed the "novel of action or events" exemplifying if with Walter Scott and Alexandre Dumas' historical novels, analysis another category: "the novel of character".

[This] novel presents, typically, one exclusive main character, who the author draws and studies lengthily and according to which the whole novel evolves. Frequently this type of novel tends to subjectivity and a confessional tone as, for instance, in Goethe's *Werther*, Benjamin Constant's *Adolphe*, Lamartine's *Raphael*, etc. Usually, the title I quite expressive as to the nature of the kind of novel, for it is, frequently, the main character's name. (2007: 685).

O Último Europeu 2284 is very similar to the definition quoted above. The Headmaster is the character that structures the whole action; he is the commentator and the main character who coordinates the utopian repeopling of the Azorean island, repeating in a future time Portuguese History, almost as if it were an anachronic historical novel.

However, one cannot say that contemporary utopian novel emerged from a complete rupture between utopia and historical novel. The model set after Walter Scott and the other authors mentioned above was object of variations in time. Writers altered the way to describe the characters, the narrators, the place, and this brought about several transformations within the genre. Language became more concise and swift, and description became less dynamic. In Real's novel, it is frequent to find that a long descriptive reference springs from the Headmaster's intention to criticize, and not so much from the mere description of a place, of characters or of events.

5 FOR A NEW FUTURE EUROPE

After having dominated the world for three thousand years, and looking at the political and social situation of present day Europe, it seems it is on the route of decadence. It is possible that, in the next centuries, Europe will not have a central role, if it cannot stand, decidedly, as a defender of two base pillars: human and ecological rights. This is the role that is left to Europe in world policy. Technological and ideological supremacy lie elsewhere. After glancing at what goes on in the world, it is not hard to imagine what will await us all in the end. There is "Arab Spring", ceaseless wars, nugatory waves, terrorism. In the name of protection

and fighting that terrorism, we are conditioned daily, accepting to change our conception of individual and collective freedom. There is a soothing acceptance regarding the loss of privacy, there is a moral decline and the reintroduction of the concept of "holy war". Our politicians and the media deal with all this lightly, even when facing pure evil and unruly consumerism.

We cannot foresee what will be the future of China, ruled by cold and cruel "mandarins". Neither can we know what will become of the American Empire with their false "infantile curiosity". These were metaphors used in the narrative to give weight to the plot and the narrative discourse.

Democracy either is living a profound crisis or became an impossibility, replaced by a modern utopia, that is, a dictatorship, apparently happy, controlled by faceless ferocious and technologically regulated societies.

The future is opened. We cannot just follow lightly. The flock mentality and the slave morality referred by Nietzsche must be replaced by an ethical responsibility for our actions, our words, our individual and collective political decisions.

O Último Europeu 2284 is an excellent novel about our individual and collective freedom. It is a utopia because it offers a positive vision of human resilience, personified in the Headmaster, who opposes a dictatorial civilization, technocratic, even if almost technologically perfect. This is governed by a hyper-rational elite, planned disregarding the principles of life and death, as a new folly exorcized centuries ago by Erasmus in the arguments exchanged with the master of *Utopia*, in 1516.

Miguel Real wants to point at some black holes that may mean the end of many European ideals that seemed to be the roots of western societies. A novel/essay startles and scares the reader, pursuing the analyses of *Nova Teoria da Felicidade* (Real 2013), *Nova Teoria do Mal* (Real 2012b), and of the *suis generis* return to Sebastianism (Real 2014).

The author does not defend any solution, and neither do Old nor New Europe seem to be political and social models worth pursuing. He expects the reader to do his own thinking about the future and presents alternative sceneries. The reader, as "other" or "new" citizen—a "rectified" citizen, compared with the ignorant hero of Saramago's *Ensaio sobre a Lucidez* (2004)– will have to vote carefully and his/our vote will decide who will govern in Europe and in the World and how happier society will be in the future. Choosing what the future will be like is everybody's responsibility, following the road selected by the author in *O Último Europeu 2284*, the one that reaffirms the feeling of continuity of the plans of transformation started in Renaissance and followed thereafter.

The future of utopia is an open one, perhaps following the ideas of French philosopher Michel Maffesoli, who defends that today there are interstitial utopias, transformed in micro-utopias[26]. There are also individual utopias, referred by Gilles Lipovestsky, who believes it to be common, the existence of "individual utopias, resulting from the acknowledgment that collective transformation may result from the action of common people, building nets"[27]. All this may contribute for a utopian network connecting all the ideas of progress, in either technology or information theory—in the concept of analysis of social networks –, the history of empires past and present, human activity worldwide. All this may help us walk towards a possible, human, hopeful future, as our utopian reasoning can imagine, create, and produce.

NOTES

Translation of the original paper from Portuguese to English by Maria do Rosário Monteiro; Revision Anabela Monteiro Nunes.

[1] "Orgulhávamo-nos do amplo progresso realizado numa dezena de anos, dotados de um plano moral excelente e de um optimismo histórico muito sólido. [...] Prolongar e continuar a civilização dos nossos pais. Grande a responsabilidade, grande a vontade, óptimo o caminho percorrido".

[2] Eduardo Lourenço received the award Prémio Camões in 1996, among several other prestigious awards acknowledging his exceptional contribution to Portuguese Culture.

[3] Cf. Medeiros (2015).

[4] "... aprendia francês e leis, a geometria analítica de Descartes, o cálculo infinitesimal de Leibniz e Newton seriam pela primeira vez ensinados em Portugal, os professores poderiam ensinar que a terra era uma esfera, mas a órbita dos astros uma elipse, como Kepler descobrira havia duzentos anos [...], tudo mudara em menos de dez anos, tudo ia mudando, Lisboa espraiara-se, geometrizara-se, racionalizara-se, industrializara-se, alfabetizara-se, modernizara-se, Lisboa vestia outra pele, burguesa, argentária, europeia, mas, qualquer que fosse a pele de que se vestisse, nunca deixaria de ser uma cidade crente, rude, supersticiosa, a cidade de S. António, pátria da inveja, sempre ansiando por uma elite iluminada que a venha salvar."

[5] Cf. *O romance português contemporâneo: 1950–2010* (2012a), Introdução à cultura portuguesa: séculos XIII a XIX (2011), A morte de Portugal (2008), O Último Eça (2006a), Geração de 90: Romance e Sociedade no Portugal Contemporâneo (2001), Narração, Maravilhoso, Trágico e Sagrado em Memorial do Convento de José Saramago (1996), G.W. Leibniz: O Paradoxo e a Maravilha (1995), and several other philosophical essays, as, for instance, Princípios de Filosofia e Quadro sobre a Ética (1987).

[6] See Pina Martins' Introductory study to the Portuguese critical edition of More's *Utopia* (2006). In 1531, Henry VIII, wanting to marry Anne Boleyn, broke the kingdom's relationship with the Catholic Church, due to Pope's refusal to concede the divorce to his marriage to queen Catherine. Thomas More refused to acknowledge Henry VIII as head of the Anglican Church. Therefore, he was condemned to death, charged with treason, and beheaded in July 6, 1535. Pope Pio XI canonized Thomas More in 1935. In 2000, Pope John Paul II declared More "patron of statesmen and politicians". Thomas More, diplomat and Chancellor to Henry VIII was a close friend of the eminent humanist and pedagogue Erasmus (1466–1536). It is attested that *The Praise of Folly* was written in London, while Erasmus was living at More's home. It was discussed between them, that important reformist text presenting a violent criticism to both 16th century society and papacy (Pina Martins 2006).

[7] "Não trabalhar, mas estar sempre ocupado—o lema de um outro filósofo, de origem portuguesa, quando as nações e as línguas eram consideradas realidades primeiras na educação de um cidadão —, tornou-se um dos lemas essenciais do nosso viver colectivo".

[8] "... ofereço-me como professor, palavra antiga que designa aquele que tem a seu cargo prioritário a educação das novas gerações a partir do tempo do desmame".

[9] "... operámos uma segunda colonização da ilha, uma colonização científica [...] depois de os portugueses terem feito, nos primórdios, uma colonização humana, geográfica, apenas de povoamento.

[10] "Não que cada um dos três represente um terço da Assembleia, como se regressássemos à antiquíssima representação proporcional própria das democracias do século XX, como se cada um figurasse uma fracção do todo, ou uma "parte" do todo. Não existem fracções entre nós, não somos uma sociedade fragmentada...".

[11] "Tudo na Novíssima Europa corria normalmente, Segundo os recursos possíveis, as necessidades básicas que já tínhamos suprido e os desejos de construção de uma sociedade o mais avançada possível do ponto de vista tecnológico".

[12] "Privilegiamos o ócio, o prazer de acordar e estar disponível, nada ter que fazer senão o que se decidir fazer, a sensação deleitosa da existência de um tempo permanentemente vazio, preenchido de um modo singular em cada dia, produto de uma escolha selecta e reflectida das nossas acções, que correspondem intrinsecamente aos nossos desejos, mesmo que realizados mentalmente".

[13] "Privilegiar o prazer em detrimento do trabalho, eis uma das nossas divisas, que os Mandarins confucionistas desprezam tanto por desconhecerem a sua excelsa virtude como por incapacidade tecnológica, por atraso científico face à Nova Europa. [...] Nós, os neo-europeus, realizamo-nos pelo prazer lúdico, livre, sem horário nem outra finalidade senão o deleite e o gosto usufruídos pela própria acção, como fazer colecções, como é o meu caso enquanto alto organizador e dirigente de museus".

[14] The Clusters are nameless cities, rebuilt in some European countries after the 2084 Great Hunger. This calamity led to the "desertion of the old cities, some with more than 20 million inhabitants, several divided in districts lacking electricity and potable water, decadent, unorganized, with minimal education and health facilities, controlled by monopolies belonging to clans. Most of the dead were thrown to dunghills, with no burials, only burnt bodies raised into pyres among the garbage". [... abandono das velhas cidades, algumas com cerca de 20 milhões de habitantes, muitas divididas

em bairros sem abastecimento de electricidade e água canalizada, decadentes, desorganizadas, com reduzidos serviços educativos e hospitalares, cobrados por empresas monopolistas pertencentes aos clãs, a maioria dos mortos atirados para monturos, sem enterro, simplesmente corpos queimados levantados em pira no seio do lixo... (2015: 39)]. Among the Clusters, the Ancient Lisbon had been one of the poorest cities of the continent. In the oriental part of the European continent, "there was Ancient Poland, a country that in the past had been continuously martyred by invasions from its Russian and German neighbours. [... a antiga Polónia, país outrora martirizado por permanentes invasões dos seus vizinhos Russos e Alemães. (2015: 111)]. "Cities once prosperous and magnificent, as Bogotá, Caracas, Buenos Aires, La Paz, Quito, São Paulo were nowadays radioactive ruins inhabited by decayed people, moribund, burdened with genetic malformations." [Cidades outrora florescentes e monumentais, como Bogotá, Caracas, Buenos Aires, La Paz, São Paulo, eram hoje ruinas radiactivas habitadas por povos decrépitos, moribundos, carregados de malformações genéticas. (2015: 110)].

[15] "Gritos, estampidos de armas mecânicas de fogo, guinchos de casais fornicadores, gemidos de moribundos, toadas de música bárbara, assente na percussão, tudo chegava amortecido aos meus ouvidos".

[16] Carlos Alberto Machado's novel *Hipopótamos em Delagoa Bay* [*Hippopotamuses in Delagoa Bay*] (Machado 2013) can be included in the mainstream of Portuguese postcolonial prose, as well as contemporary historical hybrid writing. The plot presents the anti-epic saga of the Quaresma family, telling the story of Portuguese immigrants in Mozambique. Living in Africa since the second half of the 19th century, their story is told through the prism of the Portuguese Carnation Revolution (1974), the liberation of Mozambique (1975) and the development of democracy in Portugal and Africa (or rather a credible, but unfortunate side of the phenomenon) up to the present moment of the fictional narration (c. 2010). As the protagonists feel feeble, menaced by the British and Dutch, uncertain in their relations with the Portuguese Metropolis and their new, African homeland, their mixed identity has not yet been fully expressed and leads to an inner conflict, social exclusion, and personal dramas. The novel can be read as a heterogenic, fragmentary, and hallucinatory Black Lusophone memoir (Kalewska 2015: 148).

[17] Novels as, for instance, *Um Deus passeando pela brisa da tarde* [A God wandering by the evening breeze] (Carvalho 1995), *A Quinta das Virtudes* [The Farm of Virtuousness], *Peregrinação de Barnabé das Índias* [Barnabé das Índias' Peregrination] (Cláudio 1991; Cláudio 1998), *O Concerto dos Flamengos* [The Flemish Concert] (Bessa-Luís 2002) or Real's historiographical metafiction.

[18] Some critics consider Bacon's *New Atlantis* as the foundation of another literary genre, quite close to the utopian: science fiction. Another possible ramification would be a satirical utopia, as, for instance, Swift's The Voyages of Gulliver. *Memoirs of Planets*, written by Thomas Northmore (1795) presents a positive utopian perspective of a government that achieves the status of *eutopia*, or ucronia, a future of general happiness.

[19] Fourier's idea was that communities should be able to achieve total cooperation and self-sufficiency. Some practical experiences were carried on based on Fourier's ideas: the phalanstery of Saí (1841), in Santa Catarina (Brazil), and the Cecilia Colony (1890), in Paraná (Brazil). The same happened with La Réunion (1855), in Texas, and North American Phalanx (1841), in New Jersey (USA). Marx, in *The Communist Manifesto* (1848), criticized capitalism and called the attention of the so called "communists" to avoid "the savage socialism" fearing the natural "market capitalism" since there was the danger of falling into a reproachable utopia (this was proved by the political, social and economic practices in Eastern Europe, 1945–1989). William Morris, in *News from Nowhere* (1995) still believed in a revolution that would abolish all forms of private property, which would free the human being and give him a more solidary kind of life. Among the dystopias, one can refer the narratives of such authors as H. G. Wells, Orwell, or Zamyatin (1924). During the twentieth century 60 s and 70 s there was a renewal utopian practise informed by feminism and ecology. Belong to this period the novels *Ecotopia* (Callenbach 1975), *Woman on the edge of Time* (Piercy 1976) (Fátima Vieira, *apud* (Coelho 2016).

[20] However, in the last quarter of the twentieth century, almost at the eve of the second millennium, an important Portuguese utopia was published: *Utopia III*, written by Pina Martins (1998). This long novel is structured as being the sequel of More's *Utopia*, presenting the history and actual status of the mother of all literary utopias. The question at the basis of the whole novel is, "What would More's *Utopia* be like today?" (2015: 278).

[21] "De facto, a História, a verdadeira História, só se iniciou em 2184, quando o homem se libertou definitivamente da sua animalidade, da recordação do cheiro e do sabor do sangue, do gosto da carne, crua ou cozinhada" [In fact, History, the true History, began only in 2184, when men freed himself definitely from his animalism, from the memory of the smell and taste of blood, the taste of meat either raw or cooked] (Real 2015: 55).

[22] "Hordas bárbaras armadas e aparatosas da Velha Europa, acicatadas por tambores primitivos, bandeiras hasteadas e brados dos chefes façanhudos, cercam os nossos cidadãos nas praças e invadem os nossos Depósitos Alimentícios, buscando comida, tentando dissolver o pó alimentício em água, fazendo explodir as caixas herméticas pressurizadas onde conservamos os legumes, vegetais e frutos naturais intactos".

[23] "Sem a conexão ao Grande Cérebro Electrónico, registamos apenas uma memória existencial, a recordação das acções, comportamentos e raciocínios que constituem a unidade da nossa identidade pessoal, a nossa biografia, não raro ficcionada a partir da realização virtual dos nossos desejos".

[24] "Os técnicos superiores tinham concluído que a maioria dos neo-europeus e adolescentes nunca se integrariam no Império Americano cumprindo escrupulosamente os regulamentos.

Devido à sua formação e à sua personalidade, os neo-europeus tornar-se-iam cidadãos marginais, seres contestatários, perturbadores da normalidade regulamentar, fundamento da felicidade do Império.

O seu passado privilegiado na Nova europa e de batalhadores pela sobrevivência na Novíssima Europa apontava para que a maioria se revoltasse, exigindo continuamente novas condições de vida, melhores instrumentos científicos, porventura laboratórios mais bem equipados, contactos com universidades e, até, quem sabe, condições de liberdade e de igualdade sociais—exigências não aceites pelo Império, que para eles estipulara, não uma vida de técnicos superiores, mas uma vida de cidadãos normais.

Com evidência, garantida pelos melhores técnicos de psicologia do império, a maioria tornar-se-ia, a breve ou a longo trecho, um problema social, quiçá alguns, estimulados pelos obstáculos, poderiam tornar-se líderes de futuras revoltas regionais. [...] Os militares foram encarregados de matar—sem dor, garantiu Jorge Tomás com força desproporcionada, subindo o tom das palavras—todos os adultos e adolescentes levados na nave militar. Como foram mortos, não sei.

O mais de meio milhar de crianças seria entregue pelos três militares a uma agência, que as venderia a famílias americanas que desejavam ser pais sem o sofrimento da gravidez e as preocupações do tratamento de um bebé. São famílias brancas, disse Jorge Tomás, consolando-me. As crianças ficarão bem, disse, mitigando-me a dor, assustado com a mão que eu levara ao coração.

Na nova família, os mais pequenos esquecerão definitivamente a Novíssima Europa e serão adultos felizes segundo os regulamentos.

Não valia a pena continuar a viver.

Pedi para ficar só, fui intempestivo.

Morreria nesse dia ao poente, decidi, eu, o último verdadeiro europeu.

Amanhã já não veria o sol nascer".

[25] Fátima Vieira, apud. Coelho, 2016.

[26] Fátima Vieira, apud. Coelho, 2016.

[27] Fátima Vieira, apud. Coelho, 2016.

BIBLIOGRAPHY

Aguiar e Silva, Vítor Manuel de. *Teoria da literatura.* 8ª ed. Coimbra: Edições Almedina, 2007. 817 p. ISBN: 978-972-40-0422-8.

Almeida, São José, Ribeiro, Nuno, and Real, Miguel. "Alemanha é "a âncora de uma futura Europa"." Entrevista a Miguel Real publicada dia 22/2/2015 no Jornal *Público.* Público. http://www.publico.pt/ n1686786. accessed 14/5/2016.

Bacon, Francis. *Nova Atlântida e a Grande Instauração.* Tradução, introdução e notas de Miguel Morgado. Lisboa: Edições 70, 2008. 103 p.

——. "New Atlantis." In *The Major Works.* Edited with an introduction and notes by Brian Vickers. Oxford: Oxford University Press, 2002. 457–489 p. ISBN: 9780192840813.

Bessa-Luís, Agustina. *O Concerto dos Flamengos.* 2ª ed. Lisboa: Guimarães Editores, 2002. 338, [334] p. ISBN: 972-665-386-X.

Branco, Fernando Castelo. *Lisboa seiscentista.* 4a ed. Lisboa: Livros Horizonte, 1990. 262 p. ISBN: 972-24-0793-7.

Brizuela, Leopoldo. *Inglaterra, una fábula.* Madrid: Alfaguara, 1999. 403 p. ISBN: 84-204-4163-5.

Callenbach, Ernest. *Ecotopia: the notebooks and reports of William Weston.* Berkeley, Calif.: Banyan Tree Books, 1975. 167 p.

Carvalho, Mário de. *Um deus passeando pela brisa da tarde.* 3ª ed. Lisboa: Caminho, 1995. 319, [319] p.

Chantal, Suzanne. *A Vida Quotidiana em Portugal no Tempo do Terramoto.* Lisboa: Livros do Brasil, 1986.

Cláudio, Mário. *Peregrinação de Barnabé das Índias.* 1ª ed. Lisboa: Dom Quixote, 1998. 282 p. ISBN: 972-20-1486-2.

——. *A Quinta Das Virtudes.* 2ª ed. Lisboa: Quetzal, 1991. 376, [374] p. ISBN: 972-564-101-9.

Coelho, Beatriz Dias. "500 anos depois, o sentido de Utopia não se perdeu." Público. http://www.publico. pt/n1718282. (accessed 14/5/2016).

França, José-Augusto. *Lisboa pombalina e o iluminismo.* 3ª ed. rev. e actualizada. Pref. Pierre Francastel. Venda Nova: Bertrand, 1983. 407 p.

Hutcheon, Linda. *A poetics of postmodernism: history, theory, fiction.* Londres: Routledge, 1988. 269 p. ISBN: 0-203-37141-0.

Huxley, Aldous. *Admirável Mundo Novo.* Tradução de Mário Henrique Leiria. Lisboa: Livros do Brasil, 2006. 258 p. ISBN: 978-972-38-2818-4.

——. *Brave New World.* New ed. Introduction by David Bradshaw. Londres: Vintage, 2004. xxxviii, 229 p. ISBN: 0099458160

Jardin, Alexandre. *L'ile des gauchers.* Paris: Gallimard, 1995. ISBN: 2070740307.

Kalewska, Anna. "Hipopotamy w Delagoa Bay (2013) Carlos Alberto Machado—pierwsza portugalska powieść postkolonialna o podwójnym wykorzenieniu, z Portugalii i Mozambiku." In *Współczesne literatury afrykańskie i inne teksty kultury w świetle badań postkolonialnych.* Ed. Charchalis, W. Warsaw: Aspra, 2015. 147–170.

Lukács, György. *The historical novel.* Translated from the German by Hannah and Stanley Mitchell; introduction by Fredric Jameson. Lincoln: University of Nebraska Press, 1983. 363 p. ISBN: 0803279108.

Machado, Carlos Alberto. *Hipopótamos em Delagoa Bay.* Jorge Aguiar Oliveira, fotogr.; Raúl Henriques, revisor. Lisboa: Abysmo, 2013. 288, [284] p. ISBN: 978-989-98019-3-6.

Medeiros, A. "*Memórias de Branca Dias*; uma releitura, no romance histórico contemporâneo, sobre judeus perseguidos." *Historiae* 6.1 (2015): 9–27.

Monteiro, Maria do Rosário. "Utopia III or an Ambiguous Humanist Utopia for the Second Millennium." In *Proportion, (dis)Harmonies, Identities* Eds. Kong, Mário S. Ming, et al. Lisbon: Archi & Books, 2015. 278–287 p. ISBN: 978-989-97265-0-5.

More, Thomas. *Utopia ou a Melhor Forma de Governo.* 2ª ed. Tradução, prefácio e notas de comentário de Aires do Nascimento. Estudo Introdutório de José V de Pina Martins. Lisboa: Fundação Calouste Gulbenkian, 2009. 425. ISBN: 978-972-31-1309-9.

——. *L'Utopie.* Ed. Prévost, André. Estudo e tradução de André Prévost. Paris: Nouvelles Éditions Mame, 1978. 783 p. 782 leaves of plates. ISBN: 2728900892.

——. *Utopia.* The Complete Works of St. Thomas More. Eds. Surtz, Edward and J.H. Hexter. Vol. vol. 4. New Haven and London: Yale University Press, 1965. cxciv, 629 p. ISBN: 9780300009828.

Morris, William. *News from Nowhere, or, An Epoch of Rest: Being Some Chapters from a Utopian Romance.* Cambridge texts in the history of political thought.

Ed. Kumar, Krishan. Cambridge: Cambridge University Press, 1995. xxxii, 229 p. ISBN: 0521422337.

Mumford, Lewis. *História das Utopias*. Tradução de Isabel Donas Botto. Lisboa: Antígona, 2007. 267 p. ISBN: 978-972-608-190-6.

——. *The story of Utopias*. London: Harrap, 1923. 338 p.

Northmore, Thomas. *Memoirs of Planetes or a sketch of the laws and manners of Makar. By Phileleutherus Devoniensis*. 1ª ed. Londres: printed by Vaughan Griffiths; and sold by J. Johnson; and J. Owen, 1795.

Oliver, Joe. *Uniorder; Faça você mesmo o paraíso*. Óbidos: Unified Loka Publications, 2014. 100 p. ISBN: 978-1-291-92666-8.

Orwell, George. *Mil Novecentos e Oitenta e Quatro*. 4ª ed. Traduzido por Ana Luisa Faria. Lisboa: Antígona, 2004. 326 p. ISBN: 972-608-053-3.

——. *Nineteen Eighty-four*. Londres: Penguin, 1998. 326 p. ISBN: 014027877X

Piercy, Marge. *Woman on the edge of time*. 1st. New York: Knopf, 1976. 369 p. ISBN: 0394499867.

Pina Martins, José V. de. "Estudo Introdutório à Utopia Moriana." In *Vtopia*. Ed. Pina Martins, José V. de. Lisboa: Fundação Calouste Gulbenkian, 2006. 9–99. ISBN: 978-972-31-1309-9.

——. *Utopia III*. 2ª ed. Braga: Appacdm, 2005. 475. 972–86699–62-X.

——. *Utopia III*. Lisboa: Editorial Verbo, 1989. 570. 972-22-1875-1.

Platão. *A República*. 9ª ed. Introdução, tradução e notas de Maria Helena da Rocha Pereira. Lisboa: Fundação Calouste Gulbenkian, 2001. LX, 511, [514] p. ISBN: 972-31-0509-8.

Plato. *The republic*. Edited by G.R.F. Ferrari; translated by Tom Griffith. Cambridge: Cambridge University Press, 2000. 382 p. ISBN: 9780521484435.

Priore, Mary del. *O Mal sobre a Terra; Uma História do Terramoto de Lisboa*. Rio de Janeiro: Topbooks, 2003.

Real, Miguel. *O Último Europeu 2284*. Alfragide: Dom Quixote, 2015. 277 p. ISBN: 978-972-20-5639-7.

——. *Nova Teoria do Sebastianismo*. 1ª ed. Alfragide: Dom Quixote, 2014. 262 p. ISBN: 978-972-20-5404-1.

——. *Nova Teoria da Felicidade*. 1ª ed. Alfragide: Dom Quixote, 2013. 182 p. ISBN: 978-972-20-5177-4.

——. *O Romance Português Contemporâneo: 1950–2010*. 2ª ed. Alfragide: Caminho, 2012a. 261 p. ISBN: 978-972-21-2550-0.

——. *Nova Teoria do Mal: Ensaio de Biopolítica*. 2ª ed. Alfragide: D. Quixote, 2012b. 189 p. ISBN: 978-972-20-4895-8.

——. *Introdução à Cultura Portuguesa: séculos XIII a XIX*. Pref. Guilherme d'Oliveira Martins. Lisboa: Planeta, 2011. 307 p. ISBN: 978-989-657-154-2.

——. *A Morte de Portugal*. Porto: Campo de Letras, 2008. 121, 124 p. ISBN: 978-989-625-224-3.

——. *Eduardo Lourenço; Os Anos da Formação 1945–1958*. 2ª ed. Lisboa: INCM—Imprensa Nacional Casa da Moeda, 2007. 254 p. ISBN: 9789722711999.

——. *O Último Eça*. 1ª ed. Lisboa: Quidnovi, 2006a. 236 p. ISBN: 978-972-8998-43-1.

——. *O Último Negreiro*. 1a ed. Lisboa: QuidNovi, 2006b. 398 p. ISBN: 978-972-8998-39-4.

——. *A voz da Terra*. Matosinhos: QuidNovi, 2005a. 316 p. ISBN: 989-554-233-X.

——. *Atlântico: a viagem e os escravos*. Il. Adriana Molder; fot. Noé Sendas. Lisboa: Círculo de Leitores, 2005b. 143 p. ISBN: 972-42-3494-0.

——. *Memórias de Branca Dias*. Lisboa: Temas e Debates, 2003a. 170 p. ISBN: 972-759-654-1.

——. *O Essencial sobre Eduardo Lourenço*. Lisboa: IN-CM, 2003b. 111 p. ISBN: 972-27-1276-4.

——. *Geração de 90: Romance e Sociedade no Portugal Contemporâneo*. Lisboa: Campo das Letras, 2001. 163 p. ISBN: 972-610-345-2.

——. *A visão de Túndalo por Eça de Queirós*. Lisboa: Círculo de Leitores, 2000. 214 p. ISBN: 972-42-2322-1.

——. *Portugal: Ensaio*. Lisboa: Difel, 1998. 195 p. ISBN: 972-29-0410-8.

——. *Narração, Maravilhoso, Trágico e Sagrado em «Memorial do Convento» de José Saramago*. Lisboa: Caminho, 1996. 98 p. ISBN: 972-21-0543-4.

——. *G.W. Leibniz: O Paradoxo e a Maravilha*. Sintra: Sintra Editora, 1995. 143 p. ISBN: 972-596-023-8.

——. *Princípios de Filosofia e Quadro sobre a Ética*. Lisboa: Mar Fim, 1987. 86 p.

Real, Miguel, and Oliveira, Filomena. *Vieira, o Céu na Terra: nos 400 anos do nascimento do Padre António Vieira, uma homenagem*. Setúbal: Edições Fénix, 2015. 85 p. ISBN: 978-989-667-921.

Saramago, José. *Ensaio sobre a Lucidez*. Lisboa: Caminho, 2004. 329 p. ISBN: 972-21-1608-8.

——. *Levantado do chão*. Lisboa: Caminho, 1980. 366 p.

Swift, Jonathan. *Gulliver's travels*. Ed. Rawson, Claude Julien. Edited with an introduction by Claude Rawson and notes by Ian Higgins. Oxford: Oxford University Press, 2005. 362 p. ISBN: 0-19-280534-7.

Wells, H.G. *A Modern Utopia*. Nova Iorque: Charles Scribner's Sons, 1905. 392 p.

Zamiatin, Evgenii Ivanovich. *We*. New York: Dutton, 1924. xiv, 286 p.

Part II   *Architecture—Urbanism—Design*

The utopia of paradise in architecture—gardens, countryside, and landscape in Roman and Renaissance villas

Inês Pires Fernandes

Faculdade de Arquitectura, Centro de Investigação em Arquitectura, Urbanismo e Design, Universidade de Lisboa, Portugal

ABSTRACT: A person, naturally creative, lives of utopias. Architecture is a way for humans to manifest their creativity, motivation, their life and make those utopias possible. In moments of strong urban development, people sometimes desire the restraint of country life. This way of living materializes the ideal of the relation between people and their origins. In these moments, throughout architectural history, expressions of this kind of utopia appear—the villas. Although their formal and artistic variety being product of centuries, the Villa is always the result of the owner's will for villeggiatura, for contact with nature and the leisure activities associated to it. These drives are present in the architecture of the Villas, which expands beyond the main building, includes leisure structures, gardens, and similar buildings. Beyond leisure purposes, the Villa is also equally apt for agricultural purposes, potentiating land resources. The villa project, over centuries of evolution and adaptation, transcends architecture itself and equally transforms its surroundings, by making architecture, gardens and nature a single unit, an entity in symbiosis—the maximum expression of the utopia of country life.

Keywords: Villa, Landscape, Utopia, Country, Garden

1 UTOPIA

Humans, being naturally rational and producers of cultural artefacts, build their own reality through architecture. Although being a projection, this reality does not always correspond to the ideal of functional and aesthetic perfection. Unable to fix continually the world where they live in, humans create Utopias.

Utopia, a word built from the Greek language, means "no place". Therefore, Utopia refers to a non-existent place in the real world, a fantasy created by a person to express an ideal world, place, or society.

1.1 *The garden*

The garden has always been conceived, throughout social and architectural history, as representation of paradise, the perfection of nature opposite to human imperfection.

> The image of walled garden has thus always been the setting for man's longed-for victory over human frailty and imperfection." (Schröer, p. 9).

This myth of a harmonious location gave rise to the Persian word *pairi-dae'za* meaning "walled" etymological origin of the word "paradise" (Schröer, p. 9).

According to ancient myths, the garden enclosed by walls, with a tree or a fountain in the Centre and four "rivers" is the archetype of paradise on Earth and the place of Man's origin, where he lived in peace, freedom and harmony, protected from the imperfection of the outside world—The Garden of Eden.

This archetype, described in the arts, is the basis of the *hortus conclusus*, a concept of an ideal and built nature. This model has been continuously recreated in architecture, giving origin, for instance, to the medieval cloister. Another reference to the garden and nature as Edenic places is found, for instance, in Virgil. The author idealized an earthly paradise—Arcadia—where its inhabitants lived in harmony with nature and its elements, a *locus amoenus* where they can find peace, love, and the feeling of freedom away from the harmful outside world (Schröer). This myth, unlike the Garden of Eden, has an exact location in the Hellenic peninsula.

Unlike Eden's Garden that belongs to mythology, Arcadia is the product of 1st century BC artistic culture. On the slopes of Arcadia, there were architectural and hydraulic structures that would provide leisure. The activities of its users, reported over the centuries in painting and literature, are related to the pleasure and the delight, including caves, ruins, and small temples (Schröer, p. 11). Despite its literary origin in the first century BC,

41

Figure 1. Hortus Conclusus—Manuscript "Le Roman de la Rose" British Library. http://www.bl.uk/collection-items/roman-de-la-rose [20–05–2016].

Figure 2. The Arcadian myth. Thomas Cole, 1834. http://www.explorethomascole.org/gallery/items/65 [20–05–2016].

Arcadia was recovered as reference and reflection on garden architecture since the Renaissance, and more clearly during Baroque. In the 15th century, Jacobo Sannazaro reuses the theme of Arcadia in his literary work.

The garden—walled or not—represents the desire of paradise on earth, of reconciliation and contact with nature. This does not appear in the projected garden, in its pure and natural state. It is a humanized nature, moulded according to a person's will and his/her idea of paradise, culminating in the sculpture of green elements—*Buxus* gardens.

Contrary to nature in its pure state—the forests—gardens offer proximity to nature but, simultaneously, outward-facing security which is accordance with the myth of paradise. In cultural, and architecture the garden is considered, until the Renaissance, as an autonomous entity, with specific mythic significance.

1.2 *The ideal of villeggiatura*

Humans, dependent on the culture and the society, are a product of the city. Urban life and growth had some high moments in the history of architecture. Despite being the birthplace of culture, art, and architecture, the city was, in these golden periods, also linked to accelerated rhythms, diseases, and confusion. Therefore, it gave birth to the craving for a more peaceful and healthier life, associated with the country, as opposed to the city.

Country life combines also the possibility of leisure, like walking, horse riding, hunting, among others, as well as, more rarely, agricultural work.

There were several authors throughout the history of architecture that report the benefits of interaction with countryside, including Pliny, Vitruvius, Petrarch, and Alberti.

The utopia of life in the country grew, portrayed in a house amidst a rural territory, reflecting the culture and taste of its owner and it is associated to the surroundings and nature through gardens and outdoor spaces.

2 THE REALIZATION OF UTOPIA

Humans, creators of myths and utopias, use architecture to materialize them. The utopia of a peaceful country life is registered since Roman Empire. In several moments in history, this utopia came linked to a large urban growth (Ancient Rome, Renaissance, and Industrial Revolution). The city's image, birthplace of commerce and industry, begins to be speckled by the frenetic pace but also by plagues and severe sanitary problems.

These practical reasons lead the wealthier classes to search for the country for seasonal periods of rest, associated with practices related to nature—hiking, hunting, etc.

2.1 *The country house archetype*

The country house, an ancient architectural type and the foundation of the exploitation of resources in rural areas—agriculture and livestock—is in its genesis, based on very simple and modest volumes.

The implementation is related to its function and, therefore, it is necessarily close to farmland and waterways to ensure land fertility.

The construction of the complex follows topographic shape, acting as an organic entity, integrating land. In addition to the topography, the building is also exposed to weather conditions; therefore, it is constructed to take advantage from them. (Pires)

This way country house, as a minimum entity in farmland, is structured according to very simple and modest shapes and volumes, according to the principle of fulfilment of their functions—agricultural, storage, production, etc. There are no superfluous or decorative elements, expression of a culture or an artistic movement, attempt to achieve any myth or utopia.

However, with the evolution of society and urban environment in different historical periods, the country house comes to be seen as having further potential.

2.2   *The Roman villa*

The *villa* culture arises associated with a time of intensive urban growth, during which the high social classes look for the quiet, healthy life associated with the countryside. That happened in Western culture during the apogee of the Roman Empire.

Initially, the *villa*, in suburban territory, was merely used for support and farming work, a way to increase the profits of their owners.

With the evolution of the city and the utopia of country life, the concept of *villa* entirely dedicated to the farm was blurred. At the height of the Roman Empire, the noble classes built their own *villas* in suburban territory for their pleasure and entertainment.

The purely agricultural *villa* remained. Their owners lived there permanently and dedicated themselves to its exploitation. Unlike the new emerging class, they did not share the *villa* ideology as a social and *villeggiatura* stage.

Villas gained a more classical architectural style, the result of merging the culture with contemporary architectural and decorative elements. These buildings were classified as suburban or maritime *villas* according to their location and profit.

The change of mentality has reflect in the architecture of these suburban buildings. The farming *villa* concept, with compact and simple shapes, is transformed gradually into a new villa, with more dispersed volumetry (Ackerman), according to the topography of land and its natural resources.

> To fulfil its ideological mission the villa must interact in some way with trees, rocks and fields... (Ackerman, p. 22).

However, the association with pre-existing elements exits in both types of *villas* dedicated to leisure: in the suburban *villas* integrating the main building and associated structures in the slopes,

massive plants, waterways, etc.; in the maritime *villas* recreational structures—galleries, observatories—are structured and distributed along the water line, overlooking the coast or the bays (Naples).

Accompanying this evolution, there is also progress in the compositional elements.

The change of culture is reflected in the main building volumes—from compact (influence of the city) to disperse (related with the country)—and there are new elements of relationship between entities—gantries, recreational structures, terraces, etc. (Pires).

The architectural elements of the main building show the relation with the outside. While the cubic *villa* has, in many situations, large balconies with porches—*loggias*—, dispersed *villa* may have structures such as gantries, galleries, gazebos, etc.

Despite the distinction between rustic *villa* and recreational *villa*, it is common the existence of hybrid cases (Ackerman), where the functions of recreation and farming coexist. In these cases, the building house of the noble family adopts a relevant imbedding into the farming complex. The two complexes may either be completely independent or articulated (*Villa San Rocco* and Villa *Settefinestre*).

This knowledge is the outcome of either archaeological discoveries or the product of literary texts.

The only Roman architecture treaty that survived until today was Vitruvius's *De Architettura* (*circa* A.D. 27). However, regarding the *villa* architecture, the author only refers aspects of its implantation and the arrangement of component parts.

The owners of *villas* with both functions are usually more involved in agricultural labours and they spend more time in the field—they do not share the ideology of the *villa* as a refuge but as a resource.

Those who share it, normally own a suburban *villa* where they live temporarily or even seasonally and without agricultural involvement. There may be specific yields, such as vines, but these productions do not depend on its owner. The *villas* dedicated to leisure had recreational structures that allowed a visual experience of the landscape and decorative elements—like murals on the walls of gardens. This expresses the utopian desire of relationship with nature and country life (not a physical relationship).

The *villa* concept was devastated by the fall of the Roman Empire and the implementation of the Feudal System. This era, based on farms, generates city and urban culture depreciation. In the absence or scarcity of urban life, there is no urge to dwell in the countryside. Thus, there is no record of the construction or usufruct of *villas* until the Renaissance.

Figure 3. Villa Settefinestre. Ackerman, James. *The Villa: Form and Ideology of Country Houses.* 1st ed. London: Thames and Hudson, 1990. p. 47, fig. 2.5.

2.3 *The Italian Renaissance villa*

From the fifteenth century on, with Italian Renaissance gaining a new impulse with the humanists, there was a return to the city, the arts, and consequently the *villa* culture, as refuge of cosmopolitan life. Thus, new buildings begin to emerge in suburban areas of large contemporary cities such as Rome, Venice, but specially Florence.

One of the principles of Humanism was based on the reinterpretation of classical culture, a symbol of a past apogee. However, as there were no surviving physical models for analysis and comparison, the great humanist nobles resorted to the classical literary sources so that through them they could establish a composition model for this architectural type.

The *villa* culture as myth is recovered through Virgil's Arcadia, and his references to a perfect existence associated with country life, opposing the fast pace of business, politics, industry, and the city in general.

Despite based on the work of Virgilio (1st century. B.C.), and Petrarch (sec. XIV), the concept of *villa* lacked physical and concrete architecture, implantation and composition descriptions, given the absence of real inspirational models. Thus, Renaissance *Villas* are based on Leon Battista Alberti's *De Re Aedificatoria* (written from 1443 to 1452). Alberti refers the importance of linking architecture and nature as a whole and emphasizes the importance of the gardens. These should be designed, together with the architectural construction, based on natural and cosmic proportions that guarantee the beauty of the outcome.

One must also highlight the work of Francesco Colonna's *Hypnerotomachia Poliphili* (1499). Although it is a novel, it describes the ideal garden according to the culture of this period.

Through these two works, it is clear the importance of the garden as a component and vital element of architecture itself and of the *villa* culture. The garden, consisting mainly of plants, is a reflection of the nature shaped by human endeavour (Pires, p. 24).

The garden, composed of *buxus,* was walled— still a medieval influence of *hortus conclusus*—and close to the main house, so that from it the area could be contemplated. With the evolution of architectural constructions and the increasing the sense of security, the formal garden adopts new settings and delimitations. It evolves, as architecture does, depending on the artistic culture of the time. It is easy to note the increasing existence of landscape and architectural elements that refer to the myth of perfect garden—caves, *nymphaion*, waterfalls, etc. This becomes particularly clear during the baroque.

However, there are other elements peculiar to the Renaissance *villa*, as vineyards and orchards. These, in addition to contributing to the economic viability of the property, also played a part on the land segmentation and in particular the orchards; they added the fragrant component to the enjoyment of the gardens.

Another element present in the composition of outdoor spaces is, progressively, the woods. While the gardens represent the "safe" nature, tamed by man through geometric shapes, axes and architectural elements, the woods are nature in its pure state, not shaped by people. Here they could engage in activities outside the domestic sphere as horseback riding and hunting.

The construction of Renaissance *villa* occurs mainly on the slopes oriented to the city. This location choice is used systematically in the Medici *villas* around Florence. Factors such as the existence of watercourses and plant masses were also considered, as they would bring an aesthetic appreciation and framing the set. Given the slope structure, the grounds were levelled creating platforms and terraces to maximize the construction and garden areas but also to demonstrate the power of the individual over nature, subduing her to his ideals and wills.

Both house and gardens, sited on the terraces, were subject to a prior plan, based on geometry, axes, and proportions, to achieve the cosmic order advocated by Alberti.

Despite all the restrictions, house, garden, and landscape were considered as a single entity, formed by various organisms in perfect symbiosis.

Figure 4. Villa Fiesole, Florence, Italy. http://www.vil-lamedicifiesole.it/photo-gallery.html [28–02–2016].

Figure 5. Quinta dos Marqueses de Fronteira, *Lisbon, Portugal.*

In what concerns the architecture of the house, the decline of feudalism and of the insecurity feeling, promote changes in formal architectural elements.

The boundaries of the property ceased to be built as walls, which increased and promoted the contact between the main house and the outside.

The Renaissance own language, consequence of a reinterpretation of classical culture, determined the change of compositional elements. The slain won greater dimensions and a new function—open contact with the gardens, orchards, woods, and countryside.

The mastery of perspective contributed to a new relationship between the house and the garden, making use of the sensation of depth, promoting the contact with the architecture backdrop—the landscape.

2.4 *The exportation of the villa culture*

The *villa* culture had such repercussions in the Renaissance that it was imported by other countries, Portugal being one of them. In the fifteenth and sixteenth centuries, given the commercial contacts between Portugal and the Italian Republics, as well as the cultural exchanges, the noble class and some Portuguese-speaking architects had contact with this type of architecture, adapting it to Portuguese geography, scale, and purpose. It was born *Quinta de Recreio*.

The *villa* culture and the taste for classical architecture were also imported by other European countries at different moments. It is noteworthy in

England, the Palladianism in the eighteenth century, where, besides the architectural style based on the *Palladian villas*, the gardens are punctuated by small recreational and pleasure structures—temples, waterfalls, gazebos, among others.

Treasuring the ideal of life in the countryside, there is even mention to the pictorial movement, between the late eighteenth and early nineteenth centuries. Their fundaments are the Romantic landscape painting. This movement, more aesthetic than architectural, advocated a deliberate organics of the buildings, so that they could be integrated into the landscape and surrounding natural elements. This longing for the symbiosis between building and environment, through architectural and/or vegetable elements, gives origin to a new profession—the landscape architect.

3 CONCLUSION

The *villa*, more than a utopia of the simple life in the countryside with its main advantages—escape from the frantic life and the plagues in the cities—it is a physical and actual architectural model. It embodies the country house ideal, where it is possible to rest and enjoy leisure and recreation activities associated with nature. It can also be associated with an agricultural component that ensures the property economic viability. The *villa*, despite its typological developments and variants, always represents the wish of symbiosis between human beings, architecture, and landscape. This symbiosis is ensured by architecture, with elements such as *loggias* and gantries, which allow the enjoyment of the house, gardens, and the landscape. Beside the architecture of the house, also the planned gardens ensure a harmonious transition between architecture and nature. The moulded vegetable elements, architecture, landscape, and hydraulic structures guarantee the aesthetic value of the complex and the enjoyment of nature by its owner. Finally, landscape is the last compositional element of the *villa*, vital for its implementation. Landscape is the basis of the *villa's* location, structure, and building.

The *villa*, with the buildings and its gardens, changes paradisiacal country life from mere utopia to an attainable and civilized reality.

BIBLIOGRAPHY

Ackerman, James. *The Villa: Form and Ideology of Country Houses.* 1st ed. London: Thames and Hudson, 1990.

Insausti, Pilar. "Mito y Naturaleza. Del Paraíso al Jardín Medieval". *Arché—Publicación del Instituto Universitario de Restauración del Patrimonio de la UPV.* No 4–5. 2010. pp. 227–236. In https://riunet.upv.es/bitstream/handle/10251/31072/2010_04–05_227_236.pdf?sequence=1. [accessed 25–02–2016].

Vitrúvuo. Tratado de Arquitectura. (traduzido por, Justino Maciel). 2ª ed. Lisboa: IST Press, 2006. ISBN: 978-972-8469-44-3.

Norberg-Schulz, Christian. Arquitectura Occidental. 2ª ed. Barcelona: Editorial Gustavo Gili, 1985. ISBN: 84-252-1157-3.

Pires, Amílcar. *A Quinta de Recreio em Portugal: Vilegiatura, Lugar e Arquitectura.* 1ª ed. Lisboa: Caleidoscópio, 2013. ISBN: 978-989-658-245-6.

Schröer, Carl. *Garden Architecture in Europe.* Köln: Tashen, 1992. ISBN: 3-8228-0540-8.

Virgílio. *Bucólicas, Geórgicas, Eneida.* Lisboa: Temas e Debates, 2012. ISBN: 9789896441944.

Geometry of power: Ideal geometrical shapes in military Renaissance architecture of Ducato di Urbino and Venice Republic 1478–1593

Paolo Bortot
Technology and Technical Design, ITT "M. Buonarroti", Trento, Italy
CDA, Theory and History of Architecture, University of Lisbon, Portugal

ABSTRACT: In the second half of the XV century and throughout the XVI century Europe went through the revival of the Arts and a renewed interest in human activities that will be known as Humanism in its initial stage, Renaissance as the middle phase and Mannerism in its final stage. "The Renaissance civilization" spread throughout the Italian peninsula, especially in central and northern cities, both big and small.

During the same period, the armies saw the introduction of new equipment, such as the use of gunpowder, which in turn caused the functional and formal reassessment of defensive architecture. Outposts, fortresses, and cities are redesigned to adapt to the presence of cannons and culverins that influenced the physical layout of defensive architecture.

In central Italy and in the Po plain, we meet the military works of the great architects of the time, among them, Francesco di Giorgio Martini and Michele Sanmicheli, but also the theoretical writings and drawings by Antonio Averlino, Cesare Cesariano, Pietro Cataneo, and Daniele Barbaro.

The architects mentioned above, who were at the service of the city-states, the lords, and the Republics of the time, plan and design the defence of the city but also the ideal city with its perfect geometric lines, leaving us with many testimonies of the rich and prolific era of experimentation and invention. In 1593, Giulio Savorgnan will the first Star shaped Fortress City, based in the Friuli Plain, Palma Nova, bringing to fruition the dream of ideal harmony using perfect geometrical shapes which will be reused through the century. This city will have great influence over all Europe.

The description of some military constructions made according to the new geometrical shapes, will allow us to understand the origins, development and architectural features of these buildings that incorporate technical defence, offensive force and the harmonic proportions of the reborn Vitruvian architecture also intended as a symbolic representation of the strength of political power.

Keywords: Military Architecture, Renaissance Geometry, Science of Fortification, Territory end Landscape

1 THE ORIGINS: FRANCESCO DI GIORGIO MARTINI IN THE URBINO TERRITORIES

In the second half of the XV century a civilization characterized by the presence of City-States which extended their force-based domain on neighbouring territories, develops in the mountainous regions of central Italy and in the northern plains. It is for this reason that they are studied and military constructions are developed in defence of the civil buildings functioning as military support and control of the territory.

Emblematic in the case is the story of Architect Francesco di Giorgio Martini who was initially formed as a painter at the court of Federico da Montefeltro, Duke of Urbino, and then became involved in sculpting, civil, and military architecture and the Art of War.

He belongs to that host of marvellous Renaissance artists, who's genius seems to know no limits to its expansion, and by who's work transformed not only marble and bronze but also wood, clay and even papermache, were into precious material[1]."

Francesco di Giorgio participates in the construction of the new defensive structures adapting them to resist the new firearms[2]. The walls become tilted at the base and thicker. He introduces new flanking structures such as the "Rivellino" and the "Capannato" or "Casamatta".

In the designs of the Magliabechiano Code, found in Florence, he uses anthropomorphic shapes as outlining tracks to the forts geometry adapting the shapes to the needs of the contemporary military technique, but also practically to the site's topographic characteristics.

47

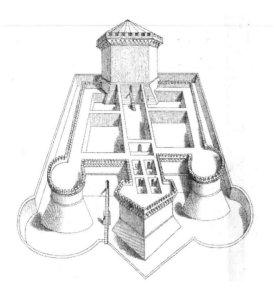

Figure 1.   Rocca of Sassofeltrio (Rocchi E., 1900).

Figure 2.   Rocca of Ostia (Rocchi E., 1900).

The form of the old wall was represented in the "Good government allegory" painted in Siena by Ambrogio Lorenzetti.

This image of the medieval city when compared with the work of Francesco di Giorgio allows us to comprehend the revolutionary magnitude of these new shapes.

In the "Rocca di Sassofeltrio"[3] (1478–1480) the square layout is characterized on one side by a central polygonal Ravelin, located between two angular circular bastions along a curtain, while on the opposite side, in the middle, we find a pentagonal tower.

Similarly, the "Rocca di Sassocorvaro" (1476–1478) has a similar but more compact layout, given the location of the inhabited centre, which is within.

In the Fortress of Ostia (1483–1486), attributed to both Martini and Giuliano da Sangallo, the design is in the form of an equilateral triangle, which has two circular bastions at the top and a triangular one. In the fort's layout we find the equilateral triangle and the square, both perfect shapes and symmetrical, but—in Sassocorvaro- a reference to zoomorphic shapes, in this case a turtle, with symbolic functions of defensive force and longevity.

Martini's work is characterized by formal research, but his solution appear often conditioned by the context.

In the Fortress of Aulla[4], erected in Tuscany in the first half of the XVI century, we find that the model of the fort is based on a square plan with four angular polygonal bastions: this model

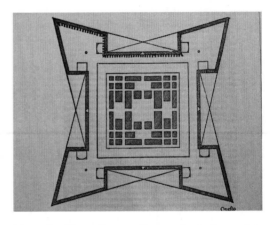

Figure 3.   Pietro Cataneo, Square city, built on plain field, 1554.

will spread throughout Europe in the following centuries.

The ideal shape of the walls of the city has only been theorized at the beginning of the XVI century. Antonio Averlino, called the Filarete, designs the ideal city[5] based on an eight-sided star inscribed in a circle called Sforzinda in honour of the Duke of Milan Francesco Sforza. In this case the attention of the architect is directed more at functional buildings, such as the Hospital, the house of Knowledge, the Buildings of Vice and Virtue, more than to military defence problems.

The case of Venice is different. The city, defended by the sea, chooses to remain without

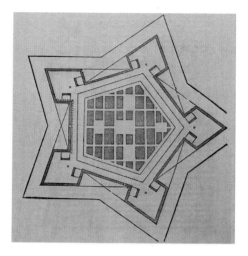

Figure 4.   Pietro Cataneo, Pentagonal city built on plain field, 1554.

Figure 5.   Bastion of the Maddalene. Verona. (Author, 2016).

walls to affirm the idea of freedom, but also the eastern imperial force through the symbols exhibited in San Marco square[6] that represent it as:

> [...] the new Constantinople (or new Jerusalem) as it is represented in the second half of the XV century by the architects Mauro Coducci, Pietro Lombardo, Giorgio Spaventa, Antonio Rizzo, or the paintings of Carpaccio. (Foscari and Tafuri 1983: 42).

In the same XV century the Republic of Venice expands to the mainland until the plains of the Po River and Friuli: "Vicenza and Verona were annexed in 1406, Udine in 1421, Brescia and Bergamo in 1428" (Delumeau 1984: 34). It is however at the beginning of the XVI century that Venice changes its ideal role identifying with ancient Rome. This program of cultural renewal becomes visible in both the Roman language of Sansovino used in the renewal of Piazza S. Marco, and in the language of the renovated Gates of mainland cities.

The strength of these cities is seen in the construction of defences with bastions and curtain walls that use forms and techniques so modern that they stand out as originally visionary and inventive.

2   THE MODERN SHAPE: MICHELE SANMICHELI AND THE VENETIAN TERRITORIES

In 1527, Michele Sanmicheli, following the precise indications of Francesco Maria Della Rovere, Duke of Urbino and head of the Venetian army, builds the bastion of the Maddalene in Verona, the first example in the world of a polygonal bastion with internal lateral flanking casamatte.

Inserted in the new walls of the city, this bastion was used with the internal changes made by the military Austrian engineer Franz von Scholl[7], until the beginning of the second half of the nineteenth century. Sanmicheli also designs the doors of the city of Verona[8].

Emblematic is the example of "Porta Nuova" (1532) built in white stones using a Doric order that represents strength and power. The semi columns and central pillars supporting a Doric trabeation with a triangular pediment. The external side, which faces the countryside, is completely covered in stone. The inside is less rich, using stone and bricks. The central element of the entrance reminds us of Greek times filtered through the Roman culture.

The work of Sanmicheli in Verona is therefore a linguistic exercise of proposing again the classical language of antiquity via optical repetition, or better of roman Verona as confirmed in the architecture of Palazzo Bevilaqua.

In "Porta S. Zeno" (1532), the experimental application of the Vitruvian language continues, but it is in "Porta Palio" (1547) where we encounter the affirmation of the Doric order's strength. In this last work, the exterior side of the city has pairs of imposing columns that support a huge trabeation—with metopes and triglyphs- in which the minor order of "lesene" is inscribed—always Doric in style- that frames the entrance. In the extremities of the external facade, the semi columns are transformed into pillars to emphasize the solution in the corner. The inner side is different: is presents the

Figure 6. Porta Nuova. External facade. (Author, 2015).

Figure 7. Porta Palio. External facade. (Author, 2016).

same Doric order but with six pairs of giant twin semi columns made with rusticated stonework.

The trabeation has the same structure of the outer facade and similarly to the outer side in the extremities of the structure the semi columns transform into pilastri with the same solution of the corner.

With Porta Palio, Michele Sanmicheli reaches, in the territories of land subject to the Venetian administration, the perfect synthesis between "munire e ornare", between resistance force and classical language.

According to the Author, though, the most modern of Sanmicheli's achievements is Fort S. Andrea (1535) built in Venice to defend the main entrance of the city. This structure is reinforced with 30 cannons in casemate, has a multilinear symmetrical layout, and is formed by two rectilinear portions separated by a semi-circular element that has at its centre a high construction, which represents a city's entrance. The columns of the central element, covered in Istria white stone, are rusticated stonework of the Doric order and anticipating the stylistic forms that we find in the inner part of "Porta Palio" in Verona. This Fort, with its purely functional plant, anticipates by approximately three hundred years the most modern forms of the XIX century by detaching itself from the tyrannical perfection of the geometry of the time.

During the same period, we find other construction presenting the application to the military architecture of the geometrical shapes of the Renaissance and modern polygonal bastions.

The Fortress of Peschiera (1549–1553) is built on the shores of Lake Garda—probably designed by the Chief of the Venetian army Guidobaldo II Della Rovere and Duke of Urbino by the military architect Giulio Savorgnan, not far from Verona (Bozzetto 1997; Concina 1983).

Even here, we can see the Renaissance tension to perfect geometric forms and the modern forms of the bastion applied to military architecture.

The fortified citadel is based on the star shape, with a regular pentagonal layout and large angular polygonal bastions. In this case, given the preeminent over watch function of this military installation, we encounter the classic language only in Porta Verona, while the one facing Milan, Porta Brescia, is completely plain, with the only function being combat.

In 1554 in Venice, Pietro Cataneo prints a detailed study of modern forms of military architecture elaborated within the work "*I primi quattro libri dell'Architettura.*"

In the eighth chapter of the first book, he writes: "of the square city located in the plain field, subject to battery fire." (Cataneo 1554: 10–12) perfecting the scheme on a square plan with angular bastions appeared in the Tuscany Brunella Fortress, demonstrating his knowledge of the flanking technique and creating perfect geometric shapes in function of the use of military force. Furthermore, Cataneo presents, in his treaty, the pentagonal regular plan, regular hexagonal and regular heptagonal equilateral layouts for forts: these last two forms presenting reinforcement bastions in the middle of the curtain walls. In his works, he also outlines the walls of every town in perspective by highlighting the importance of the geometric shapes within the definition of urban space that affirm its force on the countryside.

Cataneo also deals with "the city of the Prince with equilateral decagonal shape, located in a plain field, with its pentagonal citadels" (Cataneo 1554:

Figure 8.   Porta Palio. Interior facade. (Author, 2016).

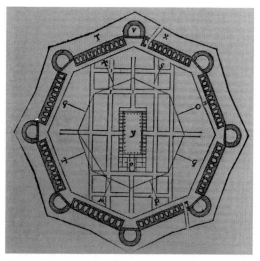

Figure 9.   DANIELE BARBARO. Fortified city, 1567.

12–13) representing it in perspective. Here we find the union between perfect geometrical shapes of the ideal city and a five-sided polygonal citadel almost to reveal the harmony via the relationship of octaves (1:2) expressed by 5 and 10 represented by curtain walls on the vertices, which are impregnable bastions.

Even Daniele Barbaro in Chapter V of the first book of his ten books on architecture published in 1567, tackles the theme "of the foundations of the curtain walls and of the towers" (Barbaro 1567: 44–52) proposing the idea of a city with a circular plan with defensive circular towers that represent a technical involution with respect to the positions of Francesco Maria Della Rovere and Sanmicheli, in these treaties we have seen how defensive strength is implemented in harmony with perfect geometric shapes which separate the space of the city from the dangers of the countryside.

Sebastiano Serlio, in his unpublished works, reminds us that the true strength lies in human behaviour, in fact:

> The prince of noble and liberal spirit, just lover of is people, fearful of Good need no fortress" as opposed to "the cruel and greedy tyrannical Prince, robber of others belongings, rapist of virgins, wives and widows [...] all fortresses in the world cannot keep safe. (Serlio 2001: 91–92).

The Prince tyrant, expresses the opposite of this ideal but also for him there are in Serlio's treaty two projects of residence: a palace-fortress with a rectangular plan defended by angular bastions, polygonal for flanking—which rehashes formal aspects presented by Giuliano da Sangallo—but also a palace with a pentagonal plan, easier according to him, to defend themselves.

The writings of Serlio lead us to reflect on the real conditions of life in the cities of that time, where war, sieges and use of force, would lead to the end of the XVI century to the realization of the debate on the theory and practice of permanent defence with the construction of Palma Nova (1593). This star-shaped fortified city, designed by Giulio Savorgnan in the centre of the Friuli plain, is the first example in the world of a star-shaped city with nine sides with defensive bastions where the curtain walls meet. This work the ideals of geometric perfection and application of new defensive theories to the permanent fortification to fruition, which is intended to serve as a base model to the subsequent military constructions throughout Europe.

NOTES

[1] "Egli appartiene alla schiera di quei meravigliosi artisti del Rinascimento, il genio dei quali sembra non conoscere confini alla sua espansione, e per la cui opera, non solamente il marmo e il bronzo, ma il legno, la terracotta e perfino la cartapesta, divenivano materie preziose." (Rocchi 1900: 6).

[2] See Nicholas Adams (1993: 158).

[3] In the historiography of the Rocca is also called Rocca di Sasso Feretrano (Rocchi 1900: Tav. I).

[4] This military building, which we do not know with certainty the designer, also called Fort of the Brunella, finds himself on a different scale in the planimetric form of the Fort of Elvas in Portugal. On the influences of the Italian School of fortifications see (Lopes 2005: 105–107).

[5] The city is described by Averlino' in its tractate... Written between 1460 and 1461. For a general framework on the subject, see (Firpo 1975).

[6] In the year 825 A.D. two Venetian merchants rubano in Alexandria of Egypt the body of San Marco and the carry in the city of Venice where is preserved in the chapel of the Doge that today is the basilica of San Marco. On facciata of the Basilica are also placed the four stolen horses to Constantinople in the IV crusade. From the same city also originates the marble group of the Tetrarchs walled in the eastern corner of the facade of the basilica.

[7] On the Discovery of Franz von Scholl see (Bozzetto 1993).

[8] See (Concina 1995).

BIBLIOGRAPHY

Adams, Nicholas. "L'Architettura militare di Francesco di Giorgio." In *Francesco di Giorgio Architetto.* Eds. Fiore, F. P and M Tafuri. Milano: Electa, 1993.

Barbaro, Daniele. *I Dieci Libri dell'Architettura di M. Vitruvio.* Venezia: Francesco de Franceschi e Giovanni Chrieger, 1567.

Bozzetto, Lino Vittorio. *Peschiera Geschichte der befestigten Stadt.* Peschiera: FRANKE, 1997.

——. Verona. La cinta magistrale Asburgica. Architetti militari e Città Fortificate dell'impero in Epoca moderna. Verona: Cassa Risparmio Verona, Vicenza, Belluno e Ancona, 1993.

Cataneo, Pietro. *I quattro primi libri di Architettura.* Venezia: Aldo Manuzio, 1554.

Concina, Ennio. "Munire et ornare: Sanmicheli e le porte di Verona." In *Michele Sanmicheli. Architettura, linguaggio e cultura artistica nel Cinquecento.* Ed. AA.VV. Vicenza: Centro Internazionale Studi Andrea Palladio, 1995.

——. La macchina territoriale. La progettazione della difesa nel cinquecento veneto. Roma: Laterza, 1983.

Delumeau, Jean. *A Civilização do Renascimento.* Tradução de Manuel Ruas. Vol. vol. 1. Lisboa: Editorial Estampa, 1984.

Firpo, Luigi. *La città ideale del Rinascimento.* Torino: UTET, 1975.

Foscari, Antonio, and Tafuri, Manfredo. L'armonia e I conflitti La chiesa di S.Francesco della Vigna nella Venezia del 500. Torino: Einaudi, 1983.

Lopes, António. *Dicionário de Arquitectura militar.* Casal de Cambra: Caleidoscópio, 2005.

Rocchi, Enrico. Francesco di Giorgio Martini Architetto civile e militare. Roma: Enrico Voghera Tipografo, 1900.

Serlio, Sebastiano. L'Architettura. Libri I-VII e extraordinario nelle prime edizioni. Milano: Edizioni il Polifilo, 2001.

Harmony as a tangible utopia

José Lopes Morgado

Centro de Investigação em Arquitetura, Urbanismo e Design—CIAUD, Faculdade de Arquitetura, Universidade de Lisboa, Lisboa, Portugal

ABSTRACT: Alberti defines harmony (*concinnitas*) in a broad comprehensive manner, however tangible, and made visible by the pleasure of the beauty (*voluptas*), the third element of the Albertian triad, linked to necessity and convenience. This paper poses the question of the relationship of architectural design and harmony, in the imagination of a possible future and the promise of a better world.

Over the centuries, architecture has shown the ability to display a harmony that survived the test of time, with inter-subjective and universal character. The timeless architectural characteristics favour an aesthetic harmony and this, in the context of everyday life architectural experience, enables a lasting ethical harmony, both in individuals and in society. Relating the parts to the whole, and with the difficult task to improve reality, harmony is a quality at once real and ideal, dreamed by utopian thinking that is our best way to combine human needs, desires, and aspirations.

This text argues that, in designing and building another world, the establishment of a certain human order is a utopian aspiration of harmony. In conclusion, in achieving the triadic harmony, that one perceives and feels every day, the tangible utopia is being achieved along with extended horizon.

Keywords: Harmony; Albertian Triad; Utopia; Architecture; Humanity

1 THE ARCHITECTURAL TRIAD

In a real perfect synthesis, architectural triad articulates the dimension of human naturalness and the dimension of architectural culture; in this reconciliation impulse, with a utopian vocation that is being held, is engendered the harmony of architecture.

In fact, as stated in the paper "Harmonia e a tríade de Alberti" [Harmony and Alberti's Triad] (Morgado 2015: 8), in a categorical holistic way

[in the] complex relationship with the operators of Albertian triad, creature and nature belong to the scope of necessity and architecture and culture to the scope of convenience. Beauty generates pleasure in man, exists in nature and architecture, and complete the other two qualities of the triad, resulting in a necessary and balanced harmonious whole. (2015: 26–27).

The utopian idea of creating a better world to live in derived from biological survival logic, life organization, genetic mutation, adaptation to the environment, natural selection, and, evolution of species. This natural impulse sublimates needs and aspirations and fits the purpose of living in architecture, in a symbiotic manner. The evolution in its relationship with environment shaped and encoded a genetic preference, setting the aesthetics of survival with characteristics observed in nature and essential for the life of our ancestors.

Therefore, in Darwinian natural selection, there are "matches between characteristics we like and characteristics that would have improved our chances of survival" (Hildebrand 1999: 10).

There is a utopia thought whenever man makes up for a lack in nature and transcends himself to change and progress; this transformation, aiming compensation of the reality, gives hope to a possible utopia with stability and the balance between nature and culture. Utopia is therefore a matter of culture.

By human transforming will, through the shelter as an adaptation, the individual improves the environment. The transition from nature to culture corresponds to the passage from necessity to convenience. Making the world habitable, individual overcomes his natural condition in architecture, overcoming conflicts between desire and reality, and determines the essential way of life conditions.

Since early Antiquity, architecture responds to three essential qualities. According to Alberti, architecture was established for the benefit of humanity. In his *De Re Aedificatoria* (2004; Alberti 2011), the triad, necessitas, commoditas and voluptas (necessity, convenience and pleasure in beauty) offers a coherent architectural unity.

The elements of the Albertian triad are the three conceptual foundations of architecture; the triad is hierarchical, from the previous necessities of daily life, to which the building is convenient and pleasantly adapted.

According to Alberti, the *necessitas*, belonging to the common order of nature, is linked to the bio-psychological unity of the human being, and implies the materiality of architecture and the constancy of the laws of physics.

So, man should not sacrifice the real necessity, underlying the desire, a second order of need; today's needs are yesterday's conveniences; the evolution creates new expectations in satisfying needs and desires; architecture has always supported the diversity of human will, which sets the built environment.

For Alberti, the *commoditas*, belonging to the order of culture, implies creature and architecture, is linked to the diversity of cultures and desires of human and societies, and refers to the use that ennobles the building, whose righteous dignity serves the common good, individually and socially.

When convenience responds to needs architecture also responds with dignity to the needs of people. This social bound with society makes possible what is felt as suitable. While the need provides uses to many, sometimes convenience only grants advantages to a few (Alberti 2011: 309).

According to Alberti, the *voluptas*, belonging to the order of the subjectivity of human psychology, is linked to the diversity of nature and culture, and refers to the pleasure of beauty, supreme end of the building; beauty is a necessary but not sufficient condition.

In Alberti, beauty is the aesthetic result of the correct and appropriate manner. In this lies the perfection contained in the utopian idea; the third element of the triad, the beauty, requires the presence of the other two vectors, for a full and desirable harmony; actually, "the third, [is] of all the noblest and most necessary" (Alberti 2011: 375).

In fact, when architecture responds to the necessity, one obtains an appropriate convenience, which confers dignity. As a result, one experiences the pleasure of beauty. Thus, the possession of beauty underscores suddenly some epiphany, a revealed fortunate coincidence; the grace of beauty is not definable, it is felt and it is durable.

We experience the built environment as a whole. In its effects, the triad is constituted a utopia ever undertaken or to be undertaken; because we are unable to think the impossible, when the *necessitas* and the *commoditas* precedes the *voluptas*, in this well balanced architectural whole it gives a certain "cathartic harmony" (Morgado 2016: 157).

2 THE TRIADIC HARMONY

In the context of the triad, the beautiful form is the sensitive part of **concinnitas** (harmony), and architecture has the potential to reveal the beauty; the effect of triadic harmony is the result of the three qualities of the triad in perfect agreement, regulating an aesthetic and an ethical dimension of architecture.

As stated in "The Harmony and Alberti's Triad" (Morgado 2015), the appropriate triad balance allows us to appreciate the architectural qualities and to improve the relationship with people; in the harmony, "the construction only makes sense if resolved at the level of necessity, the utility if provides convenience and the beauty if give pleasure" (Alberti 2011: nota, p. 24).

Harmony is unity, the result of the balance of the triad; knowing the totality of human condition, whose equilibrium provides good effects, a responsible architect may not want that the promised architecture will be a lost harmony; to make shine all the qualities is the "[a]rchitecture's harmony, its most important quality" (Pérez-Gómez 2006: 8).

With the traces of perfection of a demanding idealization, the Albertian harmony is assumed as criteria of excellence in the area of human life and for quality in architecture; the harmony creates beauty and, being cause and effect of each other in a holistic unity, beauty is concinnity and concinnity is beauty. In a coherent whole, the beauty "is the agreement and the union of the parts of a whole" (Alberti 2004: note, p. 440).

For an elegant work, it is appropriate that "everything contributes to the decorum and harmony of the whole". Thus, "beauty is the concinnity, in exact proportion, from all parts of the whole to which they belong, so that nothing can be added or subtracted, or changed without that deserves reproach" (Alberti 2011: 197, 377).

The harmony is the attempt to decipher a hidden secret in the beauty of nature, those "dimensions and correspondences of most perfect concinnity", "we just feel them. [...] we desire what is great in nature" (Alberti 2011: 584, 593).

To produce the relief that brings serenity is the intrinsic value of harmony for the potential of architecture, which prepares and promises a beneficial effect. The power of harmony is therefore the most plausible justification for the quality of architecture, which contributes to an idea of improvement, which belongs to the architectural utopia.

Harmony "[c]over all human life and all its principles"; everything in nature "is governed by the law of concinnity. Moreover, there is no greater commitment to nature than making things produced absolutely perfect. What in no way could be without concinnity: or the supreme harmony of the parties we want would"; in a true utopian hope, the Albertian *"edificatoria art* follows in a special way this same concinnity; with it, claims for itself decorum, grace and prestige: and is respected" (Alberti 2011: 593).

Human spirit seeks to contemplate beauty; harmony is "the primary purpose that art of building is chasing; [...] and it is the prize" (Alberti); according to Choay, proper to man, it "is not only an effect, it is also cause and essence" (Alberti 2004: note p. 440).

Harmony is invisible, rationally indescribable, but emotionally felt; its complexity is full of implications, with consequences in architecture; the most obvious consensual effect is beauty; the beautiful is the expression of harmony made visible. Probably architecture still lives a manifest lost canon of concinnity.

Le Corbusier was also concerned with the essence of emotion produced by all-powerful harmony. Only "the architect is able to establish harmony between man and his environment" (Corbusier 2010: 134).

When you cannot achieve perfection, your desire is to mitigate the harm; one aspires to escape the fatal consequences, establishing the circumstances of a possible harmony. For his/her own welfare, she/he wants to apply the improvement to necessary, useful, and beautiful things; and thus "enjoys in plenitude all the comforts of life" (More 2003: 62).

In harmony, the pleasure of the beauty is perfectly legitimate. In the good life, pleasure is the end of all human actions; such is the will of nature; for Thomas More, virtue is to live according to nature; outside the ideal perfection, there is the harsh reality that cannot be free from its condition.

Harmony implies the present of a human being, both psychological and sociologically. It is a consequence of what the individual is and what the society does, in a relationship between reality and dream.

Against the absence of triadic harmony, Alberti proposed "the architect to hold a redemptive task that involves providing existential well-being"; the opposed qualities are united in overcoming, by the architectural totality; with the potential of utopian character, the "aesthetic of *balance*, or agreement between the parties, it underlies the idea of harmony" (Alberti 2011: note p. 37 and 194).

3   THE ALL-POWERFUL UTOPIA

Provided by Albertian triad, the edificatory harmony is pure evidence of a praxis with the will to improve human life, which has the utopian urge to

overcome, with the architectural tribute for existential progression.

Utopia has extended the perimeter of the quality of human life, by enabling the passage of the art of building to an art of living. Anticipating the fulfilment of desires, "construct a hypothetical city, from the beginning. It is the product, apparently, of our needs"; "constructing a complete city, and making that happy" (Plato 2000: II, 369c, IV, 420c).

There are neglected aspirations in human daily life, so the architects have responsibilities to create places that have a value, with the crucial possibility of offering those little everyday amenities, the conveyers of essential joys, that make more pleasant the lives of the inhabitants.

[The soul] is always busy painting images of things that are not present [...] through [...] chimeric trait that our creative poetic power register", in "dreams [...] taken by true experience of real things. (Kant 2012: 105).

There is an aspiration of fantasy, from which emerges the ideal, with "the hope of escaping the difficulties of living" a real longing, due to the demanding "actions and [...] neglected thoughts" (Riot-Sarcey, Brouchet and Picon 2009: 10).

When designing the ideal, without focusing on the real, the architect abandons the "human environment production process. Instead of serving customers, these are converted to being their servers". In a realization displaced from reality, and not fulfilling a supportive social existence and then putting setbacks to society, one "does not solve the problem". From a desirable aspiration, guided by utopia as a project solution, a blurry ideal vision of architect is "a poor compensation for a whole lost world or a remote promise of a possible future world" (Tzonis 1977: 129, 131).

All sustainable magnification of the human being tends, by nature, to be utopian; highlighting the mitigating of daily difficulties of the common man, in architecture, this utopian substance of dreams already exists in the actual experience of global best praxis, always, as Aristotle declared, to make possible the surprising wonderful.

The operative attitude does not mean lower purposes, and is inseparable from a demanding praxis, leading to a good practice; it is useful to contrast with a well-intentioned utopia that, as the promise of a collective dream deferred, "does not take place but betraying itself" (Riot-Sarcey, Brouchet and Picon 2009: 103).

According to Michèle Riot-Sarcey, utopia accommodates to the reality of life and the necessity of human existence. Therefore, the "architect gives form to the already laid utopias, thinking to overcome its impossibility" (Rouillard).; From dissatisfaction is born the dream and a consent, to make this feasible; in this ideal way, the architectural utopia is consistently preventive and, as a whole, there is "a fundamental positivity of architecture, always conceived as a therapy or pedagogy"(Riot-Sarcey, Brouchet and Picon 2009: 26, 27).

The power of the architect's idea is in demanding for a praxis, that "translates into action that which captures in the desires" (Riviale). Indeed, in this optimistic possibility, "[e]ach epoch dreams the following" (Riot-Sarcey, Brouchet and Picon 2009: 32, 38).

Even the delayed project or not actualized has the potential to change the mind-set of the future; by an ideal principle, as of a virtuous "inadequacy of the existing order", architecture contests the real, in a symbolic opposition, projecting "an idea of what could be" (Riot-Sarcey, Brouchet and Picon 2009: 118, 119).

Imagination is the only point of contact with the future; and only man is capable for imagining. As harmony, the anthropocentric utopia not just "comes at a place that delimit it, but in the human that is liberated" (Riot-Sarcey, Brouchet and Picon 2009: 162).

What we have now is a "survival of a past not fulfilled, a kind of reminiscence of an unmet need" given this desire and the consequent failure of hope and present aspirations. If the dream is relatively impossible, however, it can be actual for "the tragic often comes from the end of a utopia" (Riot-Sarcey, Brouchet and Picon 2009: 13, 178).

In addition, utopia is right ahead of time; it is an "unfulfilled dream, but not unachievable" (Dejacques); architectural dream has the character of "premature truths" (Riot-Sarcey, Brouchet and Picon 2009: 202, 257).

Being beauty a promise of pleasure and happiness, it witnesses that perhaps utopia is achievable; and the harmony, revealing the absolute perfection of architectural synthesis is a provider of potential utopia; as a friendly utopia, the harmony exists to test, to correct and to reward architecture.

Architecture, as a kind replacement of paradise, translates the original post-sin realistic vision, compared to the original pre-sin idealistic vision. The expulsion from paradise reflects the passage of human misery to human dignity and the ability to transform; the visions of paradise reveal the state of shortage of unfulfilled desires.

Harmony and happiness are the horizons of utopia; if the adaptation and evolution are simultaneous between man and architecture, the true potential of architecture is to restructuring our ways; utopia deplores the unacceptable living conditions and architecture allows the protected man to dream in peace.

In architecture at the service of man, the design holds the promise for a better life, in a better space and in a better time, transforming the non-place in a good place.

In terms of relevance in the utopian dimension of the architectural triad, the moderated convenience does not have the same urgency of necessity, or the same seductive power of beauty, this unique thing.

In a time without utopias, even without a perfectly organized society, the triad seeks the good in humanity; the triad keeps the ideal of improving the response to necessity and perfecting the response to convenience, while it has the amazing hope of a perfect response to the pleasure of beauty.

In *Republic*, Plato imagines an ideal *polis*; associating order in the city and order in the soul.

… the person who has a good life is blessed and happy […].the government of the city, for us and for you, will be a waking reality rather than the kind of dream in which most cities exist… (Plato 2000: I. 354a, VII. 520c).

Even the magnificent island of Utopia has as ideal, the architectural triad: "eventually live […] in good peace and in the best harmony", the buildings are "conveniently arranged […] according the necessity […]. The buildings, well built and comfortable, are elegant and clean" (More 2003: 26, 73).

4 CONCLUSION: A TANGIBLE UTOPIA

Concluding, the triadic harmony is present as a true tangible utopia in the fullness of the architectural totality. In architecture, the anthropocentric core precedes and provides the beauty and harmony; then, the balanced application of the Albertian triad has the potential to generate the indissoluble whole, present in harmony, and to convey the possibility of a better world.

Architecture influences human experience of the world; thus, if architecture has a purpose, then probably it will affect humans, with the possibility to recover a lost harmony. Overcoming the transformative potential of harmony allows for a powerful architecture, with a utopian ambition.

Already present in everyday reality, in good and harmonious buildings, the pleasure of beauty appears because of an ideal solution; there is the powerful effect of "harmony […] admirable that, in a superior way, delights and holds the soul" (Alberti 2011: 172).

The architectural pleasure already qualifies its success. Under the criteria of the triad, the intelligible *necessitas,* and the sensitive *commoditas,* make visible the *voluptas*; the building responds accordingly to what is humanly intended; the social commitment to the triad creates the conditions for human happiness, here and now. To organize public happiness through the principle of equality implies the sacrifice of individual freedom; hence, in condemning the individual, there is a moral reward of virtue.

According to Le Corbusier's visionary ideal, "product of happy people and producer of happy people", in harmony, "architecture is the masterly, correct and magnificent play" (Corbusier 1977: 7, 25).

Utopia seduces with bright promises for purging the earth one day; "the greatness of the danger is largely offset by the magnificence of the benefit" (More 2003: 131).

In a cultural selection, as in natural selection, the fixed ontology of architecture favours individual ontology. Hence, ethical and aesthetic imperatives to preserve good architecture, and phenomenological experience as an aesthetic enjoyment.

Good ideas anticipate the future, causing a shift in consciousness (Damásio 2000) because we know that we really can shape architecture and configure human life, where we live can be a better place.

We conclude that architecture provides a sublime harmony that, in aesthetics and ethical level, has a beneficial effect on people. This is the aim of utopianism guiding human quest for happiness.

Of course, there are counter-utopias and even poorly designed utopias in need of correction. But the ontology of utopia is in line with the true longings of human beings, so that someday the primary utopian ideals will be a reality for the benefit of humanity.

Today there may be some kind of utopia; an "architecture enhanced in effect, whether in a dimension of harmony, either in a dimension of catharsis, see magnified the holistic force that [...] reinforces the intensification of his power; probably we already have the means to actualize the cathartic power, encouraging a triadic and conciliatory cathartic architecture" (Morgado 2015: 239).

In Aristoteles's *Nicomachean Ethics,* for man the practical horizon admits an alteration. "This is no dream, but a happy waking vision, real as day that will come true for you" (Homer 1996: XIX. 547, p. 408–549; Homero 2003: XIX. 547–548, p. 322–523).

An immense happiness awaits humanity. By degrading the dignity of the sublime soul, humanity expects nothing. This idea is prohibited throughout More's *Utopia.* Only hopeless souls and guilty consciences anticipate the end of utopia.

Fulfilled the promise of human will, the "the city we have just been founding and describing, our hypothetical city, since I don't think it exists anywhere on earth". In a certain sense, the imagination already embodies our wishes in *utopia worlds*

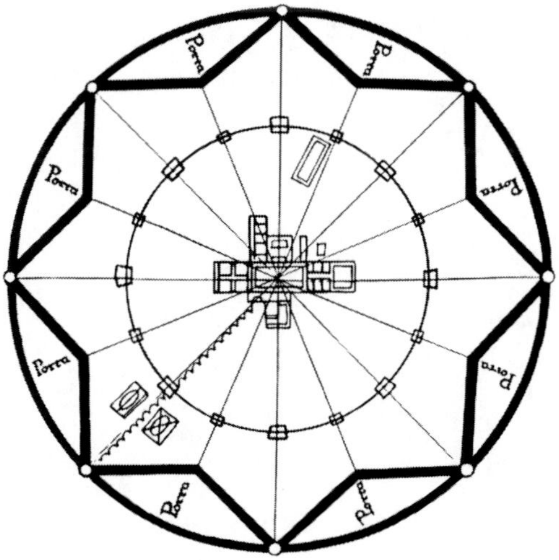

Figure 1.   Plan of Sforzinda, c. 1461–1464, Ideal city. Florence, Antonio Averlino, Filarete.

Figure 2.   Map of Utopia Island, 1516, Ideal city. Leuven, Thomas More.

and frontiers of the imaginary. "Moreover, no matter that the city exists anywhere, or it will exist". Because for now, and "here and on the thousand-year journey we have described, let us fare well"). (Plato 2000: IX, 592a–592b, X, 621d).

Offering ordered life with a reconciled and purified consciousness, the hope to enjoy the desired delights gives force to utopia; but, dazzled before an imagined possibility, reason dreams of an ideal humanity waiting for a better future; with wisdom, the lesson of Aristotle warns that the triumph of rational idealism has its dangers.

From the ancient doctrine of the Pythagoreans, Alberti's aedificatoria theory established the idea of perfection and influenced the model for the ideal city, embodied by Filarete's *Sforzinda* (c. 1464) or by Francesco Colonna's *Hypnerotomachia Poliphili* (1499). Geometry is an essential tool for architectural design, but it is not the line that makes the *lineamentum*. Alberti's lesson warns that only through the proper balance of the triad, harmony enhances certain architectural qualities.

There are very many features in the Utopian commonwealth which it is easier for me to wish for in our countries that to have any hope of seeing realized. (More 1965: 247)

Thomas More last words can be properly adapted to the question of the favourable application of the triadic harmony in a tangible utopia. It has only to do with human will and decision; and, fortunately, the sun of humankind never sets; so, for a more human existence, in this case architectural, one can say: I hope it more because I wish it.

BIBLIOGRAPHY

Alberti, Leon Battista. *Da Arte Edificatória (De Re Aedificatoria 1452)*. Translated by Espírito Santo; Footnotes by Mário Krüger. Lisboa: Fundação Calouste Gulbenkian, 2011. ISBN: 978-972-311-374-7.

——. *L'Art d'Édifier (De Re Aedificatoria, 1452)*. Translated by Pierre Caye; Footnotes by Françoise Choay.

Paris: Éditions du Seuil, 2004. ISBN: 978-202-012-164-4.

Corbusier, Le. *O Modulor*. Translated by Marta Sequeira. Lisboa: Orfeu Negro, 2010. ISBN: 978-989-955-657-7.

——. *Vers une Architecture*. Paris: Éditions Arthaud, 1977. ISBN: 978-972-564-803-2.

Damásio, António. The Feeling of What Happens: Body, Emotion and the Making of Consciousness. London: Vintage, 2000. ISBN: 978-972-564-803-2.

Hildebrand, Grant. *Origins of Architectural Pleasure*. Los Angeles: University of Califórnia Press, 1999. ISBN: 978-972-564-803-2.

Homer. *The Odyssey*. Translated by Robert Fagles; Introduction and notes by Bernard Knox. New York: Penguin, 1996. 541. ISBN: 0-14-026886-3.

Homero. *Odisseia*. Prefácio e Introdução de Frederico Lourenço. Lisboa: Livros Cotovia, 2003. 399.

Kant, Immanuel. *Ensaio sobre as Doenças Mentais*. Translated by Pedro Panarra. Lisboa: Edições 70, 2012. ISBN: 978-972-441-696-0.

More, Thomas. *A Utopia*. Translation José Marinho. Lisboa: Guimarães Editores, 2003. ISBN: 978-972-665-352-3.

——. *Utopia*. The Complete Works of St. Thomas More. Eds. Surtz, Edward and J.H. Hexter. Vol. vol. 4. New Haven and London: Yale University Press, 1965. cxciv, 629 p. ISBN: 9780300009828.

Morgado, José Lopes. *Harmonia: o Potencial Catártico da Arquitectura*. Lisboa: Caleidoscópio-Edição e Artes Gráficas, S.A, 2016. ISBN: 978-989-658-361-3.

Morgado, José Lopes "Harmonia e a tríade de Alberti." In *Proportion, (dis)Harmonies, Identities*. Eds. Kong, Mário S. Ming, et al. Lisbon: ARCHI & BOOKS, 2015. 26-39 p. ISBN: 978-989-97265-0-5.

Pérez-Gómez, A. *Built Upon Love. Architectural Longing After Ethics and Aesthetics*. Massachusetts: The MIT Press, 2006. ISBN: 978-972-564-803-2.

Plato. *The republic*. Edited by G.R.F. Ferrari; translated by Tom Griffith. Cambridge: Cambridge University Press, 2000. 382 p. ISBN: 9780521484435.

Riot-Sarcey, Michèle, Brouchet, Thomas, and Picon, Antoine. *Dicionário das Utopias*. Trad. Carla Bogalheiro Gamboa, Tiago Marques. Lisboa: Texto e Grafia, 2009. ISBN: 978-989-8285-03-4.

Tzonis, A. *Hacia un Entorno No Opresivo*. Madrid: Hermann Blume Edicones, 1977. ISBN: 847-214-114-4.

The perfect dwelling is any place in the heavens: Platonism, mathematics and music: On Kepler's thoughts and the theory of architecture in the Renaissance

Clara Germana Gonçalves
CITAD, Universidade Lusíada de Lisboa, Lisboa, Portugal
ISMAT, Portugal
Faculdade de Arquitectura da Universidade de Lisboa, Lisboa, Portugal

ABSTRACT: This paper sets out to analyse how Kepler's thoughts aligns with that of architects such as Alberti and Palladio: how that thought constructs an architectural place based on Pythagoras/Plato-derived geometry in which music plays a fundamental role.

It seeks to position Kepler as the architect of the ideal dwelling: a cosmic dwelling that was conceived analogously to the earthly dwelling of Alberti or Palladio. Like in Alberti's or Palladio's, Platonic influence is particularly evident in Kepler's work. He applies to his model of the universe the Platonic solids and calculation based on harmonic proportions while dealing with highly symbolic and aesthetic values. He also embodies in his theories fundamental aspects of humanist architecture, such as the prominence of mathematics, the mathematics-music relationship's presence, the (human-produced) work as an organism and the centrally planned design.

It is also important to position Kepler amongst those authors who paradoxically struggled between rationalism and mysticism—a struggle that was so characteristic of the 17th and 18th centuries. Those paradoxes that were still nurtured when Humanism reached maturity were manifested in the models they created.

Keywords: Humanism, Platonism, theory of architecture, Johannes Kepler

1   MATHEMATICS AND MUSIC IN HUMANISM

Some of the theories of architecture in the Renaissance refer to ideal objects, ideal spaces in an ideal space—ideal places, ideal dwelling (being them to man or to God). That ideal place—a microcosm mirroring the macrocosm—was conceived according to the laws of God. Those laws are mathematical—more arithmetical or more geometrical. In this vision of a work created by man, of the architectural work in particular, music (harmony, actually) played a fundamental role, for music was a reflection of those divine laws.

In spite of the Scientific Revolution that emerged in the early 16th century, Kepler's (1571–1630) ideas as a scientist came markedly close to that of architects such as Alberti (1404–1472) or Palladio (1508–1580). Mainly in terms of the respective notions of the ideal place conceived based on the ideal and idealised Pythagorean-Platonic geometry, where music was seen as a revelation of divine order, precisely because of its concurrence (and respective proof thereof) with said mathematical

principles. This whole conception of the world was imbued, of course, with a deep religiosity.

Humanism brought philosophical and scientific thought into line with each other; Plato and his admiration of mathematical reason ascended over the views of Aristotle. This mathematical and deductive approach, as opposed to the more Aristotelian or experimental one, gained considerable attention in the Renaissance, as Humanism preferred Plato to Aristotle, thus giving a new lease of life to the mathematical theories of Pythagoras, in which the former based himself. The Platonic notion that circles, triangles and other geometric figures were more perfect than the observed reality was perfectly matched to the Humanist mind. The same happened with the revelation brought upon by Pythagoras that makes the coincidence between harmonic consonances and the ratios between the first integers.

In contrast to traditional knowledge, which was acquired through the senses in a world of the senses and was based on intellectual activity that did not question the experience of the real as a means of arriving at the truth (which was in the cosmos and not built by man), modern science now began to favour the conceptual model over that of experiment (Pérez-Gómez 1999: 11). The truth, according to the new science, was no longer perceptible in the sensitive world; it emerged only through the medium of human action. One example of this is the Copernican heavenly system, which required man to deny what his own senses told him, given that they were patently telling him that the Sun revolves around the Earth (Remper).

However, it is also true that with the emergence of the new science, the notion of synthesis of microcosm and macrocosm, the belief in omnipresent order and harmony that had existed since Pythagoras, began to disintegrate. However, that process that led to a profound reorientation in the field of aesthetics in general, and proportion in particular (including, of course, proportion in architecture), did not determine the end of the doctrine of a mathematical universe governed in all its manifestations by a system of harmonious proportions. That doctrine was still followed by several authors throughout the 17th and 18th centuries. (Wittkower 1998: 130). Scientists—natural philosophers—such as Johannes Kepler reaffirmed the hegemony of music and thus celebrated and were part of a truth that was also shared with science. Architecture, music, and science shared a set of rules that established an analogy with the divine and symbolised its perfection. Modern science the one that "is highly mathematical in structure and argument", that "demands rigorous standards in observation and experiment", that "excludes spiritual agencies from its province and accepts a pure materialism", was "finally established only in the late nineteenth century" (Hall 1983: 4).

1.1 *Mathematics takes centre stage*

The protagonist role of mathematics in that process was decisive. The change in the way of explaining things, from verbal to mathematical, can be regarded as the greatest epistemological change during the Renaissance (Hall 1988: 28). The idea that mathematics offered a unique key for understanding the nature of things was fundamentally based on two assumptions: one, that nature was inherently mathematical, because God eternally geometrizes; and two, mathematical reasoning was the most certain that we could command (Hall 1988: 25–26). For Galileo (1564–1642), the Book of Nature was written in the language of geometry and the mathematical proof of a proposition is, in logical terms, the best that we can have (Hall 1988: 26).

> Philosophy is written in the grand book, the universe, which stands continually open to our gaze. However, the book cannot be understood unless one first learns to comprehend the language and read the letters in which it is composed. It is written in the languages of mathematics, and its characters are triangles, circles, and other geometric figures without which it is humanly impossible to understand a single word of it. (Padovan 2003: 255)

The mathematisation of nature was a burning issue—nature had mathematical origins and was subject to the laws of mathematics. The most eloquent argument for this process was advanced by Galileo, whose mathematisation of the science of the movement of real bodies provided a model for physics in general afterwards (Hall 1988: 28). Galileo arrived at the law of falling objects, for example, without conducting any experiments. The formula $s = gt^2$ is the mathematical law that governs the falling of objects regardless of what the human senses "say". The formula is the mathematical proof. Mathematics also made it possible to account for external factors, such as air resistance. It was a guarantee of having no errors.

1.2 *Aesthetics guides science*

Aesthetic judgement, which is present in Kepler's work—and is likewise a Humanist idea –, was a subjacent factor in all intellectual progress since the Scientific Revolution. There emerged a strongly aestheticizing mentality in connection to mathematics and, in particular, to an interest in regular geometric figures. Copernicus (1473–1543) was unable to free himself of the notion of epicycles and insisted on seeing uniform circular movement in his heliocentric system (according to Platonic theories, the circle was the most perfect of trajectories). In a certain sense, he appealed to the aesthetic judgement of his fellow mathematicians. He rejected the Ptolemaic system with the same reasoning other Humanists used to reject the work of the Scholastics: because it lacked beauty and unity (Bronovski and Mazlisch 1988: 129). Likewise:

> Galileo found Copernicus' proposal convincing not because it better fit the observations of planetary positions but because of its simplicity and elegance, in contrast to the complicated epicycles of the Ptolemaic model. (Hawking 2002: IX)

This faith in mathematical simplicity was so strong that it remains a constant in science up to Einstein and beyond and is, without question, intrinsic to the very idea of science (Lippman 1992: 15).

In this spirit:

> Johannes Kepler was a man who preferred aesthetic harmony and order, and all that he discov-

ered was inextricably linked with his vision of God. (Hawking 2002: XII)

2 THE IDEA OF ORGANISM

If it is true that, as Lowinsky points out (1989: 71), from Copernicus onwards the minds of both the musician and the astronomer seem to have turned towards the concept of organic unity, this was also evident in the architectural discourse. Wittkower sees Bramante's (1444–1514) plan for St. Peter's Basilica, for example, as the supreme example of organic geometry, that type of proportionally integrated "spatial mathematics" that became a distinctive aspect of the Humanist architecture of the Renaissance (1998: 34).

This idea of an organism which could be provided by proportion—evident and scientifically proven in music—was also advanced by Zarlino (1517–1590), who introduced the definition of a *corpo* for polyphonic music, now regarded as "a body, as one coherent organism" (Lowinsky 1989: 73). It was also Zarlino who, using a scientific approach, classified all musical material since Antiquity in the mid-16th century, declaring himself to be marvelled by the fact that musical consonances were determined by arithmetic and harmonic means, thus giving continuity to the tradition (Wittkower 1998: 124).

Zarlino is contemporary of Palladio who in the latter's *Quattro Libri* recommends and shows his own use of harmonic proportions. The same Palladio who defines beauty thusly:

Beauty will result from the beautiful form and from the correspondence of the whole to the parts, of the parts amongst themselves, and of these again to the whole... (Wittkower 1998: 31).

Even before, Copernicus considered is new theory as a way that could establish:

The order and magnitudes of all the planets and of their spheres or orbital circles and it would bind the heavens together so closely that nothing could be transposed in any part of them without disrupting the remaining parts and the universe as a whole. (Lowinsky 1989: 71).

For Kepler, as for Copernicus before him, the idea of the symmetrical harmony of the universe, of its form as a giant, coherent, and well-proportioned body, was of vital importance (Lowinsky 1989: 60).

Every description on the theme echo Alberti's discourse on beauty.

3 THE CENTRE, THE CIRCLE AND THE SPHERE: REPRESENTING GOD

The association of the form of the ideal temple with regular, fixed-centred geometric figures, the idea of the centralised plan that became so desired in the Renaissance, and is praised by Alberti in *De Re Aedificatoria*, was also an idea taken up by Nicholas of Cusa (1401–1464), for whom mathematics was a necessary vehicle for penetrating divine wisdom, which we should imagine through the mathematical symbol (the centre and circumference of the circle) (Wittkower 1998: 38). Only "God, who is everywhere and nowhere, is [the universe's] circumference and centre". The natural universe itself, as a contracted image of God, has a physical centre that can be anywhere and a circumference that is nowhere (Miller 2015). For Copernicus, "it is the circle alone that can bring again what has already taken place" (Lowinsky 1989: 77).

Both in the idea of the circle and in that of the sphere (or, to be more precise, the semi-sphere, which is what Alberti proposed for the roofs of temples and came to prominence in the dome forms of the Renaissance), the ratio between the circumference and the centre is always the same. It corresponds to the unison from the musical point of view—the 1:1 ratio. Naturally, the square, also representing the unison, had a fundamental importance as well. In consonance with this was also the interest in the octave (which replaced the traditional hexachords)—the 1:2 ratio. Johannes Gallicus of Mantua (ca. 1415–73) referred to it as the *perfectissima consonantia, dulcissima, modulatio*. It was the octave that encompassed in itself all the other intervals (hence its denomination—diapason). In the octave, the sounds appeared reborn (Lowinsky 1989: 76–77).

3.1 *God's wisdom*

The presence of God continued to be a fact and a fundamental factor. Despite the new scientific advances, the idea of a universe that could function without the active involvement of God was, generally speaking, unacceptable for scientists until the 18th century. Whilst it was true that, in the Renaissance, Western Christianity and its relationships with philosophy underwent profound changes.

It would certainly be erroneous to supposed that natural philosophers of around 1600 were less devoutly Christian that those of three centuries before... (Hall 1983: 8).

Prigogine, referring to Newton's age, likewise argues: although science has separated man's world and physical nature, it shares with religion the interest in finding universal physical laws that provide testimony of divine wisdom. Modern science rises from the rupture with the ancient animist alliance with nature but it establishes a new alliance with the Christian God, Universe's rational legislator (prigogine 2008: 14). Similarly, according to Voltaire (1694–1778), the great minds of the Enlightenment—all of them disciples of Newton (1642–1727)—believed in the existence of God, as the discoveries of science had made atheism impossible (Pérez-Gómez 1999: 31). Platonic geometry was the perfect representation of God and His construction.

The work of Kepler—a mystic, philosopher, geometer, and astronomer—can only be understood from the point of view of a world vision that was as purely physical as it was religious and aesthetic (Padovan 2003: 247). "[A]s a theologian and astronomer Kepler was determined to understand how and why God designed the universe" (Hawking 2002: XIV). "Kepler believed he had discovered God's logic designing the universe" (Hawking 2002: XVII). The word design has a very strong meaning here, for it is about *design*.

4 MYSTERIUM COSMOGRAPHYCUM (1596)

Kepler had, as he recalled himself, a kind of epiphany: as a lecturer in geometry in Graz he found himself reflecting on what was to be the "secret key to understanding the universe" (Hawking 2002: IX).

He would go on to construct a model for the universe in which all six planets (then known) were arranged around the sun in such a way that concentric spheres nested in a sequence of the five Pythagorean solids. For Kepler the fact that only five solids could be constructed by regular geometry explained why there could be only six planets with five spaces between them (the spaces not being uniform) (Hawking 2002: XIII). Euclid fascinated him. The absolute high point in relation to the latter's *Elements* was to be able to demonstrate that there were only five perfect solids.

Kepler was thinking as in Antiquity, philosophy comes first.

Kepler's cosmological model was an architectural model: the cosmological geometry of the ideal dwelling. A model of the cosmos, the ideal dwelling, mathematics, laws and forms, numbers and figures, arithmetic and geometry. It is also an object without accidents in a space without accidents.

In conceptual terms, one can also draw a comparison with what was happening in architecture. The problem was isolated, for example, when Alberti designed a church—a "temple" – with a centralised plan. The problem emerged when there were other factors, such as where to place the altar. Palladio's villas also had no mass, like Galileo's formulas without the resistance. In an ideal situation, things would be thus: Galileo's objects would fall at the same speed regardless of mass, and Palladio's villas would not have to adapt to the location and their walls would have no thickness. In the ideal world, science and architecture are closer (still).

Kepler transposed that model of the idealised villa—that "Rotonda free of gravity" – to the real world. That is to say, the universe in the image of man's work, which in turn, is conceived in the image of the world.

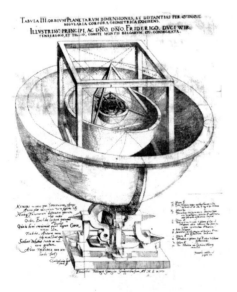

5 *HARMONICES MUNDI* (1618)

Harmonices Mundi, as Kepler's biographer Max Caspar puts it, was "a great cosmic vision, woven out of science, poetry, philosophy, theology, mysticism" (Hawking 2002: XIV).

That mysticism is not only present in Kepler's work, for it was one of the defining features of the 17th and 18th centuries.

The Counter-Reformation and baroque era was one in which oppositions that were impossible to reconcile were forced to find a way to coexist. The rise of rationalism was accompanied by the march of militant mysticism. The aristocratic cult of majesty was opposed by bourgeois domesticity.

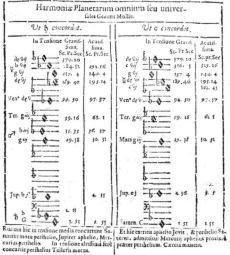

210 DE MOTIBUS PLANETARUM

Figure 2. The harmony of the planets as described in Harmonices Mundi. Harmonices Mundi. Linz: Lincii Autrial, 1619. [www.e-rara.ch].

The internationalism of Roman Catholicism conflicted with the nationalism of the Protestant sects and rising monarchies. Religious orthodoxy had to contend with freedom of thought. (Marien 2005: 359)

Kepler is a good example of the contradictions so typical of his time. How wrong his principles, thinking or conclusions were, has indeed been widely pointed out by the countless authors fascinated by his work.

Harmonices Mundi work brings together astronomy, music and geometry. That worldly music one believed in since Antiquity. That extraordinary song/dance exhibited by the planets that was beyond the human senses.

For Kepler, geometric beauty and musical beauty were a source of knowledge for astronomy. Geometry predates the world and, therefore, is divine in character. In other words, it is of God. Because everything that is part of the world is of God. Harmony in music has the same divine character. God created man in His own image and likeness, so that he could understand geometry and enjoy music, thus understanding that the world is beautiful and perfect, as the work of God that it is.

Kepler's heavenly music can only be experienced from the Sun; it cannot be perceived from the Earth:

We know the score but can never attend the performance". Its reconstitution by man is a purely intellectual act. Music "is not a human invention, subject as such to change, but a construction so rational and natural that God the Creator has impressed it upon the relations of the celestial movements. (Pérez-Gómez 1999: 29)

Music exists in nature; it is mathematical, immutable, and divine. Given these attributes there would seem to be no reason for denying the hegemony of music, particularly how its rules governed architecture, something that had been acknowledged since Antiquity.

Kepler describes the universe as if it were a piece of music:

Thus the heavenly motions are nothing but a kind of heavenly concert, rational rather than audible or vocal. They move through the tension of dissonances which are like syncopations or suspensions with their resolutions (by which men imitate the corresponding dissonances of nature), reaching secure and predetermined closures, each containing six terms like a chord consisting of six voices. And by these marks they distinguish and articulate the immensity of time.

He goes on to refer to a form of mimesis:

Thus there is no marvel greater or more sublime than the rules of singing in harmony together in several parts, unknown to the ancients but at last discovered by man, the ape of his Creator; so that, through the skilful symphony of many voices, he should actually conjure up in a short part of an hour the vision of the world's total perpetuity in time; and that, in the sweetest sense of bliss enjoyed through Music, the echo of God, he should almost reach the contentment which God the Maker has in His Own works. (Padovan 2003: 251–252)

Before Kepler, what had also led Copernicus to propose a new model of the universe was not a scientific necessity but the conviction that the universe was a mathematically ordered harmony. Thus, it should contain the same *concinnitas* that Alberti sought in architecture (Padovan 2003: 241). As Vitruvius (fl. 1st c. BC) has first referred the importance of building in accordance with the laws of music, Alberti recommends:

We shall therefore borrow all our rules for harmonic relations (*finito*) from the musicians to whom this

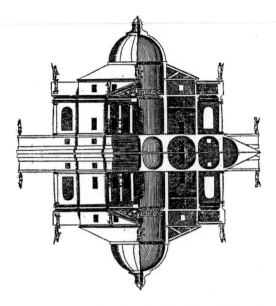

Figure 3. The Rotonda after The Four Books of Architecture.

kind of numbers is extremely well known, and from those particular things wherein Nature shows herself most excellent and completed (Alberti, apud (Wittkower 1998: 109)

For Alberti, following Wittkower:

Harmonic ratios inherent in nature are revealed in music. And the architect who relies on those harmonies is not translating musical rations into architecture, but is making use of a universal harmony apparent in music. (Wittkower 1998: 109)

6 KEPLER: THE ARCHITECT OF THE IDEAL WORLD

In *Mysterium Cosmographicum* (1597), Kepler associates the planets and their orbits to the five Platonic solids; in *Harmonices Mundi* (1619), Kepler's universe is governed by the Platonic geometry of the regular polyhedra and the numbers of musical harmony. This is a Platonic geometric world governed by musical consonances. One can speak of an architect who in addition to musical consonances also applies the geometry of

the Pythagorean-Platonic tradition. He designed the perfect dwelling—the perfect world: nowhere, in the heavens.

ACKNOWLEDGEMENT

This paper is funded by National Funds through FCT—Fundação para a Ciência e a Tecnologia under the Project UID/AUR/04026/2013.

BIBLIOGRAPHY

Bronovski, J., and Mazlisch, Bruce. *A tradição intelectual do Ocidente*. Lisboa: Edições 70, 1988.

Hall, A. Rupert. *A Revolução na Ciência, 1500–1750*. Lisboa: Edições 70, 1988. 494, [2] p.

Hall, A. Rupert *The Revolution in Science 1500–1750*. London: Routledge, 1983. 382 p. ISBN: 978-0582491335.

Hawking, Stephen. "Introduction." In *Harmonies of the World. Book five*. Philadelphia: Running Press, 2002. I-XVII. ISBN: 0762420189.

Lippman, Edward A. *A history of Western musical aesthetics*. Lincoln: University of Nebraska Press, 1992. 551 p. ISBN: 0803228635.

Lowinsky, Edward E. "The concept of physical and musical space in the Renaissance." In *Music in the culture of the Renaissance and other essays*. Edited with an introduction by Bonnie J. Blackburn; with forewords by Howard Mayer Brown and Ellen T. Harris. Chicago: University of Chicago Press, 1989. ISBN: 978–0226494784.

Marien, Mary Warner *Fleming's Arts and Ideas*. 10 th. ed. Belmont: Wadsworth Publishing, 2005. 704 p. ISBN: 978–0534613716.

Miller, Clyde Lee. "Cusanus, Nicolaus [Nicolas of Cusa]. In (ed.). Fall Edition. <http://plato.stanford.edu/archives/fall 2015/entries/cusanus/>." *The Stanford Encyclopedia of Philosophy*. http://plato.stanford.edu/ archives/fall 2015/entries/cusanus. (accessed 20/5/2016).

Padovan, Richard. *Proportion: science, philosophy, architecture*. New York: Spon Press, 2003.

Pérez-Gómez, Alberto. *Charles-Etienne Briseux's musical analogy and the limits of instrumentality in architecture*. Nebraska: The University of Nebraska-Lincoln, 1999. 70 p.

Prigogine, Ilya. *O Nascimento do Tempo*. Trad. Marcelina Amaral. Lisboa: Edições 70, 2008. 73 p. ISBN: 978–972–44–1430–0.

Remper, Gerhard. "The Scientific Revolution." http://www.solowey.net/scientific_revolution.htm. accessed 24/5/2016.

Wittkower, Rudolf. *Architectural principles in the age of humanism*. 5th ed. London: Academy Editions, 1998. ISBN: 0471977632.

Architecture and built utopia. Icons and symbols in the rural landscape of J.N. Ledoux

Domenico Chizzoniti, Letizia Cattani, Monica Moscatelli & Luca Preis
ABC Department, AUIC School, Politecnico di Milano, Milan, Italy

ABSTRACT: This paper tries to connect together the architectural interest in utopia and the ideal condition of the countryside. In the work of utopians as well as in some works of visionary architects is often exercised the idea of accomplishing an ideal dimension in planning the society or in conceiving the structure of architectural space, the so-called rural utopia.

Firstly, this paper testifies to the growing interest in the many aspects related to the concept of space in the architecture of utopia. Therefore, it intends to analyse, on architectural production, some models of spatial exploration in order to open the study of utopian literature to new lines of inquiry. Classical social utopias are among the main sources of the discipline of architecture.

Then it proposes to think of utopias not as fictional texts about future change, but as basic element in a cultural and productive process through which social, spatial and subjective identities are formed.

Finally, utopias can thus be read as textual systems implying a distinct spatial and temporal dimension; as 'spatial practices' that tend to naturalize a cultural and social construction. By examining the French context, and the period of the Enlightenment, it was considered as a case study the project of Claude Nicolas Ledoux for the Royal Saltworks, and its evolution towards the utopian dimension of the city of Chaux.

Keywords: Ideal settlement, rural ideology, visionary architecture, rural structure, Chaux, C. N. Ledoux

1 URBAN CONSTRAINS AND RURAL FREEDOM

This paper introduces some issues on the development of anti-urban ideology in the context of the French Revolution. Using some requests posed by the doctrine of progressive culture of "Philosophes", until the beginning of the political revolution of 1792, Enlightenment philosophy has traced a path in which many have followed a common trajectory. Worth mentioning social utopia, represented by Jean-Jacques Rousseau, economic research embodied by the physiocrats (among all François Quesnay and Anne Robert Jacques Turgot), and figurative experimenting interpreted by Étienne-Louis Boullée's rhetoric and Claude Nicolas Ledoux.

Due to the space given, and according to the topic of the conference, it is possible only to mention to some aspects of J. N. Ledoux's idealization of the city of. This analysis refers mainly to the formation of rural production, which aims to deal with the development of mercantile and proto-capitalist city. There is a particular attention to the case of the settlement of the Royal Saltworks at Arc-et-Senans. More precisely, the analysis centres on a number of welfare facilities, cultural and hospitality equipment settled in the mixed regime of production and research in the region of Franche-Comte.

First, however, it is necessary to contextualize briefly some reasons, which had preceded the resolution of French political revolution, which date back to a direct encounter with a certain cultural and economic emancipation of English bourgeoisie (Bergdoll 2000).

The English Enlightenment culture particularly that of the progressive bourgeoisie, moved to the conquest of some rights capable to focus, and incentive, with some public investment, the development of agricultural economy. Conversely, the French culture, a century later in 1789, with the "Declaration of the Rights of Man and of the Citizen", broke up with the foundations of the Ancien Régime, thus achieving individual liberty, collective safety, popular sovereignty, and celebrate the cult of public and communal institution. Bourgeois capitalism gave incentives to urban mercantile economy. This phenomenon has deeply affected the morphology of the city and set-up the production of the countryside. In terms of architectural culture, this trend was reflected in relevant economic investments, dedicated to the exploitation of rural resources and, at the same time, in structural transformation of the city with the reorganization of representative buildings. This new policy constrains rural environment through the deployment of all pre-industrial manufacturing activities, through a process of urbanization involving much of the rural land.

2 PHYSIOCRACY AND LIBERAL IDEOLOGY

We deal with this second aspect through the analysis of some structural experiences, with particular regard to some political and ideological constraints.

The first great insight, that was the prerogative of enlightened reformism, involves the elimination of interests and privileges of feudal origin. The investment policy in the area of production was promoted to encourage rural communities in order to organize themselves through collective structures with new settlements. In particular, in this context, the physiocratic culture postulated the return to the countryside through a process of planned and scheduled disurbanization.

To validate this drift there was the intervention of physiocratic ideology, which considered the countryside a source of prosperity, not only from the economic point of view but also as an opportunity for social reorganization. In fact, the state of prostration of the French popular class, at the threshold of the revolution is well known. The physiocrat system considered a kind of active occupation of the territory, through a series of self-sufficient communities, capable to generate economic relations with the urban settlement. On this singular aspect, the point of view of French Physiocrats Quesnay and Turgot is particularly relevant. The physiocratic ideology considered the land as a source of primary resources, essential in a modern state regulated by basic democratic institutions.

François Quesnay's famous thesis with its Economic Framework, which became the manifesto of Physiocracy, has started this polemic against feudal remnants in French economy. In that analysis there was a radical criticism of Colbert's policy, which had dominated French economy, during the Grand Siècle. Explicitly, there was expressed a criticism of economic policy detecting some controversial aspects about industrial upgrading, and also protectionist customs policy, which gave comfort to the survival of mercantilist economic forms, exclusive privilege of feudal aristocracy. This wave of liberalization of productive resources belonging to the territory was competing with the state of decay of the city, reported by Abbot Laugier, with its assumptions of urban *embellissement*:

Whoever knows how to design a park well will have no difficulty in tracing the plan for the building of a city according to its given area and situation. There must be squares, crossroads, and streets. There must be regularity and fantasy, relationships and oppositions, and casual unexpected elements that vary the scene; great order in the details, confusion, uproar, and tumult in the whole...[1]

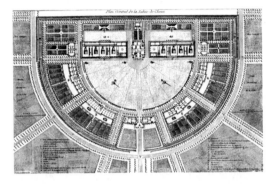

Figure 1. Claude Nicolas Ledoux, Royal Saltworks of Arc-et-Senans, 1773.

Figure 2. Claude Nicolas Ledoux, Royal Saltworks of Arc et Senans, 1773.

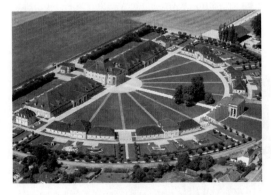

Figure 3. Claude Nicolas Ledoux, Royal Saltworks of Arc et Senans, aerial view, 1773.

Figure 4. Claude Nicolas Ledoux, General Plan of The Utopic City of Chaux, 1778.

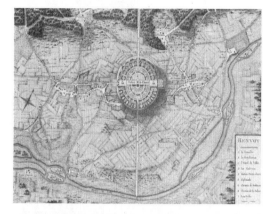

Figure 5. Claude Nicolas Ledoux, General View of The Utopic City of Chaux, 1778.

Figure 6. Claude Nicolas Ledoux, The House of guardians of the village of Maupertuis, 1778.

3 THE IDEAL CITY AS A NATURAL PLACE

The implications of this radical theory to imagine the city as a natural place resulted decisive for eighteenth century urban planning practices.

They were crucial to organize a figurative style, which inferred new words from ancient vocabulary; the words are used sometimes as reference or postulated as corollary to a compositional theme based on the principles of naturalism. Furthermore, these implications are crucial in the mystification of artefact that builds the urban or rural landscape through the discovery of decadent ruin, introducing themes esteemed by pre-Romanticism century.

These factors can be found in the Claude-Joseph Vernet's pictorial research where the ruins of a bridge builds a landscape so decadent as celebrated; on the other hand through Hubert Robert (1733–1808) who amplifies the picturesque appearance through the courtly classic reference in waterfalls of Tivoli (Chizzoniti 2007: 39–44).

Picturesque and exotic terms are also adequate to a research carried out in architecture, showing signs of a decayed landscape with repercussions in the eccentric experiments of Charles de Wailly (1730–1798), such as the park of Marigny's Château de Menars. It is the case of a Chinese pavilion (1772) placed in a highly compromised context or in the case of Françoisde Couvilliès where he proposes the architecture of a hermitage (1774) in the expanse of a princely garden, or even in the exercises on the central plans of John Soane (1753–1837), that locates circular temples in English gardens. C. N. Ledoux finds himself to work under these circumstances and in this environment. The radical nature of his creativity transforms everyday language liberating it from the superfluous of baroque involution, moving it towards a classic distillate purism, which intervenes to revive the figurative research. Emil Kaufmann noted that fragmentation of the Baroque unit and rejection of the order as the constitutive foundation of the figurative architectural party constitution marked the way to a kind of heresy composition (Kaufmann 1952).

The complaint to the ornamental excess of Marc Antoine Laugier pushed architectural research to rediscover the principles of solidity, simplicity, and naturalness: criteria of a new anesthetization. It demanded a figurative experimentation expressed in the form of free disposal of pure volumetric masses, in the combination of elementary geometric solids (Oechslin 1971). This radicalism is fully expressed in the rural setting, without those influences that generate urban circumstances in the revolutionary era. In this regard, the House for an employee of Ledoux and in particular the stereometric effect of the structure composition is a good example of that. A prism, with a rectangular base, rises to its full height renouncing the principles of subordination between the base and the stem.

Ledoux imposes a new way to regenerate an architectural theme: the country house, through the nude exhibition of architectural mass for grafts of elementary geometric solids.

4   BUILT UTOPIA. THE ROYAL SALTWORKS OF ARC-ET-SENANS

The project for the Royal Saltworks of Arc-et-Senans started with an explicit request of *Ferme Générale* as a part of the expansion process in the region of Franche-Comté, under the administrations of Trudaine, Turgot, and afterwards that of Calonne. The eastern provinces of France had been the object of significant investments, both for their strategic location—placed along the routes of important international markets—and for the availability and supply of rich deposits of raw materials (Klein 1995: 51).

The physiocrats saw in this design a real possibility for the establishment of free movement of goods, and an opening to capital flow for productive investments in the territory. The privileged sectors capable of intercepting the accessibility to capital were mainly those of specialized crafts, to give headway to the proto-industrial production. Unlike the ideological orthodoxy of Quesnay, Turgot realizes that all investment should involve particularly public institutions able to control the concentration of monetary wealth, promoting manufacturing operations for reward in the short term. Likewise, he considered possible subsequent reinvestment for improving extraction techniques geared to the longer term. This mechanism would also have triggered some economies in the agricultural sector, which should regenerate free trade activity, similar to the model of English economic development. Ledoux's assignment as responsible for the administration of *salines royales* promoted by Trudaine represented an important professional opportunity: Ledoux approach to that group of public patronage would enhance his disposition as an architect in the role of public official approval.

La Ferme Générale supervised all operations, and therefore the subsequent sale of raw materials, forcing their commissioners to exercise strict protection of public interests. Therefore, Ledoux began his project for the saltworks imposing a productive pragmatism, according to industrial economies, necessary to protect the interests of public administrators. On the other hand, the same Ledoux appears heavily involved and affected by the productive pragmatism that animated the interests of this public task, having after all a positive effect. In fact, he managed not to elude the goals of the public administration that aimed at full exploitation of raw materials.

Therefore, from these premises, the project for the saltworks of Arc-et-Senans took shape according to State interests, through public investment and did not hesitate to take on requests of the physiocratic's ideals, when this job becomes an opportunity to restore, in a wide scale plan, the circuit of major trade routes, through the reform and strengthening of territorial infrastructures. Yet, there are two feelings in the design arrangement. One that is rooted in the theme of a productive pragmatism, that for the saltworks of Arc-et-Senans; and another that aimed to be a decisive and radical modelling of an ideal city. Eventually the design evolved from a primitive industrial core towards an idealization of a new city, the utopian city of Chaux. The original ideological strain turned quickly from an ideal settlement to a social utopia, by contrasting state monopoly for the control of rural resources with the liberalization of products, and therefore the affirmation of free trade.

The ambition for this transformation is materialized in the project of an ideal city (Tafuri 1969). A kind of productive settlement was elevated to a public institution, and transformed in a project theoretically related to reforms of social order.

On this architectural design overlaps the ambition of a social order, alternative to the current city, and to the decadent urban arrangement.

Paradoxically the rate of utopia that you may encounter appears clear in some ideals requirements, when such alleged revolutionary gesture ended up. This vision is a model of urban utopia, adopted extraordinary architectural canons, since it interpreted some innovative social conditions, referred not only to public administration but also to the production and deployment of resources in the settled community.

On the other hand, the idealization of an exemplary community was emblematically achieved through symbolic exaltation of special public institutions: these new structures were offered to the community to settle a new social order.

Ledoux had endowed the cities with different forms of social buildings both for education and for entertainment. The desired reform of the production structure in the countryside resulted through modelling an ideal community, directed towards a kind of moral redemption. There are some completely innovative buildings respecting the newly established social order that reflect this form of utopia. Some examples are represented by Pacifiéré, Panarétheon, Cènobie or by Oikéma (the last one similar to the house of pleasure for Montmartre in Paris).

These organisms were thought as the new public institutions in the city, despite certain subversive pretensions concerning the figurative choice and the program of activities. The vision of the production of Arc-et-Senans Royal Saltworks turned into a moral and social innovation plan, just in some of its representative elements (Vidler 1984).

Public buildings appeared so radical in the functional program and equally conventional in typological structure, as well as in architectural design, and in the choice of the linguistic code. The discriminant of this ideological and figurative alteration is evident in the distance between experimentalism of some public facilities for the city and rural settlements (the Maison of Education, the rural school of Meilland, the House of guardians of the village of Maupertuis, etc.).

Then there is obvious conformism that refold certain paradoxical institutions located in the urban context, to ensure a new social order. In other words, this project is not adequately demonstrative of a revolutionary alternative in architecture, if only read as an abstract program of masonic ideals, opened to initiation rites that symbolically assert an idea of a new world.

This is a rather conventional way to read the work of Ledoux. Many critics have recognized a naive transposing, rather ephemeral, of the social conditioning of the doctrine of Jean-Jacques Rousseau. Others have focused on his figurative experiments, in particular on the naive symbolism declaring it artefact and caricatured.

Anyway, it is not possible to dismiss the development of Ledoux's project in relation to its presumed opportunism, manifested as a result of the political events that saw him, after having been detained, relegated to the margins of official patronage. On the contrary, it is necessary to distinguish the split personality that pushes the architect to some typological experiments, which are rooted in the productive structure of the territory, and which are confronted with objective needs of change in the relations between the state and the city, and especially between the city and the countryside.

5 THE CONSTRUCTION OF UTOPIAN LANDSCAPE

The evolution of the proposal for Chaux, from a saline to an ideal city, clearly shows the process of a planned settlement modelling theory only for utilitarian purposes to which the ideals of the agrarian reform were related.

The industrial expansion and the commercial liberalism were the basis of progressive ideology that characterized operations of structural renewal in the French economy. The Arc-et-Senans Salines should be framed in this Franche-Comté reforming process for land resources exploitation. Ledoux's program ideologically aimed to restore internal mobility within the region, towards the opening up to the Mediterranean and Atlantic trades, coming well to the North European harbours.

On an architectural perspective, his proposal aimed to a typological experimentation, which overlapped the rigid functional requirements with expressive license and compositional research, expressed also by the subversive reproduced shapes.

The first architectural plan, made «without knowing the area», is ordered according to a conventional square enclosure, which unifies the functions of production management, through a rotation of the inner court path, connecting the production pavilions to the residential ones. The second version is a part of a productive context which structurally reconnects some important communication routes into a large, open territory, hemicycle. The alleged hygienist theories cannot essentially justify the choice of a provision organized in separate pavilions along the perimeter and the diameter of a semi-circular plant, where the saline operating devices were contained.

On the contrary, this decision was prompted by the search for the character of each building, according to the principle of isolation of the pavilion, free on all four sides instead of the enfilade, each of which contains specific functions directly related to production. In this sense, the experimental character of the unanimous interweaving between production and residence is revolutionary.

This architectural plant contradicts the character of mechanical nature, of detailed searching for an architectural device, and an extravagant representation, as suggested by a stiff in critical, who claims to see a supposed landscape refinement, referring to some allusion to the picturesque or to backgrounds in the paintings of Jacques Louis David, Horace Vernet and Hubert Robert. As Panofsky argued about the iconological value of art, tends to be neutral in nature and unlikely to contain references to contemporary political events (Panofsky 1955).

The use of the Anglo-Saxon picturesque naturalism, shading the landscape etchings, which contextualize the City-Factory into an ideal place, aims to a belated moral redemption, changing ideals of a progressive thinning and decentralization of production, to those of a modelling settlement. Instead, the "exalted rationalism" of his figuration moves towards a progressive consolidation of rural urbanization through innovative typological experimentations. Production—warehouse—residence, concentrated in a single architectural structure, are characterized by a new figurative order, where it is possible to identify some characters of domesticity, as the endowment of the gardens behind the workshop-residencies, or of representation, as the rough arrangement of the ashlar on internal hemi-cycle, that marks the pavilions fronts.

The same representation of the production, as a theatrical device, subtracts itself from a simple functional and utilitarian interpretation of the general economy of the architectural plant. The geometrical combination of the hemicycle puts on the front of the scene the salt factories, which border the centre of the directional building, lined with factories, endowing the whole complex by an urban, productive, representative, and institutional coefficient.

All this stands as an alternative to some urban-ized suburbs. This settlement formed with the rise of new economies, in relation to some principles of production autonomy and independence from the city, became an example of highly specialized poly-centric production model. The subsequent evolu-tion of the project for Chaux denies this principle, taking shape with the characters of micro-univer-sality, through the progressive supply of typical urban institutions that, overpowering the original intention, reproduced by analogy the city from its internal, starting with the contradictions and the effects of alienation that were detectable in it.

It is indeed interesting to note that, for the project of Chaux, the process of building the "ideal city" departed from the supposed conclusion of a completed project itself and built within the settle-ment of saline. In the same way, some architectural units especially reproduce the experimentations deductible from the study for the village of Mau-pertuis, whose critical current enhances reductively some allegorical symbolic characters omitting, however, the degree of contextualization, in rela-tion to the structural role and foundation for a pos-sible organization of the rural landscape.

6 CONCLUSIONS

To conclude, the work we have examined has been widely known in the architecture discipline. The relation between Ledoux' social idea and the impact of a new political reform through the physi-cal aspect of the architecture is, on the contrary, rather obscure. This research tries to offer new insights by focusing on the architecture of the cit-ies, on the structure of the space and on the char-acter of the figurative language, in order to deduce

some features of the utopic and visionary explora-tion in the Ledoux's work, and how that architec-ture represents the author's political philosophy.

Utopia is a word with many connotations both negative and positive, something that is ideal, or something that is impossible to achieve. Yet the utopian dream of the ideal society and city is one of the main aspects in the discipline of architec-ture. By working on this ideal aspect of the city and with this utopic approach, Ledoux's research reveals that he made a significant contribution to architectural theory and practice and his influence carried on through eighteenth and nineteenth cen-turies till today.

NOTE

[1] Laugier, *Observation sur l'Architecture, apud.* (Tafuri 1976: 4).

BIBLIOGRAPHY

Bergdoll, Barry. "Neoclassicism: Science, Archaeology and the Doctrine of Progress." In *European architec-ture, 1750–1890.* Oxford: Oxford University Press, 2000. 9–32. ISBN: 9780192842220.

Chizzoniti, Domenico. *L'altra idea di Parigi.* Cuneo: Araba Fenice, 2007. 160 p. ISBN: 9788886771719.

Kaufmann, Emil. "Three Revolutionary Architects: Boullèe, Ledoux, and Lequeu. [With illustrations.]." 42.3 (1952): 135 p.

Klein, Bernhard. "Ledoux et les physiocrates." *Visiteur.*1 (1995)

Oechslin, Werner. "Pyramide et sphere. Notes sur l'Architecture Révolutionnaire du XVIII siècle et ses sources Italiennes." *Gazette de Beaux Arts.*113 (1971): 201–238.

Panofsky, Erwin. "Iconography and Iconology: An Introduction to the Study of Renaissance Art." In *Meaning in the Visual Arts: Papers in and on Art His-tory.* Ed. Panofsky, Erwin. Garden City: Doubleday, 1955. 26–54.

Tafuri, Manfredo. *Architecture and utopia: design and capitalist development.* Translated from the Italian by Barbara Luigia La Penta. Cambridge, Mass.: MIT Press, 1976. xi, 184 p. ISBN: 0262200333.

——. "Per una critica dell'ideologia architettonica." Contropiano.1 (1969): 31–79.

Vidler, Anthony. "The rethoric of Monumentality. Ledoux and the barrières of Paris." AA files. 7 (1984): 14–29.

Amplifying reality through quadratura: Contrappunto among corporeal and visual space

João Cabeleira

Lab2pt, School of Architecture, University of Minho, Guimarães, Portugal

ABSTRACT: Portuguese spatial production, between 17th to 18th centuries, experienced a renewal through a new feature of spatial research, the *quadratura* painting. Intended as action, the *contrappunto* among Architecture and *Quadratura* intertwines corporeal space (built) and visual space (represented) creating an apparent reality based upon the power of perspective.

The employment of perspective rules and procedures into space illusion raised *quadratura* to the condition of architectural instrument overcoming tectonic constraints. By intertwining two-dimensional images and three-dimensional reality, we witness the triumph of the perspective induction over the built space. A metamorphosis of appearances in which the projected image becomes a structural fact, transforming the perception and reasoning of the tectonic truth.

Approaching the Portuguese baroque, this new spatial achievement was introduced by the Italian authors Bacherelli and Nasoni. A path, integrating optical phenomena upon spatial configuration, followed by Portuguese authors whose *quadratura* works are judged through their capacity of blending strategies from the constructive experience with the potentialities of the pictorial essay.

Keywords: Baroque Architecture, Quadratura painting, Perspective, Illusory space

1 CONSTRUCTION VERSUS REPRESENTATION. BUILDING ILLUSION OR ACHIEVING A NEW REALITY

The synthesis provided by quadratura, painted perspectives of architecture, between built and represented space is taken by the Baroque spatial framework as a level of the spatial experience. About this architectural genre three types of discourse must be considered: a proselytizing discourse (underlying to the image's rhetorical potentiality under the logics of counter-reform and royal absolutism); a technical discourse (which calls up contents from optics and perspective science along with the theory and practice of architecture); and a symbolic discourse (referring a latent spatial induction, either by transformation of the built environment as well as the reflection of desires and ambitions materialize through representation).

Being the proselytizing discourse intrinsic to the iconographical message, the technical discourse integrates contents from the Ars pingendi and Scientia aedificandi supporting the delineation of an Architectura ficta. On the other hand, the approach to a symbolic discourse considers the reception of the image by the viewer and the subsequent validation of quadratura as deception conditioning reality.

According to this logic, the new reality provided by quadratura embraces the exterior of the mind (extra-mentis), the tangible world or the corporeal level of built reality, and its within (intramentis), the imagined materialized through representation and apprehend by the contemplation. Blending the quadratura illusion with the material world forges a visual reality. A dreamlike truth where the artifice of perspective is able to transform the corporeal relation towards the built environment.

Identifying the quadratura as an artefact capable of transforming the apprehension of form, measure, and image of the tectonic support, it can be asserted as a structural fact of the inhabited environment. In the same sense, the practice of quadratura coincides with the practice of architecture in its compositional codes (rules, grammar, architectural orders and its combining abilities) or geometrical reasoning (perspectival management of the outlined representation and recognition of optical laws), being placed within the field of spatial intervention, even if materialized through the instruments and the propositional potential of the pictorial image.

Sharing the object and purpose of its practice, architecture and quadratura may be distinguished by operating, respectively, through the organization of three-dimensional bodies, displacing and arranging the constructive substance, and the outline of two-dimensional figures that, governed by optic and projective laws, induces the viewer into the appearance of three-dimensional facts. Thus, the quadratura may be assumed as an integral and inseparable part of the inhabited space, falling within the disciplinary field of architecture. It organizes, characterizes, and defines space through the power of perspective simulation, surpassing conditions of the physical nature and imposing an apparent truth.

2 BUILDING WITH THE EASEL

Starting developed experiments aiming an architectural painting by the end of the 17th century, only at the dawn of the 18th century the quadratura, in its full technical and conceptual foundations, was hosted in Portugal[1]. Constituting a novelty for the characterization of internal spaces, this pictorial/architectural genre gets to be articulated with the built environment responding to coeval spatial and imagery requirements (Mello 1998: 97; Raggi 2004: 464).

Inaugurating this cycle, the imaginary architectures at the celling of the Monastery of S. Vicente de Fora lobby (1710, Lisbon), by Vincenzo Bacherelli, were articulated with the built surfaces imposing a rereading of its spatial dimensions, shape and image. Through observation of the image, a spatial vertical extension is evidenced, hiding the spatial (dis)proportion, and revealing a transformation of its perimeter according to five different levels.

Essentially, the visually induced composition is governed upon the constructed rectangular matrix, the corporeal level, giving rise to an increasingly complex and dynamic space: a surrounding balcony, a gallery that expands vertically the perceived space, a cornice sealing the imaginary space and a central oculus from which the Triumph of St. Augustine's above Heresy is presented to the gaze. In this sense, the physical space, with a regular and stable perimeter, is juxtaposed by the quadratura illusion, whose successive expansions and contractions streamlines plastically its limits changing the global perception of the inhabited space by the eye.

However, Bacherelli directs its concerns towards a spatial impression rather than a verisimilitude, or even obedience, with the practice and rules of the architectural science. A different understanding is followed by Nasoni, whose dedication to *quadratura* is simultaneous to his constructive practice showing operative coincidences (Bury 1956: 7; Raggi 2004: 661).

At the sacristy of Porto Cathedral (1734, Porto), the intervention of Nicolau Nasoni ranges between moments of purely decorative intent and others of

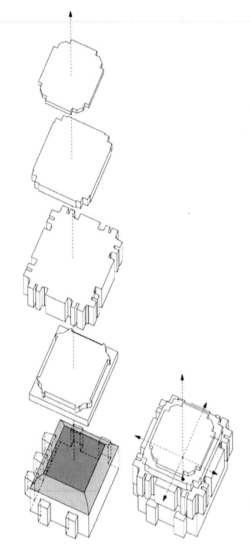

Figure 1. Interpretive scheme of the five spatial levels at the Monastery of S. Vicente de Fora lobby (1710, Lisbon), Vincenzo Bacherelli: built space and imaginary space (surrounding balcony, false gallery; upper cornice, central oculus). Cabeleira 2015, 356.

a clear spatial purpose. Here, woodcarving and *quadratura* are applied to transform the medieval structure under a modern imaginary space. A new arrangement organized at two levels: one until the cornice height, with false frames and valances integrating bays, easel paintings, the altarpiece and furniture; another over the vault where a represented parallelepiped space annuls the curvature of the built surface. If in the first level, the intention is merely decorative, integrating spatial components and its functional and iconographic program, in the second level the built space is amplified and transformed evidencing a visual metamorphosis coordinated under baroque assumptions.

Although we recognize to the works of Bacherelli and Nasoni an operational domain of perspective, the architectural composition and the manipulation of the apprehended space point out differences concerning their architectural intents and practice.

The strategy followed at the ceiling of the St. Vincent's lobby, widely spread in Portugal, seems to be confined to transmute the vault according

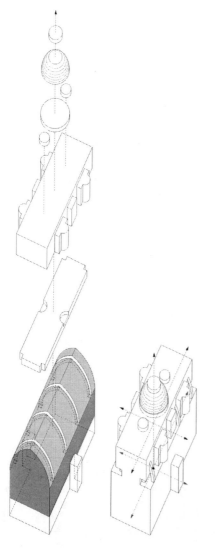

Figure 2. Interpretive scheme of illusion levels and spatial transformation at the Porto's Cathedral sacristy (1725), Nicolau Nasoni: built space and imaginary space. CABELEIRA 2015, 356.

to a wide and distant image. On the other hand, at the sacristy of Porto's Cathedral the complexity of the supporting surfaces, levels of interaction and assimilation of compositional strategies may reflect a greater spatial awareness and consequently greater ability concerning the tectonic transformation. Whether these differences could be explained through the authors' formation and background, it seems more credible to consider the disciplinary fields in which they move: while Bacherelli is specialized on painting, Nasoni gathers the practice of quadratura and architecture fostering coincident themes.

3 THREE IMAGINARY EXPERIENCES

Among the national authors who assimilate and continued the *quadratura* experience, the simultaneity among construction and representation practice is not witnessed. For these *quadratura* painters or masters of perspective, the knowledge of architecture science is based upon its image rather than a conscious mastery of its strategies and compositional rules. Still, given that the baroque space seems to be formalized under the simultaneous action of different agents, national *quadratura* masters tend to synchronize the imagined space with the constructive practice, highlighting common themes of spatial and formal research.

In this sense, three imaginary spaces of the Portuguese Baroque are considered either by the quality of the represented and built organism, revealing points of contact between the spatial research of *quadratura* and constructive practice. Clear examples of the generated *contrappunto* among visual and corporeal space, they summarize the baroque spirit and the desired phantasy.

Of octagonal plan and vault-shaped cap, the church of Menino-Deus (1711–37) emerges as a model, representative of architectural and artistic developments under the D. João V reign[2].

Here, the imaginary structure stems from the same geometrical matrix of the built nave. Even so, if some incongruities arise from the relation between constructed and represented facts (despite the equivalence of modular hierarchy some metric and alignments misfits are detected), a clear effect of spatial transformation and extension is apprehended, fostering the synthesis among built and *quadratura* proposed space. However, more evident is the coincidence between spatial anxieties and compositional procedures of the constructed and represented architectural design. Although the imaginary architectures reveal a greater formal freedom shaped by curvature of surfaces, handling and segregation of structural plans as well as the ornamental composition, that liberty aims to enhance the three-dimensionality of the illusory proposal.

Despite these variations, tensions amongst central and longitudinal axes are explored either by the outline of the built organism as well as by the illusory space. Besides these coincidences, among the transverse and longitudinal axes of the matrix rectangle, diagonal forces are considered according to sectioning of the vertices of the built rectangle, by a 45° plan, or through concave and convex curvature of the vertices of the represented cornices. Lines and curves, which foster the synthesis of both corporeal and visual features of the apprehended space. If the transverse axis is defined through modular and programmatic exception at the centre of the lateral elevations (the built pulpits and illusory balconies), the longitudinal axis, corresponding to the entry and visual approach towards the main chapel, is evidenced by false porticos and the articulation among represented and physical light.

The vertical propulsion encouraged by the *quadratura* image is characterized by a succession of perimeters. If, on one hand, the represented pilasters and vertical edges of the construction have a special role at the amplification of the perceived depth, on the other hand, surrounding cornices and galleries define levels of the imaginary construction. However, the vertical extension collides with a plan illusory ceiling at which centre a wide frame is open giving to the view, like a telescope, a vision of the transcendent. The specificity of this ceiling, concerning to coeval works, lies precisely at the replacement of a painted frame by an impressive wood frame enhancing the ambiguity between two-dimensional and three-dimensional facts.

At the Shrine of Nossa Senhora do Cabo (1701–40, Cabo Espichel) the *obra lisa* formulas of the constructive campaign contrasts with the festive spirit introduced by the baroque decorative campaigns that shaped the internal space underlining a scenic ambience supporting the religious message and program[3].

The built and represented space are organized by a transversal and longitudinal axes, giving supremacy to last one as stems from the rectangular shape of the building. Thus, despite hierarchical differences both axes define necessary symmetries to simultaneously solve the *quadratura* image projection (assisting the transference from an outlined prototype into the vault's surface), and regulation of the designed composition.

If these axes subordinate the modular variation, and value a longitudinal route, consequent to the built program and shape, an impression of diagonal tension is emphasized through transformation of the rectangular matrix. As such, at the first level of illusion a contraction of the rectangular polygon is operated through projection of the supporting brackets placed immediately above to the nave's cornice. Here, although the supporting brackets cut out the parallelepiped spatial configuration, in its distribution and rhythm, the thickness of a false slab, on which the second level of the illusory construction stands, evidences the return to the referential rectangle. Also in this second level, the surrounding elevations are arranged regularly, according to the stability of the rectangular matrix, being the transformation of the perimeter induced through the provision of free columns. Freed from its structural circumstance, the columns incite the apprehension of a rotated polygonal space, nullifying intersections of the illusory surrounding walls and bringing up continuities in shaping the spatial perimeter.

Finally, and already at the top of the false gallery, a flat ceiling is placed, on which the rotation among consecutive edges is achieved by a convex contour consequent to the introduction of oblique medallions.

The transformation of the different spatial layers is consistent with the organization of the temple and coeval compositional practices, both constructive and quadraturism, boosting the vertical axis where the major changes are operated. The space perimeter is successively compressed and amplified, the curvature of the vault surface is cancelled, the construction is elevated and the outlined set is opened by a horizontal frame that induces, symbolically, the internal space towards infinity.

In a coordinated approach among construction, sculptural and represented facts, the imaginary architectures are regulated under the modular sequence of the internal elevations. However, from a tectonic point of view the correlating logics of form and void, supporting elements and spans, are inverted. As such, the apparent weight of the illusory construction falls at the centre of the built arches opened at the nave sidewalls, overloading it perceptually. A logic which, along with the polygonal transformation and free combinatorial of the architectural ornament, resonates themes from the coeval project.

Also the church of Nossa Senhora da Pena (1705, Lisbon) is taken as a baroque lab of the age of D. João V, in which new spatial and decorative vocabularies are experience and implement. The internal built structure is characterized by a rectangular plan, oblivious to the polygonal coeval experiences, with a natural emphasis of the longitudinal axis closed by a deep main chapel. However, the stability of the built organism contrasts with the visual momentum generated by the ceiling's *quadratura* (1781) whose geometric and compositional variations enliven the perceived space[4].

Raising the imaginary architectures over the built moulding, that marks the parallelepiped space of the body of the temple, the *quadratura* enforces the perception of a vertical thrust broken by a false cornice framing a heavenly vision of the Coronation of the Virgin. Thus, above the stone moulding, false supporting brackets project the illusory construction towards the vertical axis around which the process of transformation is arranged according to contraction and dilation of the built rectangle. However, the rhythm imposed by the false supporting brackets, conditioning the represented structure, does not coordinate it with the composition of the sidewalls of the nave taking advantage of the linear abstraction of the stone moulding that, instead of participating as intermediary, functions as a frame/frontier between the physical and the visual space. A condition that leads to the perception of independent spatial entities, although arranged sequentially around the same visual vertical axis.

While *quadratura* architectures follow the symmetry imposed by the rectangular plan, only the longitudinal axis of the image is coordinated with same axis of the built structure. In fact, if the construction highlight is placed on the longitudinal sequence, dictated by form and light, the same happens with the represented space. If the longitudinal extremes of the illusory space are structured through the reinterpretation of the *Serliana* theme, side elevations are organized through tripartite composition. Unifying the imaginary perimeter a cornice of constant height conditions the impression of spatial depth and a surrounding balcony with straight and curved spans contrasts with the linearity of the stone moulding.

The effect of apparent depth is due to free columns whose simulated materiality (green marble) stands in sharp contrast to the material of the construction (white marble) accelerating the vertical impulse. The setting of free columns, with repercussions at the profile of the entablature, guides the eye diagonally inside the suggested space. A tension that values the relationship among centrality/axiality of the illusory space and may be placed in line with the spatial research of the national Baroque, managing overlaps, rotations, dilations and compressions of the built facts.

But when it comes to architectural grammar the case of Nossa Senhora da Pena distances itself from the constructive practice revealing the *quadratura* as a territory open to experimentation of architectural vocabulary and its combinatorial and formal manipulation, space rehearsal and articulation possibilities.

4 CONCLUSION

Still, we cannot claim these examples as definitive works of architecture because of the apparent compositional and metric mistakes. Inconsistencies hardly acceptable by construction, but easily surmountable by the architectural representation.

Although, if the evocation of an imaginary space is operated through quadratura, its practice arises as an architectural genre, which simultaneously allows amplifying the tangible space as well as opening it towards an open sky, in which the transcendent is revealed. A vision framed by the corporeal space, which is visually expanded through imaginary architectures.

Perspective image overcomes limitations on the nature of things, manoeuvring visually the reasoning and perception of space metric and formal nature. Therefore, structural complexity of Baroque spaces, achieved upon the intersection of corporeal and visual dimensions, synthesizes a continuous event challenging the apprehension of an imaginary universe fasten by a fictional moment which reveals the Aristotelian notion of "image-representation" as phantasy of a transcendent and orderly cosmos.

NOTES

[1] António Oliveira Bernardes experiences a 'pintura architecta' moulding and organizing architectural surfaces surpassing the two-dimensional tradition of the brutesco (Cabeleira 2015: 346; Mello 2002: 130; Serrão 1996–1997: 252).
[2] The authorship of the quadratura at the Menino-Deus Church is not consensual varying between João Nunes de Abreu and Manuel Vitorino da Serra (Mello 2002: 300; Raggi 2004: 570).
[3] The author of the temple's architectural perspectives (1640) is Lourenço da Cunha. Although, following the 1755 earthquake, the work suffered an intervention by Jose Antonio Narciso (1770) rebuilding and repainting the damaged portion. The result of different hands and times (Almada and Figueira 2002: 126; Raggi 2002: 129).
[4] Having been the quadrature (1719) attributed to António Lobo, the fact was reviewed by Reis (Reis 2006: 143) attributing the work to Luis Baptista (1781) according to renewal of the temple in the consequence of the damages caused by the 1755 earthquake.

BIBLIOGRAPHY

Almada, Carmen, and Figueira, Luís Tovar. "Igreja do Cabo Espichel: recuperação de um interior." *Monumentos*.16 (2002): 122–127. ISSN: 0872-8747.
Bury, J.B. "Late Baroque and Rococo in North Portugal." *Journal of the Society of Architectural Historians* 15.3 (1956): 7–15. ISSN: 10.2307/987760.
Cabeleira, João. "Arquitecturas imaginárias. Espaço Real e Ilusório no Barroco português." Ph.D. thesis. Universidade do Minho, 2015.

Mello, Magno Moraes. *A pintura de tectos em perspectiva no Portugal de D. João V.* Lisboa: Estampa, 1998. ISBN: 978972331402.

Mello, Magno Morais. "Perspectiva pictorum: as arquitecturas ilusórias nos tectos pintados em Portugal no século XVIII." Ph.D. thesis. Universidade Nova de Lisboa, 2002.

Raggi, Giuseppina. "Arquitecturas do engano: a longa construção da ilusão." Ph.D. thesis. Universidade de Lisboa, 2004.

———. "Dados técnico-documentais e análise crítica: reflexões em torno de um restauro." *Monumentos*.16 (2002): 128–129. ISSN: 0872-8747.

Reis, Vitor dos. "O Rapto do Observador: Invenção, Representação e Percepção do Espaço Celestial na Pintura de Tectos em Portugal no Século XVIII." Ph.D. thesis. Universidade de Lisboa, 2006.

Serrão, Vítor. "O Conceito de Totalidade nos Espaços do Barroco Nacional: A Obra da Igreja de Nossa Senhora dos Prazeres em Beja (1672–1698)." *Revista da Faculdade de Letras* V série. XXI-XXII (1996–1997): 245–267. ISSN: 0870-6336.

The utopian moment: The language of positivism in modern architecture and urbanism

Matthew Wilson

College of Architecture and Planning, Ball State University, Munice, IA, USA

ABSTRACT: Many masters of modernist design believed that sociological science and industry stood as the emancipatory powers of the 20th century. Arguably, the social and environmental implications associated with this outlook were framed by a system of thought called Positivism, which was the creation of the French philosopher, Auguste Comte (1798–1857). No work has explored the role this system of ideas played in shaping the utopian moment of early modernism, but it is often thought that Positivism was an icy, objectivist science upholding the status quo. This essay seeks to remedy this gross misrepresentation and neglect by showing that the movement of organized Positivism embraced a desire to reorganize the Victorian landscape. It begins with an analysis of the emergence of Comte's scientific-humanist ideas and their use in a participatory sociological practice; here his followers were attempting to make convivial spaces in complete regional city-communities. Thereafter, I will argue that the modern masters displayed similar patterns of language by relaying the aspiration to coordinate science and industry to improve the lives of the masses. Effectively, this essay seeks to show that organized Positivism played a role central to creating the utopian moment of early modernism.

Keywords: Victorian Utopianism; Positivism; Sociology; Modern Architecture and Urbanism; Class, Design and Urbanisation

1 INTRODUCTION

Positivism was once considered *the* philosophy of the 19th century, but design scholars have remained ambivalent about its direct and indirect influence on 20th century architecture and urbanism. Peter Eisenman writes that 'ethical positivism' produced the functionalist approach of modern design. Positivist-driven functionalism, he maintains, was essentially a 'phase of humanism' predicated on a faith in science and industry to benefit the common good (Eisenman 1996). Along these lines, Harry Mallgrave argues that Positivism informed the architectural theories of rationalism and primitivism (Mallgrave 2006: 353–354, 359, 390–354, 506–315). However, Colin Rowe and Fred Koetter quip that the bearings of Positivism led to little more than a "historical cul-de-sac" (Rowe and Koetter 1983: 21). To muddy the waters more, some assert that Positivism comprised a backward-looking ideology that made no impact on spatial design (Moore 2007; Stephen 1869; Wright 2003: 141), but Joseph Rykwert and David Smith Capon state that it offered a new social democratic humanism that aimed to 'counteract the dehumanizing influence of industrial society' (Capon 1999: 140; Rykwert 2000). Alan Colquhoun confidently concludes that the 'positivist-functionalist model' dominated modern architecture during the early modern period, spanning from the iconoclastic 1890s to the 1920s, and survived well into the 1960s (Colquhoun 2009: 167–168).

Curiously, no study has scrutinised the origins of organised Positivism in relation to the rise of modern architecture. More than any other group, the British Positivists were responsible for popularizing this comprehensive system of life, which was considered capable of ringing in a glorious epoch of global happiness and harmony.

This essay will outline how from the 1850s the Positivists acted on the utopian ideas of the French philosopher, Auguste Comte (1798–1857). Comte was famous for introducing modern sociology and a complementary creed called the Religion of Humanity. He and his followers established a network of Positivist Societies to disseminate their scientific-humanist theory and practice. Their intention here was to realise Comte's utopia called the Occidental Republic. Effectively, this paper argues that, like the Positivists, the modern masters articulated a language about coordinating science and industry to create functional, hygienic greenbelt city-communities to improve the lives of everyone, everywhere.

2 POSITIVIST UTOPIANISM, ITS EMERGENCE AND EVOLUTION

From 1817 to 1822, Comte worked as the assistant to Saint-Simon, the philosopher of 'social physics'. Inspired by the works of Condorcet, they explored the progressive implications of classifying the 'positive sciences'. Saint-Simon determined that this process could facilitate a return to 'the science on which society is founded, namely ethics' and thus 'a social state in which science will again assume a religious character' (Saint-Simon 1834: 7, 9–32, 44–38). A 'Newtonian elite', he imagined, could use their scientific authority to implement ethical social reorganisation. They would provide education, promote pacific solidarity, and coordinate meaningful work as well as resolve local, national, and international disputes. Saint-Simon's intention was to link art, science, and industry to unify humanity through peaceful industrialisation as opposed to expansionist warfare. Along these lines, he proposed transnational urban infrastructural projects similar to the Suez and Panama Canals.

Under his 'master', Saint-Simon, Comte developed scientific-historical surveys of Western society that traced the withering-away of the powers of monotheism and monarchy. Science and industry, he declared, were emerging as new modern 'spiritual and temporal' power structures (Vernon 1984). Departing from the Saint-Simonians' 'New Christianity', Comte with unshakeable confidence published his controversial Law of Three Stages, which theorised that all societies develop from a theological, to metaphysical and on to a Positive Era. A 'deliberate planning and policy' implemented by intellectuals could transform the West, opening the Positive Era of society, he maintained (Comte 1877: 538; Fletcher 1974: 29–40). Along these lines, during the July Revolution Comte introduced his new 'master-science' of sociology.

Observing the degrading and defiling conditions of Parisian proletarian life, Comte insisted, like Marx, that it was necessary to engage with the realities of 'enormous cities'. During the revolution of 1848, however, he neither endorsed a violent revolution nor the communitarian escapism Marx disparagingly branded 'utopian socialism' (Marx and Engels 2002). Comte instead declared 'moral revolution' and launched a humanist creed called the Religion of Humanity. He also produced a utopia-planning programme called the Occidental Republic, which proposed to use sociology to plan the devolution of Western Empires into 500 idyllic regional republics. Comte believed that his followers could help to realise these hygienic, modern greenbelt city-communities across the world by the 1960s. Each city-state, with a land area comparable to Belgium, would contain two million people, approximately.

Like the work of the revolutionaries of the 1790s, Comte's Occidental Republic programme included a calendar, cultural festivals, banking system, regional currencies, ethical codes, and flag system. Not least, Comte's programme called on citizen-sociologists to organise a network of new types of architecture for Positivist 'spiritual and temporal' interventions in town and country. These institutions would serve as the critical spatial agency for breaking up Western Empires and creating the republics of the Positive Era. The chief 'spiritual institution'—known as a Positivist Society Hall, Temple of Humanity, or Civic Society—would coordinate the public life of citizen groups working towards the vision of utopia on the horizon.

As the hub of the local community, the Positivist Hall was considered a catalyst for structured social change. It would function as a centre for regional sociology, institute of humanist scholarship and republican hall of social activism (Wilson 2015; Wright 1986). Part of the activities here would include cultural festivals, popular education, and social investigations for regional place making. Positivist groups would draw on Comte's designs for this and other spiritual interventions such as schools, libraries, parks, gardens, hospitals and homes. They would encourage temporal institutions—such as workshops, factories, banks, exchanges and union halls—to operate according to Comte's system of moral capitalism.

Seeking to disseminate Comte's utopian vision, the Oxford don, Richard Congreve, founded the first British Positivist Society in 1859. It attracted outcasts and the working classes as well as John Stuart Mill, George Eliot, Beatrice Webb and William Morris. By the twentieth century, a network of Positivist clubs had opened across Britain. They hosted popular lectures, technical training, arts lessons, festivals, and civic rites of passage.

Of paramount importance here was that the British Positivist Society also organised urban and rural field studies. This form of 'concrete politics' intended to train and empower conscientious citizen-sociologists for regional place making. Congreve's allies in this line of activity included Frederic Harrison, Charles Booth, Patrick Geddes, and Victor Branford. They understood the region as a unit nested between and affected by the empire, nation, city, district, and household. From the mid-1850s through to the Great War, they therefore grappled with resolving conflicts on nested spatial levels of human relations. On the international level, Comte and Congreve conducted historical and moral geographical types of surveys. These sociological investigations exposed the links between imperialism and domestic decline.

Like Saint-Simon, Comte and Congreve promoted a policy for peaceful international relations leading to the dissolution of Western empires. Thereafter, on the national level, the investigators Harrison and Booth commenced on industrial and social types of surveys. They documented poverty, overcrowding, and industrial exploitation and proposed programmes to create a national network of industrial republics. The activists Geddes and Branford subsequently led civic and rustic surveys of British regions. They aimed to link town and country into Garden City-state 'eutopias' or idyllic real places.

In the footsteps of Comte the British Positivists held that it was possible to create a world of utopian republics if, in each region, four different idealist citizen-groups agreed to rule and be ruled in turn in relation to the spaces of the city. Based in Positivist temples, schools, and studios, 'Intellectuals and Emotionals' would perform as the custodians of social and environmental welfare. These sociologists, instructors, and artists would envision programmes and plans for urban-regional social reconstruction and look after the mental and physical health of cities and citizens. Industrial 'Chiefs' would manage planning schemes as well as work in 'temples of industry'. These moral capitalists would be elected, based on merit alone, from the unionised system of trade guilds, or 'the People'.

The British Positivists' programmes proposed to coordinate the planning of new streets, parks, schools, factories, hospitals and homes following Comte's Occidental Republic scheme. They recommended the national municipalisation of industry, equal wages in town and country, and shortened working hours to enable all to enjoy civic life. Strike funds would finance mid-rise mixed use housing blocks and improved transport links. Each green-belt city-community in Britain would be replete with libraries, parks, museums and, also, 'humanist institutes', 'cathedrals of the people' or 'pantheons of humanity' (Geddes 1887; Geddes 1904b). Here schools would become the nucleus of the neighbourhood community and the means for eradicating social inequality; these spaces would shape the intellectual and emotional identity of each region. The Positivists were thus concerned with developing a participatory praxis for transforming industrial agglomerations into self-sufficient regional units.

3 THE UTOPIAN MOMENT OF THE MODERN MOVEMENTS

In this section, I will argue that during the early twentieth century the language of Positivism percolated into the work of avant-garde designers in Europe, America and beyond. As such, Positivism created the utopian moment for the modern movements of architecture and urbanism (Jencks 1985; Pocock 1975)[1].

The Positivists' aim to deliver modern life in the form of green-belt city-communities passed via Howard, Geddes, and Branford and into the work of the avant-garde. Like Comte's vision, Howard's Garden City had a size and scale compatible to the medieval city. Howard similarly suggested that the 'individualistic socialism' of trade unionism could play a critical role in producing this new social fabric. Similarly, Geddes and Branford proclaimed Comte's Occidental Republic scheme was a 'practical treatise' for post war reconstruction (Branford and Geddes 1917: 52). Through his various town-planning exhibitions across Europe and Asia, and collaborations on designing secular 'spiritual institutions', Geddes popularized aspects of Comte's work (Welter 2002: 174–250). Like Harrison, who was his 'moral teacher', Geddes recalled an important precedent: priests and guilds had united to produce the scale, character, and lifestyle of the medieval cathedral city. They praised its offering of manifold 'centres of moral and spiritual education' Geddes 1904b: 1–3, 21–35, 221; Harrison 1894: 236). Geddes' American acolyte Lewis Mumford similarly argued that the medieval city offered 'a higher standard [of life] for the mass of the population than any form of town, down to the first romantic suburbs of the nineteenth century' (Mumford 1940: 44).

Mumford's colleague Patrick Abercrombie, who is considered the 'father of British town planning', acknowledged Comte and Le Play as the founders of the modern 'French School' of regionalism. They also supported the coordination of town and country units and suggested that the school should become the centre of the modern neighbourhood community (Abercrombie 1959: 9–27, 54–25, 103–104, 126–130; Cherry 1981; Mumford 1940: 351). Abercrombie collaborated with his mentors, Geddes and Victor and Sybella Branford, in an effort to establish a national network of 'Civic Societies' to transform Britain into Garden City-states. Here groups of university-led citizen-sociologists carried out regional surveys to inform post-war reconstruction initiatives, which entailed the creation of new communities with integrated transport links, work, housing and public spaces (Abercrombie 1920; Geddes 1904a).

In America, the functionalist architect Louis Sullivan expressed a Positivist view of history, politics, and science. Science was the 'power to put forth our WILL—in place of Destiny. This is the spiritual dawning power of Democracy' (Andrew 1985; Sullivan 1961), wrote Sullivan. Following Comte, he asserted that sociology would become the 'gravitational centre of all the sciences; and a philosophy or gospel of democracy, the motive power of the world'. Sociology offered a 'salvation of man and society'. Like the Positivists he believed sociology could make democracy a 'living thing'; it would transform architecture, which had been 'made a plaything for long enough', into a 'living force'—an art for the people and by the people (Sherman 1962: 93; Sullivan 1978: 65–66; 101–102).

Sullivan's protégé Frank Lloyd Wright likewise challenged reactionary views about the machine age. In addition, like the Positivists he cautioned his followers about industrial abuses. He stated that 'the Machine [is] here to serve Humanity ... human imagination may use [it] as a means to more life, and greater life, for the Commonwealth'. The Machine would 'conquer human drudgery' and was a symbol of 'spiritual liberation', collectivism and a tool of democracy (Wright 2008: xi, 23, 97). It would alter the nature of family life and, just as radically, the appearance of cities. Along these lines, like Comte, Wright held that the environment of the modern home was an essential conduit to internalise modern civic manners and values in children.

The 'father of Dutch modernism' Hendrik Petrus Berlage aired 'almost Positivist' political and architectural ideas. He wrote about the establishment of an 'ethical settlement' founded on the spiritual concept of scientific progress. It would produce a 'new age of culture, focused on socialist principles of fraternity'. Berlage's support of Positivism also appears in the artwork in the vestibule of his most famous work, the Amsterdam Stock Exchange. Here Berlage commissioned the artist Jan Toorops to create *opera sectilia* panels that depict, in a Positivist fashion, 'the feudal order succeeded by the class society of nineteenth-century capitalism, which itself would be succeeded by a new age of harmony and religiosity' (Banham 1960: 144; Berlage 1996: 37–38, 60–33, 313). Berlage later promoted a 'religion of the new humanity' and designed a Pantheon der Menschheid, or 'Pantheon of Humanity', which drew on Comte's ideas. Such institutions, it seems, were meant to reflect the urge to forge a new post-war humanity. They formed part of greater discussions about how university elites, guilds and civic groups would usher in the 'moral equivalents of war' for everyone, everywhere (Branford 1921: 45–49, 85–91; Gould 1920).

In the footsteps of the Positivists and John Ruskin, the German Deutscher Werkbund polemicized against laissez-faire capitalism. They believed that the organisation of guild-communities could connect ethics and aesthetics. True architecture was beautiful through-and-through if it produced a vir-

tuous, memorable and meaningful way of life. As such, Werkbund leader Herman Muthesius aimed to make this group the 'sole arbiters of taste' in Germany. As if 'Positivist sociologists' they aimed to coordinate national industrial production. Here trade unions would mass-produce modern objects of a high moral and cultural value. Along these lines Peter Behrens, the Werkbund 'master of the moderns', designed the AEG's industrial identity and Turbine Factory, which represented a 'factory-temple', an 'endorsement of industrial civilisation' (Anderson 2000: 145).

Like the Positivists, the Dutch De Stijl group upheld the idea of harmonizing social love, 'social order, and social progress'. De Stijl's art, after all, represented 'dynamic equilibrium', or the balancing of opposing forces in nature and society. They aimed to establish an 'objective universal means of creation', a new visual environment for modern life (Van Doesburg 1971b). Mechanised and automated processes would decrease intellectual and physical drudgery, facilitating the growth of international emotional unity. These processes would thus obliterate the 'old consciousness' associated with individualism, ancient absolutism and nationalism. Modern art, science and machinery, as such, provided the grounds for a 'new consciousness' for a utopian, universal collectivism (Van Doesburg, Van't Hoff and Huszar 1971). De Stijl leader, Theo van Doesburg, accordingly postulated in 1924 that 'new architecture develops out of the exact determination of practical demands', and would be 'anti-decorative' and 'anti-formalist', which was a forecast Comte made during the 1840s (Van Doesburg 1971a). By the mid-1920s, de Stijl's universalist outlook had infiltrated the Bauhaus.

In his post-war expressionist phase, Walter Gropius contributed to the *Arbeitsrat fur Kunst*. This 'work council for the arts' aimed to produce art for the multitudes rather than the moneyed. Along similar lines, in the first Bauhaus manifesto of 1919, Gropius wrote that architecture was the 'crystalline expression of man's noblest thoughts, his ardour, his humanity, his faith, his religion!' (Gropius 1971). Here trade union 'guilds' would unite to build secular spiritual spaces including a 'Cathedral of Socialism'. The Bauhaus was not concerned with the 'decorative, mystical, and metaphysical'. Instead, their aim was to create a new 'anti-aesthetic aesthetic that would prize functionality'. And also like the Positivists and the logicians of the Vienna Circle, the Bauhaus was concerned primarily with technical processes for unifying architecture, painting and sculpture, to create airy and hygienic working and living spaces (Galison 1990; Gropius 1971)[2] to avert 'mankind's enslavement by machine'(Gropius 1965: 54). Architects, Gropius claimed, had been 'put into this world to remake humanity' (Gropius 1968: 61). Based on sociological knowledge, the architect was to act as the coordinator of a democratic 'comprehensive unity', from domestic to regional and national planning. Thus planning, he maintained, was to 'grow from the ground up, not from the top down by force'(Gropius 1970: 12–13, 65, 98–19, 103, 145).

The International Congress of Modern Architecture (Ciam) adopted a planning agenda based on Howard's horizontal and Le Corbusier's vertical garden city visions. Like the Positivists, CIAM 'believed that developments in industry and in the scientific understanding of human history and society were making possible a new social system based on universal human association' (Mumford 2000: 2). For these reasons CIAM, led by the 'neo-Saint-Simonian' Le Corbusier, sought to place architecture and town planning 'morally and materially' on its 'true plane', the sociological plane. They wrote about the need to redistribute the land and to plan the 'organization of the functions of collective life' across town and country (Ciam 1971b).

Along these lines Le Corbusier supported the Positivist system of syndicalism, where trade unionists assumed the means of production and elected from their body an empathetic managerial elite; Comte explained this new captain of industry would be a 'dictatorship of the proletariat'. Similar to the Positivists' thinking, Le Corbusier and the CIAM thought architecture could temper the 'ruthless violence of private interests' and warfare.

Utilising social investigation, the CIAM aimed to produce spaces that harmonised 'individual liberty and collective action'. This process was to begin with the study of the 'initial nucleus of town planning', the home. They examined it in relation to the organisation of the occupational, transport and leisure spaces of the city. Like the Positivists, CIAM was concerned with the city as a social grouping situated 'within the totality of its region' (Ciam 1971a). Increasingly, however, this process was less about humanity and more about the dictates of form and style.

4 CONCLUSION

In conclusion, the Positivists' language about coordinating science, art, and industry framed the modernists' desire to create rational, meaningful, and hygienic designs, to enable humanity to benefit from science and the machine. Positivism served as a theoretical and practical impetus to the early modernists' 'ethical functionalism'; the concept of the town and country as a planned, self-sufficient regional unit; the notion that the school is the nucleus of the neighbourhood community and the means for broader social transformation; and that

the city-community is defined by its public spaces and civic institutes.

Undoubtedly, the masters of modernism did not subscribe to the complete system of Positivism; they co-opted different aspects of its theory and practice. Positivism, after all, had one fundamental requisite. That requisite was upholding a 'higher social morality, an enlarged conception of human life' and 'a more humane type of religious duty' to the people and their city-region (Harrison 1894: 244–251, 412–255) (Harrison 1908: 262).

NOTES

[1] Here I seek to fuse the architectural and political chronologies set out by Pocock and Jencks but, in reference to the Positivists, also see (Bevir 2011; Claeys 2010).
[2] Hannes Meyer, for such reasons, discussed "anti-aesthetic" architecture.

BIBLIOGRAPHY

Abercrombi E, Patrick. *Town and Country Planning.* London: Oxford University Press, 1959.

——. "A Civic Society." *Town Planning Review* 8.2 (1920): 79–92.

Anderson, Stanford. *Peter Behrens and a New Architecture for the Twentieth Century.* Cambridge: MIT Press, 2000.

Andrew, David S. *Louis Sullivan and the Polemics of Modern Architecture.* Chicago: University of Illinois Press, 1985.

Banham, Reyner. *Theory and Design in the First Machine Age.* London: The Architectural Press, 1960.

Berlage, Hendrik Petrus. *Thoughts on Style, 1886–1909.* Santa Monica: The Getty Center for the History of Art and the Humanities, 1996.

Bevir, Mark. *The Making of British Socialism.* Princeton: Princeton University Press, 2011.

Branford, Victor. *Whitherward? Hell or Eutopia.* London: Williams and Norgate, 1921.

Branford, Victor, and Geddes, Patrick. *The Coming Polity: a Study in Reconstruction.* London: Williams and Norgate, 1917.

Capon, David Smith. *Architectural Theory.* Vol. 1. New York: John Wiley and Sons, 1999.

Cherry, Gordon E. *Pioneers in British Planning.* London: Architectural Press, 1981.

Ciam. "Charter of Athens." In *Programs and Manifestoes on 20th-Century Architecture.* Ed. Conrads, Ulrich. Cambridge: MIT Press, 1971a. 137–145.

——. "La Sarraz Declaration." In *Programs and Manifestoes on 20th-Century Architecture.* Ed. Conrads, Ulrich. Cambridge: MIT Press, 1971b. 109–113.

Claeys, Gregory. *Imperial Sceptics: British Critics of Empire, 1850–1920.* Cambridge: Cambridge University Press, 2010.

Colquhoun, Alan. *Collected Essays in Architectural Criticism.* London: Blackdog Publishing, 2009.

Comte, Auguste. *System of Positive Polity*. Vol. IV. London: Longmans, Green, and Co., 1877.

Eisenman, Peter. "Post-Functionalism." In *Theorizing a New Agenda For Architecture*. Ed. Nesbitt, Kate. New York: Princeton Architectural Press, 1996. 78–83.

Fletcher, Ronald, ed. *The Crisis of Industrial Civilization*. London: Heinemann Educational, 1974.

Galison, Peter. "Aufbau/Bauhaus: Logical Positivism and Architectural Modernism." *Critical Inquiry* 16.4 (1990): 709–752.

Geddes, Patrick. *Civics: as Applied Sociology*. London: LSE, 1904a.

———. City Development: a Report to the Carnegie Dunfermline Trust. Edinburgh: Geddes and Company, 1904b.

———. *Industrial Exhibitions and Modern Progress*. Edinburgh: D. Douglas, 1887.

Gould, F.J. "Relation of Religion to Social Life." *Positivist Review*.334 (1920): 223–227.

Gropius, W. *The New Architecture and the Bauhaus*. Cambridge: The MIT Press, 1965.

Gropius, Walter. "Programme of the Staatliches Bauhaus in Weimar." In *Programs and Manifestoes on 20th-Century Architecture*. Ed. CONRADS, Ulrich. Cambridge: MIT Press, 1971. 49–53.

———. *Scope of Total Architecture*. New York: Collier Books, 1970.

———. *Apollo in the Democracy: the Cultural Obligation of the Architect*. New York: McGraw-Hill, 1968.

Harrison, Frederic. *National Social Problems*. London: Macmillan, 1908.

———. *The Meaning of History and Other Historical Pieces*. London: Macmillan, 1894.

Jencks, Charles. *Modern Movements in Architecture*. Penguin Books, 1985.

Mallgrave, Harry Francis. *Architectural Theory*. Vol. 1. Oxford: Blackwell, 2006.

Marx, Karl, and Engels, Frederick. *The Communist Manifesto*. London: Penguin, 2002.

Moore, Stephen. "Technology, Place, and Nonmodern Regionalism." In *Architectural Regionalism: Collected Writings on Place, Identity, Modernity, and Tradition*. Ed. Canizaro, Vincent B. New York: Princeton Architectural, 2007. 432–445.

Mumford, Eric Paul. *The CIAM Discourse on Urbanism*. Cambridge: MIT Press, 2000.

Mumford, Lewis. *The Culture of Cities*. London: Secker & Warburg, 1940.

Pocock, J.G.A. *The Machiavellian Moment: Florentine Political Thought and the Atlantic Republican Tradition*. Princeton: Princeton University Press, 1975.

Rowe, Colin, and Koetter, Fred. *Collage City*. Cambridge: MIT, 1983.

Rykwert, Joseph. *The Seduction of Place: the City in the Twenty-first Century*. New York: Oxford University Press, 2000.

Saint-Simon, Claude Henri de. *New Christianity*. London: Effingham Wilson, 1834.

Sherman, Paul. *Louis Sullivan, an Architect in American Thought New Jersey*: Prentice-Hall, 1962.

Stephen, Leslie. "The Comtist Utopia." *Fraser's Magazine* 80 (1869): 1–21.

Sullivan, Louis. *Kindergarten Chats and Other Writings*. New York: Wittenborn, 1978.

———. *Democracy*. Detroit: Wayne State University Press, 1961.

Van Doesburg, Theo. "Towards a Plastic Architecture." In *Programs*. Ed. CONRADS, Ulrich. Cambridge: MIT Press, 1971a. 78–80.

———. "'De Stijl': Creative Demands." In *Programs*. Ed. CONRADS, Ulrich. Cambridge: MIT Press, 1971b. 64–65.

Van Doesburg, Theo, Van'T Hoff, Robert, and Huszar, Vilmos. "'De Stijl': Manifesto I." In *Programs*. Ed. CONRADS, Ulrich. Cambridge: MIT Press, 1971. 39–40.

Vernon, Richard. "Auguste Comte and the Withering-Away of the State." *Journal of the History of Ideas* 45.4 (1984): 549–566.

Welter, Volker M. *Biopolis: Patrick Geddes and the City of Life*. Cambridge: MIT Press, 2002.

Wilson, Matthew. "On the Material and Immaterial Architecture of Organised Positivism in Britain." *Architectural Histories* 3.1 (2015): 1–21.

Wright, Frank Lloyd. *Modern Architecture*. New Jersey: Princeton University Press 2008.

Wright, Julian. The Regionalist Movement in France, 1890–1914. Oxford: Clarendon, 2003.

Wright, T.R. The Religion of Humanity: the Impact of Comtean Positivism on Victorian Britain. Cambridge: Cambridge University Press, 1986.

Threshold and mediation devices in the domestic space approach

Bárbara Leite
Faculty of Architecture, University of Porto, Porto, Portugal

ABSTRACT: This article sets out to discuss the boundary condition between the spheres of public and private / exterior and interior in contemporary housing project, namely the intermediate territory between the two.

Therefore, it gives special attention to the theoretical framework on relations that are determined between the two sides of a boundary. There will be a particular emphasis on social sciences, taking into consideration Georges Teyssot's contributes on his essay "*A topology threshold*, as well as Arnold van Gennep's "*Les rites de passage*", who also explores the concept of "*boundary*" and "*threshold*". Then, based on the analysis of the rites of passage in human societies—as it is a culturally constructed condition -, it will be established a connection with the field of architecture, mainly focusing on Alison and Peter Smithson's notions of "*doorstep*", and Aldo Van Eyck's concept of *in-between*.

In architectural design the space *between*, although ambiguous, sets the negotiation of boundaries between private and public sphere, as principle of integration, dialogue, evasion, and transition.

Finally, we will try to assess which are the architectural elements in the domestic space—such doors, sills, windows, walls, etc. -, that convey this specificity as mediation devices and enable a forthright dialogue and negotiation between both domains.

Keywords: threshold, domestic space, mediation devices

1 CONSTRUTION OF AN IDEA OF THRESHOLD

In this paper we intend to consider on how the exterior and the interior, as well as the uses associated with them, are related and especially interconnected through the concept of threshold space, seen as an intermediate domain between exterior and interior. These two concepts, exterior and interior space, are intrinsic to the constitution of an architectural object, while distinct, they set limits and define themselves mutually, both at spatial and symbolic level.

> ... suddenly there's and interior and an exterior. One can be inside or outside. Brilliant! And that means—equally brilliant—this: thresholds, crossings, the tiny loop-hole door, the almost imperceptible transition between the inside and the outside, an incredible sense of place, an unbelievable feeling of concentration when we suddenly become aware of being enclosed, of something enveloping us, keeping us together, holding us—whether we be many or single. An arena for individuals and the public, for the private and the public spheres. Architecture knows it and uses it. (Zumthor 2006: 45–47)

As a basic principle, the architect adapts spaces through boundaries, directing and manipulating the inhabitant's experiences. Consequently, individual's space is a culturally built entity which human action materialises, either suggesting or prescribing a certain limit, which interferes with the inhabitant behaviour within that space.

Thus, for space to be understood by people, it needs to be provided with limits, since it is through this process of delimitating space that it becomes inhabitable, allowing the development of structured human practices.

1.1 Threshold concept

The threshold concept belongs to the domain of spatial metaphors that designate intellectual and symbolic operations, which can be inscribed as "passages", namely, transitions. The threshold represents an intermediate space that signals the moment of passage, the entrance.

However, it should be noted that Walter Benjamin (1982–1940), in *Passagen—Werk*[1], warns about the distinction between boundary and threshold. For him the concept of threshold (die Schwelle) should clearly differentiate itself from the concept of boundary (die Grenze). Thus, the threshold is a zone; a doorstep, a passage; leaking and filling are included in the word schwellen ("to swell")[2].

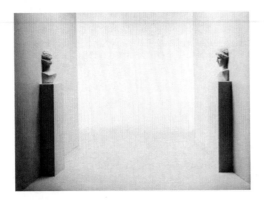

Figure 1.   Intervallo, Giulio Paolini, 1985 <http://www.fondazionepaolini.it/scheda_opera.php?id=195> [available in 2016/01/12].

Figure 2. Passage Choiseul, Paris, 1825. <https://evelyntiefenbacher.wordpress.com/2011/02/28/walter-bejamins-schwellen-wandel-ubergang-fluten-liegen-im-worte-%E2%80%9Aschwellen%E2%80%99/> [available in 2016/01/15].

Walter Benjamin, in *Passagen—Werk*, was particularly puzzled with the Parisian passages as paradigmatic examples of ambiguous spaces. These spaces assumed a hybrid character; they could not be considered a traditional inner space, neither an assumed outer one; those passages received natural lighting, since they were composed of lightweight iron roofing and glass.

1.2   *Rites of passage*

In 1909, the French anthropologist Arnold van Gennep (1873–1957) produced a research study on the transition rituals between different realities and various levels of human behaviour, which he gathered in *Les Rites of Passage* (Van Gennep 1960; Van Gennep 1978). This study contributed to a spatial reading of the society, where the ritual was the study object. There, he constructs a particular approach towards the threshold question, from an anthropological point of view, that later would have repercussions in the architectural sphere.

Through various collected examples, Van Gennep highlighted the similarity between manifestations of each individual's life cycle, of those of everyday family life, of time passing, seasons, etc. Rituals dramatized these transitional practices of social expression as if they were theatrical performances, celebrating different stages of development of one's life.

1.3   *Concept of doorstep and in-between*

Similarly, the couple of architects Alison Smithson (1928–1993) and Peter Smithson (1923–2003) built an approach to the idea of threshold. Based on studies made by the photographer Nigel Henderson and the sociologist Judith Stephen, on patterns of association between the populations of a neighbourhood in London, the Smithsons introduced the concept of *doorstep*, at CIAM IX at Aix-en-Provence, in 1953.

They were interested in promoting the relationship between domestic interior and the street. In other words the extension of house dwelling to the adjacent public space as a potentiating action for new dynamics and appropriations. The extension of the domestic space to the public arena as "a stage of social expression in which identity, social ties, a sense of security and well—being were generated"[3]

However, Aldo van Eyck (1918–1999)[4] redefines the concept of *doorstep* to *in-between*[5]. He addressed the concept of *doorstep*, recovering the sense of ritual space, that of passage:

[the door] fits us on arrival and departure, it is a vital experience not only for those who transpose it, but also for those who encounter or leave behind. The door is a place made for an occasion. The door is a mediation device, made for an act that is repeated millions of times in our lives, between the first entry and the last exit.[6][7]

Figure 3. Photographs of Nigel Henderson, London, 1951 <http://architecture-plus.blogspot.com/2010/10/pravo-na-grad-ove-2.html#!/2010/10/pravo-na-grad-ove-2.html> [available in 2016/01/15].

Figure 4. Fenêtre en Longueur. Le Petit Maison, Le Corbusier, 1925 <https://www.flickr.com/photos/462069 21@N08/5146566485/> [available in 2016/01/15].

Van Eyck suggested that the door could be considered as:

…a place which articulates and belongs to both the inside and the outside, a place where the significant aspects of both sides are presented simultaneously. The door should expand and adopt a capable form of evoking a warm welcoming, as well as materialised an invitation for a break, an invitation to remain.[7]

2 SPATIAL MEDIATION DEVICES

After the construction of an idea that could clarify the meaning of threshold, it becomes essential to understand how this concept can be materialised, as these liminal spaces, apart from being both "intermediary and intermediate, also articulate and are articulators", triggering a relationship of interdependence between the outside world and the interior[8].

The complexity of the concept of dwelling leads us to a realm populated by devices that serve as mediators between what happens inside and outside it. The various architectural resources and features, while providers of distinct qualities, prove to be essential for the intensification of this relationship. These devices—door, window, and sill—are seen as tools or elements that establish relations of mediation between spaces, but beyond this function, they are also influential in the individual's mode of action

2.1 The window, the door, the wall

Doors and windows or other type of spatial device that materialise a threshold can be designed as distinct traces of limits, as well as devices that establish the passages/transitions to the exterior. The devices can establish two types of limit, *frontier*, and *bridge*. While the first determines an enclosed space, the second refers to a space that opens to the outside.

The window, as a mediation device, implies a static position in space (as opposed to the door), a point from which one observes, contemplates and illuminates the home.

With the innovation in construction technology, windows became wider, thus observing a new spatiality where space is enlarged, integrating interior and exterior into a single frame. Nevertheless, both door and window are distinguished in terms of function, the first concerns an opening, the second a passage; however, they both constitute *interfaces* between spaces, thus allowing a physical and/or visual transition.

The door congregates in itself a willingness to move from one side to the other, as it is through it that a *breach* between spaces occurs. The door is, thus, the primary mediator of passage and articulation, conforming an element of connection between parts and creating the conditions for permanency in both domains, establishing the joint and the invitation for evasion and imagination.

The wall is also another mediation device, in the sense that it represents the physical integrity of the boundary, setting apart two worlds by its mass and thickness. As Robert Venturi states:

…designing inwards, as well as from the inside out, creates the necessary strain, which helps to make architecture. As the inside is different from the outside, the wall—the turning point—becomes an architectural event in itself. (Venturi 1995: 119)

Finally, the wall is in fact a materialised boundary, an event that, by its nature and characteristics, influences how one demarcates the inner and outer world and the relations that they establish.

NOTES

[1] *Passagen Werk* is an unfinished literary project by the German critic Walter Benjamin, written between 1927 and 1940. It reproduces a collection of texts on Parisian lifestyle in the nineteenth century, especially concerning Paris *passages couvertes* [covered passages].

[2] Walter Benjamin quoted by, Georges Teyssot (2010: 234).

[3] Alison and Peter Smithson quoted by, João Paulo Martins (2006: 255).

[4] Under the *motto* "la grande plus réalité du seuil", released in CIAM X in Dubrovnik, in 1956, Van Eyck exposes his theoretical construction, in which he emphasises the concept of threshold as a fundamental principle of the relations that architectural design establishes with space.

[5] The intermediate concept proposed, *in-between*, first appeared in the publication "The story of another idea", in the 1st edition of the magazine *Forum*, which Aldo van Eyck integrated as editorial member, alongside Bakema, in 1959.

[6] Aldo Van Eyck *apud*. João Paulo Martins (2006: 256).

[7] Aldo Van Eyck *apud*. João Paulo Martins (2006: 256).

[8] Aldo Van Eyck *apud*. (Martins 2006: 292).

BIBLIOGRAPHY

Martins, João Paulo. "Os Espaços e as Práticas. Arquitectura e Ciências Sociais: Habitus, Estruturação e Ritual." PhD Dissertation. Universidade de Lisboa, 2006.

Teyssot, Georges. *Da Teoria de Arquitectura: Doze Ensaios*. Trad. Rita Marnoto, rev. Paulo Providência, Gonçalo Esteves de Oliveira Moniz. Lisboa: Edições 70, 2010. 295 p. ISBN: 978-972-44-1615-1.

Van Gennep, Arnold. *Os Ritos de Passagem*. Trad. Mariano Ferreira; Apresentaçäo de Roberto da Matta. Petrópolis: Vozes, 1978.

——. *The rites of passage*. London: Routledge & Paul, 1960. 198 p. ISBN: 978-0226848495.

Venturi, Robert. *Complexidade e Contradição em Arquitectura*. São Paulo: Martins Fontes, 1995.

Zumthor, Peter. *Atmospheres: architectural environments; surrounding objects*. Basel: Birkhäuser, 2006. ISBN: 9783764374952.

A utopia in the real city

Eneida Kuchpil
Course of Architecture and Urbanism, Department of Architecture, Federal University of Paraná, Curitiba, Brazil

Andrezza Pimentel Dos Santos
Course of Architecture and Urbanism, Department of Architecture, Positivo University, Curitiba, Brazil

ABSTRACT: From the idea of utopia created by Thomas More in 1516, which described an imaginary island with an ideal society living in a place endowed with characteristics of great urban organization, many utopias of cities were developed by philosophers, socialists, architects and urbanists to answer the issues analysed from the observation of post-Industrial Revolution cities, rarely being realized. A negative diagnosis of society, from the description of the terrible living conditions to the evaluation of daily commuting problems and the reduction of leisure, is recurrent to proposals of utopic cities. In the formulation of an answer to a similar diagnosis, utopic cities have come close. From the review of some of these utopias, the objective of this work is to discuss the vertical cities utopic models, developed by Swiss architect Le Corbusier, leading to the formulation of the *Plan Voisin* in Paris. It is a utopia in a real city, contrary to the previous utopic cities proposals, which are characterized by the no place, without intention to relating the other real cities.

Keywords: Utopia, Vertical Cities, Utopic Cities

1 INTRODUCTION

The utopian city is based on the idea of improvement of humankind starting with the establishment of a new location gifted with characteristics of great urban organization. To talk about Utopia in the past city is necessary to mention Thomas More, who, in 1516, created the idea of Utopia, describing an imaginary island with a perfect society. In his book, the meaning of the word Utopia is presented, the "u" being the negation of the word *topos*, place in Greek. The utopian city can then be characterized as the city that does not exist, or the antithesis of the actual city. That is also the way the island developed in Thomas More's *Utopia*. Utopian island was considered the only true republic in the world where collective interests were placed ahead of personal ones. Everything belonged to everyone and there was no shortage of anything. Moreover, in this narrative, the end of private property is discussed as the solution to an ideal society, as is the egalitarian distribution of everything that was produced by society.

There are many common aspects between this island and later utopian cities proposals. Rosenau (1988) presents the definition of a utopian city: it represents a religious view, or a secular perspective, in which the social conscience of the population needs is allied to a harmonic conception of the artistic unity. It is not necessary to stress that, when executed, an ideal plan generates specific problems because of the alteration of circumstances, but its value remains intact, in that it constitutes a projection of a perfect image, an expression lived of optimist faith. The detachment of utopia's author from the society in which he lives is necessary and then, arises a world "not locatable in time and space", a world in which there are no real world vices, or "the no place", defined by More.

2 UTOPIA X REALITY

In the period between the end of the 19th century and the start of the 20th century, a profusion of utopian thoughts and proposals emerge as a reaction to a negative diagnose of society. From the description of the terrible living conditions to the evaluation of commuting problems and the reduction of leisure time due to the time spent commuting in the industrial city (Benevolo 1997).

Proposals such as those of Ledoux – through the development of complex social functions and the determination of life in a totalitarian way by the architect. Or Ebenezer Howard's – with a harmonious allocation in the countryside, of the economic advantages of big cities, in a way that suppresses all the detriments of industrial society. On the other hand, there are Robert Owen's ideas –a city with an ideal population between 800 and 1200, in an area of land of an acre per person. There was even the creation of New Harmony, in Indiana, United States, in 1825, an experience that failed after a short time. There is also the utopian socialist Henri Saint-Simon who defended alternative social structures, which materialized in utopian cities. Alternatively, Charles Fourier with the phalanstery as a counter-proposal to the industrial world he criticized. He materialized his utopia for the perfect equilibrium between the individual wishes and social expectations. The ideal communities, phalanges housed in the phalansteries, a structure with a size, if adopted generically that would replace the petty bourgeois squalor of small isolated individual houses, which then would be installed in the external interstices of cities (Frampton 1997). There is also William Morris, among many others who have contributed with some ways of getting close to the utopia vision that motivated them, allying infrastructure and sensibility to contrapose the ideal city to the actual city (Benevolo 2001).

The ideal city holds a society that cannot be corrupted, therefore is generally projected removed from the actual city. In order to maintain the standard of organization, the population is limited, with no clear solution to its growth. The autonomy of the utopian cities is also recurrent, being that there is no proposal of economic or physical relationship with the actual cities. Socialists like Owen and Fourier were even broadly criticized by their faith in the natural goodness of humankind and the natural order of society. They believed that when artificial obstacles were removed, society could then be fully realized.

3 THE FIRST UTOPIC VERTICAL CITIES

The ideal cities of the 19th century were a response to the utopian socialists for the negative diagnosis made from the observation of post-industrial Revolution cities, transformed by technological revolution and accelerated urbanization. The counterpoint to these ideal cities was the proposals of vertical ideal cities, which started from the same diagnosis, the chaotic scenario of this accelerated growth. The response to this diagnosis, however, is what differentiates them.

While the first ones propose a solution distant from the observed reality, the later appreciate some of the solutions of the existing cities and present in their plans the evolution of these proposals or even the restructuring of the existing cities from solutions of big demolitions and constructions in the urban centres for the implementation of their ideal proposals.

Observing the importance of tall buildings, Eugène Alfred Hénard (1849–1923), in his ideal city proposed the coercive construction of towers 100 to 150 meters high in medium size cities, and in big cities, of towers 500 meters high. These big towers would also serve, besides being pinpoints for pilots, as wireless telegraphy stations, for instantaneous communication between all countries in the world. These great marks opened dreams of aesthetic appearance transmitted to the great cities of the future.[1]

The tower of colossal orientation, 500 meters high, would have at its base the historical part of the city. Around it there would be a belt of eight big towers, placed in the cardinal points of a compass, 250 to 300 meters high, as a warning to pilots over prohibited areas. Each one of these towers would have a different form and would be easily distinguishable from the others.

In 1913, it was Harvey Wiley Corbett's opportunity to fantasize the streets of a future New York, with the superposition of highways, suspended sidewalks and 40 floors skyscrapers connected by elevated metro lines. The images of the future city fascinated Italian futurists.

Their debates referred to the relation between tall buildings and traffic congestion. Many newspapers presented their thoughts about skyscrapers, zoning retreats and city planning. Their ideas were usually communicated through lectures and writings. The drawings were not part of their proposals, but for them to be better understood, Hugh Ferris and Hagopian were hired to develop images of their proposals. The most well-known of these images highlighted the proposal for different levels of vehicle traffic.

Arturo Soria y Mata proposed the Linear City in 1822. Observing the traffic congestion of the traditional city, the Spanish engineer proposed an alternative to the traditional city, developed in a concentric way: the extension of the city along a single road of undefined length. At least 40 meter wide, the central road would possess afforestation and an electric railroad. The buildings would be tall and with a reduced rate of occupation, allowing for the creation of an extensive garden between them. It was the first time the imagined city was thought from the point of view of transportation system.

Another detailed study of the vertical utopian city was developed by Hugh Ferriss, a known New York perspectivist, who designed Corbett's proposal, being their image the most known proposal for different levels of vehicle traffic. In 1925, Ferriss presented a personal exposition entitled *Drawings of the Future City*, inserting himself in the field of urbanism theory. Ferriss analysed positive and negative aspects of great American buildings, mainly from New York and Chicago, he developed images about the new building volumetry from the recessions defined by New York's 1916 zoning law, and from these two analysis, developed his plan for a utopian city, that incorporated contemporary solutions.[2]

Raymond Hood's proposal for the usage of bridge structures for apartments or offices was presented as a possible solution to habitation in tomorrow's metropolis. These towers would be 50 or 60 floors high and the bridge would be suspended between the towers by cables. The base of the tower would have the boats for yachts and hydroplanes and this would allow a quick access to the private door of the apartment, through the vertical circulation that would connect directly the maritime transportation to the residential building. It presents financial advantages, besides the structural viability of the proposal.

4 LE CORBUSIER'S UTOPIAS

The most divulged and controversial vertical ideal cities were from Le Corbusier, architect born in Switzerland in 1887. Peter Hall (Hall 1995) presents facts of Le Corbusier's life, showing that his utopian cities proposals were in tune with his personal perception of the world. This is another characteristic of the ideal cities: they take the perceptions of their idealizers as a solution to all the other inhabitants. The author identifies two fundamental characteristics for the understanding of Corbusier's ideal cities: his proposal for ideal cities, highly ordered and the habitation considered as a "machine for living", answered the long period he lived in Switzerland and the fact that he comes from a family of watchmakers. Hall also makes the comparison of the watchmakers' tradition of joining many small pieces so that the whole worked in an organized way as Corbusier's city solutions.

Le Corbusier was against the ideas formulated until then for utopian cities. He was not against the proposal of urban solutions, but against the urban solutions proposed in cities projected in an isolated way, with population limits, far from the existing reality. The negation of the conventional city was contradictory to Corbusier's thinking, which describes the nonsense of living in suburbs, satellite cities, or garden cities. The necessary commuting to the urban agglomerations from the big cities would make the residents lose hours in traffic, which could be used in collective activities.

Besides the time to commute, moving energy, water and gas infrastructure large distances would generate high costs, overloading all taxpayer citi-

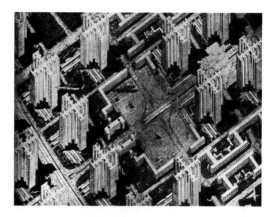

Figure 1. The Plan Voisin.http://www.fondationlecorbusier.fr/corbuweb/morpheus.aspx.

zens, since part of the time destined to work, necessary for the payment of taxes, could be utilized in leisure activities.

During the first half of the twentieth century, Corbusier developed three ideal vertical cities' projects.

In 1922, when Marcel Temporal took on the leadership of the urban section of the *Salon d'Automne of Paris*, he invited Le Corbusier to the development of an urban proposal for the exposition. In response to this invitation, Le Corbusier developed his first ideal city project: a Contemporary City for three million inhabitants.

In this project, three building typologies were proposed: 40 cruciform skyscrapers were envisioned, each 600 feet high (183 meters) in the Centre, they were designed to be occupied by technocrats, administrators, and bankers; six floor buildings *à redents* with high density in the intermediate zone and the immeuble-villas, implanted regularly in a park in the periphery. The design was based in a regular geometric layout, cut by a main vehicle circulation axis that takes to a transport centre with many levels.

The study for the city for 3 million inhabitants was developed in a model with 27 meters length. This study resulted in a proposal without precedents, different from the way of thinking the city.

In 1925, when he established the *Plan Voisin* for Paris, he created a utopia in the real city. This latter plan the proposal for a Contemporary City formed its first concrete urbanistic proposal.

In 1930, Corbusier proposes the utopian "Radiant City", in which he defends the idea of densification, which only became possible thanks to the evolution of constructive and material technology, such as steel and reinforced concrete. He makes comparisons between the necessary areas for a city with ground habitations and verticalized cities, being necessary an area eight times bigger for the distribution of a population of 1600 people in a garden city, which could be densified in a single habitation unit, covering 4 hectares. The transition from the "Contemporary City" to the "Radiant City" reflected the influence of the international contacts he maintained with other urbanists, especially in Germany and the Soviet Union.

The "Radiant City" would occupy 25 hectares and would be formed by habitation units. Since the habitation unit rises from the ground through pilotis, the land could be used in a collective way, making all the extension of the city a public space. Besides that, the roof terrace becomes a new surface for the use of residents.

Many collective activities were foreseen in the radiant city, such as maternal day cares, primary schools and the club. Corbusier (1979) believed that, with this, the city activities would be again in the human scale.

The city was established in a large park, from the concentric Contemporary City, the "Radiant City" pursued the basic idea of free circulation and vegetation inside a layout of high density, being the building types skyscrapers and collective housing blocks. This new city model allowed the separation between vehicle and pedestrian.

The new model did not deny the technological and industrial development of the existing society, but introduced solutions so that the housing and leisure conditions of industrial workers were more adequate. It is a utopian city that at the same time tries to distance itself from the previous utopian city proposals, ended up introducing the same characteristics that would hardly be established by the architect design: the ideal society, endowed with similar tastes for work and leisure, resident of the ideal city.

According to Hall in his book *Cities of Tomorrow* (1995), Le Corbusier affirmed that the evil of the modern city was in the density of its development and that the remedy, perversely, consisted in increasing this density.

5 THE UTOPIA IN THE REAL CITY: PARIS

After the proposal of the Contemporary City for three million inhabitants, in 1925 Corbusier presents his first complete urbanistic proposal—The Voisin Plan. Before presenting his proposal, he made an analysis on the serious problems observed by him in the traditional city, which were from the pedestrians' insecurity, next to the speed of the vehicles, to the typological irregularity of the constructions. He also questioned the urban legislation, which limited the height of the buildings.

He criticizes the traditional street, describing it as a road limited by sidewalks, narrow or wide, which do not offer safety to the pedestrians due to the vehicles' high speed. The question made by him was the continuity of existence of the traditional street, against to so many observed negative characteristics.

He analyses the constructions turned to the streets, which are individually of low quality, but even worse when analysed as a group, since there are many buildings different from each other. When observing the sky, the irregular silhouette formed by gables, attics, and chimneys worsens the scenario described by Corbusier.

Corbusier understood that projecting cities was a task very important to be given to the citizens. He believed that could be determined by a plan and this would be produced objectively by experts. When proposing the Plan Voisin, he faced the rejection and incomprehension of the Paris's Municipal Council, but that was comprehensible, after all, he proposed the demolition of a large Paris's central zone, besides the creation of a system of rectilinear highways.

His proposal was to demolish most of the historical Paris at North of the Seine for the construction of 18 buildings 700 feet high (213 m). Only a few monuments were to be maintained, but could be removed from their original place. For Corbusier, the new city would form uninterruptible glass extensions on the facades of the office buildings. These colossal structures would not have any vestige of masonry, only the glass would remain visible.

According to Curtis (1987), "Corbusier's health obsession in destroying the old street corridor underestimating the role of the street as a social institution, while the great overlap of road arteries did not understand the importance of territoriality and the historical memory of the previous landscape of the city". Even though the proposed city would have more inhabitants than the one to be demolished, due to densification, it would be necessary to walk three or four times less to move.

Corbusier criticized the manner in which the urban ground was occupied, with low buildings and high occupation rates. This kind of occupation generates dark streets, which, according to him, dishonoured the city. To answer the problems generated by the urban occupation in that period, he proposed the verticalization and liberation of the ground. He evaluates the height of the buildings, limited in 20 meters by Louis XIV, in the 15th Century, mostly for a technical issue, due to the masonry constructions, not so much for urbans premises. He questions the meaning of those restrictions being maintained with the technological innovations in building of the 20th Century, which would allow much higher buildings.

By densifying the population in vertical buildings, it was possible to created green areas in the city, since the big towers would utilize a small part of the total terrain. The buildings would be implanted with great distances between them, the construction of skyscrapers would happen in the centre, and they would be destined to offices: industrials, scientists and artists, 95% of the terrain would be reserved for free areas. Besides this zone, residential areas were also proposed, which would have buildings of two types: six floor buildings with luxury apartments, with 85% of the free terrain, and the modest accommodation for workers, built around the patios, on uniform squares formed by the streets, with 48% of free terrain(Hall 1995).

The pedestrian would walk not anymore on the side of vehicles, but through large parks, possible through the vertical densification of the city. Corbusier highlights, in his proposal that the streets of the new city would have nothing in common with nightmare of the streets in Downtown New York.

His proposal, which brought the destruction of the historical centre, with the conservation of only a few historical monuments, such as the Palais Royal, the Madeleine, among others, to the construction of a symmetrical system of cross-shaped skyscrapers and linear buildings was harshly criticized, but the projected was still developed and improved until 1946.

Benevolo (2001) talks about the abstract logic of Le Corbusier's plan, which classified ways according to the kinds of traffic, the buildings were linked and were implanted in a large green park and the harmony in the relations between pedestrians, automobiles, and planes.

It would be possible to watch the office skyscrapers, among the trees, when walking though these hills.

The implanted skyscrapers, spaced in regular intervals of 400 meters with a different orientation from the roads and the pedestrian paths, would be like crystals with many faces.

The Plan Voisin is the utopian proposal of a verticalized city, which, for the first time, proposes a relation with an existing city, different from the main proposals of utopian cities, which created new scenarios in new places, moving away from the problems that led them to be designed. Corbusier ended up projecting an ideal city over an existing

Figure 2. The plan voisin. http://www.fondationlecorbusier.fr/corbuweb/morpheus.aspx.

city, for him Paris was the centre of a route and, as such, could not be dislocated. This observation is essential, because it is contrary to the previous utopian cities proposals, which, characterized by being the no place, or the nowhere, do not intend to connect to a route, or to relate to the other real cities.

The Plan influenced many urban projects, even with the criticism due to the idea of destroying Paris' history. It is understood, however, that the circulation, which occurred in the great urban territory, now defined in the commercial towers, was not fully studied for the analysis of its perfect functioning. The complexity of flows and circulations necessary to the densify a commercial centre with few towers, beyond the need of energy generation, water distribution, garbage collection and sewage, would require a complex internal structure in the building, not represented by Le Corbusier in his project.

For him, with the radical proposal coming from his plane, tomorrow's Paris would be closer to waking with a new social contract, that is, a proposal of changing of society from architecture and urbanism.

6 FINAL CONSIDERATIONS

Still without scientific evidence of its existence, Atlantis was an island described by Plato in the 4th century BC, and, according to him, submerged since nine thousand years before Solon's Age. Solon was born in the 5th century and was an Athens' legislator that left a long description of the island of Atlantis, based on what he heard from men of science, in a visit to the Egyptian town of Sais. According to Plato's description, Atlantis was not just an island, since it possessed continental dimensions (Platão 2001).

In the city of Atlantis, the departure from the god's lineage removed from the human being its characteristics closer to them. The human being, gifted with imperfections, by having its human side evidenced, ends with the perfection of the city, starts a cycle of conflicts of species, greed and wrath.

Analysing the decadence of Atlantis, it is perceived that utopian cities had their decadence and extinction previously determined. Once the imaginary of the collective man is utopian, the realization of a city for the collective is also utopian. If men with a direct ancestry to gods left their human side corrupt that which the gods promised as the ideal to live, the utopian cities, with not inhabitants with direct relation to gods, would certainly be corrupted as well. By being limited even in the number of residents, they undervalued the natural question of society, which is growing and multiplying.

The proposals of utopian cities tried to combat the dehumanization arising from the industrial revolution. The overcrowded cities, the creation of workers' villages with unhealthy conditions, the dangerous streets contaminated with domestic sewage, all of this created a friendly scenario to the imagining of thinkers and architects. The human being was not viewed as an individual anymore, but as belonging to a crowd.

Regina Meyer (2007) affirms that the creation of utopias in the scope of architecture and urbanism is a way of identifying possible solution to the real problems, and a good utopia is, precisely, that proposal for a determined context in which the applied concepts can translate into possible solutions.

Le Corbusier's utopian cities brought solution in a social and technological context of the existing reality. It was not possible to deny the densification of cities and the necessity of adapt to this demand. It also was not possible to deny the evolution of the construction technology, which would allow the verticalization of cities and would realize the existence of many people in restrict spaces, without losing the necessity of green and leisure areas. Of the addressed proposals for utopian cities, maybe Le Corbusier's is the one that could be executed in its time, maybe with the reduction of number of towers' floors, due to technological restrictions.

Considering this solution, the closest to what the modern cities could be transformed into; there was a difficulty in its materialization, since Corbusier does not detail the city inside the building. In Corbusier's proposals, there are limits in detailing for the great office towers, the working spaces. That leaves a doubt on its technical viability. If the number of users of a big office tower in New York was considered, for example, there are 20 to 50 thousand daily users, which form a complexity of vertical and horizontal circulations, waste generation, water and energy infrastructure needs, etc. If for the cities there are proposals that aim to the improvement of the quality of public spaces and the organization of all sectors of society, it is necessary to analyse the proposals that seek to solve this cities inside the cities, the building spaces.

NOTES

[1] See Hénard (1911).
[2] See *The metropolis of tomorow* (Ferriss 1929) and *Architectural visions: the drawings of Hugh Ferriss* (Leich 1980).

BIBLIOGRAPHY

Benevolo, Leonardo. *História da arquitetura moderna.* São Paulo: Perspectiva, 2001.
———. *O último capítulo da arquitetura moderna.* Trad. J. E. Rodil. Vol. Edições 70. Lisboa, 1997.
Corbusier, Le. *Os três estabelecimentos humanos.* São Paulo: Perspectiva, 1979.
Curtis, W. J. R. *Le Corbusier: Ideas y Formas.* Madrid: Hermann Blume, 1987.
Ferriss, Hugh. *The metropolis of tomorow.* New York: I. Washburn, 1929. 144 p.
Frampton, K. *História crítica da arquitetura moderna.* São Paulo: Martins Fontes, 1997.
Hall, Peter. *Cidades do Amanhã.* São Paulo: Perspectiva, 1995.
Hénard, Eugène "The Cities of the Future." (1911): 345–367.
Leich, Jean Ferriss. *Architectural visions: the drawings of Hugh Ferriss.* With an essay by Paul Goldberger; foreword by Adolf Placzek. New York: Whitney Library of Design, 1980. 143 p. ISBN: 0823070549.
Meyer, Regina. Os desafios contemporâneos da metrópole: O caso de São Paulo. São Paulo: FAUUSP, 2007.
Platão. *Timeu-Crítias.* 1ª ed. Tradução e notas de Rodolfo Lopes. Coimbra: Centro de Estudos Clássicos e Humanísticos, 2001.
Rosenau, Helen. *A Cidade Idea: Evolução Arquitectónica na Europa.* Trad. Wanda Ramos. Lisboa: Presença, 1988. 201 p.

Utopia or a good place to live: Giszowiec "Garden City" in Upper Silesia

Magdalena Żmudzińska-Nowak

Faculty of Architecture, Silesian University of Technology, Gliwice, Poland

ABSTRACT: The search for "utopia" as an ideal place for life in the state of happiness and justice has been a subject of deliberations of philosophizers, writers, and even architects and urban planners for many centuries. Urban utopias were often a response to problems prevailing in cities and attempts at their solution. One of the most significant designs in the history of urban concepts of the 19th century was Ebenezer Howard's Garden City, becoming an inspiration for numerous accomplished projects.

The scope of the paper is a presentation of the earliest, and, at the same time, not well-known example of "the Garden-City" concept – "Giszowiec" housing estate in Upper Silesia. The estate was designed and constructed in 1905–1908 for workers of the coalmine owned by Georg von Giesche Erben Company.

An analysis of the design program assumptions and the process of the development of Giszowiec estate over the 100 years of its operations have been conducted, together with the evaluation of its spatial-functional, economic and social values, as well as the reasons for its decline.

Keywords: urban utopia, garden city, place, local community

1 UTOPIA: THE ASSUMED INTERPRETATION OF THE CONCEPT

The search for a happy and just political system, as well as for harmonious and unified society, goes back to the Ancient Times, i.e. to Plato's "State" dating back to the 4th century B.C.

The term "utopia" is derived from the Latin title of Thomas More's book published in 1516. This title is not explicit, and its two way interpretation as *ou-topos* – a "non-existent" place or *eu-topos* – a "good place" leads to the conclusion that the author's message was that it is an ideal and at the same time implausible place. As of to date, the concept of utopia is associated with an ideal solution, perfect but unattainable place (Szacki 1980: 14–27).

However, in such view, apart from the fantasies about an ideal, some specific rational concepts of changing the existent reality for a better one may be discerned. This aspect was highlighted by Lewis Mumford (*Story of Utopias*), who divided utopia into "utopias of escape"- enabling a flight from the drudgery and frustrations of the reality, and "utopias of reconstruction".

> In one we build impossible castles in the air; in other we consult a surveyor and architect (...) to build a house which meets our essential needs; as well as houses made of stone and mortar are capable of meeting them. (Mumford 1923: 15)

Utopias of reconstruction, which attempt to find solutions to human problems and challenges, play a significant role in the development of civilization, as concepts that designated new directions of human thought. This group consists of the majority of the so called: "urban utopias" which appeared at the turn of the 19th and 20th centuries. They all searched for new forms and principles for the functioning of cities faced with the crisis of unrestrained industrialization in the 19th century. As avant-garde and visionary solutions, they were a driving force behind progress.

The first half of the 19th century witnessed the emergence of the concepts of utopian socialists: Charles Fourier (Phalanstery) and Robert Owen (New Harmony) based on common property and collective work in agricultural and industrial sectors. The dynamics of the development of civilization led to more technologically advanced concepts: Arturo Soria y Mata proposed a linear layout of cities (Linear City), the skeleton of which was a system of transport. Antonio Sant'Elia designed only a limited number of tower blocks (*Città Nuova*); whereas, Tony Garnier—a modern industrial city (Une Cité Industrielle), with separated zones of: occupancy, work, services, and leisure, to improve the living conditions of its inhabitants.

Implemented, predominantly partially, utopian concepts confronted with the reality were sources of successes and failures, but always contributed to the verification of diverse ideas and development of urban planning concepts. One of the most resounding widely disputed and followed city healing concepts was the "Garden City" advanced in 1898 by Ebenezer Howard.

2 "GARDEN CITY": EBENEZER HOWARD'S CONCEPTS

Howard's ideas were somehow an emanation of reformatory impulses of his current times, when Europe suffered from worsening living conditions in city settlements, the example of which was London (according to Howard); yet, the elements of his theory may be tracked down to previous concepts, i. e. Claude-Nicolas Ledoux's "Ideal City of Chaux" from the Enlightenment Period. According to Howard's definition, the "Garden City" was supposed to be a city designed in consideration of healthy, hygienic occupational and living conditions and industrial activity. The scale of the city was supposed to be small, to protect the development of a local community, and to secure an organic state of equilibrium between the city and its surroundings.

The "Garden City" was to be a satellite of a big industrial city. The layout based on a circle, had centrally located services surrounded by park and cultivation zones. The successive zones were to accommodate housing, and the outer ring would be allotted for industrial establishments.

Howard presented the theory of the three "magnets" that draw inhabitants by the attributes of occupational conditions. The magnet of the city is culture, work and technology; the magnet of the country is nature and ecology; the magnet of the "Garden-City" are opportunities for physical and spiritual development of human beings, building up a healthy society, and provision of economic grounds for the comfort of living (Czyzewski 2009: 151–159).

Howard put a great emphasis on social processes and the economics of the designed settlement, which was more important than a spatial form of a given development. He dedicated his concept not to utopian idealists, but to rationally calculating investors who had to see profits from their financial outlays (Hall 2002: 95–97).

The first implementations of Howard's ideas were Raymond Unwin and Barry Parker's design of Letchworth Garden-City (1903) and Hampstead's garden-like precincts of London (1907) and, several years later, Welwyn (1920).

3 GISZOWIEC: MINING GARDEN-CITY

3.1 *Design assumptions*

Giszowiec housing estate was designed in 1905, and initiated in 1906–1908. The estate is one of the earliest implementations of the Garden City concepts, but remained almost unknown. This was due to its location, in Upper Silesia, and complicated history of those lands in the 20th century. At the time of its construction, the estate belonged to the eastern part of Germany. In 1922, the plebiscite was held in Upper Silesia. As a result, Giszowiec became part of the Polish Republic. World War II (1939–1945) was the time of the German occupation. After the War, Upper Silesia returned to Poland, but with the new political and social system of the Polish Peoples' Republic. The change in the political reality was crucial to further fate of Giszowiec estate.

The estate was constructed by Georg von Giesche Erben mining company (inheritors of Georg Giesche), and its direct initiator was Anthon Uthemann, at the time the Company director. He secured the architects Emil and Georg Zillman (from Charlottenburgh- which is nowadays a district of Berlin), who created the spatial concept of the estate and designed its buildings. The design was preceded by in-depth studies of the land formation, types of traditional land developments and the style and customs of inhabitants of Silesia, to create the most amicable and familiar habitation conditions. All buildings, despite their reference to historic architecture, were equipped with state-of-the art technological novelties, thanks to which the estate was very modern, and, in many aspects, experimental at that time (Szejnert 2007: 33–46).

The plan for the estate was 750 × 1000 m surrounded by forest greenery. The name "Giszowiec" is derived from the surname of the founder of the coal company: Georg Gische (1653–1716) and is a Polish rendering of the name: Gischewald (Giesche forest). The layout of the estate clearly refers to Howard's Garden- City concept, which is manifested not only in the concentric plan, but also in a distinct separation of the central zone from the agglomeration of buildings in the public and commercial zone. There is also clearly defined boundaries of the further growth of the estate and systematic grounds for its economy. In accordance with Howard's principles, the estate combined positive aspects of the country and the city into one organism—the variety of its functions and the provision of work constituted the city-like profile of the settlement, whereas, the houses connected by arable allotments—its rural profile. The assumed program and system solutions of the estate guaranteed its economic, administrative, social, and cultural sufficiency, making it an independent satellite of the parent company (Gische coalmine).

3.2 *Construction of the estate*

The decision on delineating the site for the settlement and commencement of design works was undertaken in 1905. In June the construction began, and, in October 1908 the first settlers inhabited the already finished buildings; whereas, the almost complete end of the works was in December 1909. In 1910, Herman Reuffurth, professor at the Royal School of Building Crafts in Katowice, published a book devoted to a detailed description of the guidelines and design layout of the estate, its construction process, as well as to the technical, economic, and social principles of its functioning. The author emphasized that the rate of works was consistent with good workmanship and low costs of the investment (Reuffurth 1995: 74).

The functional program of Giszowiec estate entailed: flats for about 600 workers' families accommodated in semi-detached houses. The semi-detached houses, according to the designers, met the postulate of an independent flat at the concurrent optimization of the costs of construction and maintenance. Each family had a garden of about 1100–2800 m^2, with the obligation of its cultivation and an outbuilding utility for animal breeding. The assumption was that miners' wives would look after the household and the garden, staying within the boundaries of the estate, which would secure proper care of children without the need to establish kindergartens.

The architects designed a settlement that referred to traditional forms of Silesian houses, yet completely adjusted to meet the requirements of the building and technological provisions in force at that time. For example, an advanced technology of timber impregnation enabled use of shingle roofing in observance of the tradition, but, also in fulfilment of fire-resistance requirements. The construction of housing for miners' families was not expensive, but the grouping of a variety of forms (40 types of houses) translated into an unusual visual attractiveness of the estate and rejection of monotony. The workers' houses were one-floor. They had high-pitched roofs with various arrangement of the entrance vestibules. They were heated by coal delivered free of charge by the coal company (Reuffurth 1995: 20–31).

In comparison with workers' dwellings, the houses for administrative staff, teachers, and a physician were more extended and had higher standard. Furthermore, the estate also offered accommodation for unmarried engineers and miners. It provided a canteen, dining room and bathing rooms.

The public utility program was also relatively comprehensive. The main public facilities were grouped around the market square and included: District Forestry Office- main administration and estate management situated in a park in the northern part of the central square, an inn with supporting facilities located in a big part bordering with the central square from the eastern side, a compact

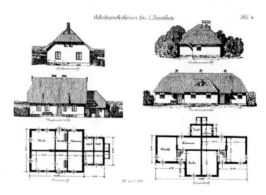

Figure 2.   Exemplary designs of workers' houses (source: Archive).

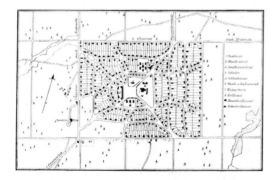

Figure 1.   Original plan of Giszowiec estate (archival materials). 1. Workers' Inn. 2. District Forestry Office. 3. Shops. 4. Schools. 5. Hostels. 6. Laundry and Bath House. 7. Water Tower. Duty and Tax Chamber. (source: Archive).

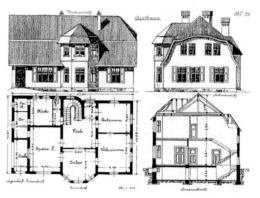

Figure 3.   Design of a physician's house (source: Archive).

Figure 4.   District forestry office-facade (source: Archive).

frontage of shops from the western part, and, the closing southern part with three schools.

The inn was the most splendid building in the entire estate. The costs of its construction were high, but the founders' intention was explicit: to provide the inhabitants with a pleasant place for leisure and a respectable building for celebrations, receptions and other cultural events (Reuffurth 1995: 53–59). The building contained a restaurant and guest rooms. The main hall with a view was used for receptions, parties, and theatrical performances. In the park there was a band shell.

There was a complex of shops along the central square offering a full range of commercial services. An extensive, over 100 m long sequence of facilities included: textile shop, grocery, green-grocery, fruit, butcher's, bakery, pharmacy and post office outlets. Such compact settlement on the western frontage of the market square referred to urban architecture and emanated a sense of appropriate prestige.

The occupants of the estate also had an opportunity of baking their own bread and pastry in freestanding "baking ovens" located at every street, to facilitate the continuation of homemade baking traditions. There were three schools for Catholic children and one for Evangelical Church. A huge church for 3500 people was also erected to be of service to the two neighbouring estates, as well as a narrow-gauge railway to transport workers to Gische coalmine.

From the earliest years of its functioning, Giszowiec estate was almost completely self-sufficient. It generated its own electricity, which was supplied to houses free of charge. In addition, it had its own telephone dispatch centre, with connections to the public utility buildings and to some to the administrative buildings. It had a water supply system—with a water tower constructed at the highest point of the estate. Street standpipes served as water supply points and as fire hydrants. There was also a fire brigade functioning at the estate.

Giszowiec estate had a central heat generation unit, which supplied heat energy to the majority of the public buildings, launderettes and bathing house. The launderettes were equipped with boilers for the laundry, centrifuges, dryers, and electric linen presses. All inhabitants were obliged to use public launderettes, as doing laundry in the houses was prohibited. Such system substantially reduced the consumption of water and protected the housing from moistness.

There was also a biological water treatment plant, which processed manure for the cultivation of gardens. Furthermore, the estate had its own ice generation system for food cooling. As far as health protection was concerned, a separate hospital building was built in the vicinity, with technical facilities for contagious illnesses. It should also be emphasized that all service facilities were exceptionally clean and hygienic (Reuffurth 1995: 60–69).

3.3   *Micro-society: Local community*

The estate was predominately inhabited by qualified miners who constituted the main part of Gische coalmine crew, as well as the biggest social group of the estate. They settled down and had families with many children—in accordance with the Silesian tradition. Other occupants represented the mining management staff, office and administrative workers and teachers. Due to very good living conditions, social care, continuity, and job security, not to mention the sense of familiarity, the occupants were not only satisfied but also proud of Giszowiec estate. In a short time, it became the place of origin of a local community or, more precisely, a network of local families deeply rooted in the place of their occupancy. Friendships and love relationships were formed, the occupants jointly organized wedding receptions, baptisms of numerous children were born to the world, and they also held celebrative funeral receptions.

Women met while laundering and baking bread; whereas men played in a mining orchestra. There was a local theatre, choir in memorial of Frederick Chopin, and other political and social organizations. A local newspaper called: "Górnoślązak" ("The Upper-Silesian") was edited.

All inhabitants participated in masses, religious celebrations and other forms of parish activity, as parish church played an important role in the formation of the community. One of the most significant events was Saint Barbara's Day, who is the patron of miners, when processions accompanied by the mining orchestra, parties and celebrative holy masses were held.

In 1919, one of the occupants of Giszowiec established a photographic atelier, which still exists. It has become a centre for documenting the life and most important events held at the estate, as well as of portraying its inhabitants. The archive of the atelier is a priceless chronicle of this site.

The local community created a specific ethos of work, respect, and human solidarity. Many preserved documents and reports testify the authenticity of inter-personal relations and ties with the place of occupancy. Some multi-generation families of the original settlers have survived to date, despite the tribulations of the difficult history of Upper Silesia in the 20th century: two world wars, uprisings, changes of country allegiances, political and social transformations. The families stay in touch and cultivate the memory about the estate, collect artefacts, thus confirming the philosophy of a long-lasting preservation (Szejnert 2007).

Figure 6.   District Forestry Building listed as heritage—the current condition (Photo by the Author.

3.4   *Giszowiec: Life and death of the place*

The estate survived without damages the period of World War II and the first post-War years. The process of spatial degradation and destruction of the community began in the 1960s, when coal began to be extracted from a new "Staszic" mine located near Giszowiec. In 1969, the state authorities of the Communist regime undertook the decision to liquidate the estate due to the need of providing mass housing for miners employed at the new mine. Hence, the estate settlements were systematically demolished and new 11 floor concrete slab tower blocks constructed. The demolition works were discontinued following the decisions of conservators dated 1978 and 1987, on listing the urban and spatial layout of Giszowiec as examples of the heritage of Katowice Voivodship, which prevented only one third of the historic estate from utter destruction.

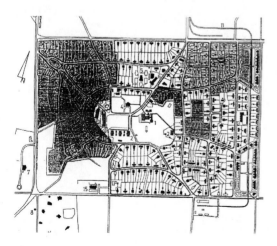

Figure 5.   Giszowiec—layout- the current condition (demolished buildings are marked in grey color).

In the early 1970s more and more occupants of Giszowiec were forced to relocate to the slab concrete tower blocks as their houses had been demolished. "Modernity" that took over the estate was a result of the policy of the Polish Peoples Republic authorities towards Upper Silesia. Such policy led to irreparable damages of material heritage. Wide-spread demolition works in Silesia entailed buildings dated to the turn of the 19th / 20th century, of which procedure Giszowiec is just one example (Kaszuba 2002). The Communist regime also attempted to change the post-German name "Giszowiec" to "Staszic housing estate", but, fortunately, despite the administrative decision, the old name prevailed and was continuously used until its official restoration in 1990, when political transformations began. In 1999, the coalmine transferred the property rights to the buildings of the historic part of Giszowiec to Katowice Municipality. Some of the buildings were sold to the families that had been occupying them for years. Nowadays the entire unique settlements of the estate are under permanent conservatory protection. The urban layout of Giszowiec was listed as heritage, as well as its separate buildings.

4   RECAPITULATION: SUCCESS AND DECLINE OF GISZOWIEC ESTATE—CAUSE/EFFECT ANALYSIS

Giszowiec estate was an implementation of Ebenezer Howard's postulates of combining: "the health of the country with the comfort of the city". Such utopian-sounding concept became completely real in the discussed estate. Giszowiec estate combined dreams with the reality.

What was the phenomenon of its success?

- Its founders were perfectly aware of the fact that there are some basic aspects that secure a durable functionality of the city and its further growth: spatial, economic, and social elements, which must occur and develop harmoniously. Their synergy constitutes the mechanism of urban growth.
- The key to success were rational assumptions of the design based on reliable analyses. The future vision of Giszowiec was grounded on strong foundations of tradition and history, creating a sense of preserving the continuity of growth.
- Rational and efficient economic and social mechanisms of the functionality of the estate were applied: jobs for the occupants, good wages, and system of services, social care, and places for leisure.
- The estate was perfectly managed, there was discipline, but also, the sense of security and satisfaction, which was an effective tool supporting its growth.
- The introduction of unusually advanced elements of technological modernity was "discreet"—it enhanced utility and functionality of the estate, without distorting the harmony and continuity of spatial development.
- The program assumptions were based on respecting the dignity of the occupants- miners, who were to find their real home. The assumptions based on truth and honest intentions rendered a good result.

Giszowiec offered friendly space, economic foundations for further growth, and conditions for developing the local community. The emerged social capital- a community of the occupants, their work, talents, involvement- have all created value over the decades of the estate's existence and contributed to its continuous development.

What was the reason for its decline?

The reasons behind the decline and destruction of Giszowiec were external. They did not stem from the assumed design principles or management system, but from their negation. The estate functioned as a healthy organism in which all its particular elements made up an indissoluble network of connections and relations. Poland's new political reality after 1945, the political and social command, and control doctrine of the Communist authorities destroyed a subtle network of sustainable relations functioning at the estate. The changes in economic conditions, centrally undertaken decisions concerning the development of industry, caused a dramatic increase in the number of inhabitants and emergence of new, inadequate forms of collective housing. Thus, it was not only the spatial structure of the estate that was destroyed, but also its social capital. The political doctrine assumed an intentional rejection of the "idyllic" image of Giszowiec regarded as heritage of the previous political system. It was only in the 1990s that the remaining historic buildings of the estate became subject to strict conservatory protection.

BIBLIOGRAPHY

Czyzewski, Andrzej. *Trzewia Lewiatana: miasta ogrody i narodziny przedmieścia kulturalnego.* Warszawa: Panstwowe Muzeum Etnograficzne, 2009. 238 p. ISBN: 978-83-88654-81-7.

Hall, Peter. *Cities of Tomorrow.* 3th ed. Oxford: Blackwell Publishing, 2002. 553 p. ISBN: 0-631-23252-4.

Kaszuba, Elżbieta. "Dzieje Śląska po 1945 roku." In *Historia Śląska.* Ed. Czaplinski, Marek. Wrocław: Wydawnictwo Uniwersytetu Wrocławskiego, 2002. 716 p. ISBN: 978-83-229-2872-1.

Mumford, Lewis. *The story of Utopias.* London: Harrap, 1923. 338 p.

Reuffurth, Hermann. *Giszowiec - Nowa górnośląska wieś górnicza, Spółki Górniczej Spadkobierców Georga von Gieshe.* Translated by Bronisław Machnik. Katowice: Kopalnia Węgla Kamiennego Staszic, 1995. ISBN 83-03-03669-6.

Szacki, Jerzy. *Spotkania z Utopią.* Warszawa: Wydawnictwo Iskry, 1980. 218 p. ISBN: 83-207-173-2.

Szejnert, Małgorzata. *Czarny ogród.* Krakow: Wydawnictwo Znak, 2007. 547 p.

From the garden city to the red village: Howard's utopia as the ground for mass housing in Soviet Russia

Irina Seits
Department of Aesthetics, Södertörn University, Stockholm, Sweden

ABSTRACT: Beautiful urban utopia by E. Howard promised an ideal city model that combined comfort of the city life with rural paradise. Experiment resulted in fascinating suburbs around the world, yet none of them proved sustainability of the garden city model. The concept greatly influenced on the 20th century urban theory and stimulated various mass housing solutions.

Present paper traces destiny of the garden city concept in Soviet Russia. A guiding idea for constructivists, it inspired new type of housing—zhilmassivs—semi-autonomous estates within bigger cities that combined independent infrastructure with close connection to peer towns. State watched Avant-Garde experiments and used their outcomes to set control over population, which resulted in appearance of the "red villages": settlements near industrial centres that neither allowed tenants to master land nor use benefits of urban life.

To prevent the raising of autonomous self-maintained households, garden city concept was declared a utopia that propagated bourgeois lifestyle and that was hostile to the ideas of collectiveness promoted through official housing policy. Yet the poor living conditions in the red villages were explained by the failure of the garden city concept and functionalist practice *per se*, in spite of the fact that those settlements had little relation to both.

Tracing the history of the garden city concept in the early Soviet Russia, I reflect on how and why it could happen.

Keywords: Garden city, urban utopia, Soviet Russia, Avant-Garde architecture, housing estates.

1 GARDEN CITY CONCEPT IN MASS HOUSING CONSTRUCTION OF THE 1920–30'S IN RUSSIA AND EUROPE

In the interwar period, the construction of mass housing in Europe was taken under state control and handed over to Avant-Garde architects. In spite of all financial and political challenges, it acquired mass character.

One of the major concepts that influenced on development of mass housing in the young Soviet State was the idea of the Garden City by E. Howard. Yet, none of the Russian regions received a garden city that fully resembled Howard's model. Among those districts, where idea of the garden city was realized in a most precise way could be mentioned the Sokol village in Moscow, built by architects N. Markovnikov, brothers Vesnins, I. Kondakov, and A. Shchusev in 1923.

From the very beginning, those estates that resembled Howard's urban utopia in both Russia and Europe kept close connection to the big city metropolis. Same could be said about Letchworth and Welwyn garden cities laid out by E. Howard in the UK; Hellerau garden city in Dresden initiated by businessman Karl Schmidt-Hellerau; the Gartenstadt Falkenberg in Berlin by Bruno Taut; and Södra Ängby residential district in Stockholm, built by Edvin Engström already in the 1930 s as they were not autonomous from their peer cities either.

European experiments that resembled Howard's models were not sustainable towns that were able to define their own organization and infrastructure, to offer occupation and services for their inhabitants and maintain other development patterns that were crucial for Howard's garden city concept.

On the contrary, those estates were completely depended on the big cities' infrastructures. Their existence was provided by the communication system of the peer towns, which turned them into sleeping suburbs rather than sustainable independent settlements. Moreover, even though the life of their tenants was organized more rationally and healthier in Howard's sense, those suburbs were the types of residential areas that Howard argued against, since they complemented big cities, making them more sustainable and resistant to elimination, which was Howard's final goal.

Figure 1. Gartenstadt Falkenberg in Berlin. Arch. Bruno Taut, 1913–1916. Photo by the author.

Still, those programmatic estates, generally referred to as garden cities, were certainly designed under strong influence of Howard's ideas. They became early platforms for the functionalist experiments that defined later solutions for mass housing construction and outlined its further development deep into the 20th century. All estates mentioned above were extremely different from each other in size, planning, type of housing, social status of their inhabitants, infrastructure and overall architectural appearance. They were built under rather different political, economic, administrative, and social circumstances. What they had in common was the period of construction and the reference to the concept of the English garden city combined with the functionalist approach.

Such combination resulted in development of new types of housing in Europe of the 1920s: Siedlungen in Germany and Zhilmassivs in Russia and that were creatively processed in other European countries in the 1930 s and later on.

In order to trace and analyze the origins of the modernist housing estates in Russia, it is crucial to recall social, political and economic environment in the country that shaped the landscape for their evolution. Far not all settlements that were constructed in the country at that period could be referred to the heritage of functionalism.

2 CONSTRUCTIVISTS SEARCHING, STATE WATCHING: DEVELOPMENT OF THE NEW HOUSING POLICY IN SOVIET RUSSIA

In Soviet Russia of the 1920s the state was watching carefully, first with sympathy, later with concern, but always with great interest after the development of constructivist theory and practice. In the beginning constructivists were given front pages of mass media to propagate their ideas; they were provided with commissions and financing to build.

The results of their work were closely studied by state from all aspects of economy, ideology, politics, timing and efficiency. In the 1920s the new state had not yet developed any unitary policy of how to solve tremendous shortage in housing. The government had no other means to develop housing policy besides watching different experiments that were being realized within modernist movement. Its apologists welcomed Bolshevik Revolution with great enthusiasm and were ready to break with the past not only in the art of building, but in all other aspects as well.

Moreover, constructivists were not simply the minstrels of Revolution, but highly trained and experienced professionals that were able to offer fast and efficient solutions. Thus, theoretical and practical searches by constructivists were moving along the 1920 s together with their critical watching and evaluation by state.

Once the vertical of power had grown more or less sustainable, constructivists with their purely scientific approach to construction of reality and aim to raise a liberated man of the future were not demanded anymore. Constructivism faced severe critique and was later prohibited along with the theories that directly or indirectly influenced on its development, including the concept of the garden city.

At the same time, both the method developed by constructivists and elements of the garden city concept continued their existence in Soviet mass housing construction until the end of the regime. Yet, it was already in the second half of the 1920-s when the notion of the garden city disappeared from the officials' lexicon, state commissions and press.

3 FROM THE GARDEN CITY TO THE RED VILLAGE

First, the term of the garden city was replaced with the "garden village", then with the "workers' village" or "red village". That way any reference to the Howard's model was eliminated.

The "workers' villages" already by the end of the 1920 s were equally and endlessly distant from the reference both to the garden city and constructivism. They remained the unfortunate creation of the Soviet State housing policy of the 1920 s to 1930 s where workers were forced to live in a perverted space that resembled something in between the town and the village in such a way that it implemented only the shortcomings of both.

The typical "workers' village" was constructed around a certain industrial centre (a factory, a mine or a power station). According to the electrification plan of GOELRO [ГОЭЛРО], the State Commission for Electrification of Russia established in the 1920, which aimed to provide the country with fast economic and industrial growth, the new industrial centres were usually built in a distance from other metropolitan or even populated areas. Thus, the new "workers" or "red" villages were often organized far from the traditional settlements.

Workers were moved to the apartments in the blocks of houses deprived of direct access to the land even though it could be in a huge excess. After a housing committee had announced a competition for a new workers' village it could happen that the contest was won by a project, which offered construction of individual houses (cottages or villas) as its major type. Yet such approval was a seldom case, since the state favoured multi-apartment blocks rather than individual houses. Even if independent villas were built, they were inhabited by several families that shared the kitchens and other facilities in the house. Thus, the communal living according to the "one room-one family" principle was always secured.

The green areas in workers' red villages were used not for individual farming, but only as public areas. Tenants did not have private gardens to grow food, which was natural for living in the country. They could apply only for the seasonally rented kitchen gardens to grow vegetables, and those gardens were kept under close regulation by the factory administration that provided workers with housing. That way the very sense of the countryside with the living in close connection to land was purposely eliminated.

The disconnection from the land and industrial labour that was a major feature of the town life could be assumingly compensated by the infrastructure that provided citizens with living necessities. Yet, the town-like infrastructure was hardly ever built in the workers' villages due to economic and logistic reasons as such settlements were usually distanced from the older towns.

The density of the villages where people were forced to live in small apartments or even in communal flats in 2–3 storied blocks was not caused by the lack of land or high construction costs, as it could be a case in Western Europe. As the many reports by the architects of that time show, the cost of the construction of the small 1-storied single-family houses, when combined with the low cost of land, was financially more efficient than construction of the multi-storied panel blocks (Meerovich 2007).

It was state policy that discouraged workers from living in separate cottages, hidden behind the private gardens, which would dramatically lower the control over the workers living practices and limit penetrability of the communist ideas (Meerovich 2007: 144).

The tenants were provided with the housing by administration of the plant where they worked. Private or even cooperative ownership of land and housing, as it was offered by Howard, was unthinkable. Dwelling was given to a worker for as long as he was employed at the factory. One could not simply live in the "red village" without having a job at the plant. Housing became the means of manipulation over the worker, as it was the major stimuli to keep him at his place of employment.

The idea of communal living when family and private life was exposed to the community was heavily propagated in opposition to the bourgeois' living in the villas where the exploitation of men by men was inevitable to maintain the living.

4 CONTRADICTIONS BETWEEN CONSTRUCTIVIST METHOD AND STATE HOUSING POLICY

It is significant, that in the housing estates, where one family occupied an individual villa, most of constructivists had seen no contradiction to the socialist ideology. Along with their Western colleagues, Soviet architects argued for construction of single family cottages proving their economic efficiency in the low populated regions (Meerovich 2007: 144).

The counterargument could be the high costs of communications that were to be brought separately to each house, yet it was hardly the case, since the practice was different. Both in the settlements with individual villas and apartment blocks gas and hot water (often cold water as well) were provided not to each house or flat, but to the shared properties such as collective banyas (bathhouses) and kitchens. They could be either separate buildings on the territory of the workers' village or spaces in the living blocks that were shared by tenants.

Even when developing new types of buildings such as collective banyas and collective kitchens, constructivists did not mind individual bathrooms and kitchens in the apartments. The special space for them was given in most of constructivist Zhilmassivs. It was another issue that bathtubs and showers never found their way to the bathrooms.

The decisions to keep flats without facilities were taken by the housing committees that altered original projects by constructivists for economic as well as ideological reasons. In contrast with state housing policy, constructivists offered new way of living, where a man, and first a woman, could choose to socialize and educate herself rather than to do cooking and washing in the kitchen after the end of her working day. The collective banyas and kitchens were means of liberation of women from kitchen slavery. On the contrary, collectivization of the everyday living practices that was realized through the state policy only hardened life of the workers as cooking, washing and laundry remained their personal duty. At the same time, since kitchens, laundries and washrooms were intentionally separated from the living quarters it took more time and efforts from tenants to maintain them.

The difference lied in the very goals that constructivists and state set for their search of the ideal mass housing solution. The architects planned their estates with intention to ease life of the tenants, to liberate them from everyday routines that took their time and energy from working on their life improvement. They managed to keep construction costs as low as possible, while building rational living space. On the other hand, state was interested in holding the worker attached to the site of his employment and in controlling his life in all aspects in order to maintain sustainability of its own power.

The economic, political, and social situation of the 1920s forced the government to use any means to keep huge masses of people under control. Industrialization and collectivization caused enormous migration within the country. Driven by civil war and hunger, workers were leaving cities and their poorly paid jobs at the plants for the country where they hoped to grow food on the land to survive.

In the 1920 s the urban population in Russia was decreasing dramatically, and major leverage of the factory's administration to keep a worker at his place was to provide him with the housing.

The development of workers' settlements was nearly fully given in hands of the factories' administration. Living meters became the means of manipulation, punishment, and stimulation of a worker. To provide employee with the private living in an individual cottage, even if it was the cheapest and the smallest, was the best thing to lose control over him.

Families that shared apartments or at least some facilities of every-day use (such as kitchens and bathrooms) had to open up the most intimate living practices to the public. No space or time was left to spend moments of leisure apart from the community. Family and religious holidays were replaced with the abundance of the "Red Days of Calendar"—different celebrations of communist ideology in the workers' clubs, which attendance was mandatory.

In fact, the life of an inhabitant was organized in a way that would not allow him to leave his plant as he was chained to it by the living space that was given to him only for as long as he worked there. The holder of a living unit could be punished by the forced eviction if he didn't meet requirements for a good worker and tenant which would deprive him of the means to survival.

Those practices that secured an employee at his working place were fixed and strengthened through different legal procedures such as registration at the place of living (*propiska*) that limited one's mobility. Many workers did not have passports to record even the place of registration, and thus could not move around the country legally. The "labour books" kept records of the worker's previous and current jobs as well as of his behaviour and personal characteristics.

5 RED VILLAGE—THE BLAME ON CONSTRUCTIVISTS? THE REASONS OF IGNORANCE TOWARDS THE HERITAGE OF ARCHITECTURAL AVANT-GARDE IN RUSSIA

The type of the "workers' red villages" was yet specific as it was developed around the newly formed industrial sites. In older metropolises, the situation was different, and there were other means in use that provided workers with housing.

The type of the workers' red village, though it was very spread in Soviet Russia, did not belong to the Avant-Garde heritage. Still those workers' villages that had been massively built in the 1920 s and 1930 s were often identified with constructivism among general population. Constructivists were unfairly blamed for those inhuman living conditions in the workers' villages that nearly majority of population had to live in, since a room or an apartment in a red village was often the only alternative to the barrack.

Due to the fact that constructivism as a theory and practice was officially denounced in Soviet Russia, there was little known among general population about what it actually was and what kind of buildings belonged to constructivism as architectural style. In the 1930 s to 1950 s, while masters of Avant-Garde were still alive, constructivism was heavily criticized, and later it was neglected and forgotten. That might be one of the reasons why there is such a low respect and little recognition of Avant-Garde heritage in Russia, which results in its fast dilapidation and destruction, as it receives support neither on administrative level nor among general population.

The districts that resemble true constructivist style and that are usually called Zhilmassivs—are nowadays located in central parts of the grown cities, where land is the most expensive, and thus they are extremely vulnerable to the replacement and deconstruction.

The tenants who still occupy the houses built by Avant-Garde architects are often unaware of their high artistic and cultural value, since after ninety years of neglect those buildings remain in poor condition. The usual argument for the destruction of constructivist housing heritage that is articulated by administration, interested in replacement of the low-rise buildings with the modern skyscrapers, is that "people do not want to live in the houses without bathrooms". Residents often support the efforts to remove constructivist houses, willing to improve their living conditions. However, neither it was ninety years ago, nor it is today that the absence of facilities in the buildings is the fault of constructivists. The original projects included all modern facilities.

For the preservation of Avant-Garde heritage it is important to realize that mass housing in Soviet Russia in the 1920 s did not consist of purely constructivist solutions. The leading type of housing in Russia by the start of the WW II and long after it remained the wooden barrack with the corridor system, where a family occupied a room, and all facilities such as shared kitchens and bathrooms were located either in the end of the corridor or outside of the barrack.

6 THE ORIGINS OF SOVIET ZHILMASSIV. EARLY ZHILMASSIVS IN LENINGRAD AS THE FUNCTIONALIST ALTERNATIVE TO THE GARDEN CITY

The more sustainable and innovative type of mass housing in Russia that kept relation to the concept of the Howard's garden city and was its most successful realization was the newly developed type of a zhilmassiv.

The first Soviet zhilmassiv was raised in Petrograd in the Narva District on the site of the densely populated workers settlement close to the Putilov (later Kirov) Plant, which was the largest factory in the country. Zhilmassiv in Traktornaya street became the very first platform for a social experiment in mass housing that combined the concept of the garden city with the functionalist approach, and which managed to overcome the hardships of existing social and financial conditions.

The idea was not only to provide workers with shelter, but also to construct new social space with certain ideological resemblance. Architects A. Nickolsky and A. Gegello developed new types of buildings that reformed and rescheduled life of a dweller. In his everyday living a worker was forced to go through certain buildings, which function was to take care of his needs: he would sleep in a room of a zhilmassiv (apartment block), then go to eat in a factory-Kitchen, then work at the plant, wash in the collective banya (bath-house) and spend leisure time in the Palace of Culture. The politically, mentally and socially controlled routine was constructed with its rhythm and trajectories

Figure 2. Zhilmassiv in the Traktornaya street in Leningrad. Photo of the 1930-s, from: Kirikov, B., Schtiglitz, M., Leningrad Avant-Garde Architecture, St. Petersburg, 2008.

that a worker could not avoid: from food that he ate to the suit that he put on for the concert that he had to see. Thus the district infrastructure served also as a mechanism of social and mental reformation (Seits 2012).

Architects adjusted modern technologies of contemporary housing constructions used in Europe (such as German Siedlungen) to the Soviet reality and state commission. In the first Leningrad zhilmassiv, architects partially realized the idea of line building that was widely used in German cities. The low-rise three-storey blocks with individual flats were direct references to the garden cities.

The intimate character of the district was close to the atmosphere of the Howard's urban utopia. The Traktornaya street was designed as a green alley, which added to the calm and garden-like character of the estate.

Yet the first Russian zhilmassivs could not be called garden cities or directly compared to German Siedlungen: they were neither independent villages nor suburbs, but rather the estates that were deeply integrated into the structure of Russian cities. They became inseparable parts of the peer towns and yet developed their own autonomous infrastructure.

Due to the deep ingrowths into the existing urban structure, Russian zhilmassivs were more distant from the independent character of the English garden cities, than German functionalist suburbs.

The paradox lied in the fact though, that Howard argued against suburbs, and his garden cities were to be self-maintained structures, independent from the bigger towns. German Siedlungen were designed as sleeping residential estates, where people only lived and spent their leisure time. The tenants had to go through the buffer zones that separated living quarters from the places of work

Figure 3. Palevsky Zhilmassiv in St. Petersburg. Arch. A. Zazersky, N. Ribin, 1925-1928. Photo by A. Nickolaev, from the author's collection.

or areas where other services such as medical care, schools, and shops were located.

Such separation of spaces drastically contradicted with the model offered by Howard, where all sites that a tenant lived through during the day were to be placed within the borders of the same garden city, producing no zoning of the space.

The idea of the garden city was taken to closer consideration by Russian constructivists and was realized in zhilmassivs that possessed very well developed infrastructure. There was no necessity left for a tenant to go beyond the borders of an estate, as her working place, living quarter, necessary services, and entertainment were located within the territory of the zhilmassiv.

The original structure and basic idea of zhilmassivs were their well-organized infrastructure that provided tenants with alternative to the traditional city life, where distances between functional zones were stealing a lot of the lifetime.

Unlike garden cities, zhilmassivs were not meant to keep private lives of their tenants away from the community. On the contrary, their structure

exposed living practices of inhabitants with intention to improve them in the process of reformation of a former peasant into a perfect resident of the future through the means of architecture.

Residents' living space was not divided into the working, sleeping and leisure zones anymore, it became homogenous. All spaces, including that of a plant, became the tenants' living space. The everyday routines were sectioned within that space. In this enchaining model state saw the strings of control, while constructivists—the potential for improvement and reformation of men's living.

As the state strengthened its power, it swept out all humanistic elements of the garden cities that remained within zhilmassivs, such as separate apartments, small size of the low-rise buildings and the spacious green yards—the last reminding of private gardens. State prohibited the notion of the garden cities, constructivism was abandoned, and the concept of zhilmassiv as a model for the mass housing construction was perverted into the workers' red villages. The latter incorporated everything that both the Howard's utopia and constructivism were fighting against.

The dehumanized living space of the red villages possessed little relation to the garden cities as well as to the Avant-Garde housing estates. Yet the responsibility for the unfortunate settlements was put on those, who argued for the most rational and comfortable living for all.

BIBLIOGRAPHY

Meerovich, M. "Rozhdenie I Smert Sovetskogo Goroda-Sada [Birth and Death of the Soviet Garden City]." *Vestnik Evrazii*.1 (2007): 119–160.

Seits, I. *Urban memory places of 1920-s in St. Petersburg: reconstruction of everyday life in new socialist districts of Narva Gates and its re-enactments in the present.* Cities & Societies in Comparative Perspective. 2012.

Utopian visions in the Portuguese city: An interpretation of Planos Gerais de Urbanização

José Cabral Dias
Faculty of Architecture, University of Porto, Porto, Portugal

ABSTRACT: One can have many approaches to utopia. While focusing on ideas aimed at transforming a particular context into a non-achievable reality, we are also referring to utopia. Within the context of dictatorship (28th of May 1926–25th of April, 1974), the Portuguese city, which was planned through *Planos Gerais de Urbanização* (General Urbanization Plans), was also a realm of utopia. This paper will have a non-conventional approach to those plans. It is expressed by a break with the former context. Aspiring to a new urban reality, which was focused on modernity, the town planners intended to break up with tradition. Accordingly, the car should be seen as a contribution for shaping urban space through innovative principles. Demolitions and new areas of expansion would be the most visible face of it, among a wide range of features for transforming the pre-existent context. However, even taking into account the inadequacy of Portuguese city to its own time, the Government was in opposition with complete ruptures. Instead, the adequacy to financial resources was in the centre of its concerns. In other words, the reality was Government's work scope in contrast to non-achievable goals. The architects, on the contrary, set their own goals well beyond reality.

Keywords: Portuguese city, Planos Gerais de Urbanização, Utopia, Estado, 20th century urban concepts

1 INTRODUCTION. MOVING ON TO A NEW REALITY

The concept of utopia can be used to define a dream that is either yet to be materialized or difficult to achieve. It can also be used to refer either to a fantasy, or a strong desire, or even to a place or an unattainable idea. This will be the scope of this article. It will be framed by the *Estado Novo* City[1] in the way the town planners/architects faced it. In other words, the scope of this paper will be framed by the city planned through *Planos Gerais de Urbanização* [General Urbanization Plans].[2]

Such city was conceived as an impossible place, considering the country that Portugal was at that time. The approach to this idea, considering the Governmental perspective, shows that the Dictatorship intended to build a completely new urban world upon an unrealistic range of settlements, from the north to the south, and from the smallest to the biggest ones. Over more than 400 urban settlements with more than 2500 inhabitants should be the subject of a new plan, including all the places with either touristic, leisure, climatic, spiritual, historic, or artistic relevance.[3] Moreover, this purpose should be achieved in solely three years after the approval of the Decree-Law nº 24802, which took place in 1934. These figures are surprising within

the Portuguese reality. Portugal was then a country facing deep problems affecting almost all its infrastructures and urban equipment. It is important to notice that the Minister of Public Works, Duarte Pacheco soon intended to modernize Portugal[4] by building new roads and sea ports, developing energy production and its distribution as well as agricultural hydraulics, and by improving hospital facilities, building school and university buildings, law courts, city halls, and post and telegraph offices. In the urban context, which is the scope of this paper, many of such measures converged, in 1938, for the definition of both the *programa de melhoramentos urbanos* [urban improvements program], and the *programa de melhoramentos de águas e saneamento* [water improvements and sanitation program] (Costa 2012: 101, 109–113, 141).[5] In other words, it is possible to confirm that *Planos Gerais de Urbanização* took part in a wider program to transform profoundly the country and its urban world. Actually, this program was defined also under the dependence of the Minister Duarte Pacheco. In the one hand, the city councils faced the obligation to submit each plan to the ministerial administrative services through a very slow and bureaucratic process in order to obtain their approval (Lôbo 1995a: 46). On the other hand, the Minister on his own previously defined the group of town planners that were able for urbanistic issues. Even when it became clear that this group was insufficient to satisfy the wide number of plans that were being designed in 1943, Duarte Pacheco was assertive: the municipal authorities could invite new architects in order to enlarge the initial group if they wanted to, however each of them should be approved by himself (Lôbo 1995a: 41). Even if the aim of the Decree-Law n° 24802 had been defined in order to provide city councils with effective urbanistic instruments, this fact miscomprehends the reality. The truth is much more complex: all the urbanistic system that was created by the Portuguese dictatorship was deeply centralist, and was part of governmental policy (Lôbo 1995b: 34–49). Duarte Pacheco faced those urban studies as a medium to organize the construction surge met by Portugal during the consolidation of the political regime, says Nuno Portas (Portas 1973: 727). According to this author, such plans were part of a process undertaken in order to ensure that the inner cities of the country and the ones located in the province would be developed over large avenues like Lisbon (1973: 727). In other words, such process defined a strongly controlled urbanistic program. It was implemented in order to give a territorial face to the political regime, and, accordingly, it should be rapidly spread all over the country by both a solid and unrealistic number of settlements.

In a more emphatic way, the architects intended to go well beyond the impossible reality sought by the Decree-Law n° 24802. This approach, regarding the architects' intentions, addresses an idea of progress, which adopted the automobile as a symbol as well as it spaces as a metaphor of a new era. Such intentions contradict the most known interpretation of *Planos Gerais de Urbanização*, which sees on them an instrument to the affirmation of tradition. Actually, they can be seen as the expression of alternative ideas focused on modernity and progress.[6] Even though the urban form is not always clear about this issue, the descriptive documents made on each plan do not deceive. Their authors struggled for building new urban meanings by the exploitation of innovative spatial concepts and mechanisms. They intended to break up with traditional spatial organization. In this approach, the automobile must be understood, therefore, as a medium to express a different way of thinking about urban issues. It was a symbol of a new era rather than solely a subject on itself. Considering the Portuguese urban context and taking into account the architects' goals, the car expressed, actually, an idea of utopia: it provides, in part, the framework for *Planos Gerais de Urbanização*. This paper is actually about that conceptual understanding behind the car. Accordingly, the aim of this article is to identify and explore such concepts.

In summary, when the Government intended to transform all the country in order to express its political power over the territory, there was no possibility to achieve such goal. Besides the solid number of settlements to be transformed, the architects, (even the most conservative ones) tried to bring modernity and progress into urban planning. This will be the focus of the next section.

2 MOVING ON TO CITY SPACE. DRIVING INTO UTOPIA

Planos Gerais de Urbanização are usually understood as a space of promenades, squares and small houses, which were designed on the basis of formalistic and traditional urban solutions. Accordingly, those plans are seen as the result of Government's ambition of showing political presence all over the country as it was seen before. However, when launching the program of *Planos Gerais de Urbanização,* the urban scenario faced by the Minister Duarte Pacheco[7] was disqualified. The national urban network was undone. Towns and villages presented expressive signs of degradation (Dias 2011).

As a testimony of a degraded reality, the way the buildings, street network, urban equipment and infrastructure are described on the technical reports of the *Conselho Superior de Obras Públicas* (CSOP—Superior Council of Public Works) does not deceive.[8] The undemanding, not intentional, spontaneous, and careless urban growth of Portuguese towns and villages as well as their poverty are clearly pointed out: alleys, as well as unhealthy and too much occupied blocks with low oxygen and light generally formed the urban space. It was also formed by degraded buildings, street widths and layouts incompatible with modern car traffic, as well as improperly installed urban equipment. This deprivation, regarding the precarious state of such equipment, was increased by the one that did not exist at all. Even the major cities needed to be reformed.

The architects intended to set up almost a zero level of urbanity. They intended to do so sometimes with pragmatism, often with apologetics. The car was then a symbol of a contemporary living framework in accordance with its own time. Regarding the effort for updating and improving urban space, the urban plans tended to shape new paradigms aiming at breaking up with both tradition and pre-existing territorial organization. However, this trend met opposition by CSOP, for both realism and conservatism. The ambition of transforming the urban context in a utopian basis—in the sense that architects proposed a radical reform of the cities and small villages by going beyond governmental goals—was an idea with no possible future. That will be seen bellow.

In an objective way, architects misrepresented the perception of reality when they attempted to express new urban concepts. The misfit between the way in which car was seen and the physical medium in which it should move was complete. In other words, architects were too much optimistic. In modern times, the car was intentionally accepted as a symbol of an era marked by progress as well as by greater freedom in discovering territory on an individual experience. However, that was a trend with no success over Portuguese territory. As it was said above, the Portuguese urban context was deeply degraded. The intended motorized revolution was not possible over the precariousness of, simultaneously, buildings, squares and streets, nor with both the lack of sewage systems and water network failure. It was not even possible with the lack of urban equipment of any kind. In summary, it was not possible when everything was, apparently, undone, and, simultaneously, the concerns should be different. Looking outside the urban world, it is also clear that such revolution was neither achievable within the context of a country that was still balancing its financial situation, nor in the framework of a clear lack of economic development. Still in 1944, Parliamentary—Secretary of State, for Trade and Industry, José Ferreira Dias, was assertive against the specialization of the economy, especially if it was based on the agricultural activity and on country ruralisation (Rosas and Brito 1996: vol. 1, 466). Reinforcing this argument, Portugal did not have changed significantly in the early 1960s. As Minister for Economic Affairs (1958–62), Ferreira Dias said that the Portuguese economy had to change solidly, even considering some longing for rural economic basis. Consequently, the country expected a serious reform of the industry, he said.[9]

The difficulties imposed by the lack of economic development might explain the slow development pace of the plans. Besides that, the lack of flexibility of such plans explains that in 1946 (Decree-Law nº 35031) all of them were classified as preliminary plans: In a manner of fact, this decision intended to adapt each plan to the financial conditions of the city councils at any time. In other words, this measure intended to face real needs, and, therefore, to allow the adaptation of each plan to changing circumstances (Lôbo 1995a: 47). In short, the particular conditions of Portugal were not the ideal ones to try a radical movement towards a new urban condition. The circumstance of taking into account the car as a metonymy of progress and modernity was, therefore, a key to failure. The architects would experience this evidence on their own. Nevertheless, the automobile would

move on according to the intention to transform the national urban context. Even if either manifestos or pure theoretical conceptualizations were set aside, the car was also revealed as a symbol and a myth in Portugal. This trend, which refers to Banham (Banham 1977), sets the car as a solid contribution to build a new idea of urban territory.

Regarding the words above, the first feature for car effective movement was always embodied by a radially urban form (there were few exceptions, and they were always located in the most modest and smaller villages). This spatial feature aimed at driving the long-distance traffic away from the city core.

[Such option] is really in the spirit of the Road Plan, by moving away these routes intended for large and rapid inter-urban circulation from the urban settlements, and thus defend them from the proximity of buildings of any nature". (Ramos 1949: 9)

The separation between interurban and intraurban traffic founded, therefore, the accurate mechanisms. New spaces for free and speedy movement were supposed to cross urban design. *Variantes* [by-pass Roads], *circulares* [ring roads], *vias periféricas, anéis periféricos, anéis de circunvalação, artérias periféricas, vias de cintura,* [peripheral roads], and *vias tangenciais à periferia do aglomerado* [tangential roads to urban settlement] are new nomenclatures which were spread from the north to the south in order to support the new vehicle (Dias 2011: 483). Such features gathered with other mechanisms in order to achieve urban segmentation (Dias 2011: 483). Indeed, the zoning was used as a new reality. Residential space and buildings were organized with independence with respect to car traffic. Street hierarchy was a feature for fluidity of car movement. In other words, those mechanisms were put together as part of an urban design based on the separation between pedestrians and automobiles. The antinomy between pedestrians and automobiles was, by itself, a strong feature for a new era: it only made sense in the presence of the car. The car omnipresence set a new mind-set concerning the urban space. The road circulation broke through the existing urban fabric, and the narrow winding streets gave rise to new and wider spaces. The car had a role, therefore, to define all the urban spaces, from major roads to parking places inside private lots (Aguiar 1945: 9). Accordingly, the very notion of road network was clarified (Franco 1957: 14). New paradigms concerning efficiency, pleasure and speed, which were early referred by Le Corbusier[10] in international context, were part of the equation also in Portugal. New signs of a different way of looking at urban issues using foreign references were present from

Northern to Southern Portuguese territory. The *Neighbourhood Unit,* the *Athens Charter,* and the *Garden City* with small houses were used all over the country in different plans, even if they were not used in extensive areas nor in their pure form. In addition, the *Parkway* was also a new international contribution for organizing Portuguese city space in a pleasant and fluid way that was metaphorically expressed by Giedion in *"Espacio, Tiempo y Arquitectura"* (Giedion 1982).

The terminology is itself an expression of a mind-set change. Grande circulação [great circulation], circulação acelerada [fast circulation], grande trânsito [large traffic], viação acelerada and trânsito acelerado [fast traffic], movimento acelerado [fast movement] or even movimento interno intenso [internal intense movement] are all concepts which took part among the hierarchy and separation of different kinds of traffic. In the same basis, vias inter-urbanas [interurban roads] make a contrast with vias intra-urbanas [intra-urban roads]. Similarly, vias de distribuição [roads for traffic distribution] were in opposition with vias de penetração [roads to enter the city)] as well as vias de atravessamento [roads to cross the city] were different from vias locais [local streets] and also from artérias de serviço and artérias de circulação local [local circulation roads]. Following the same principles, vias de circulação [circulation roads] were in opposition with vias de habitação [residential roads]. Similarly, vias de acesso ao centro [roads for centre access] were different from becos [alleys]. Vias arteriais [major traffic roads] and artérias radiais [radial roads] completed the traffic hierarchy which was, dichotomously, divided between trânsito arterial interno and trânsito arterial externo [major internal traffic and major external traffic] (Dias 2011: 485).

All the words above show a way of thinking in accordance with territorial changing conditions. In addition, other features contributed also to modify the meaning of urban landscape: nós de circulação (circulation nodes) e praças de giração (roundabouts), vias de sentido único (one-way streets), cul-de-sacs, viaducts and tunnels—for pedestrians or automobiles—joined the ideas to change urban hierarchy, and both the shape and the meaning of urban space (Dias 2011: 485).

In accordance to those changes, the streets would be useless if they were not able to support car circulation—since the car was the medium that gave meaning to urban space (Dias 2011: 485). Accordingly, street enlargements and alignment corrections complemented such changes. These changes are much more significant as it is known that many measures resulted in strong destructions within urban fabric (Dias 2011: 485).

As a consequence of the new way of perceiving city space, cosmopolitanism overlapped rurality (Dias 2011: 492). Real needs were put aside in the struggle with the car. The idealization was mainly the context in which the plans were designed. Reinforcing this idea, the man at car's steering wheel had changed his condition: he was a traveller (Dias 2011: 484). Moving, besides circulating, became pleasure, according to a modern idea that was emphasized by Giedion as it was seen before.[11] In summary, it is possible to say that the urban space reflected optimism and apologetics. The closeness to good national and regional roads was a cause of happiness, David and Maria José Moreira da Silva said in 1944 within the design process of Matosinhos urban Plan (Dias 2011: 486).

Nevertheless, CSOP had a solid different way of looking at the urban context. In pre-existing areas, there was no space for radical transformations. The city of Faro is a fine example of a generalized attitude, which is clarified with words of reproach:

[The proposed design] does not deserve agreement by the Council (...), since the adopted solution implies almost full reform of the city. (Públicas 1946a: 19)

The expression of the official point of view intends to be doctrinaire:

It is believed that authors' attitude of each plan should avoid a radical overhaul in the old core of the settlements, apart from keeping the architectural or picturesque character, since it is unaffordable. The large achievements in modern and rational manner should preferably be reserved for the extension zones. (Públicas 1946a: 19)

Such reproach is made by *Direcção-Geral dos Serviços de Urbanização* [General Directorate of Urban Development Services] and receives full agreement of CSOP: "The Council absolutely agrees (...)" (Públicas 1946a: 19).

It is true that CSOP was not always against the architects' options. It is even true that the so-called conservative architects tried also to go beyond the former context they found in the city. The *Plan d'Amenagement et Extension de Coimbra* designed *by* Etienne de Groër in 1944 is a role model of it. The unique drawing that is inserted in the descriptive document of the Urban Plan is clear: the graphical representation of the automobile is the medium chosen to give full meaning to *Avenida de Santa Cruz* length. As a metaphor of modernity, the new *Avenida de Santa Cruz* was a 400 meters long avenue, which was supposed to cross Coimbra downtown. Therefore, the historic relevance of urban fabric should be submitted to mechanized movement. Indeed, this avenue shows an apologetic and aesthetic conceptualization: it can be found in it a desire to link the space of the city with its own time. It is true that a rhetoric statement was also behind the option of opening this avenue. However, the intention of highlighting the historic and urban relevance of *Igreja de Santa Cruz* [Santa Cruz Church] is directly related with the car movement towards its facade. As it was said above, it is true that CSOP accepted de Gröer's proposal.[12] However, this does not deny all the arguments. In line with the ideas expressed in this paper, *Avenida de Santa Cruz* has never been built, and the blocks that had been demolished persisted as an urban void that was a fair ground and a car parking, even after 25th April 1974. In summary, this avenue became unfinished as a symbol of the difficulties for transforming the city upon modern ideas. This argument is in accordance with the approach undertaken in this paper.

3 CONCLUSION. ARRIVAL IN A REALISTIC FIELD OF ACTION

The refusal of the proposals by CSOP resulted from the awareness of their incompatibility with municipal financial resources (Dias 2011: 463). Actually, such refusal shows a permanent and major concern in the assessment process of the plans. Despite the unachievable ambition of reshaping almost all the urban context, it also shows a realistic approach: CSOP had always been in opposition with the more speculative or essayistic architects' attitude. Accordingly, the maximum use of existing streets was also imposed by CSOP in opposition to new unnecessary extensions solely motivated by design reasons (Dias 2011: 464).

It should also not be forgotten that conserving existing streets, as far as possible, is a contribution for maintaining the local character within the settlement. Great changes should be avoided, even when the traffic demand impose them (...). (Públicas 1946a: 11)

CSOP fights, therefore, against breaking up with the past and tradition. This is a very relevant idea. It shows one more possible way of understanding the antinomy that separated the Government and the architects. The architects intended to completely transform reality and to build an entirely different world (Dias 2011: 463). Differently, the Dictatorship intended to preserve tradition and to assuring the feasibility of each proposal. Another statement of CSOP is solidly relevant. It highlights these arguments by expressing the opposition between reality and oneiric idealizations (Dias 2011: 466):

Indeed, it is important to (...) foresee and to organize the territory; however, guidelines expressing grandiosity should be avoided, since they are, in most cases, incompatible with local resources and needs. (Públicas 1946b).

Such statement undoubtedly shows an opposition to the architects. The reality was the Government's scope of work. Thus, the car could not able to impose a new universe. There was no space for a radical and innovative life scenario. CSOP could never support the architects' goal for they were too ambitious. As it was said, the reality was the Government's context of work. In contrast, urban design was set well beyond reality.

NOTES

[1] *Estado Novo*—New State—is the way the dictatorship established in Portugal by the 1933 Constitution called itself. Cf. (Rosas and Brito 1996: vol. I, p. 315).
[2] The Decree-Law n. 24 802, dated 21th December 1934, established the legal framework that instituted *Planos Gerais de Urbanização*.
[3] Vd. Article 2nd of Decree—Law 24 802.
[4] The foundation of the new Ministry of Public Works and Communications results from Duarte Pacheco's intentions to transform the Portuguese territory. Cf. (Costa 2012: 98).
[5] Cf. (Santa-Rita 2006).
[6] Vd. DIAS, José Cabral. Episódios Significativos de Espacialização Urbana a Partir do Automóvel: [os Planos Gerais de Urbanização; 1934–1960]. Porto: Faculdade de Arquitectura da Universidade do Porto, PhD Dissertation, 2011. The aim of this PhD research was to assess the impact of the emergence of car within the context of urbanistic ideas in Portugal
In the scope of such research, there were studied 26 plans covering from both the larger Portuguese cities to the smaller villages, and from the North to the South (excluding Lisbon and Porto): Coimbra, Braga, Viseu, Castelo Branco, Portalegre, Leiria, Tavira, Tomar, Caldas da Rainha, Matosinhos, Olhão da Restauração, Espinho, Marinha Grande, Lourinhã, Santo Tirso, Nisa, Fundão, Mafra, Felgueiras, Macedo de Cavaleiros, Fonte Santa-Monfortinho, Santa Cruz, Sul de Braga, Vila Nova de Gaia, Zona Litoral entre Espinho e Vila Nova de Gaia
The covered period was defined between 1944 and 1956. The research was supported by the descriptive documents and technical drawings made by the architects in each plan, and by the Technical Reports made by *Conselho Superior de Obras Públicas* (Superior Council of Public Works) in the assessment process of such plans.
[7] Portuguese Minister of Public Works between 1932 and 1936 and later between 1938 and 1943.
[8] The *Conselho Superior de Obras Públicas*, which was reorganized by Duarte Pacheco in 1933, was a technical and advisory body responsible, through his 4th section, for coadjuvating the Ministry of Public Works in the assessment process of the plans. Cf. (Costa 2012: 110). The aim of those reports was to assess each one of the

plans made within the framework of *Planos Gerais de Urbanização*. They were written by *Conselho Superior de Obras Públicas* (Superior Council of Public Works). Regarding the preexisting condition of both the cities and the small villages, the rapporteur's general opinion pointed out a status of poverty and decay. This is a testimony given by the review of the 26 plans studied in the scope of the PhD research referred in endnote 9. Vd. DIAS, José Cabral, Op. Cit.

[9] Ferreira Dias apud (Rosas and Brito 1996: 478).

[10] Regarding this issue, Le Corbusier gave us a new way of looking at the link between the territory and the car. In the pages of a central book of his own authorship, he identifies an important symbol of that link, the Parkway, with a new pleasure, aiming at driving with aesthetics concerns. Cf. (Corbusier 1995: 80, 91–92)

[11] Vd. (Aguiar 1945: 9).

[12] It can be said that CSOP exceptionally accepted de Gröer's proposal. The reason can probably be found in the significant role given to *Igreja de Santa Cruz* in the urbanistic scheme. *Igreja de Santa Cruz* took a central role on Portuguese cultural and territorial definition in Middle Ages, and since Coimbra was the first capital of the Portuguese Kingdom, it became pantheon for the first and second kings of Portugal. In other words, the proposed urbanistic solution would also have a role to emphasize a symbol of Portuguese national identity with rethoric expression. That role would be pleasant to a conservative and nacionalistic regime like *Estado Novo* was, and it must explain the acceptance of the wrecking ball. Cf. (De Gröer 1944).

BIBLIOGRAPHY

Aguiar, João. *Memória Descritiva do Ante-Plano Geral de Urbanização de Castelo Branco*. Lisboa: Arquivo do Serviço de Estudos de Urbanização da Direcção-Geral do Ordenamento do Território e Desenvolvimento Urbano, 1945.

Banham, Reyner. *Theory and Design in the First Machine Era*. 7th ed. London: The Architectural Press, 1977.

Corbusier, LE. *Maneira de Pensar o Urbanismo*. 3ª ed. Translated by José Borrego; ti. original Manière de Penser l' Urbanisme. Lisboa: Europa-América, 1995.

Costa, Sandra Vaz. *O País a Régua e Esquadro: urbanismo, arquitectura e memória na obra pública de Duarte Pacheco*. Lisboa: IST Press, 2012.

De Gröer, Etienne. *Plan d'Amenagement de Coimbra*. Lisboa: Arquivo do Serviço de Estudos de Urbanização da Direcção-Geral do Ordenamento do Território e Desenvolvimento Urbano, 1944.

Dias, José Cabral. "O contraponto: o CSOP—a apreciação dos planos e expressão do regime político na cidade." In *Episódios Significativos de Espacialização Urbana a Partir do Automóvel: [os Planos Gerais de Urbanização; 1934-1960]*. Ed. Dias, José Cabral. PhD Dissertation. Porto: Faculdade de Arquitectura da Universidade do Porto, 2011. 433–479.

Franco, Lima. *Memória Descritiva e Justificativa da Revisão do Ante-Plano de Urbanização de Leiria*. Lisboa: Arquivo do Serviço de Estudos de Urbanização da Direcção-Geral do Ordenamento do Território e Desenvolvimento Urbano, 1957.

Giedion, Siegfried. "Espacio-Tiempo en la Urbanística." In *Espacio, Tiempo y Arquitectura*. Ed. Giedion, Siegmund. 6ª ed ed. Translated by Isidro Puig Boada. Madrid: Editorial Dossat, 1982. 769–972.

Lôbo, Margarida Souza. *Planos de Urbanização à Época de Duarte Pacheco*. Porto: FaupPublicações, 1995a.

—— "Aparecimento, expansão e declínio dos Planos Gerais de Urbanização." In *Planos de Urbanização à Época de Duarte Pacheco*. Porto: FaupPublicações, 1995b. 34–49.

Portas, Nuno. *A Evolução da Arquitectura Moderna em Portugal: uma interpretação*. História da Arquitectura Moderna. Ed. ZEVI, Bruno. Lisboa: Arcádia, 1973.

Públicas, Conselho Superior De Obras. *Parecer 1703: Ante-Plano Geral de Urbanização de Castelo Branco*. Lisboa: Biblioteca e Arquivo Histórico do Ministério das Obras Públicas, 1946a.

——. *Parecer 1720: Projecto do Plano de Urbanização da Vila de Felgueiras*. Lisboa: 1946b.

Ramos, Carlos. *Memória descritiva do Ante-Plano de Urbanização da Vila Fundão*. Lisboa: Arquivo do Serviço de Estudos de Urbanização da Direcção-Geral do Ordenamento do Território e Desenvolvimento Urbano, 1949.

Rosas, F, and Brito, J. M. B, eds. *Dicionário de História do Estado Novo*. Coord. e Pesquisa Iconográfica Maria Fernanda Rollo. Vol. vol 1. 2 vols vols. Lisbon: Bertrand Editora, 1996. 527 p. 522 colns. ISBN: 972-25-1015-0.

Santa-Rita, José de Santa. *As Estradas em Portugal: da Monarquia ao Estado-Novo (1990–1947)*. Lisboa: Edições Universitárias Lusófonas, 2006.

Staging modernism: Case study homes and suburban subdivisions

Anna Novakov

Art and Art History Department, Saint Mary's College of California Moraga, California, USA

ABSTRACT: Between 1945 and 1966, California Arts and Architecture magazine commissioned dozens of modern architects to design technologically efficient homes for post-war Americans. The project addressed the post-war need for affordable housing while also displaying the latest in American ingenuity on the pages of the magazine. Most of the realized single-home projects, which were actually quite expensive, were in southern California and embraced the mild Mediterranean climate with transparent glass walls, lush gardens, open floor plans and the views of the Pacific Ocean. The completed houses, situated in the heart of the film industry, were staged, photographed, filmed and animated as models for a new post-war Utopian lifestyle. Case Study #8 (The Eames House, 1950) in Pacific Palisades and Case Study #22 (The Stahl House, 1960) were the most photographed and promoted projects.

Keywords: California, Case Study, Film, Post-War, Magazines

As early as the 1940s, architectural magazines began showcasing designs for small, inexpensive housing for people living alone, married couples or young families. The designs were focused on creating healthier and less expensive lifestyles for people of various social classes. Architects who had professionally come of age during the Depression were focused on the role of architecture as a vehicle for social change. This approach to building was very much influenced by the *Bauhaus* and formed the basis for the early modernist housing designs. The houses were curated by the architects and photographed and open to the public in a fully staged form—attracting more than 350,000 visitors. Shelter magazines, such as *Good Housekeeping*, started promoting suburban house designs and new home prototypes as early as the 1940s. Even though experimental designs for small homes were developing across the United States, they took on a particularly innovative tone in the Los Angeles area—where the mild climate encouraged a form of leisure living that way focused on play and sports. The shortage of wood as a building material in Southern California, pushed architects to design houses out of alternative materials.

In Los Angeles, there was an extraordinary amount of architecture within easy reach. Some of its specific lessons were that a good house could be made of cheap materials, that outdoor living was as valued as indoor spaces, that a dining room was less necessary than two baths and glass walls. (Morgan 2012: 166)

The post-war period was one bursting with optimism and idealism. Pierre Koenig (1925–2004) recalled that as a young Case Study House architect,

Everybody wanted to produce answers to housing problems. Everybody was going into mass production ... social problems were being addressed. It was an exciting period of time and all kinds of things were being tried. (Jackson 2007: 9)

The climate and clients in Los Angeles were particularly appealing to many innovative architects who wanted to follow Mies' lead by connecting indoor and outdoor living.

As a resort and a place of transitory employment for a Bohemian class [Los Angeles] supported an unusual number of bachelor dwellings, temporary quarters and weekend cabins. These small structures with their programs fixed and lifestyles fluid offered wonderful opportunities for architects to experiment and test ways to simplify [their designs.] (Gibbs et al. 2012: 159)

This combination of a mild climate and large number of interested clients formed a fertile environment for modern individual and the later development of tract homes. This space for innovation was made possible by the *Case Study Home* project that began seven months after the end of World War II. It started as a social program by John Entenza (1905–1984), the magazine's publisher and his assistant Susan Jonas who was the link to foreign architects and magazine editors.

Following the Bauhaus dictum that architecture must push modernism forward, the ambitious CSH Program shaped and honed an idea of modern design—the influence of which continues to reverberate in Los Angeles and beyond. (Peabody, Bradnock and Singh 2011: 9)

The boundaries of the city were ever expanding into the suburbs and offered

> … a desirable place of health and recreation, balmy weather and vast uncharted areas of undeveloped land, southern California was available to be designed by the car. (Craig 1986: 26)

Entenza began his career in the movie business working for Metro-Goldwyn-Mayer in the experimental production department. This early experience taught him the value of encouraging and promoting photogenic homes. After buying *Art and Architecture* magazine, Entenza launched a program that addressed the issues of America's post-war economy. The first designs for houses were small with two bedrooms, two baths, and no spaces for servants. The intent was to create a new lightweight lifestyle for post-war families. During the post-war era

> designers and manufacturers used materials and technologies developed or honed during wartime—molded fiberglass and plywood, synthetic glues and plastics—to create stylish furnishings that were accessible and affordable in a way that most earlier modernist designs had never been. (Barter 2001: 16)

One of the earliest completed Case Study houses was the 1949 home and studio of creative partners Charles (1907–1978) and Ray Eames (1912–1988), constructed from industrial steel and plated glass. It is a Utopian paradise that utilizes the latest technology inside the house as well as the spectacular coastal view from Pacific Palisades. It was also a showplace for the Eames' Bauhaus inspired "moral campaign to improve the world through good design" (Peabody, Bradnock and Singh 2011: 9) The house grew in an additive manner "with frame and cladding not separated, but working together, and that it possesses wit, a quality extremely rare in architecture" (Jackson 1990: 183).

The Eameses embraced the parameters of the *Case Study* call for post-war housing and utilized industrial material in the design of their "ground-breaking *CSH 8* … [a home and studio that] consisted of two steel-framed cubes covered in panels of glass and colourful plaster." (PEABODY, BRADNOCK and SINGH 2011: 9) They designed an innovative live and workspace utilizing the technology of industrialization, perfected

during the war, and applying it to domestic spaces. Julius Shuman (1910–2009) photographed the house in 1958, for *Life* magazine. The image was unusual in that it was curated to

> show Charles and Ray Eames surrounded by their belonging: rugs, throws, plants, toys and objects collective from their travels around the world. (Stevenson 2005: 34)

The images in *Life* magazine were significant as partners to the role of television within the post-war home.

> Though television … was entirely ephemeral … tens of thousands of Americans preserved millions of copies of Life and other photo weeklies... Life's talented and well-equipped photojournalists typically shot the same events as television cameras and the photo magazine also occasionally printed excerpts from news footage as stills. Thus it became a universal archive of television imagery, which often ended up on in posters and collages, private albums and scrapbooks. (Walsh 2004: 2)

The Eames own film about their Case Study #8 house, *Home: After Five Years of Living* (1955), was created using fast cutting—a rapid-fire technique that utilizes our optical ability to process images of nature and urban spaces, these films created proto-digital film experiences. The film was assembled using thousands of slides that they took while living in the house. The images, presented as a film, embraced both the still images of popular magazines and the growing importance of moving television images.

Charles Eames, like Entenza, had associations with the film industry—working as a set designer for MGM in the early 1940s. This experience was influential on the way that he reproduced and promoted his architecture and design.

> All Eames architecture can be understood as set design. The Eamses even present themselves like Hollywood figures, as if in a movie or an advertisement, always so happy, with the ever-changing array of objects as their backdrop. (Colomina 2001: 22)

Creating design for the masses involved being able to visualize the modern home through the media of the time. Staging of the projects for reproduction was one of the ways that the Eamses became so recognized for their work.

Case Study House #22, situated on 1635 Woods Drive (above the Chateau Marmont Hotel) in Los Angeles, was designed by San Francisco-born Pierre Koenig as a family home for C.H. Buck Stahl a former football player, his wife Carlotta and their children Bruce, Sharon and Mark. The 1960 house was a dream home for the Stahls who purchased the land in 1954 for $13,500 and still own the property today. Three years after purchasing the lot, Stahl hired Koenig to design the house. Koenig, a professor at the University of Southern California and an early assistant of Greek-born Raphael Soriano (1904–1988), was a lifelong proponent of prefabrication and the social responsibilities of architecture. Soriano had designed and executed Shulman's house, begun in 1947 and completed three years later, in the Laurel Canyon neighbourhood of the Hollywood Hills. The professional connection with Shulman would prove significant in the photographic documentation of Koenig's projects.

In designing CSH #22, Stahl insisted on retaining the 270 degree unobstructed view from the mountains to the ocean. Stahl and Koenig understood the site as "an eagle's nest in the Hollywood hills". The two-bedroom 2200 square foot house ultimately cost $34,000 with an additional $3,600 for the pool and bare some resemblance to a bird in flight—a reference to the eagle's nest metaphor. "As in most Keonig houses, water became a part of the floor plan". (Morgan 2012: 176) The three large-scale plate glass walls—a remarkable engineering feat—define the house as a transparent pavilion. The heated concrete floors provided the Stahl children with ramps for skateboarding and the dramatic cantilevered living room looking out over the lights of L.A.

The iconic 1960 photograph of the house by Shulman was staged using two female students (one from UCLA and one from Pasadena High School). He believed,

models should be used with great care in architectural photography. The architectural design is the important element of the photograph; disturbing, outdated fashions and over powering human forms are merely distracting and have no importance. Therefore placement should be based on normal scale and for demonstration of architectural elements". (Stevenson 2005: 35)

In Shulman's photograph,

the glass living area of this dramatically spare house [is] silhouetted against the nocturnal cityscape, a photograph that captures a wide swath of the 240 degree expanse visible from the house's transparent living room … Transparency reveals a glamorous and desirable lifestyle as much as it reveals a stunning vista. (Whiteley 2003: 11)

Nearly all the published commentary on the photograph, which was made using a seven-minute exposure, positions it as a Utopian image that depicts the promise of the California lifestyle and of architectural modernism for residential applications.

CSH 22 was a photogenic project used as the backdrop to numerous films, television shows, music videos, and even video games. The film *Smog* captured Case Study #22, just two years after its completion. Unlike Shulman's photograph of the home, Italian director Franco Rossi's 1962 cinematic depiction of the house brings up darker, more Dystopian associations. The 1962 film takes place during a 48-hour period when Vittorio Ciocchetti, an Italian attorney lands in Los Angeles for a two-day layover. He meets Italian ex-pats who extol the virtues of the city and who drive him around empty streets shrouded in air pollution. Ciocchetti says:

Oh this is paradise on earth! Who lives in this paradise? In all my walking, I didn't meet a soul.

The film, a Dystopian view of Los Angeles, focuses on the isolation of the individual within the car-reliant urban space. As the first post-war European film to be shot completed in the City of Angels, it has something of a documentary tone—focusing on newly built houses (such as the Stahl), the recently completed Los Angeles International Airport, and the oil wells of Culver City. The economic divide between the opulence of the Stahl house and the deserted streets of Hollywood are an indication of the deepening schism between post-war social classes. Both the photograph and the film focus on the private home as an emerging Utopian dreamscape held together with plate glass, steel and spectacular views of the city and ocean.

By the 1960s, the *Case Study* program evolved from single family dwelling in the 1940s and 1950s to designs for entire communities. The magazine recognized this suburban expansion as a trend and highlighted Fordist prototypes for new forms of mass-produced housing. The 1960s, a seminal decade, saw the dawn of the Kennedy era that began with so much optimism and resulted in such a state of chaos. This political and social shift was reflected in suburban developments that served as an escape from the problems of the time as well as a way of creating a fully contained living environment that was insulated from many of the urban turmoil of the time. The expanding metropolis marketed itself through images of leisure and play—a lifestyle associated with warm climates and the Sunbelt. They also offered many white Americans, who were resistant to change, with the promise of a homogeneous society devoid of diversity. This new vision of the city as a suburban, insular, planned community increased the access to a leisure-filled life to middle class whites who profited from the decentralization of urban centres.

The exodus to the suburbs was a search for an imagined Arcadia that was promoted through advertisements—stories about places that have not even been built. Made possible through the Federal-Aid Highway Act of 1956, freeways were built in remote areas where land was affordable and available. The American Dream was marketed to people who were mobile, had cars and were ready to move.

Television and print advertising marketed the suburban lifestyle as a post-war Utopia for many Americans.

Television and Life undoubtedly influenced each other, creating, vivid, highly focused and dramatic images that told a powerful story in an instant and became icons of the events themselves. (Walsh 2004: 9)

Images of smiling, happy families living in their own piece of California paradise were potent images of post-war luxury now available on a mass scale.

By the late 1950s commercial television had found its way into an estimate 85% of all American homes—households where television was watched for some five hours a day... The aesthetic texture of early television—the foggy, often distorted black-

and-white images or the garish, bleeding colors; the incidental static, ghosting and interference—had almost disappeared. (Walsh 2004: 9)

Television shows, such as *Leave it to Beaver* and The Adventures of Ozzie and Harriet, provided weekly templates of emerging American lifestyles—prototypes for suburban houses, interior décor, the latest appliances, fashion and family dynamics. The programs, highly anticipated family entertainment, became barometers for a middle-class lifestyle—an image to be endlessly emulated and purchased through new consumer goods.

BIBLIOGRAPHY

Barter, Judith A. "Designing for Democracy: Modernism and Its Utopias." *Art Institute of Chicago Museum Studies* 27.2 (2001): 7–105. ISSN: 00693235.

Colomina, Beatriz. "Enclosed by Images: The Eameses' Multimedia Architecture." *Grey Room* - (2001): 6–29. ISSN: 1526–3819.

Craig, Lois. "Suburbs." *Design Quarterly*. 132 (1986): 29–30.

Gibbs, Jocelyn, et al. *Carefree California: Cliff May and the romance of the ranch house*. Santa Barbara, Calif. /New York: Art, Design & Architecture Museum, University of California In association with Rizzoli International Publications, 2012. 275 p. ISBN: 9780847837823.

Jackson, Neil. *Pierre Koenig, 1925–2004: living with steel*. Hong Kong: Taschen, 2007. ISBN: 9783822848913.

——. "Metal-Frame Houses of the Modern Movement in Los Angeles: Part 2: The Style That Nearly." *Architectural History* 33 (1990): 167–187. ISSN: 0066622X.

Morgan, Susan, ed. *Piecing together Los Angeles: an Esther McCoy reader*. 1st ed. Valencia, CA: East of Borneo in collaboration with the Art School at California Institute of the Arts, 2012. 391 p. ISBN: 9780615528236.

Peabody, Rebecca, Bradnock, Lucy, and Singh, Rani, eds. *Pacific standard time: Los Angeles art, 1945–1980*. Los Angeles: Getty Research Institute and the J. Paul Getty Museum, 2011. xxi, 330 p. ISBN: 9781606060728.

Stevenson, Rachel. "Living Images: Charles and Ray Eames "At Home"." *Perspecta* 37 (2005): 32–41. ISSN: 00790958.

Walsh, Peter L. "This Invisible Screen: Television and American Art." *American Art* 18.2 (2004): 2–9.

Whiteley, Nigel. "Intensity of Scrutiny and a Good Eyeful Architecture and Transparency." *Journal of Architectural Education* 56.4 (2003): 8–16. ISSN: 1531–314X.

Figure 1. *Smog,* 1962. Director: Franco Rossi. 35 mm film. University of California, Los Angeles Film and Television Archive.

An appropriation experience of the empty space

Andre Soares Haidar & Daniela Getlinger
Faculdade de Arquitetura e Urbanismo, Universidade Presbiteriana Mackenzie, São Paulo, Brazil

ABSTRACT: Growing cities often leave empty spaces on historic and consolidated areas. The city of São Paulo (Brazil) experienced during the late 20th century a growth movement towards an every expanding area, leaving decaying buildings and public spaces in the city centre. After decades of lack of investment on spaces for public interaction, the city experienced an emerging debate over who was entitled to use those spaces, from parks to streets and avenues.

This paper studies some of the spontaneous events that need the appropriation of public spaces for them to exist. Focusing on one of the most important São Paulo landmarks, the Paulista Avenue, it is developed a system to list residual spaces left on the urban fabric of the area. Studying concepts of public intervention, public space, and architecture heritage developed by authors as Bernard Tschumi, Jan Gehl, and Marta Bogéa it is then proposed an intervention on one of the most representative residual spaces left by real estate speculation at Paulista Avenue.

The intervention proposed is the building of a new museum, the Sexual Diversity Museum, to exemplify the importance the avenue has as the stage for the fight for LGBT rights in Brazil. The final object is a building that is successful in its intention to open some of the last heritage buildings in the region to the population and creates a new public square.

Keywords: space, occupation, connection, museum, heritage

1 INTRODUCTION

During the year of 2015, the city of São Paulo experienced a surge in debates about public spaces and who is entitled to use them. New cycle lanes, bus lanes were built and avenues were closed in favour of public use during the weekend. Streets and avenues that were previously the domain of car are appropriated by pedestrians and, through them, acquire new uses and functions.

Imbued with the same objectives to discuss the city, it is intended throughout this paper to see how the architecture can be thought and designed to free the body and its occupation of space. Starting with readings from different architects, trying to understand processes that appeared to question the functional pragmatism and spatial development program in architectural design. Analysing the experiments of intervention and project seeking to create ruptures, disjunctions in the program, and freedom of manipulation and appropriation of space by those who use it, with special emphasis on the movement of the individual forced by architecture, we seek to develop a hypothesis of intervention in the territory that prioritizes the city and the user rather than the closed and restricted space.

2 PAULISTA AVENUE—HISTORY AND URBAN DEVELOPMENT

The current avenue that emerged from deforestation of the old forest of Caagaçu is very different from its origin as birthplace of coffee barons and the industrial elite. The study of its history is fundamental for understanding its evolution and its current conformation as a business and cultural centre in the metropolis.

The avenue was created as a high-end luxury region, being inaugurated in 1891. The choice of location was key at the top of the central ridge of the city geography, with a privileged view, proximity to the city centre, where businesses were located and with remnants of native vegetation.

During the first years of existence until the early twentieth century, the avenue was very sparsely populated. This fact is explained by the region lack of basic infrastructure such as electricity, water, and sewage until the turn of the century. This situation will only be reversed in 1900 with the arrival of the tram, which accelerates the occupation of lots with the first houses, totalling 50 at that time.

Initially the result of a real estate project, the fact that the avenue had been occupied by wealthy population and being conformed as an elegant neighbourhood, explains why it was the target of privileged investment by government; providing services to the area long before other parts of the city, such as the aforementioned tram in 1900 and the asphalt in 1908.

Understanding the ease with which the municipality made investments in the region makes it possible to understand how, from the beginning of the avenue, it was considered a key element in the city and how, later, that higher availability of infrastructure would attract other investments and use profiles for the area.

In 1936, the city hall enacted a law authorizing the construction of residential and commercial buildings on the boulevard. Residential towers began to be built and to replace the old mansions. The towers came as representation of a new architectural movement considered as a "national product"; modernism, in contrast to the eclecticism models present in homes. Modernism that represented progress and was evident in the early buildings, initially sought to piggyback on the ideal of the mansion, being composed of large luxury apartments. However, gradually buildings with smaller apartments were built, providing home to a new middle class.

It was the Conjunto Nacional building, proposed by Argentine businessperson José Tjurs and designed by architect David Libeskind, who led the change of use in the avenue. Foreseeing the migration of sectors of the economy from the city centre to this region of São Paulo, it was designed combining services and shops, with apartments and offices.

A sequence of office buildings were built during the 1960s, however, it was the construction of MASP (Sao Paulo Art Museum) by architect Lina Bo Bardi at the end of the 1950s that created a landmark in the avenue, introducing another use, and beginning with what was to become a cultural hub in the city.

The real rise of Paulista Avenue came along with the decay of the downtown area. The demand placed by cars, especially when they became important elements in the city during the 60 s and 70 s, exerted pressure for space in the centre that this could not meet. There were few parking spaces and the mesh of streets hindered the access of vehicles to the region. The search for wider areas to facilitate the flow of cars boosted the output of cultural facilities and corporate headquarters from the central region to Paulista Avenue.

Given the high values of land and the ever-increasing pressure from the real estate market, heritage bodies were unable to act at the speed necessary to protect many of the remaining mansions on the

avenue, which were demolished and converted to skyscrapers.

In 1991, the centenary of the avenue was celebrated with a series of mostly aesthetic oriented reforms, such as the change of visual communication (signs and street names), change in newsagents and other public facilities, these projects were concomitant with the opening of a new subway line. In 2008, it was carried out a change on the sidewalks, with special floors that allowed the movement and orientation of people with visual disabilities. In the year 2015, it was made an intervention to create a bike path, connected to a municipal system.

The concentration of culture equipment at the region extends to the present day, taking advantage of the area increasing accessibility, with bus lines and subway (from the 1990s), with examples of libraries, museums, art galleries and cinemas.

Paulista Avenue got its first museum installed in the 1960s (MASP), but until today new headquarters of cultural institutions are moving to the area. The road conforms as a cultural centrality, and its easy accessibility and visibility (being home to several radio and television networks) help to attract many investments.

The current situation of the Paulista Avenue is as a multifunctional space. It is not the modern era functionalist ideas that govern it. When analysing land use in the region it is evident that the avenue differs from its surroundings, predominantly residential occupation. There is not a single use that is imposed as a vocation to the area, allowing the development of the various programs that are currently found in it, from retail and basic services to headquarters of multinational companies, judiciary, and central bank offices, through bank headquarters, residences and various cultural facilities.

A key aspect when understanding the Avenue, which is influenced by the multiplicity of existing uses in the area, is the network of paths and shortcuts that are formed in the region, developing alternatives to the flow of pedestrians on the sidewalks, breaking the grid of urban design and significantly increasing areas for the free movement of pedestrians. To make an analysis of these territory characteristics it is necessary to

Figure 1. Nolli map showing the permeability of ground floor for pedestrian use in Paulista Avenue area. The selected block is highlighted. Source: Authors production.

use the criteria developed by Giambattista Nolli and his Rome survey in the eighteenth century, when he developed a method that did not analyse the city from its aerial view, but from its ground floor. Observing the Nolli map it is possible to observe how large private sector enterprises, such as department stores and malls, or smaller ones, such as restaurants, bars and street shops, encourage greater pedestrian circulation expanding public space and making sidewalk enter the interior of the blocks.

2.1 Development of the selected block

The Paulista Avenue as it is today is the result of a long process of territory construction. The first land division and its subsequent subdivisions must always be taken into account to understand the operations of the real estate market, over more than one century of history.

Some of the blocks are examples of this process. At one extreme, we have the situation of blocks whose owners have managed to keep their property unchanged to the extent that land speculation allowed, and had the ground sold at its entirety. At the other extreme, we have large lots that were divided over the years among family descendants or when financial needs demanded.

The block studied in this work is an example of this second situation. Composed originally of just one lot, only with one mansion built at the Paulista Avenue corner with Ministro Rocha Azevedo Lane by the René Thiollier Family (Villa Fortunata). Over time, in the early decades of the twentieth century, it was subdivided into smaller lots, reaching a maximum of 12 lots. However, when skyscrapers started to be built in the Avenue this process was reversed, when smaller lots were bought to allow building constructions.

The current situation of the block is unique in comparison to the remaining of Paulista. Inside it are located a public park with the remaining vegetation from the old Villa Fortunata (now demolished), two heritage houses, the Italian consulate, a modern office building with a shopping arcade on the ground floor, a small shopping arcade and a 1991 office building.

Perhaps the most iconic element in the court is the Joaquim Franco de Mello residence, designed by Portuguese engineer Antonio Fernandes Pinto for the Franco de Mello family. The mansion, built in 1909, is one of the few remaining houses from the first residential occupancy period of the avenue. It was listed in 1992, which guaranteed its existence to the present day.

The project surroundings, comprised of some of the blocks from Paulista Avenue with higher pedestrian permeability, emphasizes the need to develop a project proposal that seeks to achieve the same goals; i.e., provide an alternative for the movement of pedestrians.

In the year 2014 it was undertaken a competition for the design of the Sexual Diversity Museum in this court. It was while working in the office Pauliceia Architecture and Restoration that I had first deeply studied the area, which motivated me to undertake this project academically.

2.2 Social movements occupying the avenue

The Paulista Avenue obtained a unique character in the city panorama, surpassing its condition as being just a road or a commercial and residential area; but certainly becoming a stage for different events and demonstrations that affect the dynamics of the metropolis.

The first decades of occupation were characterized as the representation space of São Paul's elite, with the emergence of events for this class. Noteworthy are the Carnival Parades, with its apex during the 1930s. Although very restricted, those parades already denoted a vocation for public use of the avenue.

With the change of the occupation from residences to offices and services throughout Paulista Avenue, came the first union demonstrations, with increasing frequency from the 1980s, with the organization of protests by professional categories that link this city space with the activities of the tertiary sector. One can also say that the visibility generated by the presence of several radio and television networks spread over the region also served as an attractive for these events.

Along with the political protests, the creation of events and celebrations in the area grew, based on the same principles of visibility and accessibility of the avenue. The first major event was created in 1924, by Casper Libero Foundation, the annual Race of São Silvestre. The government also began organizing festivities in the region, starting in 1991 with the avenue's centenary celebrations and other dates such as Christmas and New Year's Eve.

Always taking into account the visibility by media channels and the accessibility of the road axis, the avenue is occupied annually by various manifestations of social groups. This is the case of events such as the Gay Parade, which has come to gather around 3 million people, marking the region a stage for the struggle for equality and against prejudice.

The Paulista Avenue is configured today as the unquestionable space of popular representation in the city of Sao Paulo and the incorporation of architectural spaces for these occupations has become an inseparable fact.

3 SPONTANEOUS OCCUPATIONS AND POPULATION FLUXES IN THE CITY

With the emergence of modern architecture, a need arises to order the spaces of the building and its contents, creating shapes and patterns that become models for the use of these spaces. Such templates repeated over the course of history generate a conditioning for the use and experimentation of spaces. The advent of so-called "functionalism" sharpened up a search for correspondence between content and form, creating a model that could be universalized and, therefore, the possibility of intervention and manipulation of these spaces by individuals who used them weakened.

In this context, it is clear that the development of Avenida Paulista would only focus on the automobile. When the first skyscrapers were built, the logic should be the application of modern ideas, with the avenue use restricted to the passage, the flow as large as possible, of vehicles and pedestrians. However, it was not what happened. A large number of mixed-use projects were designed with large open areas and galleries on the ground, responsible for creating spaces that break with the strictly functionalist ideals, emerging spaces that enable the population to use them without barriers.

Based on this example about Paulista Avenue, questions arise about how and why such breaks in an established model of organization occur.

The architect's role in this scenario of transgression is based on the search and intention to create conditions so that such events and spontaneous gatherings may occur, promoting the interaction between the "defined and non-defined, the design and non-design, intelligible and not-intelligible" (Tschumi 1996). These contrasts and ruptures are made possible through the establishment of tensions that question the hierarchical organization of the spaces and programs.

Marilia Solfa is an author that says, from a study of Tschumi and artists such as Matta-Clark and Hélio Oiticica, that the perception of space is fundamental to understanding the body's behaviour, ordering, and logic. Such understanding is crucial when thinking about the transgression of order, because every human body that is inserted into the space can violate it, as well as the order of the space may violate the body that it is inserted.

The city of São Paulo has several examples of this violence. The growing impermeability of public spaces to pedestrians, always focused on the increasingly rapid flow of vehicles, generates a lack of civility, lack of meeting in public spaces and the privatization of living held behind walls and indoors. Some authors have dedicated themselves to the study of how to reverse this situation. Jan Gehl is an architect and urban planner studies the relationship between public space and the urban society. His analysis begins with the identification of elements that attract pedestrians, focusing on the relations between movement and stop of the population within the common parts of the city.

His research also indicates that the existence of the city is governed by the movement of its people and itself (the city) is responsible for creating this movement, because before becoming a fixed living point, the metropolis began as a meeting place, attracting those seeking work, trade, etc. ... in short, the city "is the road correlative", it exists only with the existence of a circuit network.

The appreciation of the metropolis is constituted by the relationship between mobility and permanence; because if the city is known for its ridged face (architecture) it is, in fact, the flow of people that gives it life. As a result, it is clear that accessibility and circulation are the key aspects to define a place. The tools to generate this accessibility are also studied; it is not enough to build a new square and just expect it to be busy. There are elements necessary for this to occur; as shading, spaces to sit, open spaces and a possibility of creating visual.

A completely different situation is the adaptation of existing spaces in the city, for new uses and new possibilities of occupation.

The architect Marta Bogéa, studious of the city flows (with a specific focus in the city of São Paulo) and communication and semiotic theories, analyses some fundamentally artistic interventions carried out on the urban fabric, but that went beyond its intentions on to make interventions in the territory. Marta Bogéa further explains that the relationships, created between these interferences and the consolidated city fabric, are as new connections that complement an urban space logic already developed. These elements that get support from the environment are considered mobile, because they have different speeds than the vicinity (the city occurs at a slower processing time), but simultaneously reconfiguring the time the stable element moves. In this regard, the author claims that these movements and appropriations of space "enable transformation without dissolution" of the consolidated city.

The challenge for architects, regarding the need for intervention in the city, emerges in the question of how to transform the movement of people in spaces, so powerful in shaping the city, in a pause that generates the shift needed to transform the place? In a way, creating generic spaces with features that allow free occupation by events and by body movement ensure that the particularity is installed. Special attention has to be allocated to the creation of these spaces, in order to avoid becoming just places of movement and passing by, but connections (a "between" in relation to two points).

Analysing the studied territory again, Paulista Avenue one observes that this region differs from its immediate surroundings by having an accommodation capacity for the pedestrian, which accesses the area through a unique transport infrastructure in the metropolis and especially for its wide sidewalks. However, it is the high concentration of cultural facilities, shopping centres, coffee shops, and restaurants that guarantee an extended occupation beyond the working hours and weekends.

4 MEMORY AND MONUMENTS

In Brazil, there are governmental mechanisms and organizations that are "responsible for preserving, disseminating and supervising the Brazilian cultural heritage, and ensure the use of these assets by current and future generations". This means heritage preservation institutions have the power to value assets tangible and intangible, with the intention of protecting elements of society, from architecture to a popular party or work of art, which keep the memory of a collective.

The city of São Paulo has examples of listed buildings that vary from parks and nature reserves, through the first milestones of colonial urbanization, to the buildings that represent aspects of Sao Paulo's expansion from the late nineteenth century until the present day.

Paulista Avenue is a special section of the city, when taken into account the attention given by preservation organizations. In this road axis are found remnants of vegetation that existed before urban occupation, examples of the first homes to be built in the area, buildings displaying the advancement of modern architecture, and representative buildings of the current contemporary architecture.

The most significant thing about this large number of listed properties in a single road axis is the recognition of the importance that the various examples of architecture and occupation have for the history of the city and its memory. The presence of these assets ensures the region a unique character, a real story line that shows how the city from 1889 to 2015 was.

The government has an important role over the definition of what is considered heritage, since social memory is a fundamental part of a project of society, this way giving importance to those buildings and areas that are capable of transmitting the history:

The cultural heritage can be thought as a support for social memory, [...] as an external stimulus that

helps reactivate and revive certain traits of the collective memory in a socio-territorial formation. (Mesentier 2005)

The above quote makes clear the role that listed buildings in the city have as memory support, as always-present landmarks of the society evolution. Such assets survival over time is critical because social memory is not built briefly, but over the course of many generations interacting with those assets and transforming the individual memory in the collective history. As well as the preserving of the listed buildings, the total disappearance of a historical record (the architecture of a time) is harmful because it also disappears with the ability to realize that the social universe subject to a continuous process. However, it is not enough to preserve a heritage property, but also ensure proper access to the property. Therefore, developing a project that attempts to open of some of the last remaining listed mansions at Paulista Avenue is an extremely relevant issue when discussing about the city that we are designing.

5 THE PROJECT: SEXUAL DIVERSITY MUSEUM

This project, supported by the studies done, begins with the questioning of what, for us inhabitants of the twenty-first century would be the ideal city, with a special focus on the public spaces for all. The project of the Sexual Diversity Museum attempts to materialize this idealism, creating a space appropriate for LGBT representation, but not unique to this group. Due to pedestrian flows that cross the project, it will be a space for everyone, occupied by all.

This project, supported by the studies done, begin with the questioning of what, for us inhabitants of the twenty-first century would be the ideal city, with an special focus on the public spaces for all. The

Figure 2. Access by Paulista Avenue, the Joaquim Franco de Melo house, and the park are connected to the project.

project of the Sexual Diversity Museum attempts to materialize this idealism, creating a space appropriate for LGBT representation, but not unique to this group. Due to pedestrian flows that cross the project, it will be a space for everyone, occupied by all.

The museum can be considered the reality of what in the past was a utopia. A space on the main avenue of the city where the LGBT community may manifest, where they can meet and have truly a landmark in the urban scenario. An area of socialization through culture, which integrates and presents to visitors and passers-by the difficulties of a group long excluded and discriminated.

After selecting the lots for the museum's implementation, it was important to project a building that could unite all de different heritage elements and that could connect de court to the rest of the avenue.

The two listed mansions on the court (the Joaquim de Mello Franco Residence and Vicente de Azevedo Residence) were integrated into the museum complex. The first, facing Paulista Avenue was restored and opened its main floor for permanent exhibitions. The second will be used on the upper floors as the museum's management and on the ground floor as coffee shop.

The design theme led the concept of interior spaces. The large central hall serves as a place to see and be seen and conveys the social struggles for LGBT visibility. A network of walkways connects the different floors crossing the space; generating new points of contemplation to the expositions, the museum surroundings, and even visitors.

The program includes exhibition halls, restaurant, and an observation deck facing Paulista Avenue. There are also areas for the collection, library, and classrooms. Other service and auditorium have been placed underground.

The central void space motivated the structural conception. The building has four concrete cores, in which a large metal structure is supported, responsible for overcoming the central span and conforming the two museum bodies.

6 CONCLUSION

This paper started from the premise of investigating the flows of people and spontaneous occupations of space, with a particular interest on Paulista Avenue in São Paulo, identifying open spaces in the region and observing how their occupation does not have a single character, providing ground for different interests and uses, since its origin as the birthplace of São Paulo's elite.

The development of a representation place for the LGBT community at Paulista Avenue would undoubtedly need to answer to the demand for public spaces that already exist in the region, creating a place of connection and integration. Therefore, the project was designed to meet these demands. Its ground floor is not only an open space but also a conversion point for the different flows coming from all sides of the court, sheltering them and protecting them. Thus, the project is successful and fits perfectly to the regions context.

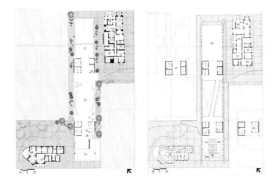

Figures 3 and 4. Ground and First floors plans.

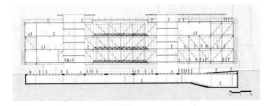

Figure 5. Main project section.

BIBLIOGRAPHY

Bogéa, Marta. *Cidade Errante*. São Paulo: SENAC, 2009. 248 p. ISBN: 9788573599114.

D'Alessio, Vito. *Avenida Paulista: a síntese da metrôpole (Avenida Paulista: the synthesis of metropolis)*. 1ª ed. São Paulo: Dialeto—Latin American Documentary, 2002. 120 p. ISBN: 978-8588373051.

Gehl, Jan. *Cidades para Pessoas*. São Paulo: Perspectiva, 2013. 264 p. ISBN: 9788527309806.

Mesentier, Leonardo Marques de. "Patrimônio urbano, construção da memória e da cidadania." *Revista Vivência* 28 (2005): 167–177.

Tschumi, Bernard. *Architecture and disjunction*. Cambridge, Mass.: MIT Press, 1996. 268 p. ISBN: 978-0262700603.

"We dream of silence": Adriano Olivetti and Luigi Cosenza in the Pozzuoli factory, Naples. Utopia and reality in the workspace

Raffaella Maddaluno

Departamento de Artes, Humanidades e Ciências Sociais, Faculdade de Arquitectura,
Universidade de Lisboa, Lisboa, Portugal
CIAUD—Centro de Investigação em Arquitetura, Urbanismo e Design, Lisboa, Portugal

ABSTRACT: Adriano Olivetti chose to build another factory, in addition to that of Ivrea, in Pozzuoli near Naples and commissioned the design to an old associate the engineer Luigi Cosenza. Following a conversation in which they determined the terms and objectives, Cosenza went to Ivrea to study the production cycle and understand the principles upon which the industrialist based his working philosophy. The project began in 1951 and a site was identified on a hill with a privileged view of the Bay of Naples. The building was designed according to the centrality of workers and workspaces were marked by a relationship of symbiotic exchange with the outside world and the landscape. The cross floor plan of the factory allowed for an unforced adherence to the ground and the language used by the designer locate the work within the most significant examples of architecture of the modern movement in Naples. The classicism of Cosenza, intensely able in communicating with the Neapolitan artistic tradition, converses peacefully with the utopian vision of Olivetti who aims, through this project, to lead men of the south towards modernity in the workplace without becoming a slave to the mechanization of the assembly line. The Pozzuoli factory is thus part of a broader project of the industrialist, which, through its activities and civic engagement to the problems associated to the territory, managed to turn every decision into a success and of high artistic quality.[1]

Keywords: Adriano Olivetti, Luigi Cosenza, Pozzuoli Factory, Industrial Architecture, Italian Modernism

1 ONE SUNNY MORNING: GENEALOGY OF A REENCOUNTER

An architectural project, any type of architectural project, is the result of a slow and gradual reduction of a distance: a distance between man and nature, a distance between intuition and concretization, a distance between client and designer, a distance between utopia and reality.

The process of this reduction usually begins from a stage, a scene. The Gulf of Naples: an embrace between the land and the sea. A Sorrento Terrace: the silence of the citrus groves. Two characters: the industrialist Adriano Olivetti and the engineer Luigi Cosenza. Therefore, the story begins the early 1950s.

Adriano Olivetti is in Naples as he wants to build a factory in the south and he is discussing it with Luigi Cosenza, with whom he shares an ethical vision and special attention to working conditions. The two are reflecting on the Italian post-war situation and the need for a form of beauty in the workspace as a comfort for the soul. As a consequence of this exchange Luigi Cosenza went on to Ivrea, to study the organization of the first Olivetti factory, to understand relationships and the centrality of the production areas, to become familiar with the production process, to understand the role of workers within the Olivetti working environment.

Client and designer thus cancelled their initial distance: a long and fruitful dialogue begins, animated in the way that conversations between great personalities are animated. Adriano Olivetti had already met Luigi Cosenza during the war when commissioning a territorial plan for the Campania Region, a very similar design exercise to that performed by the BBPR Studio in the Valle d'Aosta.[2] The war and the complexity of the nature of the territory itself disrupt the plan, yet not the will to resume dialogue in a more favourable moment.

2 "AN INDUSTRIAL PLANT"[3]

We must work in this field with great sensitivity, in the sense that being too rigid towards the functional can lead to results in manners and a formalism no less distant from our modern aspirations in architecture than the old disorderly and presumptuous "factories" (...) Among the fields in which architecture operates, that of industrial buildings is undoubtedly the most attractive in terms of the rational orientation of modern techniques. The consistency of geometric relationships and spatial proportions can express themselves most clearly in this sector. However, it is also the sector in which it is easiest to slip back into cold geometric compositions, back to the dry academy. (Cosenza 1950)

The factory project of Pozzuoli was entrusted to Cosenza in 1951 and lasted for four years. The methodology adopted revealed a clarity of purpose and a mastery of design methods that marked the experience as a happy occasion in Italian architectural history. Cosenza drew up the idea for the project, developing it out of a sequence of sketches organized through 59 cards.[4] The design path borrowed the language of a film sequence, where each scene is described in detail through decisive lines, colours that identify functions, pathways, and captions that complement and enrich.

They were the result of Cosenza's visit to Ivrea and the necessary need to bring order, through drawing, to a flow of ideas and then be able to respond to aesthetic, functional and production requirements. The project progressed towards a more detailed definition: the distance between insight and realization that was slowly being cancelled. They were drawings in which everything is schema, graphs, diagrams, distributions, and flows. Surely, they were not conceived as disposable products, but were probably the result of a final draft of reflections, notes, and inspections. Nothing is known of these preparatory graphical notes, a fate common to many of the sketches of architects of the period; indeed, the cult of the original architectural sketch to display and be preserved is a recent phenomenon (De Seta 2006: 30).

The plot on which the factory stands is rectangular, bordered by Via Domitiana, the main Roman road that connects Naples to Rome, and by Via Fascione; while on the left side, a possible area for future expansion was immediately identified. From the sketches, one begins to understand just how much and in which ways the distribution values of flows (workers, materials, management, and the public) determine spaces. The compositional concepts on which the entire project was structured include, firstly, a cross floor plan ensuring that the factory adapted itself to the ground, embracing the slopes rather than working against them; the second is the will to ensure exposure and the most natural lighting for workstations. All of this guarantees the possibility of a continuous view even beyond the walls of the building, thus connecting with the wonder of the landscape of the Gulf that opened up in front of the generous sequence of

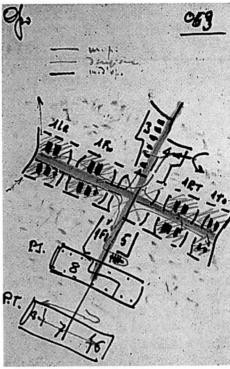

Figure 1. (*Continued*)

Figure 1.   Sketches of Luigi Cosenza's project (www. luigicosenza.it).

Figure 2.   The gardens with small artificial lake; Pietro Porcinai's Project (www.luigicosenza.it).

Figure 3.   Aerial photograph of the factory (www.luigi-cosenza.it).

windows. The gardens, designed by Pietro Porcinai (1910–1986), regulate this opening towards the exterior, mediating and combining spaces. (Fig. 2) Marcello Nizzoli (1887–1969) developed the colours of the interior and exterior walls. It is a type of design that is not a consequence of the assembly line, but is arranged according to independent bodies, connected to each other and with the external environment. This greater freedom of composition derived from the fact that processing systems would enter into a crisis during the 1950s, allowing greater flexibility in positioning machines in their spaces. The composition of the factory is organized on two floors with a north-south orientation. Each wing has an element that concludes the axis: respectively, the west has changing rooms, the south has management offices, the east has spaces dedicated to heat treatments, and the north an area designed for future enlargement.[5]

The factory, in its original configuration, consists of four groups of buildings: the one that faces Via Domitiana, a group of support buildings, a group of buildings that make up the actual factory along with offices and a section dedicated to warehouses. A system of covered walkways and parking lots completes the functional composition. An area of the building is intended for the exhibition of products and the training of young apprentices. Along the route used by workers to enter the factory, there is also a library. Subsequently, beyond a heated passageway, we find the canteen, divided into two large serving areas with a capacity of around 500, connected with a staircase to an upper gallery with tables designed for use post-meal. The canteen serving areas have terraces offering views. The entire complex is made of reinforced concrete and concrete tiles, the floors are tile except for the management and office area which are green Verona marble mosaics with the door and window thresholds in Trani marble. The stairs used by the workers are made of iron sheeting lined with galvanized iron.

Is there not, beyond its apparent rhythm, something more charming, a destination, a vocation in the life of a factory? We can answer: there is an order in our everyday work, in Ivrea, as in Pozzuoli. In addition, without prior knowledge of this order, it is futile to hope for the success of the work we have undertaken. Because a plot, an ideal plot beyond the principles of business organization has informed over many years, inspired by the thought of its founder, the work of our society. (Olivetti 2012)

3 "AN INDUSTRIAL OLIVETTI PLANT"[6]

Olivetti products seem almost illuminated by the exact proportions and love with which an object should be manufactured, by the love, with which one carries out their duty, by the love of their work. He placed the responsibility of man with man himself. He wanted to realize the dream of a new society on earth, not postponed by unspecified deadlines. (Corbusier 2006: 156)

In 1908, Camillo Olivetti founded the first Italian manufacturer of typewriters in Ivrea. He had five children, including Adriano, who would receive a free, secular education. Adriano's first work experience was in his father's factory. His character was forged by ideals related to workers' struggles in Turin and the experience of World War I, in which he participated as a volunteer. He graduated in engineering from the Polytechnic of Turin but continued to maintain his political views that caused persecution by the Fascist regime and forced him to flee to France. He returned to Italy but was officially marked as subversive.

The experience that would shape his final vision of the conditions of workers and concepts of the organization of a factory was his trip to the USA in 1925. Because of this trip, he would propose to his father the organization of the company through the introduction of several innovations: the decentralized organization of personnel, Directors according to functions, the rationalization of time and assembly methods and the development of commercial networks in Italy and abroad. Yet his desire was not simply the development of the business, he had still higher ambitions.[7]

During WW II he would never cease to innovate, learn, share, and inform, waiting for opportunities that would result as profitable in order for the dream of a workable utopia to happen. Moreover, it did happen. He could apply the principles he had formulated during a period in Switzerland (1944–1945) in his theoretical treatise *"L'ordine politico della comunità"* published in 1945. After the war, the link between territory, enterprise, and workers was solid, and became a powerful tool for organizing and planning. Ivrea would become the hub for ideas arising from a group of philosophers, sociologists, urban planners, architects, writers, and economists. Olivetti founded the Community Movement and gave life to the Edizioni di Comunità publishing house. The magazine *Comunità*, which he founded in 1947, began to advance the most modern cultural debates in post-war Italy, all marked by the need for modernization of the country and social equality. In 1948 at the Ivrea plant, a Management Board was established, for a number of years the only example in Italy of a joint body with advisory powers of a general nature in the allocation of funding for social services and assistance. In 1956, Olivetti reduced working hours from 48 to 45 hours per week without any loss of pay, several years in advance of national labour contracts. Adriano's civic engagement, which would also witness a political turning point with his election as a Deputy in 1958, was to be accompanied by his careful and constant work in the development and modernization of the Olivetti company; indeed, the 1950s saw the firm's great international expansion and Olivetti's signature will always be associated with an image of artistic and industrial quality and excellence.[8] The architects and artists who made up the great Olivetti family were chosen carefully and every piece of work was considered as unique. Even today, Luigi Figini (1903–1984) and Gino Pollini (1903–1991) consider Olivetti shops and factories think of the Carlo Scarpa (1906–1978) shop in Piazza San Marco in Venice from 1958, or the one entrusted to BBPR on 5th Av. in New York in 1954 or, indeed, the factory in Ivrea itself (1934–1940). Each project of the industrial plant was accompanied by the construction of auxiliary services, such as cafeterias, schools, kindergartens, housing for workers and configurations entrusted to architects such as Marco Zanuso (1916–2001), Ignazio Gardella (1905–1999), Mino Fiocchi (1893–1983) and Eduardo Vittoria (1923–2009). Adriano Olivetti's commitment to urban issues saw him cover several positions of significant prestige and commitment such as his accession to the National Urban Planning Institute of which, in 1948, he became a member of the Executive Council; the founding of the journal *Urbanistica*; his appointment in 1956 as an honorary member of the American Institute of Planners and that as Vice President of the International Federation for Housing and Town Planning. In 1959, he was elected President of the UNRRA-Casas for post-war reconstruction in Italy.[9] Only his death, on February 27 1960, would interrupt his dream, the continuance of which was delivered into the hands of people who, although having grown up surround by the Community ideas of Olivetti, were not sufficiently capable of continuing their materialization.

Real talent, able to open up new pathways, is born already equipped with the tools to be able to travel and to be able to manage changes of direction. Olivetti was a multiplier of directions. His writings were always centred on an anthropocentric view of the world and on the price one had to pay for a presumed and compulsory modernization. His every action in the local area or the organization of a workspace always aimed for the possibility of creating the conditions for a more worthy and more integral life with an essential reunification with nature. He writes in his work

As the chemical composition of a body is powerless to resolve the mystery of life, so the figures and numbers, statistics and calculations of our ministers, the hundreds of billions of state investments will not be enough to advance a nation, if you are not ready to live a new life based on the deep human roots that are found in the villages, in the communities, close to nature and the landscape, and that is only expressed in the life of municipalities and provinces. (...) In the civilization of the land over millennia, the farmer looking at the stars could see God, because the land, the water, the air express the continuity of vital energy, because water not only serves to wash the body, but it also affects the soul just as a baptism purifies the heart. (...) For this, the modern world, having enclosed man in offices, factories, living in the cities among the asphalt streets and the raising of the crane and the noise of engines and the messy tangle of vehicles, somehow resembles a vast, dynamic, loud, hostile prison from which eventually we must escape.[10]

As noted by Gregotti, in the preface to the publication referred to above, it is possible to identify three key points in defining the profile of Adriano Olivetti. The first is a sense of moral duty in all of his actions springing from an enlightened spiritualism shared and refined through building relationships with Piero Gobetti (1901–1926) and with the thought of Simone Weil (1909–1943). The second is the discovery, through the work of Frank Lloyd Wright, of the relationship with nature, the possibility of a foundation of an ideal village, antithetical to the notion of the city organized by functions by the Modern Movement in Europe. The third is the idea that disciplines such as architecture, urban planning, sociology and philosophy should share methods and objectives for a common good, namely that of the territory. This vision of a communion of knowledge and tools reached levels of modernity never subsequently attained, where even workplaces and factories, historically considered conglomerations of alienation, could be incorporated within this ideal. Olivetti himself considers the factory in these terms

Figure 4. Assembly line view in Pozzuoli factory (www. luigicosenza.it).

Moreover, as our economy comes from the factory, we will speak first of the factory, in terms of the Community factory. What is this Community factory? It is a workplace where righteousness dwells, where progress dominates, where we may shed light on beauty, around which love, charity, and tolerance are not names and voices devoid of meaning. (...) The joy in work today denied to the great majority of modern industrial workers, will finally return when the worker understands that his effort, his labour, his sacrifice—which will still always be a sacrifice—is materially and spiritually tied to a noble and human entity that he is able to sense, measure and control, because his work will serve to strengthen that community, real, tangible, where he and his children have their life, their ties, their interests".[11]

4 "AN INDUSTRIAL OLIVETTI PLANT IN NAPLES"[12]

Luigi Cosenza introduced the Modern Movement to Naples, but his architectural production can be considered, without fear of falling into forced definitions, as belonging to the Neapolitan culture and tradition. Franco Purini, in a text devoted to the architectural poetics of Cosenza, identifies certain components, referred to as *invariants*, when aiming to define Neapolitan architectural culture. The first is related to a strong presence of classicism, which assumes characteristics closer to a mythological essence, rather than to a monumental sense that we may witness in Rome. In the tradition of the latter city, the concept of the classical is identified with the Ruin, an archetypal element to be decoded; in Naples, full of archaeological remains from the ruins of Pompeii and Herculaneum, the classic becomes a figurative object, a vehicle of sensory experiences due to the explosive nature of the earth, always in constant tension and in a condition of permanent impermanence. The second characteristic is added as an explanatory afterthought to the notion of the classic when we talk of Naples: a landscape of hills and coves, of breaths and sudden dizziness. The architecture in Naples is the bodily encounter between nature and artifice, and those who wish to get involved must possess both the strength of Pluto and the gracefulness of Persephone.

How did he do it?" asks Alberto Campo Baeza, "How did that devil Bernini ensure that cold marble could give off so much warmth? As in architecture, there is always in sculpture, there must be, a moment of strength. The algid point, the divine point, in which the outstretched hand of Pluto grabs the soft flesh of the left leg of the worried Persephone. It is undoubtedly a highlight in the history of Sculpture through the ages. Moreover, I repeat, this came to my mind clearly during my visit to Cosenza's Olivetti. It awakened my admiration for its location on the Via Domitiana hill. With the same wisdom of Bernini. It is no coincidence that Bernini and Cosenza were born in Naples. (Campo Baeza 2006: 214)

Luigi Cosenza is classical, and with the same mastery of the classics, devotes much time to reducing the distance between artifice and nature. He is familiar with every slope of the hill where the factory will be placed, just as the Greeks knew their Acropolis, and works during the breaks in this age-old dialogue between man and the Neapolitan territory, refusing to reduce everything to a single radical surface, a highly frequent attitude within the modern movement. The third component involves a child of the Enlightenment legacy: it is that concrete precision; it is the attention to simple and complex detail, which is born almost as a need to be set against the imposing presence of

Figure 5.   The canteen serving areas (www.luigicosenza.it).

nature. It is the rationality of simple volumes, the almost never superfluous application of functions to forms, consequence of the tradition of holiday homes in the Gulf Islands and the utopian Enlightenment of Ferdinando Fuga (1699–1781).

All of the works of Cosenza are therefore Neapolitan and have in common the assumption of the role of manifesto, despite their classicism. For example, for the factory in Pozzuoli, the choice of a cross-floor plan also reveals a classical attitude, yet releases the shape from the sacredness of the symbol. It thus becomes secular as importance is taken away from its centre; notoriously the most powerfully sacred and symbolic place in religious buildings, in this case simply the node of the two axes that define the structure of the various spaces. The rigor of his architecture, not exhausted only in the distribution of spaces and functions, allows for absorbing the unforeseen and mutation typical of architectural design. His method has ancient roots and is typical of those who possess a sense of reality, firmly anchored to place, yet without losing the visionary insights of one accustomed to living by the sea. His collaboration with Bernard Rudowsky (1905–1988) with whom he designed Villa Oro,[13] develops and refines an unquestionably Mediterranean architectural language which, however, never falls into localism, is traditional without being folkloristic. Yet the real great design utopia of these spaces is that the worker can, while performing their job, always allow the eyes to wander in the direction of the windows, along the sinuous lines of the Gulf. What an immense form of freedom, in a working environment traditionally marked by mechanical rhythms and obsessive focus on the tiny component pieces that pass you by for a lifetime, never stopping for a moment to consider where they will end up or what they will go on to constitute. Here, the pieces pass you by the same but at least tired eyes have something to admire, every so often.

Thus, in front of the most beautiful Gulf of the world, this factory is elevated, in the idea of the architect, in respect of the beauty of the place and so that beauty may be central to our everyday work. We also wanted nature to accompany the life of the factory. Nature was in danger of being repudiated by such a great building, in which the closed walls, air-conditioning and artificial light would have tried to transform man, day by day, into a different being from that which had entered, albeit full of hope. The factory was therefore conceived on a human scale so that they could find release, rather than suffering, in the workplace". (Olivetti 2012)[14]

We left, in little more than a generation, an ancient civilization of farmers and fishermen. To this civilization, which is still the civilization of the South, the light of God was real and important, family, friends, relatives, neighbours were important, the trees, the earth, the sun, the sea, and the stars were important. Man was working with his hands, exercising his muscles, drawing the means of life directly from the land and from the sea.

5 "A POST-WAR INDUSTRIAL OLIVETTI PLANT IN NAPLES"[15]

When, four years ago, the construction of this facility was decided, the battle initiated at the Ivrea factory to become an international enterprise was already in full swing. The problem of the South had already entered our soul some time previously in all its painful magnitude (...) yet the issue was not our setting-up in the South. It was, rather, the deviation, demanding and sudden, that could distract us from the hardest fight we had undertaken in Europe, in the Americas, in South Africa. We willingly accepted the new burden. It was an act of faith in the future and the progress of our industry, but above all a conscious homage to the needs of these regions. Moreover, it was not only a monetary contribution, but also a genuine sacrifice by our workers. Because all of Italy is affected by the painful disease of unemployment. (...) Can industry provide us with aims? Are these to be found simply in the profits index? (Olivetti 2012)

Following WW II, the southern question took on dramatic tones indicating the need for resolutions as a matter of urgency. The civil commitment of Olivetti in the desire to restore dignity not only to work spaces but also to domestic spaces was demonstrated on several occasions in which his role was indisputable, such as in the case of the construction of new districts as a result of displacement away from the Sassi in Matera. It is as if the war had suddenly drawn back the curtain of false progress and had revealed social situations that could not be shown to those who, from the other side of the Atlantic, had attempted to establish the price of a covenant: the Americans.

Olivetti arrived in Matera, a city presenting serious sanitation issues, in 1950, the same year that the *Lettera 22* was released in Ivrea. This city became the symbol of his strong sense of the redemption of the South after the war, the same sense that had led him to set up a factory in Pozzuoli.[16]

Reduce the distance: this time the social, between North and South. In a constant tension that unfolds, the utopian greatness of Olivetti is outlined, in his ability to fit the dream to the reality, respecting the languages of every territory, understanding the nature of each place and the creative individuality of each action. Only then, in compliance with individual personalities, is universal value bestowed upon all of his works.

I would not say by this that our discipline postulates impossible revolutions and ventures along treacherous paths of utopia. It merely acts according to the precept that says not to leave behind, working meticulously, day by day with fatigue, a faith in other larger and perfect accomplishments, but also requires that we do not overlook, due to a faith in these, the obligation to everyday work. The continuous exchange between practice and the ideal is nevertheless the rule for our conduct also at this stage. Studying and experimenting in the living social body, always meeting new difficulties yet also learning to equip ourselves with new tools and master their use.[17]

NOTES

[1] Note to the title: "Noi sogniamo il silenzio" is part of a speech delivered in Turin during October 1956 at the Sixth National Congress of the Institute of Urban Planning and published for the first time by Edizioni di Comunità in the anthology Città dell'Uomo in 1959 under the title "Urbanistica e libertà locali".
[2] BBPR is the acronym that indicates the group of Italian architects formed in 1932 by: Gian Luigi Banfi (1910–1945), Lodovico Barbiano di Belgiojoso (1909–2004), Enrico Peressutti (1908–1976), Ernesto Nathan Rogers (1909–1969). The Urban Plan of the Valle d'Aosta is from 1936.
[3] The division into paragraphs of this text refers to a text of Cosenza himself and the paragraphs are textually cited with the following comments by the same author. "Evidently it is not the search for a solution to economic problems, unions, politicians in opposition to workers and the various industrialists; but of modern research in the face of issues of form and content", (L. Cosenza 2006: 22).
[4] The cards, 59 in all, are identical in their dimensions (23 cm × 15 cm), are in Bristol paper with a drawing on one side and the captions and some written notes by Cosenza in black marker on the other.
[5] As early as the 60 s, Olivetti in Pozzuoli was presented with the possibility of increasing its workforce from 1200 to 2700, due to the expansion of European and American markets. Cosenza studied and provided various solutions, constantly reflecting on the existing scheme of the cross with perpendicular arms. In this second phase, the method was limited to the coordination of solutions previously developed during an early stage and debated at length with the technical Directorate at Ivrea. This was presented as a testing phase for the functional verification of the time that had elapsed (in both the Ivrea factory and in the Pozzuoli factory) for all the elements. Processing had now become electronic and less mechanical, the excessively long waiting times in the canteen required a new organization of space in the dining halls; open spaces are reduced; assembly assumes a primary role with respect to the workshop, as well as storage deposits for processing. But "...the main issue is the new role of an managerial apparatus substituting Adriano Olivetti who had recently passed away; then decides Ivrea, disinterested in the historical experience of a factory built

according to the human scale principles of the ideology of Adriano Olivetti and the political and architectural beliefs of Luigi Cosenza" (G. Cosenza 2006: 164).

[6] "A good opportunity to open a dialogue with Adriano Olivetti, who has also been searching for years for a new relationship between man and environment, architecture and tradition, climate, construction materials, between form and content" (L. Cosenza 2006: 22).

[7] "There was a model from across the Atlantic; there was a push to an almost inexorable move towards a new state of bigger, more efficient things where many workers would find a reason for existence. My father hesitated, hesitated, because - and I said so for many years and for long periods - because the large factory would destroy man, would destroy the possibility of human contact, and would have led us to considering the completely human make-up as a mechanical make-up. Every man as a number. Yet the open pathway unfolded anyway. (...) The scientific machine was set in motion, technical departments grew, new products were studied, were put into production, were sold. Each year, architects studied enlargements. There was something beautiful in this; there was a certain pride in seeing the old red brick factory out of these large modern windows. And gradually the factory emerged as it is currently". (Olivetti 2014), first edition by Cadeddu Davide (Ed.), La riforma politica e sociale di Adriano Olivetti (1942–1945), Quaderni della Fondazione Adriano Olivetti, 2006.

[8] "Between the end of the 1940s and the 1950s, Olivetti introduced several products destined to become real objects of worship due to the beauty of their design, their technological excellence, and functional quality, among them the Lexikon 80 typewriter (1948), the portable Lettera 22 typewriter (1950) and the Divisumma 24 calculator (1956). In 1959 the Lettera 22 was chosen by an international jury of designers as first among the hundred best products of the last hundred years". www.fondazioneolivetti.it.

[9] http://www.fondazioneadrianolivetti.it/lafondazione.php?id_lafondazione = 1.

[10] Cosenza, Luigi, "Noi sogniamo il silenzio", cit. note 1

[11] Olivetti, Adriano, *L'industria nell'ordine delle Comunità*, a text that appeared for the first time in 1951 in the Movimento Comunità pamphlet "Tecniche delle riforme" and was subsequently published with the same title in 1952 in the anthology Società Stato Comunità.

[12] "A contribution to the efforts of Italian architects, the architects of Olivetti in the study of building traditions and the environmental conditions of Campania for the rigorous analysis of the issues of distribution, static solutions, the connection between interior and exterior spaces, the rooms used at work, the cultivation of the body and the spirit, water, housing, size and rhythms as a function of motion and circulation issues" (Cosenza 2006: 22)

[13] On Villa Oro see (Maddaluno, Cosenza and Rudofsky 2002).

[14] Adriano Olivetti delivered the speech to the workers of Pozzuoli on April 23, 1955 at the inauguration of the new plant. In 1959, Edizioni di Comunità published it in the anthology Città dell'Uomo. It was recently published in Olivetti Adriano, Ai Lavoratori, Comunità Editrice, Roma/Ivrea, 2012.

[15] "A precise reference to the tasks of the designer on finance and production, in the present state of anarchy of national and local production, the growing threat to local production, the consequent need to contain the costs of the plant and operate within the limits of functionality" (L. Cosenza 2006: 22).

[16] On the situation in Matera see (Bilò and Vadini 2013).

[17] Cosenza, Luigi, "Noi sogniamo il silenzio", cit. note 1.

BIBLIOGRAPHY

Bilò, Federico, and Vadini, Ettore, eds. *Matera e Adriano Olivetti, Conversazioni con Albino Sacco e Leonardo Sacco*. Roma: Fondazione Adriano Olivetti, 2013. 278 p. ISBN: 978 88 96770 21 4.

Campo Baeza, Alberto. "La fuerte mano de Pluton sobra la tremula carne de Persefone." In *Luigi Cosenza La fabbrica Olivetti a Pozzuoli*. Ed. Giancarlo, Cosenza. Naples: Clean Edizioni, 2006.

Corbusier, Le. "Ricordo di Adriano Olivetti." In *Luigi Cosenza La fabbrica Olivetti a Pozzuoli*. Ed. Giancarlo, Cosenza. Naples: Clean Edizioni, 2006.

Cosenza, Giancarlo, ed. *Luigi Cosenza La fabbrica Olivetti a Pozzuoli*. Naples: Clean Edizioni, 2006.

Cosenza, Luigi. "Nascita di una fabbrica, ANIAI, 10 July 1955." In *Luigi Cosenza La fabbrica Olivetti a Pozzuoli*. Ed. Giancarlo, Cosenza. Naples: Clean Edizioni, 2006.

——. "Architettura industriale." In *Esperienze di architettura*. Ed. Cosenza, L. Naples: Macchiaroli, 1950.

De Seta, Cesare. "L'origine del progetto." In *Luigi Cosenza La fabbrica Olivetti a Pozzuoli*. Ed. Giancarlo, Cosenza. Naples: Clean Edizioni, 2006.

Maddaluno, Raffaella, Cosenza, Luigi, and Rudofsky, Bernard. "Villa Oro a Napoli 1934–1937." *D'architettura*.17 (2002): 158–167.

Olivetti, Adriano. *Le fabbriche di bene*. Rome/Ivrea: Edizioni Comunità, 2014.

——. *Ai Lavoratori*. Rome/Ivrea: Comunità Editrice, 2012.

Radical and utopic: The mega-structuralist formalization of house and city

Ana Marta Feliciano & António Santos Leite
Project Department, Faculdade de Arquitectura, Universidade de Lisboa, Lisboa, Portugal

ABSTRACT: The events that marked the period after the World War II marked a particularly optimistic moment, but ambiguously also critical in regard the question of behavioural norms and values of society. Reflecting the decisive changes underlying it, this time changes in what regards the realization not only of problems inherent in the functionalist city (urban model that had driven the major reconstruction initiatives of post-war urban structures), but also assists the dissolution of a specific and topic sense of city. Particularly, the previous findings can explain that, after a first exploration period (primarily associated with a visual universe linked to the image of the machine and a certain fascination for a technological universe), the projects tried to attend an important paradigm transition. It would now search for a generic and radical utopian understanding of the city assumed as living organism, expression that would be materialized in a diverse set of proposals for new structures for contemporary dwelling.

Keywords: megastructure, japan metabolism, utopic architecture, post-war city, living city

Today we each recognize the existence of a new spirit. It is manifest in our revolt from the mechanical concepts of order and in our passionate interest in the complex relationships of life and the realities of our world...[1]

Presenting itself as one of the first perceived realities, contributing to the transformation of society and for a vision of the world interconnected in a progressive global village, the period after the World War II, with its unavoidable demographic boom, has come to imply a new global awareness of the consequences of that population explosion. Given this finding, discussed in numerous scientific studies but also with repercussions in the most widespread means of communication, a new critical consciousness became emanate as part of the thinking of new utopian and generic spaces.

Thus, the utopic development of future scenarios, promoted by the demographic 'boom' and global vision of the post-war society, would be the imaginative engine to the new elementary principles and a set of new urban proposals. Given the alarming atmosphere contained in that prefigured world, the technological progress was present in various fields of human activity. It lend a feeling of euphoria and positive belief in scientific progress, of which the space projects were paradigmatic symbols. They seemed to present themselves as an alternative to a generic future for humankind; opening new perspectives that should questioned the traditional boundaries, the aggregation modes and the physical and territorial contexts of urban organization.

Figures 1, 2, 3, 4. Aerial view of Levittown, Pennsylvania, 1959; "Life" Magazine, January 1955; Raimund Abraham, "Universal City"; "Aerial Perspective of the Earth's Horizon and axiometric", 1966.

Consequently, and in addition to the possibilities explored in the new territories, the generic problem of planetary population density seemed present relevant issues: the concentration or density of new urban structures, their capacity of regeneration and transformation; scenarios that from this moment would approach an idea of urban agglomeration as a living organism.

Figures 5, 6, 7. Team 10 in Otterlo Meeting 1959 (CIAM '59), 1959; Pueblo pre-Columbian, The Cliff Palace, Mesa Verde, Colorado; 'Ponte Vecchio', Florence.

Developing this sense of global and utopian thinking about the city as a natural organic entity, the 50s and 60s promote investigation on the theme of the city and dwelling. Supported by progressive knowledge and historical consciousness, attempt to create a set of associations and analogies with some past period's achievements, as well as the emergence of renewed interest in the generic study of architecture and urban aggregations of vernacular character.

Following the radical nature of this cultural environment, members of Team X would try to consolidate the intention to design spaces for cultural people, characterized by the sense of aggregation and a new desire for consumption. On the other hand, this period would also be associated with new research conducted by Japanese architects. They were working on new forms of urban structure resulting from a new '*homo movens*' emerging in the structures of post-industrial society. Sharing a number of common ideas with Team X members and due to the possibility of dialogue, the change of thoughts materialized in meetings that mark the 50s and early 60s. The group of Japanese architects also had the intention to rethink culture and interpersonal relationships, determined by inherent cultural differences resulting from anthropological spaces.

Responsible for one of the first attempts of characterization and theoretical systematization of this universe of search, Fumihiko Maki, a student of Kenzo Tange, in an early period of his research devoted to the problem of urban genesis, sought the work in "Investigations in Collective Form"[2],

definition of a generic concept of Group Form. Specifically, according to this author, the concept could be associated with an urban situation generated by a repetition and agglomeration of anthropological buildings, similar and standardized, articulated in groups or conformations characterized by a clear structure and a varied formal silhouette.

Considered as one of the first attempts for a theoretical definition of 'mega form', which later evolved into the generalization of mega-structure, Fumihiko Maki sought to clarify and define the concept of mega-structure as

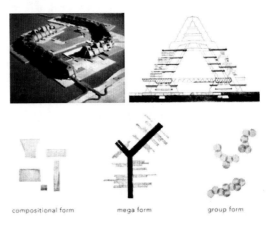

Figures 8, 9, 10. Kenzo Tange, model and section of "Community for 25000 in Boston Harbour", 1959–60; Fumihiko Maki, "approaches to collective form (...): compositional form, mega form, group form", in "Investigations in Collective Form", 1964.

Figures 11, 12. Le Corbusier, perspective drawing for "Fort l'Empereur", Argel, 1931; model of "Plan Obus for Algiers", 1932.

... The large frame in which all the functions of the city or part of the city are housed. It has been made possible by present day technology. In a sense, it is a man-made feature of the landscape..."[3]

And came to underline one of its most important features:

... A mass-human scale form which includes a mega form and discrete, rapidly changeable functional units that fit within the larger framework..."[4]

The increase of dimensional, structural and functional difference of these two moments of the megastructure, recovered some of the concepts previously developed, in 1931, by Le Corbusier's project "Fort l'Empereur" in Algiers. It had the possibility of filling the main structure with different subunits, and seemed now to delegate in the future inhabitants of the structure, the possibility of construction and real final definition of the basic cells of the project. This concept, together with the proposed dialogue between a hierarchically dominant and permanent structure, containing a set of subordinate accommodations, ephemeral and open to spontaneous local participation, would be a controversial vision within the architectural discipline.

After a brief period, four years after the attempt of Fumihiko Maki to theorizing the concept of megastructure, Ralph Wilcoxon sought in the introduction of his essay "Megastructure Bibliography", to define the concept of megastructure as:

... Not only a structure of great size, but (...) also a structure which is frequently: 1) constructed of modular units; 2) capable of great or even 'unlimited' extension; 3) a structural framework into which smaller structural units (for example, rooms, houses, or small buildings of other sorts) can be built—or even 'plugged-in' or 'clipped-on' after having been prefabricated elsewhere; 4) the structural frame work expected to have a useful life much longer than that of the smaller units which it might support..."[5].

In this systematization of the idea of megastructure, and in the creative action that characterized the artistic and architectural thinking during this period, we can detect in the works of many mega-structuralists, an intention to propose urban structures open to various dimensions of future ownership. Indeed, these new structures sought to sketch a complex interplay between modern technological society and the spontaneous construction of its own urban cluster through a natural process, a clear biological evocation of cyclic and continued growth, contraction, and reconstruction. Using examples from vernacular architecture or built

Figures 13, 14, 15. Mario Fiorentino, section and actual views of 'Corviale' building (1 km length), Roma, first phase 1972–1974.

structures of the past and with the background of problems inherent in most cities during the postwar period, these architects wanted, therefore, to rethink the emerging conflicts between design and spontaneity, macro-scale and micro-scale, the permanent and the ephemeral.

In addition to the assumptions underlying the concept of megastructure, the research triggered around this background becomes responsible for an unusual proliferation of images. These not only valued the conceptual assumptions of megastructure idea, also stimulated a focusing around the image of the concept itself, transporting visual allusions that could not be expressed in words. Therefore, the idea of megastructure as utopian and prospective urban proposal began and in addition synthesized the four Wilcoxon's definition points, to be associated to specific visual universes.

Frequently referenced throughout this conceptual process, is the complex structure of 'Ponte Vecchio', in Florence, characterized by the convergence of several functions: articulate along its pedestrian route a set of commercial spaces, promoting the simultaneity of this mixed spaces crossing simultaneously with the living spaces. In fact, the diversity of different spaces and dimensions inherent to these differentiated areas were in analogy with megastructure, with the hierarchical rhythm of the bridge itself and its supports, which in turn could join a broader level of service: the movement of the watercourse itself and pathways that could be established.

Indeed, the complex structure and the diversified spaces underlying the different levels of megastructure would carry also a visual universe with points

Figures 20, 21, and 22. Japanese lineal village, s.d.; "agricultural city", Kisho Kurokawa, 1960; Kisho Kurokawa, axonometric and view of Nakagin Capsule Tower, Tokyo, completed in 1972.

Figures 16, 17, 18, and 19. Japanese lineal village, s.d.; "agricultural city", Kisho Kurokawa, 1960; Kisho Kurokawa, axonometric and view of Nakagin Capsule Tower, Tokyo, completed in 1972.

of contact with numerous industrial structures. Actually, these concerns tended to rediscover an imaginary, which is inevitably linked to the huge complex of refineries, bases, and sea fortifications, which in addition to harbouring a variety of functions inherent in all of its community life, also had a powerful visual image composed of several connected functional units.

This need for discovering these complex products of human creation, in particular mechanisms merged into a whole compact and a great diversity of tasks, would be also responsible for a symbolic revisiting of structures such as aircraft carriers, submarines or liners, whose direct analogy with the inherent complexity of urban structures and their habitat forms had already been set out by Le Corbusier in the heroic age of the Modern Movement. In fact, the 'big transatlantic' understood as a container of a complex and autonomous human community, would be, during this period, under renewed attention through proposals such as Paolo Soleri's works and its 'arcologies':

> [...] Compactness and definite boundary; the functional fullness of an organism designed for many, if not most, of man's needs; a definite and unmistakable three-dimensionality..."[6].

Thus, we can argue that the mechanistic inspiration and a certain formal imagination proved, in the first phase of megastructure, clearly recognizable in the huge collages that tend to demonstrate the architectural ideas using technical illustrations worked at different scales, offering amazing urban images, which were evident both in generic aerial views as in the detailed design of small units. Significant, is that these small units tend to be presented as possible assemblies of components of a huge machine, that could be inserted in the general context of the proposal, giving strong, technical and realistic appearance.

Despite the interest initially developed around analogies between the machine and the city, the development of the concept of megastructure would however focus its investigation not so much in the exploration of this concept—or visual image—compact container identifiable, but towards the discovery of the internal organizational structures inherent to various bodies. Indeed, reflecting a broad development in different areas of knowledge, the research carried out by mega-structuralist movement would advocate an investigation operating an unconstrained urban aggregation, an aggregation of functional elements, operated not so much by a bounding entity but through intentional exposure of new spatial and dynamic aggregation, unifying its various elements:

> [...] Well before the first manned flights, von Braun and Bonestell had driven the sleek, projectile-shaped

Figures 23, 24, 25. Soleri in front of a supersonic jet-port and city that might be built on the mud flats of New Jersey 'Arcology' (architecture plus ecology), 1969; Hans Hollein, section and photomontage of "Aircraft Carrier Projects", 1964.

Figures 26, 27, 28, 29. Space station McDonell Douglas phase 2, 1970; Jean Jaques Cousteau, "Pre-Continent II", Port Sudan—Red Sea, 1963; Cover of "Space Oddity" the David Bowie, 1969; cover "Life" magazine, special edition, 1969.

spaceship image out of currency and replaced it by images of open frameworks filled with spherical pressure-vessels, command modules, and power packs. When Armstrong set foot on the moon, the module from which he descended must have looked as familiar as home to the megastructure generation..."[7]

The fascination developed around the concept of the 'transatlantic' as an universe of analogies (enunciated by Le Corbusier during the golden age of the Modern Movement), which resulted in an image characterized by the compact nature of a form, would during this period, result in the

influences of scientific and technological world, progressively replaced by a new sensibility which implied recognition of the constitution principles of complex structures that made humanity's new universe. A new kind of aggregation and visual articulation conform now the design of complex structures of space projects, characterized by a clear hinge of autonomous components coupled in a visually expansive composition.

Thus, according to the new spirit of the times, the concept of megastructure would be, at this stage of evolution, associated with this new sense of diverse and unconstrained aggregation:

[...] As against the international style's classicizing view of technology and machinery as neat smooth regular solids of anonymous aspect, the younger mega-structuralists clearly saw technology as a visually wild rich mess of piping and wiring and struts and cat-walks and bristling radar antennae and supplementary fuel tanks (...) all carried in exposed lattice frames, Nasa-style..."[8].

Paradigmatic was the conquest of new spaces or new territories, shown in the generic sense of a new exploitation of oceans and their consequent potential, as well as attending to the beginning of the exploration of new territories advocated by space projects, which would definitely marked the arrival of humans on the moon.

In a further move to this achievement of macro-scale, this prospect towards a journey to distant areas of space, scientific research developed also towards the discovery of micro-scale, towards the unveiling of physical structures of various bodies and towards an understanding of structure and metabolism of living organisms. These researches were also responsible for a new vision and understanding of the universe, with consequences in the form of projecting urban and architectural structures. In fact, the daily view of this extraordinary set of achievements and transformations would eventually mark the emergence of a new thinking on the different spatial supports of modern people and the specific needs of the post-war society lifestyle. In fact, despite the intrinsic demand for a greater spontaneity and a possible involvement of individuals in shaping their own habitat, the generic and utopian development of the concept of megastructure, as advocated in numerous architectural projects, would be closely connected with the desire to create a new order or principle for the city. Away from a primary and purely mechanistic concept of order, generated simplistically by economic activities or traffic conditions, the concept of megastructure seek to establish itself essentially as a new general order or principle intelligible according to a true architectural dimension.

Figures 30, 31, 32, 33, 34, 35. Kiyonori Kikutake, "Marine City"—perspective and studies for mobile houses, 1958–1963; Kisho Kurokawa, growing sketch's for "Metamorphosis 65", 1965; Kenzo Tange, Tokyo Bay project, 1960.

Undeniably, the events that marked the post-war made it a particularly optimistic moment, but ambiguously also critical regarding the question of society's behavioural norms and values. Consequently, reflecting the decisive changes underlying it, this time change respecting the understanding not only of the problems inherent to the functionalist city, model that had driven the major reconstruction initiatives of post-war urban structures, but also assisting the dissolution of a specific and topic sense of city. Specifically, the previous findings can explain that, after a first exploration period, primarily associated with a visual universe linked to the image of the machine and a certain fascination for a technological universe, the projects tried to attend an important paradigm transition, that would search for a generic and radical utopian understanding of the city assumed as living organism, expression that would be materialized in a diverse set of proposals for new structures of contemporary dwelling.

NOTES

[1] Team X, "Instruction to Groups, Dubrovnik 1956", in Joedicke, "Documents of Modern Architecture: CIAM '59 in Otterlo", p. 14; in Sarah Deyong; "Memories of the urban future: the rise and fall of the Megastructure"; in AA.VV. "The Changing of the Avant-Garde: Visionary Architectural Drawings from the Howard Gilman Collection"; ed. The Museum of Modern Art; New York, 2002; p. 25.

[2] Supported by an article published by Fumihiko Maki and Masato Ohtaka in spring 1960 under the title "Group-Form", the concept of 'Collective Form' developed a more significant systematization in the work of Fumihiko Maki," Investigations in Collective Form" (ed. Washington University orig;. St. Louis 1964). Realizing the need, in methodological research, not only of a critical observation and analysis as well as a use of strategic projective instruments, the previous work is supported by three types of systematic conceptual approach, the concepts of 'compositional form', 'mega form' (or megastructure), and 'group form'; "... We become accustomed for so long to design the buildings as separate entities that currently suffer from a mismatch in spatial design languages, able to create meaningful environments. (...) collective form' represents groups of buildings—segments of our cities. 'Collective form' is however not a collection of separate buildings unrelated, but rather a set of buildings with motifs to be together..." (Fumihiko Maki, "Investigations in Collective Form"; Washington University, St. Louis, 1964, p. 5); authors note.

[3] Fumihiko Maki; "Investigations in Collective Form"; Washington University, St Louis, 1964; p. 8.

[4] Fumihiko Maki; "Investigations in Collective Form"; Washington University, St Louis, 1964; p. 9.

[5] Ralph Wilcoxon, "Megastructure Bibliography"; in Reyner Banham, "Megastructure, Urban Futures of the Recent Past", Thames and Hudson Ltd, London, 1976; p. 8.

[6] Paolo Soleri; "The City in the Image of Man"; Boston, 1970; p. 14; in Reyner Banham, "Megastructure, Urban Futures of the Recent Past"; Thames and Hudson Ltd; London, 1976; p. 22.

[7] Reyner Banham; *"Megastructure, Urban Futures of the Recent Past"*; Thames and Hudson Ltd; London, 1976; p.22.

[8] Reyner Banham; *"Megastructure, Urban Futures of the Recent Past"*; Thames and Hudson Ltd; London, 1976; p.17.

Viennese utopia—the expression of revolutionary architectural thought

Aleksander Serafin
Institute of Architecture and Urban Planning, Faculty of Civil Engineering, Architecture and Environmental Engineering, Lodz University of Technology, Lodz, Poland

ABSTRACT: The paper concerns the question of utopia in the context of artistic and architectural achievements that were introduced by Austrian neo-avant-garde.

Revolutionary utopianism of the Expressionism initialized a visionary creative activity that is especially involved in social communication. Progressive Viennese community in the 60s and later has continued certain tracks. The key representatives of Austrian neo-avant-garde introduced the architectural thought based on a modern technology. However, their works created a kind of paradox, as some of them presented an architecture as elitist issue. That could be concerned as a negation of primary social values. Particular performative projects introduced by Viennese collectives presented further technological inspirations as a background for autonomous forms that radically opposed the tradition. However, visionary Austrian designers used utopia primarily as the form of expression. Their manifestoes looked at architecture as an expressed opinion.

The both presented myths of "the technology" and "the autonomy" are introduced in this paper as a utopia. However, it should be noted that they are useful for the development of the contemporary thought that refers to a global culture. This solution ensured a wide range of new forms and a sort of resistance against the prevalence of the modernity.

Keywords: architecture, culture, expression, technology, Vienna

1 INTRODUCTION

The standard comprehension of the term "utopia" evokes the vision of the ideal solution that is unattainable. The question of this concept may gain slightly different meanings in various domains. Utopia is inherently transgressive because its essence is based on crossing disciplinary and conceptual boundaries, together with ordering and separating existing modes of thoughts (Howells 2015: 25). However, Nathaniel Coleman maintains that utopia always requires an architectural frame because the problems of evolving form refers to the way people appropriate space (Coleman 2011: 2). Moreover, the issue of utopia can be seen in the context of the entire city. The evidence may be set in Viennese projects by Camillo Sitte and Otto Wagner (Whyte 2004: 286).

Utopia may be considered as an effect of the form concept crisis. Coleman claims that

architectural invention is akin to utopian projection and utopia harbours the potential to rescue architecture from aimlessness, obsessive matter-of-factness, or a non-critical embrace of global capitalism." (Coleman 2005a: 6)

The social factor seems to be fundamental also. Utopia corresponds to impetuous ideals of the Romanticism, but on the other hand, it refers to the social "organic work". Theodor Adorno noticed that most of the fulfilled utopian dreams assumed a peculiar character of the spirit of Positivism (Bloch 1988: 1). According to that point of view, it can also refer to a social as well as an architectural revolution.

2 ESCALATION OF AN EXPRESSION AS A WAY TO THE NEW SOCIAL QUALITY

The great Expressionistic though may be considered as the one responsible for a contemporary appearance of architectural utopia. Originally, Expressionism, like any artistic movement, set the goal to transfer a social content in the form of visual values. A strong underlying political purpose was present in many aspects. On one hand, the beginning of the trend is marked by a search for a new style indirectly referring to the fanciness and the tradition of Gothic, but on the other hand, the leftist movement adapts it. As primary avant-garde of the Eastern Europe was engaged in the creation of revolutionary utopianism (Margolin 1997: 22), also the German-Austrian trend gained subversive character. The belief in creating a better world based on the interaction between artistic and social aspects. Some influential thinkers, like Ernst Bloch presented expressionism as utopianism, which is revolutionary (West 2000: 103), but he proclaimed the new reality supporting by metaphysical rendition. One of his visions is, as he wrote, the "implementation of the central concept of utopia" (Bloch 2000: 3). However, this philosophical background has become favorable conditions for the development of Austrian neo-avant-garde after a few decades.

3 NEW VISIONARY APPROACH

Urban context may be concerned on many levels. The city is the combination of the ever-changing entities transformed by ideological, political, economic, and social factors (Álvarez Layna, Maidana Legal And Vélez Cipriano 2015: 88). That is why the performance can also be seen as a factor of the overall development, including architectural one. First trend in Austrian contemporary art that became famous in the world is a Viennese Actionism. Its role seems to be extremely political and concerned essentially cultural issues in spite of social aspects. However, other movements that were born in Austria at that time proposed different unique visionary solutions.

4 ELITIST VERSUS EGALITARIAN PARADOX

Social associations with utopia represent a sort of paradox, as long as they are associated with the searching for the ideal social system. It seems that such a dispensation should be based on democracy, but on the other hand, leading representatives of this movement enforce an elitist. Hans Hollein, who could be concerned as influential Austrian visionary designer, claimed that the architecture is a matter of elites (Weiss 2011: 267). Together with him, Walter Pichler manifested:

> Architecture is not an integument for the primitive instincts of the masses. Architecture is an embodiment of the power and longings of a few men [...]. It never serves. It crushes those who cannot bear it. Architecture is the new law of those who do not believe in the law but make it. (Pichler and Hollein 1971: 181)

Even the reminiscence of the cult of power in the meaning Friedrich Nietzsche's thought is identifiable. Hollein's graphics "Superstructure above Vienna" from 1960 clearly demonstrates the dominance of designed stone structures over the urban landscape. This project highlights the imperfection of the utopian character of the social challenges set itself by a primary *avant-garde*.

5 UTOPIAN TECHNOLOGICAL THOUGHT AND ASPECT OF SUBCONSCIOUS

Viennese designers introduced also some grotesque values in architectural form, so not only human but even "the machine" could be a new elite for Pichler (Jencks 1987: 66). New doctrine was supported by surrealistic grasp. Renata Hejduk emphasized that

> Hollein went on to state that much of this work could be seen in relation to the strong Freud-based Austrian heritage of operating psychologically on the individual user's mind and consciousness as well as on the immediate environment while introducing ideas of individuality. (Hejduk 2006: 45)

Processes such as rest and action had to be related to a new technology.
Manfredo Tafuri wrote that

> The presentation of the historical *avant-garde*, and in the 1960s and 70 s of the neo *avant-garde*, of presenting work as a critical experimentation of the articulation of the language, should thus be measured against the reality of concrete and productive work on the new possibilities of programed communication. (Tafuri 1976: 161)

Selected projects by Viennese collective Haus Rucker Co (Fig. 1) are the expression of both the technological mind and the interest in sub consciousness of an individual. Projects of this team were "adequate medium for easy and provisionary, transistor, and flexible architecture" (Weibel 2005: 572). However, the manifesto "Absolute Architecture" seems to be more radical. Pichler claims: "machines have taken possession of it and human

Figure 1. Haus-Rucker-Co (Laurids Ortner, Günther Kelp, Manfred Ortner), "Environment - Tranformer". Source: http://www.ortner.at (used with the consent of Ortner & Ortner Baukunst).

Figure 2. Hans Hollein, "Media Linien", Munich, Germany. Source: phot. by Aleksander Serafin ©.

beings are now merely tolerated in its domain" (Pichler and Hollein 1971: 181), but at the same time Hollein explains: "we make an architecture that is not determined by technology but utilizes technology, a pure, absolute architecture" (Pichler and Hollein 1971: 182).

Project designed by Hollein that represents the supremacy of a modern technology is Media Linien (Fig. 2). It consists of the long pipes in different colours that run through the Olympic Village in Munich. First, it was designed as the media and orientation system. This precast structure could have been built instantly. The utopianism lies in the fact that today the structure fulfils its role only in limited range. Despite it, Hollein as well as Pichler and Haus Rucker Co had been one of the earliest importers into architectural thinking of Information Age. They were able to predict the impact of technology on the contemporary culture. The success of all mentioned designers demands lays in the fact that an architecture was treated as a sort of criticism. At the same time, it was expressive design for the experimental and utopian city environment.

6 ADDITIONAL MYTH OF AUTONOMY

Utopian Austrian Neo-expressionism assumed the possibility of creating an autonomous work. Andrew Benjamin noticed, "autonomy should not be understood as involving architecture's separation from the social or the political" (Benjamin 2007: 40). Instead, autonomy could skip the cultural context and focus on the work itself. Evaluation of architecture would be the result of its own internal operation and therefore in terms of its

own self-conception (Benjamin 2007: 41). The very expression of the work could replace the symbolic values.

Two performative projects called Hard Space and Soft Space by the Austrian collective Coop Himmelb(l)au may be therefore considered as an introduction of the contemporary autonomic architectural form. The phenomenon is based on the complementarity of the two separate and different experiments (Serafin 2014: 157). In the field of architecture, this concept is extended to the meaning of a futuristic vision that is entering into discussion with its context. Next project called Restless Sphere refers to a form that protects against the environment rather than adapt to the surrounding (Fig. 3). Coop Himmelb(l)au claims:

The more constricting the demands on architecture become, the tougher architecture has to respond. Till it breaks. (Prix 2005: 40)

The artistic revolution done by the collective was directed against "political incongruous aesthetics" (Prix 2005: 40). Recently Wiltrud Simbürger noticed that the pneumatic architecture is in relation to the new political utopianism of that time (Simbürger 2015: 159).

The autonomy and independence of established symbols seems to be a directive also for other architectural group called Missing Link. The team represented by Adolf Krischanitz, Otto Kapfinger and Angela Hareiter challenged with broad spectrum of activities, inter alia action-art, where all

Figure 3. Coop Himmelb(l)au, "Restless Sphere". Source: phot. by Peter Schnetz © (used with the consent of Coop Himmelb(l)au Wolf D. Prix & Partner).

Figure 4. Adolf Krischanitz, Kunsthalle Wien, Vienna, Austria. Source: phot. by Aleksander Serafin ©.

Figure 6. Ortner & Ortner Baukunst (Laurids Ortner, Manfred Ortner), Museum Moderner Kunst Stiftung Ludwig Wien (MUMOK), Vienna, Austria. Source: phot. by Aleksander Serafin ©.

7 CONCLUSION

Edward Rothstein wrote: "utopia is not an impossible place, or at any rate, it is generally not supposed to be" (Rothstein 2003: 3). Controversy is an integral part of that language of forms. That is why it is difficult to consider the topic in the category of "positive-negative". According to Lars Lerup, influential Tafuri's conclusions on utopia were full of despair (Lerup 2001: 22). However, the utopia is an integral part of the architecture. It is also necessary for the development of global culture. The portrayed myths of the technology as well as the autonomy turned out to be a utopia. A controversial issue that refers to it is the extreme approach of a so-called "all or nothing". Nathaniel Coleman wrote: "utopia turns mean, pathological, when the model of superior situation, which it puts forward, must be fully realized" (Coleman 2005b: 2). Visionary Austrian architects used utopia primarily as the mean of expression. What is more, they used it as an indispensable tool for the creation of new spatial forms. They were in opposition to the formula of modernism that turned out to be a doctrine, which is a menace to the diversity of artistic language. It could be said that neo-avant-garde was not so radical in creating utopia like the modernists. Although "by the early twenty-first century we can still see the ultimate legacy of modernity as part of a strongly ideologically loaded discourse" (Álvarez Layna, Maidana Legal and Vélez Cipriano 2015: 88), the tracks developed by neo-avant-garde are recognizable after all. What is more, they are internally different.

Figure 5. Haus-Rucker-Co (Laurids Ortner, Günther Kelp, Manfred Ortner), "Oase Nr. 7", Hamburg, Germany. Source: phot. by Dennis Conrad (Creative Commons License).

the media were employed equally (Weibel 2005: 574). Multidisciplinary experiments eventually led to architectural minimalism. However, the aspect of a technology has been treated only superficially. Kunsthalle Wien may be considered as an example of this approach (Fig. 4).

Missing Link's architecture reduced a form to the level of geometric minimum that was necessary for the preservation of its expression, but the latest works of other movements' participants gained a completely different character. For example, Coop Himmelb(l)au has some projects that referred to the original performance with spherical forms but largely their autonomous architecture was built based on deconstruction, what could be regarded as a direct reminiscence of Expressionism. It should be noted, however, that Haus Rucker Co was able to save the sphere theme. Not only the exhibition at the Museum of Arts and Crafts (Fig. 5), but also the projects like the building of the art gallery MUMOK takes the advantage of spatial model developed earlier (Fig. 6).

BIBLIOGRAPHY

Álvarez Layna, José Ramón, Maidana Legal, Andrés, and Vélez Cipriano, Iván. "Owenite architecture and urban rationality: notes and reflections on class and utopia." In *The individual and utopia. A multidisciplinary study of humanity and perfection*. Eds. Jones, Clint and Cameron Ellis. Farnham: Ashgate, 2015. 85–98.

Benjamin, Andrew. "Passing through deconstruction. Architecture and the project of autonomy." In *Critical Architecture*. Eds. Rendell, Jane, et al. Abingdon: Routledge, 2007. 40–47.

Bloch, Ernst. *The spirit of Utopia*. Stanford, Calif.: Stanford University Press, 2000. ISBN: 0-8047-3764-9.

——. "Something's missing: a discussion between Ernst Bloch and Theodor W. Adorno on the contradictions of utopian longing (1964)." In *The utopian function of art and literature. Selected essays*. Ed. Bloch, Ernst. Cambridge The MIT Press, 1988. 1–17.

Coleman, Nathaniel. *Imagining and making the world. Reconsidering architecture and utopia*. Vol. Peter Lang. Oxford, 2011. 349 p. ISBN: 3034301200.

——. *Utopias and architecture*. Abingdon: Routledge, 2005a. 352 p. ISBN: 0-415-70085-1.

——. *Utopias and architecture*. Abingdon: Routledge, 2005b.

Hejduk, Renata. "A generation on the move: the emancipatory function of architecture in the radical avant garde 1960–1972." In *Transportable environments*. Ed. Kronenburg, Robert. Vol. 3. Abingdon: Taylor & Francis, 2006. 40–52.

Howells, Richard. *A critical theory of creativity. Utopia, aesthetics, atheism and design*. 1st ed. London: Palgrave Macmillan, 2015. 216 p. ISBN: 978-1-137-44616-9.

Jencks, Charles. *Ruch nowoczesny w architekturze*. Translated by Morawinska, Agnieszka; Pawlikowska, Hanna. Warszawa: WAiF, 1987. 554 p. ISBN: 83-221-0225-9.

——. *Modern Movements in Architecture*. Penguin Books, 1985.

Lerup, Lars. *After the city*. 2nd ed. Cambridge: The MIT Press, 2001. ISBN: 0-262-62157-6.

Margolin, Victor. *The Struggle for Utopia: Rodchenko, Lissitzky, Moholy-Nagy, 1917–1946*. Chicago: The University of Chicago Press, 1997. 267 p. ISBN: 0-226-50516-2.

Pichler, Walter, and Hollein, Hans. "Absolute architecture." In *Programs and manifestoes on 20th-century architecture*. Ed. Conrads, Ulrich. 3rd ed ed. Cambridge The MIT Press, 1971. 181–182.

Prix, Wolf Dieter. "The tougher architecture, 1980." In *Get off of my cloud. Wolf D. Prix. Coop Himmelb(l)au. Texts 1968–2005*. Eds. Kandeler-Fritsch, Martina and Thomas Kramer. 3ª ed ed. Stuttgart: Hatje Cantz, 2005. 40–41.

Rothstein, Edward. "Utopia and its discontents." In *Visions of utopia*. Eds. Rothstein, Edward, Herbert Muschamp and Martin E Marty. Oxford: Oxford University Press, 2003. 1–28.

Serafin, Aleksander. "Expression as the basis for redefining order of the form in the urban space." In *Space reloading*. Ed. Gibala Kapecka, Beata. Vol. 2. Cracow: ASP, 2014. 153–165.

Simbürger, Wiltrud. "Utopia's Bubbles: Pneumatic Architecture of the 1960s and 1970s as a Vehicle for Urban Exhibitionism." *Architecture and Culture* 3.2 (2015): 2013–2015. ISSN: 2050-7828.

Tafuri, Manfredo. *Architecture and utopia: design and capitalist development*. Translated from the Italian by Barbara Luigia La Penta. Cambridge, Mass.: MIT Press, 1976. xi, 184 p. ISBN: 0262200333.

Weibel, Peter. *Beyond Art: A third culture: a comparative study in cultures, art and science in 20th century Austria and Hungary*. 2th ed. Wien: Springer, 2005. 617 p. ISBN: 3-211-24562-6.

Weiss, Klaus Dieter. *Workmanship. Working philosophy and design practice 2000–2010*. Basel: Birkhäuser, 2011. 327 p. ISBN: 978-3-0346-0481-9.

West, Shearer. *The visual arts in Germany. 1890–1937. Utopia and despair*. Manchester: Manchester University Press, 2000. 243 p. ISBN: 0-7190-5279-3.

Whyte, Iain Boyd. *Modern German architecture*. The Cambridge companion to modern German culture. Eds. Kolinsky, Eva and Wilfred Van Der Will. 3rd ed. Cambridge: Cambridge University Press, 2004. 282–301.

Lemons into lemonade: Materializing utopian planning in Providence, Rhode Island (RI)

Carlos Balsas

Geography and Planning, University at Albany, Albany, NY, USA

ABSTRACT: Utopian planning is uncommon. Powerful and long-lasting urban visions are difficult to conceive. Fruitful collaborations are required to implement components of utopian visions. The US city of Providence, Rhode Island (RI), has redeveloped its downtown over the last three decades in hopes of making the city more attractive and competitive. The city's urban renaissance was envisioned by, among others, Mayor Vincenti Cianci and accomplished through the collaborations of numerous stakeholder agencies and individuals. How can community members work together to materialize a revitalization vision? In this paper, I analyse three propositions on utopian and collaborative planning and place making. I argue that the scale of the revitalization strategy, the individual commitment and the responsibilities of various stakeholders influence the results of collaborations to a great extent. The Providence case demonstrates that coordinated planning and continued efforts are required to materialize the utopian vison of a revitalized city.

Keywords: Utopian planning, urban revitalization, public-private partnerships, territorial equity, Providence, RI

1 INTRODUCTION

Recently, there have been extensive efforts at understanding and improving collaborative planning within contexts of globalized and neoliberal uncertainty (e.g. Angotti 2008; Healey 1997; Short 1996). Many of these efforts have revealed the existence of formal and ad-hoc partnership arrangements involving multiple people and entities (Pierre 1998; Walzer and Jacobs 1998). In this paper, I use a qualitative storytelling approach to research the reasons why different socio-economic and geopolitical sectors work together to define, promote, and implement urban revitalization strategies.

How can these different sectors with perhaps competing aims and opposite rationales collaborate in the realm of city revitalization? At first, attempting to join these two sectors seems almost like a lost cause. However, there are reasons for both the public and the private sectors to work together in order to attempt to resolve common problems. I argue that effective city revitalization partnerships need the support and expertise of both sectors, and that there are reasons for the success (and potential failure) of these collaborative arrangements (Fig. 1).

The US City of Providence, Rhode Island (RI) is utilized to reflect on the critical elements of collab-

Figure 1. 5th Century A.D. Mediterranean Mosaic. (Author's archive).

orative planning. Recently, Providence has implemented a series of audacious planning projects that have led to its internationally acclaimed "urban renaissance" (Leazes and Motte 2004). With a collective vision and a very strong mayoral leadership, the City of Providence was able to establish collaborative partnerships with the state's economic, social, cultural, and environmental agents to implement its vision of a revitalized city. This paper is a hybrid and methodologically eclectic piece between an ethnographic and a policy evaluation study. I attempt to not only describe and understand a social and organizational phenomenon, but above all, I strive to reach findings useful to decision-makers and scholars. The methods used in this research included formal and informal interviews, meeting observations, literature reviews, extensive document analyses, and discussions with scholars and practitioners.

This paper has three parts. The first part reviews the basics of utopian, collaborative planning, and place making, while clarifying the role of revitalization partnerships, mostly in the United States. The second part introduces the city of Providence and presents the research findings at the state, city and neighbourhood levels. Themes range from public participation in collaborative processes and the difficulties in establishing and sustaining partnerships, to the effectiveness of their outcomes. Part three closes with some concluding remarks.

2 UTOPIAN, COLLABORATIVE PLANNING AND PLACE MAKING

Utopian planning utilizes extraordinary visions of distant imaginary places and social arrangements to stimulate collective memories (Pinder 2002). Usually, one person conceives such visions. Sir Thomas More's (1516) *Utopia* and Edward Bellamy's (1888) *Looking Backward* are only two of a handful of idealized places and societies. Furthermore, powerful visions also have important physical and moral dimensions. Architects and designers have conceptualized visions of the good city, mostly in physical terms. Examples include Ebenezer Howard's "garden city", Frank Lloyd Wright's "broadacre city", and Le Corbusier's "*ville radieuse*" (Fishman 1982/1994). Friedmann (2002) has written about the good city in the context of human flourishing and of a fundamental set of rights, freedoms, and responsibilities, such as the right to shelter, fair employment, and decent living conditions, as well as service to one's community. Central to both physical and social conceptualizations are responsible global city governance arrangements, to which Barber (2013) has alluded to in "If Mayors Ruled the World".

The incremental implementation of utopian visions requires a whole array of collaborations. Those collaborations are best conceptualized as a group of individuals or organizations working together to resolve an issue at interdependent levels. City revitalization schemes aim to enhance the social, political, physical, and economic value of a certain geographical area of a city. Public-private

partnerships for city revitalization are not easy arrangements to establish and maintain, particularly due to their high number of players, and quite often, competing aims and operational philosophies. Table 1 shows a set of principles and puzzlements about collaborations.

It is generally assumed that the public sector is bound by "social rationality", which aims to improve public goods for all, and that the private sector operates with a "market rationality", which aims to maximize profits for few individuals or organizations. Public sector organizations are generally slow moving, rigid and with restricted resources for the numerous investments they are accountable for. On the other hand, private sector entities can be flexible and efficient because their decision-making processes are not subject to the same level of public scrutiny. In this sense, partnerships are effective mostly when both sectors are able to move in the same direction (Paumier 2004). Even if each sector commits different efforts, the initiative remains only as long as a certain balance is maintained.

Place making comprises mostly a set of urban design techniques aimed at improving the physicality of specific places (Bunnell 2002; Castello 2010; Lang 1994). Augmenting the innate characteristics of places, either natural and or man-made and turning them into amenities for human enjoyment

Table 1. Fundamental principles and doubts of collaborations. (Author's conceptualization).

	Principles	Questioning the assumptions
Political	Create political influence	Will the public or the private sector set the agenda?
	Create support networks	What and whom are these networks formed to support?
	Enhance credibility	Whose believability?
Economic	Encourage cost sharing	Who benefits from the arrangements?
	Leverage scarce resources	Where are funds to come from?
	Increase efficiencies of service delivery	How is efficiency to be measured?
Operational	Develop collaboration and teamwork	Who constitutes the team?
	Help build capacity	With what purpose?
	Get activities done	What activities and who decides on the priorities?

requires knowledge, vision, and resources (Bohl 2002; Rossman and Rallis 1998). Recent place making attempts with traction include a myriad of urban regeneration programs all over the developed world and the new urbanism movement in the United States (Duany, Plater-Zyberk and Speck 2000).

3 ROGER WILLIAM'S, CIANCI'S OR TAVERA'S PROVIDENCE?

Providence will forever be associated with Roger Williams, the founder of the Rhode Island colony in 1636—possibly inspired by More's utopia written more than a hundred years earlier and the puritans' search for religious freedom in the new world. Fig. 2 is a triptych depicting Providence's evolution from its riverfront land plotting inception in the mid-17th century to the 19th century settlement structure already with segments of the river's embankments protected by landfills. Why study the city of Providence? The city of Providence was chosen because it has been subjected to a range of planning projects, which have contributed to its internationally acclaimed urban renaissance. In fact, Providence, a city of approximately 150 thousand inhabitants, has been transformed in the last three decades from a somewhat depressed community to a revitalized city with a wide panoply of arts, culture and consumption oriented destinations.

The state's size and the metropolitan area's 1.6 million people contribute to its abundance of activities, strong commitment to historic preservation and quality of life, to the point where Providence was even distinguished with the "City Livability Award" by the United States Conference of Mayors in 1994. The riverfront acreage of real estate in the newly redeveloped central area is particularly impressive when thinking about the New York City—Boston corridor (Fainstein 1994).

Before the shopping malls were built in nearby suburbs, downtown Providence was the only place to shop, visit restaurants and find entertainment activities (Bunnell 2002). As in other cities in the late 1950s and early 1960s, much of Providence, especially downtown, was eroded, erased, or transformed by the crashing waves of urban renewal (Frieden and Sagalyn 1989/1991; Hall 1995; 2002). Although Providence was founded and prospered due to its waterfront location, the city turned its back on the waterfront in the 1950s. The city literally covered over part of its waterfront with parking.

By the 1980s, suburbanization and economic decline had closed all of the downtown's major retail stores (West and Orr 2003). However, Providence

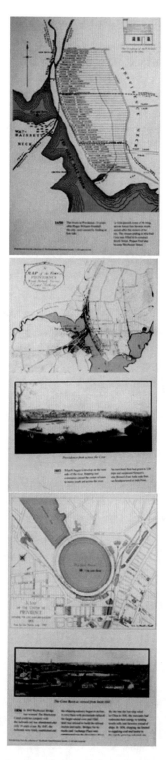

Figure 2. Providence's Triptych 1650–1803–1856. (Images of a mural in downtown Providence, author's archive).

has been transformed. Massive redevelopments have taken place on the waterfront in the 1990s and early 2000s. The historic retail core has undergone a transformation into a somewhat vibrant arts and entertainment centre with an emerging residential community while retaining its very special character (Birch 2009; Bunnell 2002).

Fig. 3 shows downtown providence and its public works projects in 1989 with the confluence of the Moshassuck and Woonasquatucket rivers to form the Providence River, together with the newly landscaped public spaces Waterplace Park and Riverpark bordering the state capitol (Fig. 4).

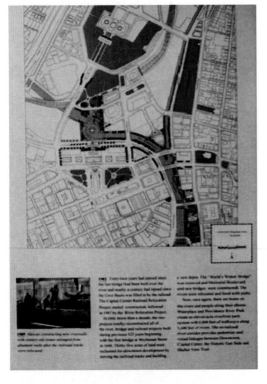

Figure 3. Board of Downtown Providence in 1989. (Mural in downtown Providence, author's archive).

Figure 4. Providence Waterplace Park in 2000. (Author's archive).

The Riverwalk Park is a linear stretch, which goes from the Route 1 and Park Street bridges to Crawford Street Bridge. The park by the Rhode Island School of Design (RISD) is the most landscaped stretch of the downtown Providence river greenway. Multiple public art sculptures have been placed in Memorial Park to celebrate the city's cultural history.

With perhaps the exception of Vancouver, the Capital Centre/River Relocation Project has been recognized as one of the most audacious downtown waterfront undertakings anywhere in North America (Punter 2003). Its concept was to remove the decking that covered the Providence River and to make the waterfronts available for public enjoyment. Along the reopened rivers, a number of commercial projects have also been built. Apartment buildings were built nearby, tourism related activities were moved into the former railroad station area and the new convention centre and hotel is pulling visitors to the area. Together these investments have spurred a new regional shopping mall on the edge of the CBD. Providence is believed to be now a completely different city. A city that, with a collective vision and a very strong leadership, including that of a half-dozen terms mayor Vincent Cianci, was able to establish various collaborative partnerships (Stanton 2003).

Since then the daylighting of rivers and streams has been emulated in other US cities, such as Yonkers—NY and Kalamazoo—MI, and abroad in as faraway places as Seoul, South Korea (Lee and Anderson 2013). The apparent success of the newly created public spaces is visible in the animation programs and artistic installations on the embankments and in the rivers themselves, which have attracted a high number of visitors, especially late in the afternoon and at dusk on weekend summer nights. Barnaby Evans' Waterfire installation was a particularly popular attraction (Ellin 2006). Angel Taveras's mayoral term was initiated in 2010, giving rise to a stronger emphasis on multiculturalism and ethnic diversity issues in the city (Orr and Nordlund 2013; Sandercock 1998). Also recently, additional office and condominium towers were built on the waterfront, which finally may have spurred the renovation of the old monumental Mason Building—left unfinished since the depression era—into an upscale hotel.

4 PROPOSITION I: STATE AND CITY LEVEL PARTNERSHIPS

Partnership arrangements at the state and city levels are usually top-down efforts between agencies. Partnerships are commonly recommended economic development strategies, even when they are not utilized to the full extent of the law. Plans and scientific studies present the virtues of collaborative planning among city and state agencies. However, the implementation of partnerships usually proves difficult and with a greater number of stumbling blocks (Ryan 2006). During my interview with a local economic development specialist in the early 2000s, Astride Stevenson (pseudonym), three themes emerged related to public participation in urban contexts, the effectiveness of collaborative outcomes, and the value of partnerships.

The first theme was the need to include people of limited resources and of traditionally segregated ethnic groups in open and democratic planning processes. When I questioned Stevenson about the advantages of public participation within partnership arrangement, she replied:

> Public participation is important, particularly to define the major objectives of a certain program, but many times it is a waste of time and money because we could have just come up with those objectives or plans by ourselves in the office without spending hundreds of dollars hiring people to coordinate the meeting and mediating conflictual interests.

Another interesting perspective arose when I asked her about how easy it was for her to engage other organizations in the creation and implementation of partnerships, to which she replied that community leaders very often see her organization as the "mean to access money" and they tend to cooperate easily (paraphrased answer).

On the other hand, when I asked her to reflect on the outcomes of the revitalization programs she developed and implemented in the past, she answered quite pragmatically: "I believe we were successful in turning lemons into lemonade" (direct quote).

Among other ideas, Stevenson also shared with me

> How boring it was to go to certain meetings at night, after a day of work, and coordinate partnership arrangements with people that were not interested in being there at all

Finally, when I asked her to comment on the following statement:

> Working together to maximize community and regional resources is the most efficient use of time and materials in any scheme", she replied promptly: "no! It is not true! Collaborative planning is difficult and not efficient at all.

5 PROPOSITION II: LOCAL LEVEL PARTNERSHIP ORGANIZATION

At the sub-municipality level, I conducted an interview with the director of a local neighbourhood organization created in 1995 that aimed to improve an area centred on an important avenue which connects downtown Providence to the suburbs and adjacent towns in the metropolitan area. The office was located in a renovated building centrality located in the neighbourhood. My interview with Ana Lopez (pseudonym) was very promiscuous with themes ranging from the type of daily work performed, to the difficulties in establishing and maintaining collaborative processes, and the value of partnerships.

In my interviewee's opinion, one of the major problems to implementing revitalization programs and to obtaining more resources for her organization, and consequently to the neighbourhood, was the almost powerless neighbourhood political recognition of the area. In her words, she said:

> This is very political, this is a very small community but it is very political, what they [the politicians] look at is the voting power of the community. There is a large Latino-American population and they are very aware of what is going on here. You have a lot of people that are not registered to vote, only about two per cent of Latino population are registered to vote and this also affects how much interest they devote to the neighborhood.

I immediately related this statement to my prior knowledge of politics in the city and Mayor Cianci's strong leadership, which had unveiled an impressive downtown transformation. I also reflected on what had been happening in other legacy cities, where most of the political attention, and financial resources, in cities such as Philadelphia, Pittsburgh, and Buffalo, was being devoted to the central cores and not to their inner-rings, often decaying, and dilapidated neighbourhoods, mainly because of their identities. Usually, downtown revitalization has had priority because of symbolic power, if no longer due to their local economies (Marshall 2003).

Another theme that emerged during the interview with Lopez was the difficulty in establishing and maintaining collaborative processes. At the centre of a partnership, arrangement there needs to be a clear and achievable purpose. However, this implies, as Lopez so well put it, "identifying the issue and how to implement it, how the outcome is going to be channelled, and who is going to be responsible for it". This finding demonstrates that collaborative efforts are not easy, but if they are to succeed, more workable and consensual solutions need to be found. Besides identifying issues and reaching consensus, unresolved conflicts among partners seem to have an important role in shaping the outcomes of collaborative processes. Again, in Lopez's words: "there are a lot of hard-headed people, to say the least, and there are a lot of egos, bumping heads to each other". These are just some other examples of unforeseen conditions that can limit partnership accomplishments.

The final theme to emerge from my interview with manager Lopez was the value of partnerships. Several scholars have characterized the planner as an enabler, a facilitator or mainly a technical advisor helping others to identify their problems and to create their own solutions (Cullingworth and Caves 1997/2009; Kelly and Becker 2000). This seems to be also applicable to local level partnerships. As long as elected officials and interest groups are committed to working together on resolving common problems and the possibility of duplication and costly competition is reduced and resources are shared, the various benefits of collaborative processes are likely to be achieved (Barber 2013). Lopez confirmed this belief in local knowledge with the sentence: "community process is slow but very effective".

6 PROPOSITION III: EQUITY AT THE CORE OF URBAN GOVERNANCE

Social equity and social justice are noble goals in most planning literature, if not well treasured life guideposts and even universal and philosophical humanistic beliefs (Mullin 2003). When economic development specialist Stevenson faxed me a notice of a neighbourhood meeting in the City of Providence, I did not believe it to be relevant to my research. Its location outside of downtown Providence made me question the various items on the agenda. However, several expressive terms, such as revitalization, commercial, and neighbourhood needs and improvements resonated with me and I attended the meeting. In fact, this was an appropriate introduction to the planning dilemma of "territorial equity" in allocating investment and fair treatment between downtown and the "rest of the city".

The meeting took place at the West End Community Centre, a structure six or seven blocks southwest from where I had my interview with manager Lopez. The room was similar to a large cafeteria with a seating capacity of approximately hundred twenty (Fig. 5). The first person whom I met was a young, friendly, Caucasian man, Robert Windsor (pseudonym), the director of a Community Development Corporation (CDC) responsible for an organic farm and various urban gardens, mostly in abandoned lots.

Figure 5. Small Business on the West Side. (Author's archive).

The meeting was chaired by a city planner, Elise Guilot (pseudonym), who started the meeting by explaining its purpose and saying that it was the first time that the City of Providence was conducting a focus group in order to help improve the neighbourhood. Then she introduced the consultant, Roberta Katz (pseudonym), and informed meeting attendees, that independent consultant Katz was going to conduct the meeting to reduce any potential conflicts of interest in the public participation process (Smith 1961/2000). Planner Guilot started by briefly reviewing the agenda, which had been distributed previously to participants at the beginning of the meeting.

Three major themes surfaced during the meeting: territorial equity among neighbourhoods in the city, collaborative planning, and obstacles and priorities in neighbourhood revitalization.

Regarding the first theme, it is important to clarify what many participants seemed to have realized at the meeting; the "leveraging potential" of funding which depends of where and when the money is invested. For instance, one million dollar spent downtown is different from the same amount invested in another neighbourhood. Downtown was perceived to be capable of, through revitalization partnerships and other types of collaborative arrangements, leveraging additional investments.

The need to avoid competition for scarce public funding and to collaborate in order "to get things done" was a significant belief espoused frequently during the full duration of the meeting.

However, at a certain point, the director of a CDC devoted to neighbourhood revitalization, *Jean Strong* (pseudonym), voiced the following comment: "collaboration kills me, with so much collaboration we can't do anything!" This statement was important to reflect on the advantages, limits and the precepts of successful collaborative planning. Fortunately, this is not the case in all collaborative arrangements. Nonetheless, this is something that participants need to be aware of before committing to any partnership arrangements. In other words, this proposition also substantiates that it is frequently easier to diagnose problems, than to agree on the strategies to eradicate them.

This is a brief list of obstacles and difficulties identified by meeting attendees in order to revitalize the neighbourhood:

- "scarce financial resources and technical assistance in the city to help the CDCs" said Strong,
- "the fact that even minority middle class professionals leave the city to the suburbs because they do not want to raise their kids in a possible environment where there is easy access to drugs" said *Michael Elm* (pseudonym),
- "absentee landowners" mentioned one resident,
- "land and housing costs" stated another resident,
- "building and fire codes contribute to expensive housing mortgages and insurances costs" further explained an African-American resident.

The following were among the top priorities for the neighbourhood:

- "we need more comprehensive planning, we need a holistic approach!" Windsor said,
- "I want to know more about the history of my street" said the African American resident, explaining that "it brings respect… to the neighbourhood and creates a sense of place",
- "revitalize the main streets," suggested the planner, immediately following up with the argument that "we need more groceries and small shops, don't you think!?" and the enumeration continued…

7 THE LIMITS OF UTOPIAN PLANNING: LEMONS INTO LEMONADE, ANYONE?

The initial motivation of my research was to examine the reasons why and how administrations intervene in city centres (Redstone 1976). As the research progressed, I expanded my initial questions to include also issues of territorial equity and social justice.

This paper has three main sets of propositions. The first concerns state and city level partnerships. The second sheds light on the role of local level partnership organizations in advancing urban utopias (variously conceived). The third set of propositions is centred on the question of territorial equity, to which unplanned grassroots community gardens, family owned grocery stores, and small viable businesses, are critical components (Carlsson 2008).

Similar to processes described by Angotti (2008), and after listening on the news that New York City wanted to sell community gardens—cared for mainly by Hispanic individuals—to real estate investors, I found myself reflecting on what I had observed recently in a meeting in a poor part of Providence. The director of a non-profit organization and several other meeting participants vehemently conveyed the dilemma of preserving neighbourhood community assets for collective use to planners. Community planning is conducive to utopian planning so long as physical improvements are complemented by social and moral conceptions of the good, *sine qua non,* the just city (Barnett 2003; Friedmann 2002).

In these concluding remarks, I reflect on the value of leadership for the implementation of urban utopias (Ben-Joseph 2005; Olsen 2003). Public and private sectors are known to have different, in many cases diametrically opposed goals, procedures and rationalities.

By their very own nature, opposites are extremely difficult to mix. For instance, water and oil are immiscible—meaning that they cannot be easily mixed. History has shown that the same may occur with public-private partnerships in cases where partner organizations lack a clear purpose, cannot agree on a unified vision, in themselves tend to commit non-equitable resources to the collaboration, have a history of prior conflicts, share information unwillingly, or finally when there is minimum interaction among community leaders (Pierre 1998; Walzer and Jacobs 1998).

Partnerships are collaborative efforts with binding responsibilities for both the public and the private sectors. In this sense, collaborations can only succeed when representatives from both sectors are able to agree on how to materialize the urban utopia. Each sector can devote and commit different human and financial resources; however, the effort works only as long as a certain equilibrium is maintained. In synthesis, the tenets of successful partnerships involve broad membership mirrored in broad constituency of stakeholders that are likely to create utility from the commitments allocate to the arrangements (Healey 1997; Marshall 2003).

A bottom-up approach that emerges from the deeply rooted needs of the community (and is supported by practices and solutions arising from local knowledge is likely to be more effective than a despotic top-down approach, which mostly turns "lemons into lemonade".

In fact, in his study of sixteen US downtowns, Ford (2003) concluded that Providence ranked in fourteen place, only above Phoenix and Charlotte, in a list, which was topped by Seattle, Portland, and Baltimore (Fig. 6). The rise of the creative city

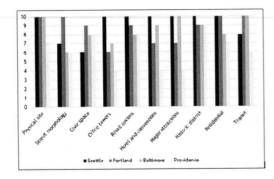

Figure 6.   Ford's (2003, p. 279) Downtown Ratings of Selected US Cities.

movement since then, together with the negative consequences of the global financial crisis, a.k.a: great recession, may have altered this ranking considerably (Florida 2014).

An effective communication process, a sense of direction and a multidimensional understanding of urban revitalization increases everyone's knowledge of what is a possible, thereby reducing possible ground for conflict (Orr and West 2002). Partnerships involving real collaborative decision-making processes, rather than only perceived "ways to access money" are more effective in allowing individuals to reach consensus and maintain their compromises. To conclude, the need for resources and to welcoming volunteers is conceivably the most sought after contribution to the success of partnership arrangements and visionary utopias.

ACKNOWLEDGMENT

Part of this research was conducted with a doctoral grant from *Fundação Para a Ciência e a Tecnologia* (FCT).

BIBLIOGRAPHY

Angotti, Thomas. New York for Sale: Community Planning Confronts Global Real Estate. Cambridge: The MIT Press, 2008.

Barber, Benjamin. If Mayors Ruled the World: Dysfunctional Nations, Rising Cities. New Haven: Yale University Press, 2013.

Barnett, Jonathan. Redesigning Cities—Principles, practice, implementation. Chicago: Planners Press, 2003.

Ben-Joseph, Eran. The Code of the City: Standards and the Hidden Language of Place Making. Cambridge: The MIT Press, 2005.

Birch, Eugénie. "Downtown in the "New American City." The Annals of the American Academy of Political and Social Science. 626 (2009): 134–153.

Bohl, Charles. Place Making: Developing Town Centers, Main Streets, and Urban Villages. Washington D.C.: The Urban Land Institute, 2002.

Bunnell, Gene. Making Places Special. Chicago: Planner Press, 2002.

Carlsson, Chris. Nowtopia: How pirate programmers, outlaw bicyclists, and vancant-lot gardeners are inventing the future today. Oakland: AK Press, 2008.

Castello, Lineu. Rethinking the Meaning of Place: Conceiving Place in Architecture-Urbanism. Surrey: Ashgate, 2010.

Cullingworth, Barry, and Caves, Roger. Planning in the USA: Policies, issues, and processes. 3rd Edition. London: Routledge, 1997/2009.

Duany, Andreas, Plater-Zyberk, Elizabeth, and Speck, Jeff. Suburban Nation—The rise of sprawl and the decline of the American Dream. New York: North Point Press, 2000.

Ellin, Nan. Integral Urbanism New York: Routledge, 2006.

Fainstein, Susan. The City Builders: Property, politics, & Planning in London and New York. Oxford: Blackwell Publishers, 1994.

Fishman, Richard. Urban Utopias in the Twentieth Century: Ebenezer Howard, Frank Lloyd Wrigth, Le Corbusier. 5th Edition. Cambridge: The MIT Press, 1982/1994.

Florida, Richard. "The Creative Class and Economic Development." Economic Development Quarterly 28.3 (2014): 196–205.

Ford, Larry. America's New Downtowns: Revitalization or Reinvention. Baltimore: The Johns Hopkins University Press, 2003.

Frieden, Bernard, and SAGALYN, Lynne. Downtown Inc.—How America Rebuilds Cities. Cambridge: The MIT Press, 1989/1991.

Friedmann, John. The Prospect of Cities. Minneapolis: University of Minnesota Press, 2002.

Hall, Peter. Cities of Tomorrow. 3th ed. Oxford: Blackwell Publishing, 2002. 553 p. ISBN: 0-631-23252-4.

——. Cidades do Amanhã. São Paulo: Perspectiva, 1995.

Healey, Patsy. Collaborative Planning, shaping places in fragmented societies. Vancouver: UBC Press, 1997.

Kelly, Eric, and Becker, Barbara. Community Planning—an introduction to the comprehensive plan. Washington, D.C.: Island Press, 2000.

Lang, Jon. Urban Design: The American Experience. New York: Van Nostrand Reinhold, 1994.

Leazes, Francis, and Motte, Mark. Providence, the Renaissance City. Boston: Northeastern University Press, 2004.

Lee, Jong, and Anderson, Chad. "The Restored Cheonggyecheon and the Quality of Life in Seoul." Journal of Urban Technology 20.4 (2013): 3–22.

Marshall, Alex. How Cities Work: Suburbs, sprawl, and the roads not taken. Austin: The University of Texas Press, 2003.

Mullin, J.A. "Bellamy's Chicopee: A laboratory for Utopia?" Journal of Urban History 29.2 (2003): 133–150.

Olsen, Joshua. Better Places, Better Lives—a biography of James Rouse. Washington D.C.: The Urban Land Institute, 2003.

Orr, Marion, and Nordlund, Carrie. "Political Transformation in Providence: The election of Mayor Angel Taveras." In 21st Century Urban Race Politics: Representing Minorities as Universal Interests. Ed. Perry, Ravi. Bingley: Emerald, 2013. 1–12.

Orr, Marion, and West, Darrell. "Citizens' Views on Urban Revitalization: The Case of Providence, Rhode Island." Urban Affairs Review 37.3 (2002): 397–419.

Paumier, Cy. Creating a Vibrant City Center. Washington D.C.: The Urban Land Institute, 2004.

Pierre, Jon, ed. Partnerships in Urban Governance. New York: St. Martin's Press, 1998.

Pinder, David. "In defence of utopian urbanism: Imagining cities after the 'end of utopia." Geografiska Annaler 84B.3–4 (2002): 229–241.

Punter, John. The Vancouver Achievement: Urban planning and design. Vancouver: UBC Press, 2003.

Redstone, Louis. The New Downtowns: Rebuilding business districts. New York: McGraw-Hill Book Company, 1976.

Rossman, Gretchen, and Rallis, Sharon. Learning in the Field. Thousand Oaks, CA: Sage, 1998.

Ryan, Brent. "Incomplete and incremental plan implementation in Downtown Providence, Rhode Island, 1960–2000." Journal of Planning History 5.1 (2006): 35–64.

Sandercock, Leonie. Towards Cosmopolis—Planning for Multicultural Cities. Chichester: John Wiley & Sons, 1998.

Short, John. The Urban Order—an introduction to cities, culture, and power. Oxford: Blackwell Publishers, 1996.

Smith, Herbert. The Citizen's Guide to Planning. Chicago: Planners Press, 1961/2000.

Stanton, Michael N. The Prince of Providence: The True Story of Buddy Cianci, America's Most Notorious Mayor, Some Wiseguys, and the Feds. New York: Random House, 2003.

Walzer, Norman, and JACOBS, Brian, eds. Public-Private Partnerships for Local Economic Development. Westport: Praege, 1998.

West, Darrell, and ORR, Marion. "Downtown Malls as Engines of Economic Development, Community Spirit, and Political Capital." Economic Development Quarterly 17.2 (2003): 193–204.

Utopia in Terra Brasilis: Lina Bo Bardi's design for an ideal future

Myrna de Arruda Nascimento & Eneida de Almeida
Project Department, Faculty of Architecture and Urbanism (FAUUSP), University of São Paulo (USP), São Paulo, Brazil
Faculty of Architecture and Urbanism, University of São Judas Tadeu (USJT), São Paulo, Brazil

ABSTRACT: This paper discusses Utopia's concept using an intrinsic strategy familiar to fictional development dilemmas of a dreamed place. First it manifests criticism to a reality that is to be overcome (or transformed); and then discusses the qualities of the new scenario, designed to fit the intervention's intention, and achieve the goals that throw individuals into the unknown, arguing for the unusual.

From literary construction of the plot, to architectural constructive intervention, the idea of planning a place for a new spirit of social conviviality, democratic and generous, is achieved in the Leisure Centre Factory SESC Pompeii. The leaving of Italian post-war scenario toward Brazil, fertile soil to plant innovative uses in abandoned spaces, redefining the destination of pre-existing building and promoting changes in neighbourhood social relations and society routine, became an opportunity, a challenge, and a hope to the Roman architect Lina Bo Bardi.

In Pompeii Factory scenery, faced with the options to preserve or to innovate, Lina has chosen both and others, giving São Paulo city an indelible mark: a timeless territory in constant metamorphosis, always updated and subject to reinvention use by the ones who adapt it to various purposes. Utopia of utopias, the pioneer experience inspired designs of multiple nationalities, and witness the production of genuine and transcendent knowledge: most devout to spontaneity than to scientific rigor.

Keywords: Utopia, Lina Bo Bardi, Architecture, SESC Pompeii Factory Leisure

1 INTRODUCTION

In "The Garden of Forking Paths" (1941)[1], detective story by the Argentine writer Jorge Luis Borges, the main storyline character, to unravel the parable engendered by his ancestor, reveals to the reader the complex circumstance of the argument upon which the fiction is built[2]: time is a phenomenon that is infinitely bisected in countless future, and not always the choice of an alternative eliminates the others.

The notion of a labyrinthine temporality is familiar to dreamers and explorers when facing the incognito, ambiguities and complementarities present in chimeric elaborations of places and human organizations, similar to those found in Thomas More Utopia, whose memory is evoked (in this Congress) 500 years after its first edition.

When presenting the Brazilian edition of *Utopia*, in 1980, the essayist Afonso Arinos comments the double aspect of the book (More 2004: 35–37), either seeming to be a fantasy, or a historical report, "a side of satirical fiction and other of realistic libel. Behind the mists, cruel facts. This ambiguity is perhaps its enduring interest"[3]. The critic also points out that the voyagers' writings and letters, especially the narratives of the Florentine Amerigo Vespucci (1454–1512), served as inspiration for More, and one can find in the description of the Utopia Republic much of what we would like to have in our contemporary cities.

Recent publications and celebratory editions of Lina Bo Bardi's centennial in 2014, and the overwhelming contributions that distinguished her career and professional experience in Brazilian metropolises, enable us to identify curious affinity between the Roman inventive idealist and the theme of this meeting.

Also Linda, on her own way, involved with simultaneous and different activities, jobs and dreamed plans with which she divided her attention and devotion, was used to amuse herself challenging and questioning the regular temporal logic, unable to cover the complexity of her ideas articulation, her design proposals and the spatiality generated from her projective work[4].

A contumacious pioneer, Lina transcends cultural limits opening connections between fields of knowledge and cognitive streams often uneven and adverse[5].

The contradiction presented in her personal references, as well as identified in her professional production, has never been interpreted as a negative sign or fragility: created in the Fascist Italy and a World War II survivor, Lina sought in America a place to grow and play a new type of humanism, able to harmonize her Illuminist heritage with references from popular culture and simplified and accessible technologies[6].

2 FROM EXPERIENCE TO SCIENCE GENERATION

The experience gained throughout childhood, recognized by independent attitudes and living among ambiguity, perfected during university education, became protagonist of a unique production of knowledge and an unprecedented knowing about the science of architectural making, regarding to her particular expression and circumstance, personally and professionally. The decision to defend applied knowledge, evident in her utilities projects executed with rigor, come from careful observation of social and human dynamics. The attention to behaviours and vernacular cultural heritage characterizes the work of Lina Bo Bardi, updating it and recognizing its relevance as timeless legacy, notably marked by an approach that interweaves art and life, and transits between the new and the old, as if they were continuous and reciprocal conjunctures (Fig. 1).

According to Lima (2013), it is noticeable the presence of approaches and interpretations in Lina Bo Bardi's work, in line with Edoardo Persico (1900–1936)[7], a critic with a visionary thinking, whose collaborator, Giuseppe Pagano (1896–1945) decided to put the Rural architecture as a central theme of the Milan Triennial (1936), suggesting with the exhibition's title, "Continuity and Modernity", the underlying dilemmas presented in comparisons between the economy and simplicity of country houses, and the rationalism and functionality tectonics of the emerging modern architecture in Italy[8].

Figure 1.   SESC Fábrica Pompéia, multi-use space. By Myrna Nascimento, nov 2014.

From experiences of special approaches and her familiarity with the amphibologies, we believe that history understood as "living memory" is probably the essence of Lina Bo Bardi's architecture. Its synthesis is recognized as a whole made of three main features: her seeking to overcome the historical division between "old" and "modern"; her effort on building a continuity between past and present, and the relationship between the universal and the particular frequently noticed on her work. It is not, however, continuity and restoration of that condition from more distant times when the intervention on the pre-existence obeyed only the injunctions and conveniences of the present. Instead it is an activation process of careful and indispensable critical exercise in relation to the testimonies of the past that not only guide conservation, but mainly determine judicious transformation of the goods to be submitted for intervention.

At the same time that her understanding of history shows up in tune with the design of the New History, her words also confirm the perception that the contemporary landscape imposes a critical review of orthodox modernism, especially in what concerns the questioning of the dominant antihistoric vision that permeates the cultural environment of the first half of the twentieth century[9].

We emphasize, therefore, the attention to architecture as a cultural fact, as "organism fit for life", which is equivalent to professional acuity in drafting enforceable and enduring architectural programs, in relating to the reality it faces, as well as attention to the user, reconciling one's everyday physical activities with its own rich symbolic universe.

It is also possible to recognize Lina Bo Bardi's creative strategy affinities with the curious paradox, according to academic criteria applied to traditional educational institutions, where modern sayings procedures intend to classic principles of generating ideas, subjecting them to criticism and review when facing the seduction imposed by new situations, scenarios and challenges.

Risking to experience unprecedented paradigms and develop her projects in a unique, inaccurate and unpredictable perspective, Lina set processes and methods, dare proposing devices and new perspectives of work, designed in other realities, but not yet observed in the Brazilian context. In this sense it is essential to recover evidence of this questioning attitude in her adventures abroad. One such experience took place just before moving to Brazil, alongside Carlo Pagani and Bruno Zevi, with the launch of the magazine—Culture della Vita, a leading journal for the time, however, short-lived: one total of nine published figures (the first in February 1946), the first seven with fortnightly, the last two as a weekly publication (the last in June of that year).

In the immediate post war period, the innovation proposed by the editors was to address the themes of architecture, housing and art, among others, in order to extrapolate a technical focus, discipline, aiming to establish a close dialogue with readers, and to propose an improvement of daily life[10]. In the first issue, dedicated to the topic of reconstruction of the war-torn country, the editorial invites readers to take part in this collective work of rebuilding[11].

This first edition presented a report entitled "Research on housing" based on an interview with the magazine journalist given by a resident from the centre of Milan, in order to portray the characteristic ways of life of those difficult years of war.

The questions were divided into two types: those concerning the habitability of accommodations: the area of the internal space, the solar insolation conditions of the main environments, possibility to enjoy the scenery; and those relating the everyday habits: contact and coexistence between neighbours, receive visits, use of common areas, go out to ride, and others. Specific questions were raised about the habits of children and places where they gathered to play. Deserves attention the comment that one of the interviewees, although dwell in a tiny apartment, owned a 2,000-volume library. Asked about what would make him happier at home, the mother replied without hesitation: "I would like, as well as additional services privacy (bathrooms were collective, located externally to accommodation), another room where my son could study in peace and entertain friends."

Is not hard to find in this interview the intimate desire to mobilize architects and other professionals to rebuild a city, aiming to make it ideal. In this attitude we find the source of unrest, and the preference to know and, more than to learn, to respect and incorporate into her plans and work foreign cultural information, exploring the peculiar simplicity and spontaneity of users as a *"materia prima"* for the spaces she designed, a procedure that will feature incisively the legacy of Lina Bo Bardi throughout Brazil. Her words about the second time she visited Factory Pompeii's place should be quoted: "a Saturday, the atmosphere was different: no longer the elegant and lonely hennebiqueana structure[12], but a cheerful audience of children, mothers, fathers, and elders walking from a pavilion to another"[13].

Therefore thinking about the design of projects from rates evidenced in the observation of events and/or urban environments to be transformed (created, rethought, reworked), and the behaviour and preferences of the audience for whom it is planned, considering the pre-existence of historical and cultural interest, is a posture that surpasses the anti-historicist one, frequently associated with the modern movement.

This innovative and bold strategy, finds in the Pompeii Factory Leisure Centre plan the most legitimate expression of finding fertile and available ground to build a utopian, timeless territory in constant metamorphosis, always updated and subject to reinvention use by the community and cultural agents who exploit and adapt it to various purposes.

The term "reconstruction", mentioned in the Italian publication aforementioned, referring to action in setting destroyed by war, deserves attention because, in the field of heritage preservation, reconstruction is considered mistaken in the sense of going back in time, eliminating distinction between the original material and the restored elements, and assuming a connotation of "remake", while the term "reuse" is more suitable.

Lina Bo Bardi's project corresponds, strictly speaking, to a requalification of action, reuse or recycling, to the extent that it recognizes the dignity and the documentary value of the pre-existing architecture, and through an elaborate transformation enhances and gives significance to the pre-existence, reviving it and reinventing both the program and the quality of the space. (Fig. 2)

Figure 2.   Eight pre-stressed concrete walkways link the two towers, crossing gaps of up to 25 m; a wise proposal to create a useful path and solarium deck in a non-edificated area. Beneath the deck ran a channelled stream called Córrego das Águas Pretas, Black Waters Stream. By Myrna Nascimento, Nov. 2014.

3 FACTORY POMPEII: CITADEL OF FREEDOM

Bruno Zevi, Italian historian and critic of architecture, co-founder with Lina Bo Bardi magazine "A—La culture della vita", writes in the journal "*L'Espresso*" an article on the SESC Pompeii entitled "Factory of signs" noting that the challenge facing the architect was to present "a complex of socio-cultural services in an 10 million urban gear, with the "connotation of a chaotic mixture of signs and tumultuous in perpetual mutation [...]"; and in addition, he highlighted Lina's purpose of "keeping intact the existing volumes, reconfiguring it, but in order to transform the introverted structures in bright extroverted spaces. [...]"[14].

Lina Bo Bardi's experience is different from the usual procedure and is somehow pioneering in that context, for there was not even consensus on the value of industrial buildings, nor restrictions to the demolition of the brick warehouses.

It is significant that the tipping of the factory set by the cultural heritage preservation agencies had occurred only after Lina Bo Bardi's intervention (Conpresp, 2009; Iphan, 2015—respectively to the municipal and federal responsibility. It can be said that the intervention is now considered a national reference in the field of heritage preservation as a valid alternative to the nature of *viollettiano* interventions, likely to eliminate the passage of time signals to recover the unity of style or primitive condition, identifying value only on the origin of the object of intervention. Instead, the Factory of Pompeii is an affirmative intervention that is distinguished from the pre-existing, enabling not only differentiation but also the reversibility, as the increases are autonomous and independent of the original structure. Therefore, it is commendable to recognize the relevance of the action that distinguishes the new from the old, and assumes the right to introduce new elements sparingly and with deference to the existing architecture to enable a new appropriation of this architectural space (Fig 3).

The philosopher Eduardo Subirats, among some meanings he attributed to the factory of Pompeii, evokes the intervention as "concrete utopia of a warm architecture, able to transform the alienated labour sphere of the factory"[15] in place "where it makes room to pleasure, to beauty, to the playful". Comparing the recreation centre with a community dedicated to games and pleasures, Subirats suggests to call it a phalanstery (Wainer and Ferraz 2013: 83).

A kind of seed one wants to spread in São Paulo city, the SESC Pompeii is the idealized space, identified by the Mandacaru flower[16] [19], and made of iron profiles, placed on the concrete walkways between the buildings. The metaphor of the plant,

Figure 3. "The freedom of the artist was always" individual "but true freedom can only be collective. A freedom aware of the social responsibility that overturning the boundaries of aesthetics", Lina Bo Bardi [http://lina-bobarditogether.com/pt/2012/07/17/about/). By Myrna Nascimento, Nov. 2014.

unpredictable and unusual, that is born and grows wild, enhances the fantastic character of Lina Bo Bardi's design: a definite metamorphosis in São Paulo urban landscape.

NOTES

[1] Published in Fictions (1944), the first work of the Argentine writer translated into English, honoured in 1961 with the International Prize for Literature. "[...] Bajo árboles ingleses medité en ese laberinto perdido: lo imaginé inviolado y perfecto en la cumbre secreta de una montaña, lo imaginé borrado por arrozales o debajo del agua, lo imaginé infinito, no ya de quioscos ochavados y de sendas que vuelven, sino de ríos y provincias y reinos [...]. Pensé en un laberinto de laberintos, en un sinuoso laberinto creciente que abarcara el pasado y el porvenir y que implicara de algún modo los astros. Absorto en esas ilusorias imágenes, olvidé mi destino de perseguido. Me sentí, por un tiempo indeterminado, percibidor abstracto del mundo". (Borges 2004: 475).

[2] "Dejo a los varios porvenires (no a todos) mi jardín de senderos que se bifurcan. Casi en el acto comprendí; el jardín de los senderos que se bifurcan era la novela caótica; la frase varios porvenires (no a todos) me sugirió la imagen de la bifurcación en el tiempo, no en el espacio. La relectura general de la obra confirmó esa teoría. En todas las ficciones, cada vez que un hombre se enfrenta con diversas alternativas, opta por una y elimina las otras; en la del casi inextricable Ts'ui Pên, opta—simultáneamente—por todas. Crea, así, diversos porvenires, diversos tiempos, que también, proliferan y se bifurcan." (Borges 2004: 477).

[3] "Utopia comes from Greek, unreal place; the name of the capital, Amaurote, means non-existent city; King, Ademos, expresses the idea of a chief without people, and so on"(More 2004: 36).

[4] "But linear time is an invention of the West, time is not linear, it is a wonderful tangle where, at any time, solutions can be chosen and spots invented, without beginning or end." (Bardi 2013).

[5] "(...) Instead of universal values, her complex work and writing unveil the roles of plurality, otherness, and

instability in the constitution of modernity. Instead of agreeable forms, she strived to embrace the contingency and spontaneity of life. She spent her own life in transit, navigating the contingencies of different locations and worldviews. Her ventures established a weaving path in and out of modern culture, materialized in the communication between innovation and tradition, abstraction and realism, rationalism and surrealism, and naturalism and history, as well as between revolutionary impulses and melancholy" (Lima 2013: Preface, p. xii).

[6] "The 1936 exhibition rejected the early culturalist views of environment and *architettura minori* (vernacular architecture) by presenting rural building techniques as the authentic tradition for rethinking modernization.[...] For all their critical advancements in combining rational logic and rural architecture, rationalist designers' desire for revision soon met official resistance" (Lima 2013: 16).

[7] "Just before his premature death in 1936, Persico [...] asserted that the purpose of abstract aesthetics, technical simplification, and sanitation, though essential in addressing the realities of mass society, was not to serve efficiency but to welcome human life and to help society define and share collective values. Persico planted the seeds of a self-critical realism that would bud a decade later among Lina Bo's generation during the efforts of post-war reconstruction. (Lima 2013: 16).

[8] "Despite her praise of the spontaneity of nature and uneducated culture, she remained attached to Enlightenment principles such as rational logic, scientific knowledge, and free will. [...] Her sense of humanism was pragmatic. It acknowledged the constraints of specific situations and aspired to justice. She was especially fond of the improvisation and authenticity of simple people, whom she often saw as deprived of material means but rich in creative capacity" (Lima 2013: Preface, p. xii)

[9] "[...] The modern architecture is, like all human activities, the product of man's experience in time, and that there is no break between the so-called "modern" and history as it is the "modern" firstly the product of the story itself, through which you can only avoid the repetition of experiments overcome. It is the story [...] as a living thing and present, revived in its fundamental problems endowed with transmissibility and fruitful teachings [...]" (Bardi 2002: 5–6). New edition of the text presented by Lina Bardi to FAUUSP, when she ran for Theory of Architecture chair in a competition in 1957.

[10] "The call is a reality that we do not know; it is like shouting in the desert. Even worse is to refer to a reality that is believed to know having it seized by the statistics. The true reality is harvested only from the mouths of those who had lived it. The others are false realities, unable to practice attention. This failure explains the fact for which, notwithstanding the splendour of urban theories, reality remains so squalid and miserable. We know the truth about housing only by asking tenants, the more casual the choice of people to be interviewed, more close we will be to the real search. We have more interest in spontaneity than in scientific rigor". (Bardi and Zevi 1946).

[11] "It is necessary for people to concern themselves with their own homes, the schools themselves, the plans of cit-

ies, factories, public buildings, and means of transport, the sports buildings, and public spectacles. It is necessary that every people participate consciously in the architectural and urban solutions for the country itself."

[12] Reference to François Hennebique, a pioneer in concrete structures design.

[13] "Children ran, young people played soccer in the rain that fell from the cracked roof, laughing with the ball kicking on the water. Mothers prepared barbecue and sandwiches at the entrance of Clelia St.: nearby it was running a puppet theatre, full of children. I thought that everything should go on like this, with all that joy [...] No one changed anything. We found a factory with a beautiful structure, architecturally significant, unique, no one modified... The architectural design of the SESC Pompeii Leisure centre came from the desire to build another reality".

[14] "Renouncing professedly the mythology of classical beauty, this socio-cultural centre of Sao Paulo plays the cards of dissonance with boldness and spontaneity. Without intellectualism, provides a desirable environment model, stiff with humanity and poetic fantasy". (Zevi 2013).

[15] "[...] The intent of this dignifying and recreational architecture resides that revolutionary tradition set by the called utopian socialism. A kind of the new loving world of Fourier, translated into an industrial civilization of the tropics [...] a kind of *locus amoenus*." (Wainer and Ferraz 2013: 83).

[16] "Mandacaru quando flora lá na seca, é o sinal de que a chuva chega no sertão" ("Mandacaru when flowers in the dry, is the sign that the rain started in the backcountry". Verse of the song "Xote das meninas", Luís Gonzaga, Brazilian musician, 1953.

BIBLIOGRAPHY

Bardi, Lina. "A Fábrica da Pompéia." In *Cidadela da Liberdade: Lina Bo Bardi e o SESC Pompéia*. Eds. Wainer, André and Marcelo Ferraz. São Paulo: Editora Sesc-SP, 2013.

Bardi, Lina Bo. *Propaedeutic contribution to the teaching of the Theory of Architecture*. São Paulo: Institute Lina Bo Bardi and PM, 2002.

Bardi, Pagani, and Zevi, A. *Cultura della Vita*.1 (1946)

Borges, Jorge Luis. *Obras Completas*. 15ª ed. Vol. I (1923–1949). Buenos Aires: Emecé, 2004.

Lima, Zeuler Rocha Mello de Almeida. *Lina Bo Bardi*. Foreward by Barry Bergdoll. New Haven: YALE, 2013.

More, Thomas. *Utopia*. Preface: João Almino. Trad. Anah de Melo Franco. Brasília: Universidade de Brasília, 2004.

Wainer, André, and Ferraz, Marcelo, eds. *Cidadela da Liberdade: Lina Bo Bardi e o SESC Pompéia*. São Paulo: Editora Sesc-SP, 2013. 31 p.

Zevi, Bruno. "L'Espresso, Roma: may, 1987." In *Cidadela da Liberdade*. Eds. Wainer, André and Marcelo Ferraz. São Paulo: Editora Sesc-SP, 2013.

Small utopias. Influence of social factors on spatial development of ecological settlement structures

Anna Szewczenko

Faculty of Architecture, Silesian University of Technology, Gliwice, Poland

ABSTRACT: The paper is focused on the issues concerning shaping and developing ecological concepts of settlements, perceived as a process of spatial and social growth, based on the need of creating a new social order in the face of the necessity of going back to nature, and maintaining the state of equilibrium between the built and the biological environment. In each of such realizations some ideas of utopian visions may be found, targeted at creating ideal conditions for life, based on a certain concise axiological order. The author of this paper declares the thesis that the trend of sustainable architecture, targeted at reaching the state of equilibrium of the ecological, economic and social factors is a continuation of the idea of a new system of values that should motivate action against profound social conflicts and degradation of the natural environment. The selected concepts and their implementation were analysed, demonstrating a harmonious connection of the anthropogenic and natural factors, which became one of the fundamental assumptions. The analyses and the author's experience from participating in the project: "Eco-Zone Settlement" led to the conclusion that the determinants of such solutions are: joint objectives and professed values of local communities, as well as the processes that enable their functioning in connection with other urban structures.

Keywords: sustainable development, ecological settlement, local communities

1   INTRODUCTION

The problems and threats that city inhabitants have to face nowadays lead to a continual search for ideal solutions that could measurably improve the quality of life. Utopia is a foundation of various urban development ideas, as it proclaims the creation of a new social order and the assumption of commonly shared ethical values [Pańków, 1990, p. 170]. A considerable number of theoretical concepts of future cities entail solutions aimed at ecological problems associated with the connections between social and spatial concepts. Thus, they are closely related to utopian ideas. Visions of virtual cities and *eco-city* concepts are based on the Sustainable Development principles, often aspiring to the idea of utopian metropolis. They are successive attempts at creating an ideal city, perceived in terms of architectural and aesthetic solutions. Undoubtedly, they look for new solutions as far as the equilibrium between the natural and the anthropogenic environments are concerned: referring to a certain degree to the utopian desire for harmony between man and nature—which is a key issue in the majority of ecological concepts of cities and housing estates. According J. Wines: The objective of (green settlements—the author's com-

mentary) "is to take joint responsibility over the environment by all communities on the grounds of profound philosophical, psychological, and cultural awareness" [Wines, 2008, p. 11]. Therefore, the following thesis may be formulated: the trend of sustainable architecture that attempts to reach the state of equilibrium of ecological, economic, and social factors is a continuation of the ideas of a new society whose system of values motivate the prevention of profound social and spatial conflicts. The question arises: What are the possibilities of implementing such assumptions and what is their spatial scale?

Multi-scale urban layouts of space create definite values, but also reveal the mechanisms of social behaviours and relations with space. The scale of modern cities and their multi-functionality results in the fact that it is difficult to find durable social standards obliging all inhabitants and reconciling all interested parties, especially in view of an increasing role of neo-liberalism and relativism. However, there are examples of smaller settlements, implemented by and for smaller social groups and functioning in accordance with the superior values system. This is particularly true as far as the relations between local communities and the natural environments are concerned, which makes them similar to utopian assumptions. Moreover, the concept of the functionality of such separate communities is grounded on social relations based on a new social order and prioritizing the well-being of the community over specific individual benefits.

2 ELEMENTS OF UTOPIAN ASSUMPTION IN ECOLOGICAL URBAN STRUCTURES

The search for solutions applicable to ecological cities in compliance with the Sustainable Development principles is nowadays one of the most important directions in modern architecture and urban planning. This search is reflected in a variety of sustainable architecture trends: from low-intensity technologies based on natural building materials and passive energy-efficiency systems, to high-tech solutions and active energy-efficiency systems. Nowadays, the common background is the awareness of the depletion of natural resources and energy sources, as well as the degradation of the natural environment, which impose the reconsideration of the relations between the natural and the built environments and between the biological conditions and the society. Innovative solutions of modern architecture are focused on proficient management of sites, water, energy and building materials efficiency, durability and flexibility of solutions in consideration of recycling and biodegradation, renewable energy sources, advancement of energy-efficient construction technologies.

The starting point of all concepts of ecological cities was the assumption that the city constitutes an eco-system, perceived as a number of biological, social, and cultural dependencies. In the majority, the social concept of such cities refers to the return to nature. The indicated fundamental ethical values were responsibility, frugality, and privacy of inhabitants. An example of such approach is the Dutch *biopolis* dating back to the 1960s [Gutowski, 2006, p. 112–113]. Other *eco-city* concepts did not assume the creation of new city structures, as they emerged at the time of growing ecological problems, revitalization of post-industrial zones and degraded city quarters. Urban space was treated as a synergy of the ecological elements, social responsibility, attitudes of average people, and the *non-violence* movement [Gutowski, 2006, p. 117]. The concept of *eco-city* had a number of long-term objectives leading to the return to life consistent with nature. It indicated the need of social changes in the economy, social ecology and grass-root ecological movements, as well as bioregionalism, in accordance with which the equilibrium was achieved by identifying people with the natural environment of their occupancy area. Urban transports including bicycles, ecological solutions in the architecture of buildings were supposed to create an ecological style of urban life. A certain continuation of this train of thought is the current phenomenon of *Transition Towns*, which is developing mainly in Great Britain. This is a concept of cities assuming a wide range of activities aimed at energy conservation, perma-culture farming, and profound social changes—cooperation that builds social ties and social involvement [Godlewski, 2011, p. 11].

The concepts of eco-cities have an impact on the current development of cities, especially in the face of growing ecological threats. They are certainly reflected in modern urban planning trends. There are also examples of wide-range activities, such as *Healthy Cities Project* initiated by the World Health Organization. Specific stages of this project were implemented in terms of the Sustainable Development principles and healthy life style.

Ecological housing estates or urban quarters are opportunities for verifying, in practice, some model solutions. Among many examples of ecological estates, the following should be distinguished: Järn estate near Stockholm based on efficient settlements built of natural construction materials. It is, at the same time, a cultural and educational centre (functioning from 1964) that inspires many pro-ecological activities. It houses Waldorf schools, biodynamic households, shops, and restaurants that use organic food production, a hospital applying an anthroposophy approach in patients' treatment, alternative-banking systems (Fig. 1). The entire estate is based on Rudolf Steiner's philosophy [www.kulturcentrum.nu].

Another example is BedZED energy-efficient British housing estate founded in 2002, which enabled the testing of an experimental technology and standard solutions for Ruralized housing systems [Arczyńska, 2010, p. 81].

Similar objectives were accomplished within the framework of "Eco-towns" British government program. Such comprehensive activities indicated the need of undertaking initiatives in

Figure 1. Buildings in Järn estate near Stockholm based on Rudolf Steiner's philosophy. By: Author.

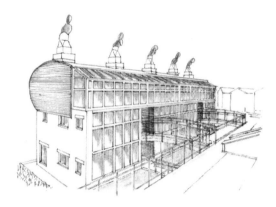

Figure 2. BedZED energy-efficient British housing, 2002. By: Author.

Figure 3. EVA-Lanxmeer quarter in Culemborg, view of ecological houses. [www.groenblauwenetwerken.com].

favour of education, as well as marketing and financial measures aimed at changing social attitudes, habits and respect of common good, as decisive criteria for undertaking every-day decisions. The first steps towards such direction were economic incentives supported by the government program.

Another interesting example of an ecological estate is EVA-Lanxmeer quarter in Culemborg (Holland). Its layout emerged as a 24-hectare social and ecological settlement situated on former farming sites and surrounded by the protection area of potable water. The quarter designed for about 1000 inhabitants consists of 250 flats, 40 000 m² of office and service floor, ecological farm, information centre, biological renewal centre, congress centre, catering outlets and a hotel. The design is focused at promoting Sustainable Development and reflects an integrated approach to sustainable urban planning. Its objective is to integrate technology and innovativeness with the natural environment and behaviour patterns that promote the care about natural resources consumption in everyday life.

The design was formed following extensive social participation, in the course of which a "coaching-model" was selected as a way of cooperation with future users of the estate. The users took active part in workshops and the planning process from the earliest stages of the design. The creation of the quarter was initiated by EVA foundation in 1994. The concept was devised in participation of a group of scientists and specialists representing different fields. Some of them were interested in moving into the designed quarter. The background for the involvement of future occupants was the changes that occurred despite social divisions and hierarchy of influence, yet in favour of creating a structure of a cooperation network.

The network was rapidly expanding and together with it, the number of potential inhabitants who held frequent meetings and co-created the design assumptions. The worked out model provided the appointment of a leader—who, as a chief designer was supposed to supervise the implementation of the design in cooperation with the project management team.

The decisions were undertaken by means of cooperative participation of the specialists and the interested parties, in accordance with the spirit of an "open design" that considered the level of awareness and knowledge of the society. The future inhabitants participated in several workshops, which were opportunities for a dialogue on planning decisions. The authorities of Culemborg, in accordance with the Sustainable Development principles and in consideration of the social awareness of the wide range of the inhabitants' interests, also cooperated in the project [Van Timmeren, 2007, p. 8–9].

3 THE CONCEPT OF "ECO-SPHERE ESTATE" AS A MODEL OF THE SYNERGY OF ANTHROPOGENIC AND NATURAL ELEMENTS. CASE STUDY

Conclusions from the author's own studies certify the importance of a leader who initiates and coordinates the activities associated with planning and social processes in the implementation of the above-mentioned projects, another example of which is "Eco-Sphere Estate" in the Silesian Botanical Garden Complex in Mikołów. The main assumptions for creating the Estate were [Kojs 2011]:

- A structure grounded on a symbiosis of the anthropogenic and natural environment elements,
- Ability of long-term functioning on the bases of renewable and non-renewable energy sources,

- Implementation of the postulates of Sustainable Development in terms of ecology, economy, society, architecture, aesthetics, space, and culture.

The functional assumptions of the Estate followed the adaptation system reflecting a tree in the process of adaptation. Hence, the Estate is subject of the regulation of the laws of nature and creates anthropogenic conditions for the stream of the processes. It was also assumed that the local community of size about 150 families would be formed by participation in all stages of the process—so that the final shape of the estate provided motivation for taking part in the spatial reality on the level of undertaken activities. The dynamics of the functioning of the Estate was based on the permanency of supreme *objectives and values* recognized by the local community. The size of the neighbouring group of inhabitants should enable the formation of deeper ties (about 25–150 families). At the same time, the Estate was to have a developmental nature, to enable further growth of other settlements. The elaboration of the conditions for the continuity of the natural and anthropogenic systems involved the following solutions [Szewczenko, Czachowska, 2011]:

- Preservation of the existing biological habitats and their enhancement in the form of a park, botanical garden, recreation greenery zones, to supplement social spaces,
- Introduction of the functions of education, exhibition, experiments, services—as elements supporting the growth of social relations,
- Separation of road traffic from the pedestrian traffic with the demarcation of the zones of accessibility,

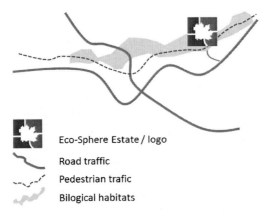

Eco-Sphere Estate / logo

Road traffic

Pedestrian trafic

Bilogical habitats

Figure 4. Scheme of the "Eco-Sphere Estate". Worked out by: A. Szewczenko, A. Czachowska.

- Construction of buildings from natural materials, also as passive energy-efficiency systems.

Although the concept has not been implemented, it united specialists representing many fields of science and technology: civil engineering, architecture, economics, culture studies, social psychology. The barriers to its implementation were the absence of a leader, a person that would be a *spiritus movens* of the Project, and, undoubtedly, the lack of cohesion between the specialists' objectives and future users' objectives—only few specialists expressed an interest at settling down at the Estate. Furthermore, the assumptions of the Estate were somehow opposite to the concept of a high standard of living, perceived as founded on the application of advanced technologies. It is sometimes essential to change the lifestyle, which may be an obstacle to undertake the decision of inhabiting such type of settlement.

4 CONCLUSIONS

One of the philosophical approaches to the concept of utopia assumes the use of values and behaviour patterns to model the social life [Pańków, 1990, p. 78]. If the local community inhabits its particular territory through the generations—ties with the place of residence become part of the community's culture. Moreover, the homeliness becomes the value of such place initiating the relationships between the place and local community [Pawłowska, 1996, p. 7]. Whereas in case of the newly formed settlement structures the relationship with the place of residence is forming due to the uniqueness the cultural landscape. Most of the ecological concepts of housing settlements present homogenous and spatially readable forms. Furthermore, this uniqueness is a result of participation of their future inhabitants, motivated to create a harmonious relation with the natural environment.

Conception of ecological settlement entailed profound changes in the social attitudes towards the biological environment, which is the starting point of building a new quality of social standards. This integration of spatial and social factors results from the scale of habitat that is the foundation of strengthening ties with living place and responsibility for it. In practice, this concerns those estates or urban quarters, the scale of which enables the creation of a cohesive local community that may influence the processes of changes occurring in the surroundings. However, what is the main concern? Utopias are often islands.

BIBLIOGRAPHY

Ańków, Irena. *Filozofia utopii*. Warszawa: PWN, 1990. ISBN: 8301099992.

Arczyńska, Monika, Pancewicz Łukasz. *Miasto oszczędne—utopia, idea społeczna czy polityka miejska?* Czasopismo Techniczne (Technical Transactions). Kraków: Wydawnictwo Politechniki Krakowskiej, 2010, 14, (107). Serie: Architecture.

Gutowski, Maciej. *Przestrzeń marzycieli. Miasto jako projekt utopijny.* Warszawa: 2006.

Godlewski, Paweł. *Transition Towns. Miasta przełomu.* Green2. 2011, Nr 5.

Kojs Paweł. *Ekosferyczna osada—czy możliwa jest symbioza antropogenicznych i przyrodniczych siedlisk?* Prezentacja w ramach konferencji "Współczesna architektura krajobrazu—trendy, technologie i praktyka" pt.: "Różnorodność ekologiczna—teoria i praktyka", Kraków 2011.

Pawłowska K.: *Idea swojskości w urbanistyce i architekturze miejskiej.* Kraków: Wydawnictwo Politechniki Krakowskiej, 1996.

Szewczenko, Anna, Czachowska Agnieszka. *Osada Ekosferyczna. Miejsce—ludzie—środowisko.* (non-published material). 2011.

Van Timmeren, Arjan, Kaptein, Marleen, Sidler, Dick. *Sustainable Urban Decentralization: Case EVA Lanxmeer, Culemborg, the Netherlands.* ENHR 2007 International Conference "Sustainable Urban Areas". Rotterdam: 2007, p. 8–9.

Wines, James. *Zielona architektura.* Warszawa: Taschen, 2008. Green Architecture. ISBN: 978-83-89192-47-9.

www.groenblauwenetwerken.com

www.kulturcentrum.nu

Architectural imprints and performing the imaginary

Višnja Žugić

Department of Architecture and Urbanism, Faculty of Technical Sciences, University of Novi Sad, Novi Sad, Serbia

ABSTRACT: This paper proposes the idea of performing the Imaginary within the urban environment, through a specific ephemeral spatial phenomenon—The Architectural Imprint. The imprint, understood as an example of an architectural *Non-Design*, emerges because of the urban transition of a city, where single-family houses are replaced with multi-family housing buildings.

In this non-linear and temporarily stretched process, a demolished house leaves a two-dimensional imprint on the sidewall of a new neighbouring building, exposing fragments of a former life to public space of the urban environment. This imprint functions as *spatial evidence*—a displayed projection of one's identity, which triggers a dynamic production of narratives within time-space gaps of new urban structures' development.

Publicly exposing the private—physical traces of the demolished house's interior—the imprint appears as a specific kind of *scriptible* architectural text. Being determined by the juxtaposition of numerous oppositions within one single phenomenon: private-public, temporary-permanent, built-demolished, and real-imaginary, it reveals its essentially heterotopic nature.

The spatial system of an architectural imprint fulfils a specific role in its physical context—it appears as a co-creator of produced meanings, and thus can be discussed as a paradigm for performative architectural space. Namely, the performance achieved by an architectural imprint emerges through the mechanism of correlation and results in the production of meanings, which always remain between real and imaginary.

Keywords: architectural imprint, imaginary worlds, spatial performativity, non-design

1 INTRODUCTION

The articulation of an idea of the Architectural Imprint is deeply connected to processes of building, and the way they are perceived on the level of urbanity. The process of creation of a single building demands a certain duration, of several months, sometimes years. On the level of townscape, the existence of this process is never negligible. Similar to the transition of the entire society, the morphological transition of a city as its consequence is rarely happening at once. Considering the perceived scenery of the city, the individual processes of construction are changing almost on daily basis. In that way, a morphological transition is built by a collection of ephemeral, unique images. Understood as connected time sequences, these images demonstrate an increased potential for being open to what Diana Agrest refers to as *mise-en-se'quence* (Agrest 1998: 198–213). Here, the urban environment is understood as a set of fragments, rather than a closed "object of reading", where the analysis focuses on the conditions of content, rather than the content itself. *Mise-en-se'quence*,

therefore, refers to a *situation* which "makes possible the inscription of sense in a free and highly undetermined way" (Agrest 1998: 207). This possibility of the inscription, in the context of an architectural imprint, is ultimately conditioned by its physical and spatial circumstances.

2 SPATIAL OPPOSITIONS

An imprint, as a phenomenon, which is essentially architectural, is determined by the series of contradictions. Namely, an imprint emerges from the demolishment of an existing architecture; it exists in this newly formed spatial gap within the urban tissue and disappears with the construction of a new building. While architecture as such begins with construction and ends with demolition, the imprint demonstrates a specific kind of architectural paradox—its very creation through an inverse process (Žugić, Marić and Mirkov 2010). At the same time, the question of materiality of the very imprint reveals new layers of contradictions. While the imprint is primarily determined by the demolished house—morphologically and on the level of meanings it absorbs—its very existence is conditioned exclusively by a neighbouring building. In that way, it is impossible to read the materiality of an imprint independently of any of the two components that determine it, while at the same time, it essentially does not belong to either of the entities. The imprint does not belong to a non-existing house, but it functions as a specific form of the *memory* of the house. Being partly defined by the physical void left after the demolishment of the house, and by the borderline between the two individual building sites, it becomes an explicit materiality of an architectural absence, rather than a mere witness of empty space (Fig. 1 and Fig. 2).

Establishing a gap in urban tissue is not only physically determined, but also temporarily. The process of construction of a new building continuously transforms the structure of the very imprint, as well as the possibilities of its perception. The phenomenon emerges in a time gap between the life of two distinct architectures, and in that way, its character often becomes extremely ephemeral.

Figure 2. Architectural imprint in Lisbon, 2010. Photo credit: Žiga Zupančič.

Since architecture, in its most conventional concept, presupposes a category that is essentially permanent this makes yet another contradiction specific for the analysed spatial system.

Therefore, it is possible to determine main characteristics of an architectural imprint: it is a temporary, liminal, dynamic, and metaphorical architectural presence within the time-space rift of the morphological transition of a city. These specific physical circumstances typical of an imprint, function as the main impulse for the interweaving of Real and Imaginary worlds, once the situation is faced by an observer, "putting into play a force analogous to that of the unconscious" (Agrest 1998: 208).

3 IMPRINT AS AN ARCHITECTURAL NON-DESIGN

In the substantially unstable nature of an architectural imprint, it operates as a perfect example of a phenomenon which Diana Agrest marks as *non-design*: "a semiotically heterogeneous object in which many different signifying matters and codes intervene" (Agrest 1998: 208). Observing an architectural imprint as a non-design, it is understood as a text,

Figure 1. Architectural imprint in Lisbon, 2010. Photo credit: Žiga Zupančič.

which appears in a dialogically structured system of intertextual relationships. On the level of urbanity, the text of an imprint constantly correlates with other texts from its environment. Being present in the web of surrounding texts (which belong to the realm of urban landscape), an imprint appears as a kind of *intervention* into its semantic context, while simultaneously this very context, with its "semantic volume" (Agrest 1998: 209) perpetually shifts the possibilities of its reading.

Within the complex system of ever-changing perception of the townscape, the process that unfolds is deeply determined by an uncertain play of meanings. The meanings in question are triggered by a scriptable architectural text of an imprint: the text "giving no indications as to how it is to be read, demanding that its reading be in effect its rewriting" (Hays 1998: 198). With "no unique producer" of those meanings, "the predisposition of non-design to openness implies permeable limits and an always fluctuating or changing specificity" (Agrest 1998: 207).

By exposing the traces of someone's private life directly to viewers within an urban environment, and, therefore, possibly exposing someone's identity, the non-design of an architectural imprint offers the possibility of a productive reading—the reading which is not a passive consumption of a previously defined or unique sense, but the one which calls for additional inscribing and imagination. Being reduced to a specific kind of *spatial suggestion*, rather than a system that transfers the accurate facts, the quality of produced narratives depends equally on several factors. Namely, meaning that emerges is a result of a synchronic correlation between the initial potential given by the materiality of an imprint, its dynamic contact with the urban environment, and *the observer* that finds him/herself facing this spatial situation. This process results in forming a specific *liminal space* of the observer, which "provides a spectator with an illusion of the Great Idea, spanning beyond its limits" (Zekovic 2014: 70–82). The *great idea* of the situation of the described *mise-en-se'quence* refers precisely to the produced Imaginary, and is directly dependent on both the existence of an imprint, and at the same time, on the "ability of the spectator to feel, recognize and employ all relations that occur in the given space-time framework" (Zekovic 2014).

In relation to the typology of architectural spaces, an architectural imprint represents a certain form of heterotopia (Foucault 1997: 330–336). Resulting in the creation of *liminal space*, which always exists on the borderline between Real and Imaginary,[1] the heterotopia of an architectural imprint is "capable of juxtaposing in a single real place several spaces, several sites that are in themselves incompatible" (Foucault 1997: 330–336).

Existing as a real and imaginary place at the same time, it operates in the juxtaposition of the present, concrete, physical and material, and the absent, abstract, immaterial and inscribed. Resulting in a *heterotopia of illusion*, it always reaches beyond what is visible. Therefore, the imprint can be understood as an essentially dynamic phenomenon, both physically and semantically.

4 PERFORMING THE IMAGINARY

In this complex system of production of meanings, the imprint, as an ephemeral state of architecture, fulfils a specific and important role. This role is directly related to producing the Imaginary, and generating a temporary function of this spatial entity—the performative function.

Namely, according to the theory of functions in architecture (Dinulovic 2012), every architecturally articulated space is capable of fulfilling not one, but many different functions. Depending on the changes of its primary context, architecture is also capable of generating new ones, explained by Umberto Eco as *secondary functions* of architectural space (Eco 1997: 173–195). The key question in considering possibilities of defining and recognising new spatial functions refers to analysing the roles that space fulfils in a specifically defined problem. In the problem of articulating the phenomenon of architectural imprints and its effects on experiencing the urban environment, the architectural space fulfils the task of *production of meanings*. The meanings that emerge in correlation with the imagination of an observer are not prescribed meanings—the architectural space was not initially built with an idea to represent the meanings in question. Instead, the meanings of an architectural imprint are being established and realised at the very moment of their reception. This means that the meanings are not previously defined and reflected, nor are they transferred or represented, but *produced*. The described qualities of the phenomenon open the possibility of analysing it as the problem of performing arts. According to Ana Vujanović, performing arts is the "term that focuses on the problem of performing, as the process of establishing, production and realisation of meanings" (Vujanovic 2004: 131). In that way, it becomes an "operative term which can be applied to any artistic discipline, practice or act when it centres the problem of performing or when it is theorised from the aspect of that very problem" (Vujanovic 2004: 130).

Relying on this theory, it is possible to define a specific performative function of architectural space, which is, in this research, demonstrated by the idea of architectural imprints as its paradigmatic example. The architectural entity of an imprint performs through a specific mechanism of correlation—a physical and textual juxtaposition of its materiality and the surrounding texts. Precisely the imprint, as a gradually vanishing spatial evidence absorbed with traces of a former life, but also, a fragmentary frame, which allows for loading of meanings, directs this very production towards imaginary narratives that may have never existed. Here, the imprint, with its delayered, exposed, and reinterpreted memories of past times, appears as a *co-performer* of the Imaginary. Functioning as the unreachable dreamlike places, the worlds that unfold through the co-performance of an imprint appear as potentially personally determined micro-urban utopias.

The established performative function of this space is also ephemeral in its character. It is not related to a static, stable, and existing narrative, but a rather labile, dynamic narrative-in-process. In that way, the space of an architectural imprint becomes essentially performative and acts as one of the key components of establishing the temporary coexistence of the Real and the Imaginary in the public space of a city.

5 CONCLUSION

Architectural imprint is an ephemeral state of space and time in architecture, materialised as a two-dimensional projection of an imagined private life. Being at the same time a concrete, existing, and material physical fact and an illusionary space of limitless possibilities of inscribing and interpretations, it operates in the mode of a constant and overall tension. Within the urban landscape, an imprint, therefore, functions as a Non-Design—a fluctuating text constantly intervening in its surroundings and opening the possibilities for a productive reading.

Through the mechanism of correlation, the imprint establishes a synergy between the absorbed meanings of its physical entity, the circumstances of its appearance on the level of urbanity, and the imagination of an observer facing this spatial situation.

The role that the architectural imprint fulfils in this synergic relationship refers to a specific kind of

performative function of architecture. As a result, the imprint operates as a co-performer of the Imagined worlds, ephemeral and fictional, and at the same time authentic and personally determined.

Overlapping, layering and facing diverse and opposite characteristics within one single phenomenon, the Imaginary, performed by an architectural imprint, enriches the experience of a built environment, setting in motion an unexpected urban drama.

NOTE

[1] This interpretation of liminal space is given by Jelena Todorović, PhD, a Full Professor of History of Art, as well as a Visiting Professor at the PhD study programme of Architecture, University of Novi Sad, where she leads a course titled Liminal Spaces and Infinity of Illusions.

BIBLIOGRAPHY

Agrest, Diana. *Design versus Non-Design*. Architecture Theory since 1968. Ed. Hays, K. Michael. Cambridge: The MIT Press, 1998. 198–213.
Dinulovic, R. "The ideological Function of Architecture in the Society of Spectacle." In *International Conference Architecture and Ideology, Proceedings.* Belgrade: Board of Ranko Radovic award, Association of applied Arts Artists and Designers of Serbia (Ulupuds), 2012.
Eco, Umberto. "Function and Sign: The Semiotics of Architecture." In *Rethinking Architecture. A reader in Cultural Theory.* Ed. Leach, N. London: Routledge, 1997.
Foucault, Michel. "Of Other Spaces: Utopias and Heterotopias." In *Rethinking Architecture. A Reader in Cultural Theory.* Ed. Leach, N. London Routledge, 1997.
Hays, K. Michael, ed. *Architecture Theory since 1968.* Cambridge: The MIT Press, 1998.
Vujanovic, A. *Razarajuci oznaciteljile performansa.* Beograd: Studentski kulturni centar, 2004.
Zekovic, M. "Liminal Space in Art: the (In)security of our own Vision." In *Dramatic Architectures: Places of Drama, Drama for Places.* Eds. Palinhos, J. and M. H. Maia. Porto: Centro de Estudos Arnaldo Araújo CESAP/ESAP, 2014.
Žugić, V, Marić, M, and Mirkov, B. *Lifeprint. In Personal Infrastructures—SMIBE Short Film Competition.* Chicago: SMIBE, The Society for Moving Images about the Built Environment, 2010.

From Rotor to utopia? On Rotor, Fourier and the relation between architecture and utopia

Francesco Deotto & Marcela Garcia
Département de Langue et de Littérature Françaises Modernes, Université de Genève, Geneva, Switzerland
Département d'Histoire de l'Art et Musicologie, Université de Genève, Geneva, Switzerland

ABSTRACT: The aim of this paper is to consider from a utopian point of view the activities and theoretical positions of Rotor, a collective of architects and designers founded in 2005 in Brussels, which, for the moment, has never explicitly written about utopia.

First, we underline how this group is deeply engaged with a question that in the last decades has become more and more crucial for utopian theories and projects: the question of sustainability. We will analyse one of its most recent exhibitions devoted to this topic, *"Behind the Green Door"*, as well a series of examples that show how their work can change our relationship primarily with construction materials, but not only with them.

Then, we consider the relationship between this approach and the work of Charles Fourier. We suggest that despite their apparent distance, their proximity becomes manifest when, instead of the stereotypical notion of *phalanstère*, we look closer into Fourier's texts and into the crucial role he gives to details.

In this perspective, we show how Rotor is valuable to rethink the relationship between architecture and utopia, in coherence with the most recent critical reflections on both of them.

Keywords: Rotor, Fourier, Sustainability, Details, Exhibitions

1 ROTOR IN NUTSHELLS

Rotor was created in Brussels in 2005 as a collective bringing together professionals from different fields—mainly architects and designers—to work in common projects[1]. In many ways, its configuration and work method stands as a new paradigm for architecture firms in a 21st century already marked by world financial crisis such as the one in 2008 and by the decline of the cult of the "starchitect". First of all, Rotor won't define itself as an architecture firm *per se*, seeing that they work in other fields besides architecture, such as research as well as industrial and exhibition design. As opposed to the common model of an architecture firm in which there is a main architect or group that represents it and the rest of the employees that anonymously orbit around it, Rotor functions in a horizontal structure and avoids glorifying any of its members.

The collective has worked in standard architecture projects such as the remodelling of the building for the Galleries Lafayette Foundation in Paris, or the project for the *Mode & Design* (MAD) museum extension in Brussels[2] [2]. However, its most renowned projects are its researches and exhibitions, especially the latter ones, which have allowed it to gain recognition in the rising field of architecture exhibition academia. This stems from its novel approach to exhibition design, which it defines as an alternative to museographic displays based on the presentation of documents and materials pertaining to architectural projects, and to "phenomenological" displays that favour the immersion of visitors in a different space, as it is the case with installations. Rotor, instead, wishes to discuss a topic or ask a question through the presentation of everyday life objects. Another important aspect is that its exhibitions aim at being as objective as possible, avoiding value judgments and letting the visitor construct its own opinion of the subject in question. This was evident in Rotor's contribution for the Venice Architecture Biennale in 2010, which consisted on the presentation of used materials such as an old carpet or wall panels extracted from demolitions. All objects were accompanied by a label with their description and nothing more. Though minimalist in its appearance, this exhibition was actually the result of Rotor's extended research and analysis on the use and wear of construction materials, available in the catalogue (Rotor, 2012).

Rotor was subsequently selected for curating the 2013 Oslo Architecture Triennial, titled *"Behind the*

Green Door, A critical look at sustainable architecture through 600 objects". This exhibition focused on the question of sustainability through the display of a selection of diverse objects deemed sustainable by their creators. Once again, it avoided a prescriptive point of view. It invited the object's authors and different specialists to comment on the objects, but the collective itself, rather than imposing its opinion on them, was above all interested in encouraging the visitors to form their own judgment on the topic.

2 UTOPIA, SUSTAINABILITY AND REUSE

"*Behind the Green Door*" is particularly interesting from a utopian point of view because it questions a topic that is not only at the centre of many political and social debates in the last decades, but also of contemporary thinking on utopia. Indeed, as Jean-Louis Violeau has recently stated, ecology is one of the central themes that composes the utopian horizon in our current context (Violeau, 2013, 77). In relation to this question, this exhibition is important because it constitutes an extremely detailed and diverse catalogue of the different approaches in which architects nowadays are attempting to create sustainable architecture. More precisely, it shows the limits of the discourse developed around the term "sustainability", an "essentially contested concept" (Rotor, 2014, p. 14)[3] that is not outlined in a clear way and that anyone can interpret differently. Rotor underscores the fact that sustainability cannot be really defined in a scientific way, because it is a political matter that must be confronted as one, without being masked by pseudo-scientific theories that also keep the lay public away from the debate, as it has been so far. In the exhibition's catalogue, we can see that each project uses the concept of sustainability according to its own rhetoric, with arguments on its favour but also with certain limits. An evident example is Sir Norman Foster's project for Masdar City, located in Abu Dhabi's vicinity. A so-called self-sufficient city that one could find rather close to the stereotypical utopian city (but which obviously does not exhaust all forms that utopia can have in architecture) that incorporates the most recent innovations in terms of ecology. Yet, once we look closer it is difficult not to see that the city keeps outside of its walls a series of elements from which it benefits but that do not fit the sustainable category, its airport for example. Rotor even notices that Masdar is in reality an Abu Dhabi neighbourhood instead of a truly independent city[4].

Even though characterized by their perceptive remarks, members of Rotor do not suggest that we abandon altogether the concept of sustainability. They explicitly refuse the cynical perspective according to which there are no solutions to current climate, social, and political issues, nor any possibility of change. On the contrary, against any resignation, they invite the public to delve deeper into current initiatives and to think of new ones.

For instance, in relation with a fundamental feature of Rotor's practice—its working with used materials and on existing buildings—we could then mention three related projects. The first is the *Vade Mecum for Off-Site Reuse*, a manual that begins with the many benefits of reusing materials, which include those in the perspective of the global environment. Its main characteristic is to offer very pragmatic steps that professionals in the building sector, as well as in theory any reader, can undertake to discover, extract, and reuse differently abandoned materials[5].

"Opalis", the second project, is an online platform that offers an inventory of companies in Belgium that deal in different ways with used materials, with the purpose of facilitating access to their services[6].

"Rotor Deconstruction", finally, is another more recent website in which members of Rotor resell to the general public materials that they themselves have recuperated from demolition sites: generic materials, but also "more exceptional" ones, such as rare pieces by important Belgian designers, many of them not available in the market anymore[7].

3 ROTOR, FOURIER AND THE "PETITS TRUCS"

Such a way of working could give the impression that Rotor's approach has nothing to do with a utopian project, at least if we refer to the stereotype that associates utopia and monumentality. However, we want to show that things are more complicated and we can prove it by taking one of the most quoted figures when it comes to utopia and architecture: Charles Fourier[8]. On the one hand, in Rotor's projects there is nothing that comes close to the images we have of Fourier's *phalanstères*, that's to say the gigantic collective housing building that Fourier envisioned for 1200 people. But on the other hand, Rotor and Fourier are closer than one could think if we approach the latter from a point of view that starts from one of the essential aspects to define his idea of utopia: the role of details.

As Pierre Macherey recently highlighted (Macherey, 2011, 311–317), but also as Roland Barthes had already noticed[9], Fourier's way of thinking on utopia is based essentially on his extreme attention to details. First, in the sense that

Fourier's work is composed primarily of details. In *Le nouveau monde industriel et sociétaire* there are, for example, meticulous descriptions of pear culture, of the proper art of eating soup or of the infinite varieties of cheese (Fourier, 182). Then, because details are essential to the passage from the so-called Civilization, which Fourier despises, to the world of Harmony. Indeed, according to him, the creation of the harmonious world does not depend on a sovereign decision, on a political revolution, or on an institutional or legislative project. It depends on activities that pertain to our daily life and which constitute exemplary actions, which Macherey describes as *"petits trucs"* (small things) that can change everything, not through a break following a decision, but through a form of "spontaneous contamination", without the intervention of a centralized power (Macherey, 2011, 454).

In Rotor's projects we don't find the affirmation that if we follow its model we will attain the ideal world. However, its approach is undeniably close to Fourier's in the sense that it gives value to small actions and everyday objects. First, because Rotor is interested in the reuse of elements destined to demolition, a practice that architects have systematically underestimated. Second, because when the collective works on a building, it gives extreme attention to every single component and object in it, taking the effort to imagine new uses for them. For instance, the former marble façade elements from the Social Sciences Library in the Université Libre de Bruxelles were transformed into tables[10]. Finally because, as seen in the pavilion for the Venice Biennale that exhibited materials from everyday settings, Rotor encourages the public to change their attitude and their practices: first towards construction materials, but also, in a more general sense, towards all objects that surround us.

4 CONCLUSION

In conclusion, we want to come back one last time to the open character of Rotor's exhibitions and activities, which don't pretend to provide any definitive solutions and are the signs of an ongoing research. We can indeed observe that in this sense Rotor's approach is perfectly aligned with one of the characteristics that utopia has taken in contemporary times. As Miguel Abensour has frequently highlighted, the most interesting utopian texts, starting from the second half of the 19th century, are characterized by their abandonment of pedagogic and monologic postures (Abensour, 2013). In other words, authors such as William Morris increasingly turn

away from precise descriptions of ideal models. They are interested instead, without giving the reader solutions from a presumed knowledge, in inspire in him a utopian impulse, that's to say to make him think of the possibility of a different world.

In this same way, we do not want to restrict to Rotor's approach the relationship between architecture and utopia in the contemporary world. However, two aspects in the work of this collective must be highlighted again: the first one is the critical approach that Rotor implements towards projects such as Masdar, which can be read as continuations of an outdated idea of utopia in architecture. The second one is the way that Rotor encourages a new relationship to objects and materials through a deeper attention to details. And even though utopia may take unpredictable ways, it seems to us that these two elements are essential to foresee the relationship between architecture and utopia in the world to come.

NOTES

[1] The founding members are Maarten Gielen, Tristan Boniver, and Lionel Devlieger. According to their website (http://rotordb.org), as of February 2016, other partners are Michaël Ghyoot, Benjamin Lasserre, Melanie Tamm, Renaud Haerlingen, Lionel Billiet.
[2] For more information on the Galleries Lafayette project, see http://rotordb.org/project/2013_LAF. On the MAD museum, see http://rotordb.org/project/2012_MAD_Brussels.
[3] Rotor refers to a series of cases that W.B. Gallie investigated in his essay Essentially Contested Concepts (1956).
[4] For more information on Rotor position on Masdar project, see Rotor, 2014 (pp. 80, 117, 332–334) and Gielen—Devlieger, 2013.
[5] Available at http://rotordb.org/project/2015_Vademecum_Deconstruction.
[6] http://opalis.be/.
[7] http://www.rotordeconstruction.be/.
[8] See Vidler, 2015, for a recent comprehensive study on Fourier's relationship with architecture.
[9] "Peut-être l'imagination du détail est-elle ce qui définit spécifiquement l'Utopie (par opposition à la science politique)" (Barthes, 1971, p. 110).
[10] These tables are available in the following Website: http://rotordb.org/project/2008_TableManon.

BIBLIOGRAPHY

Abensour, Miguel. *L'homme est un animal utopique.* 2nd ed. Paris: Sens & Tonka, 2013. ISBN: 9782845342156.
Barthes, Roland. *Sade, Fourier, Loyola.* Paris: Seuil, 1971. ISBN: 9782020055112.

Fourier, Charles. Le nouveau monde industriel et sociétaire ou invention du procédé d'industrie attrayante et naturelle distribuée en séries passionnées. Paris: Bossange, 1829.

Gielen, Maarten—Devlieger, Lionel. Pockets of Sustainability. An introduction to the Oslo Architecture Triennial 2013. In *architecture Norway. An Online Review of Architecture* (www.architecturenorway.no/questions/cities-sustainability/pockets-rotor/), 2013.

Macherey, Pierre. *De l'utopie!* Le Havre: De l'incidence éditeur. 2011. ISBN: 9782918193104.

Rotor. *usus/usures. Etat des lieux—How things stand.* Brussels: Editions de la Communauté française Wallonie-Bruxelles, 2010. ISBN: 9782960087833.

Rotor. About Our Exhibition Ambitions. In *OASE Journal for Architecture*. 2012, n° 88. Nijmegen: NAi Publishers. ISBN: 9789462080140.

Rotor. *Behind the Green Door. A Critical Look at Sustainable Architecture through 600 Objects.* Oslo: Oslo Architecture Triennial, 2014. ISBN: 9788299937016.

Vidler, Anthony. Fourier l'architecte. (Translated by Roger, Philippe) In *Critique.* 2015, n° 812–813. Paris: Editions de Minuit. ISBN: 9782707328380.

Violeau, Jean-Louis. *L'utopie et la ville. Après la crise, épisodiquement.* Paris: Sens & Tonka, 2013. ISBN: 9782845342194.

Postmodern questions about *Progetto e utopia*

Jorge Nunes

Departamento de Artes, Humanidades e Ciências Sociais, CIAUD, Faculdade de Arquitetura, Universidade de Lisboa, Lisboa, Portugal

ABSTRACT: The work of the Italian historian Manfredo Tafuri in the domain of history of contemporary architecture stands out by the radical way it addresses the critique of ideology of modern architecture. This theme was consolidated definitively in 1973 with the publication of *Progetto e utopia,* a book that reveals the most combative side of Tafuri and stands out since then as one of the most radical interpretations of modern architecture. Over the past three decades, it was being read and criticized, with many authors drawing attention to the contradictions and prejudices underlying the thought of the author. Although its methods of analysis have since been abandoned, revisiting its theses have now a renewed sense given the economic crisis that devastates the world since 2008, Europe in particular, showing how the "crisis" analysed in this fundamental work of historiography of contemporary architecture retains all its topicality.

Keywords: Tafuri, Utopia, Crisis, Postmodernism, Modern Architecture

The work of the Italian historian Manfredo Tafuri (1935–1994) in the context of history and architecture criticism embody a vast work, which stands out by the number of publications, the temporal scope and especially the radicalism that addresses certain contemporary issues, including criticism of the ideology of modern architecture.

The contemporary cycle begins with the first essays published in *Contropiano* magazine and the first edition of *Teorie e storia dell'architettura*, to stand out definitively with the publication of *Progetto e utopia*. It extends after the 1970s, with works dedicated to the revision of the architecture of the twentieth century, and ends with the publication of *La sfera e il labirinto. Avanguardie e architettura da Piranesi agli anni '70*, the author's final reflection on contemporary architecture.

The critique of ideology had a first formulation in *Teorie e storia*, but it is the essay *Per una critica dell'ideologia architettonica*, published in 1969, in the first issue of the magazine *Contropiano*, which definitely appears as a "working hypothesis to be deepened and consolidated in future". The goal was to "criticize the guidelines of architectural ideology, rereading the history of modern architecture in the light of the methodological tools provided by a critique of ideology understood in the most rigorous Marxist assertion" (cf. Tafuri, 1985: 9). An expanded version of this essay was published four years later, in 1973, in the form of a book with the title *Progetto e utopia. Architettura e sviluppo capitalistico*.

Progetto e utopia reveals the most combative side of Tafuri and remains, still today, as one of the most profound and radical analyses of modern architecture. If *Teorie e storia* is "a positive statement of the critique of ideology", as stated Jean-Louis Cohen, *Progetto e utopia* testifies the passage "to the negative charge of the crisis of the ideology" (Cohen 1995: 51).

Progetto e utopia retraces 200 years of contemporary architecture, considered from the earliest formulations of the architectural theories of Laugier to the experiences of the vanguards in the late 1970s. The book's central thesis is that the course of modern architecture cannot be understood independently of capitalism's infrastructure and that its development takes place within the parameters of this economic and social model. Tafuri understands that there is a structural analogy between the laws of monetary economy that regulate and govern the entire capitalist system, on the one hand, and the movements of the architectural avant-garde, on the other.

The aim of the book is to show that this analogy reflects an ideological subservience of modern architecture, which manifests itself even in positions that appear explicitly to reject the model of bourgeois and capitalist civilization. It is this servility that leads the Italian historian to denounce the ideology of contemporary architecture as a "bourgeois intellectual discipline" (Rosa, 1995: 33), considering its intellectual project as a "process of crisis". Starting from the Enlightenment, architecture has been gradually imprisoned by the strategies of the capitalist rationalism, becoming one of its ideological instruments. It is in this sense that Tafuri speaks of crisis: a crisis of the ideological function of architecture, which makes it irrelevant to the point of questioning its disciplinary survival:

> What interests us at this point is to define the functions that capitalist development took from architecture. That is, that it took from the ideological prefiguration. From this point of view we are almost automatically led to the discovery of what may also seem the "drama" of architecture today: to be compelled to return to a pure architecture, to a kind of private utopia or, in the best cases, to a sublime uselessness (Tafuri, 1985: 10).

The radicalism of *Per una critica dell'ideologia architettonica* and *Progetto e utopia* motivates many authors to argue that the ideological emptying imposed on modern architecture by capitalist development implies, in a way, his "death" as a discipline, as Tafuri acknowledges in the introduction of *Progetto e utopia*:

> The conditions were created to consider our article as a tribute to an apocalyptic attitude, as a 'poetic renunciation', as extreme denunciation of a 'death of architecture'" (Tafuri, 1985: 9).

This radical approach won a wide audience throughout the 1970s, both in Europe and in America. However, the complexity of the texts and the erudition of his intellectual constructs led often to misunderstandings and misinterpretations. Not always his readers realized his intentions, and not always Tafuri was willing to correct these misinterpretations that, with time, were assimilated and assumed as *faits accomplis*, to the point that today integrate architectural culture.

These interpretation problems are not new and are mainly related to how the work of Tafuri was disclosed. In 1995, in a special issue of *Casabella* magazine dedicated to the memory of the Italian historian, Giorgio Ciucci, his *compagnon de route* in the School of Venice, attributes the fragmented nature of the reception of his work to the unsystematic way that his books have been translated and published outside Italy (Ciucci 1995: 13). This happened particularly in the English-speaking world, where these readings and interpretations were based mostly in "poor or even disastrous translations" often published without respect for the original sequence of the Italian editions.

The reception of Tafuri is particularly problematic in the cultural and academic context of the United States, as shown by Joan Ockman (1995). In America there has always been difficult, or even

inability, to follow the transformations of Tafuri's thought throughout the 1970s, and the works after *Progetto e utopia*, past largely unnoticed by American readers. This, as suggested by Ciucci, may be related to how his books were made available in English, but it should be, above all, because *Progetto e utopia* is his work more widely read in American territory.

One of the most important American readings of Tafuri's work appeared in the next decade with the essay *Architecture and the Critique of Ideology* written in 1982 by the critic and literary theorist Fredric Jameson (Jameson, 2000). This text, in spite of the repairs that it motivated in the following years, is one of the most important interpretations of Tafuri's thought due to the ability to penetrate an extremely dense and complex thinking, and, above all, by trying to position Tafuri in the debate on post modernism which dominated almost all disciplinary discussions at that time. The influence of this text is still detectable in the authors who in recent years have addressed the work of Tafuri, as evidenced by the critical readings of Hilde Heynen and Ignasi de Solà-Morales, referred to below.

Architecture and the Critique of Ideology tries to overcome what Jameson understands to be Tafuri's "political impasse", saying, in a tone of controversy, that the pessimism of the Italian historian is determined by the same problems that gave rise to the affirmative and celebrative postmodernism of Robert Venturi (Ockman 1995: 67).

Jameson analyses the Italian historian's work from three perspectives: the first concerns the Marxist context in which this work was produced, with particular focus on *Progetto e utopia*; the second refers to the historiography, and more particularly to narrative history, whose problems and methodological dilemmas determine some of the fundamental concepts of Tafuri work from late 1960s. The latest perspective fits Tafuri within the postmodernism whose doctrinaire principles were in heated discussion at the time when the text of the American critic was drafted. This last issue is the most controversial since, as Jameson relates, Tafuri never recognized in his periodization of historical narrative, the emergence of a post-modern age (Jameson, 2000: 444–45).

Jameson begins with the "crisis" that affects history as narrative since the late nineteenth century, to reconfigure the terms of the critique of representation discussed by post-structuralism, the current of thought that underlies many of the theoretical discussions of post—modernism. In his view, this crisis is rooted in a contradiction difficult to solve: first, "the representation of history tends necessarily to suggest that history is something you can see, you can witness, in which you may be present"—an inadmissible idea for Jameson given the theoretical precepts of post-structuralism. On the other hand, "the story is always the count of one or more stories, being intrinsically narrative in its own structure."

This dilemma arises mainly among authors who, as Tafuri, are committed to a dialectical view of history. Jameson points out a structural feature of this vision that in his opinion is the final condition to which must be "painfully" submitted all those wishing to prosecute the dialectical thinking: "this feature is the sense of need, the necessary failure, the closure, the hopelessly insoluble contradictions, and the impossibility of the future" (Jameson, 2000: 446).

Tafuri's negativity is, in the understanding the of the American critic, a requirement of dialectical thinking; the account of the "growing asphyxia caused by capitalism" and the "settlement of all aesthetic," conveys a "paralyzing and asphyxiating" sense which sees any architectural or urban innovation as mere "worthlessness." Tafuri sees history as "a closed process" increasingly closed by a "totalitarian system" in growing expansion, confirming the view taken by Jameson that dialectical history, "analytical" and "rigorous" is a "stoical resignation of action", almost a "Hegelian resignation of all possible future", representing ever, and under any circumstances, "the story of a failure." The Tafuri's statement about the impossibility of contemporary architecture to achieve something more than "sublime uselessness" (Tafuri, 1985: 10) should be read in this light: instead of being an "opinion" or a "carefully considered position" is a "formal necessity" of the dialectical historiography feeding his work since the second half of the 1960s (Jameson, 2008: 450).

After considering that the pessimism of the Italian historian can be understood as a systemic feature of dialectical thinking, Jameson fits the thought of Tafuri in the discussion of the "end of ideology" promoted in the United States in late 1950s. This discussion advocated the end of classical capitalism, heralding the emergence of a new "post-industrial" social order that the new left currents came to designate as Society of Consumption, Consumer capitalism, Society of the Spectacle.

This new condition of the contemporary world, dubbed Late Capitalism by Jameson gives rise to "new forms of struggle and resistance," as the great upheavals of the 1960s, however, it is also accompanied by a wave of pessimism and disenchantment towards a system connoted with totalitarianism and for which you do not see escape. A system in which the revolt no longer means the emergence of new forces of resistance and a radically different future, but only "a mere inversion" within the system itself, "a one-off change of this or that characteristic" of "a system no longer dialectical in

its strength, but merely structural(ist)" (Jameson, 2008: 451).

The remarkable attribute of this position is the insistence on what is after all a classical notion of Marx, namely that a revolution and a socialist society will only be possible when capitalism runs out, somehow, all its possibilities, becoming a world and global system. The pessimism of Tafuri has in this idea one of its theoretical foundations arguing that there can be no qualitative change in any dimension of the capitalist system, for example, architecture or urbanism, without first having an eradication or "revolutionary" transformation of this system (Jameson, 2000: 452–453).

This insistence on the "asphyxiating" and totalitarian character of Late Capitalism, is also shared by Massimo Cacciari, Tafuri's accomplice in the School of Venice, as pointed out by Hilde Heynen. On the negative thinking of Cacciari—writes this author—the reality is defined based on the conviction that every form of "synthesis", any attempt to reconcile the contradictions, is illusory. Any theory that promises an emancipated society becomes, therefore, impossible. Moreover, any critical system that seeks coexistence with the reality of capitalist civilization is always seen as a crisis phenomenon, which in the end always turns out to confirm the system itself. The only acceptable attitude towards this assumption seems to be an attitude of resistance that promotes a completely disillusioned understanding of its own existence. This resistance does not end any positive definition because it, as Cacciari asserts, would inadvertently take nostalgic or utopian desires, and be, once again, doomed to failure (Heynen, 1999: 145). What remains is the Negative Thinking, that is, the refusal of any dialectical thought.

Heynen following Jameson says that both Tafuri and Cacciari notions of history are monolithic at some degree. The two members of the School of Venice understand modernity as a closed system, within which the policies and cultural practices have no impact on the course of historical development: "In their positions there is little room for practices that may have an effective critical influence on the direction of liberation and emancipation" (Heynen, 1999: 144–145).

Despite all this radicalism, Jameson finds that in the study of specific cases, Tafuri—"the darkest and the most relentlessly negative of all thinkers"—has sometimes a subtler and more ambivalent approach. The book *Architettura contemporanea*, written "in the most favourable atmosphere of 1976" traces a very different picture of the historian's "end of history" dominant in the works of the final years of the 1960s and beginning of the 1970s. According to Jameson, the ambiguity revealed in this work supports two positions simultaneously:

a totally negative and a more Gramscian one, more open and "realistic" (Jameson, 2000: 455).

This ambivalence pointed by Jameson has been one of the most controversial and more fertile topics in the reception of Tafuri's work. Not only because it questions the intransigence image that is traditionally associated with it, but above all because it puts relativism in the centre of a thought that has always opposed all relativisms. The historian Jean-Louis Cohen refers to this problem in a 1995 essay, concerning one the misunderstandings more associated with the work of Tafuri:

> One of the most common misunderstandings consists of viewing his work as simply announcing the death of architecture. (…) Although he articulates a degree of historical pessimism as to the destiny of architecture, it is no less evident that he also proposes to mobilize the forces of resistance and, at the very least, a critical view of modernization. It would be rash indeed to view this as signifying the irremediable failure of architecture as a sphere of human production (Cohen, 1995: 53).

This topic is also addressed by Hilde Heynen, which, moreover, supports her views in Jameson, to analyse the changes operated in *Architettura contemporanea*. As Heynen tries to show the ambivalence detected by Jameson lies less in an easing of Tafuri's doctrinal position, and more in a contradiction that seems to affect the work of the Italian historian since the mid-1970s. This contradiction is related to the incompatibility between the concept of Marxist truth and the concept of truth defended by post-structuralism, with both having active presence in the work of Tafuri. Marxism, in the strict sense of the word, defends the existence of something we can call "objective" reality that allows us to make a clear distinction between ideology and genuine theory: authentic theory explains objective reality, while ideology only gives a distorted and mystified image. Heynen argues that a similar assumption seems to be based on *Progetto e utopia*, which requires a clear and monolithic interpretation of modernity relatively "as a blind historical force that, though clearly scheduled, is not programmatic in the sense that it does not allow us to achieve any conscious project of emancipation and liberation" (Heynen, 1999: 145).

In *Architettura contemporanea*, however, the emphasis is placed from the beginning in the multiple nature of architecture recent history: "obviously the intersection of all these different stories never ends in unity," says Tafuri (Tafuri apud Heynen, 1999: 145). The influence of post-structuralism, which has as its starting point the notion that reality is a subjective construction that cannot be understood except through socially defined catego-

ries—and so distorted—is here more pronounced in the opinion of Heynen. Modern life has such a plurality of layers; the truth is so diverse so that it becomes impossible to portray it thoroughly. The inevitability of this fragmentation produces the negative thinking. This notion is, however, problematic because it is not easily reconcilable with the Marxist conception of a "true theory". To Heyden, this contradiction is permanent and still remains today as a major epistemological problem of contemporary theory.

Tafuri, moreover, is fully aware of the problem. In *Il Progetto Storico*, the introductory chapter of *La Sfera e il Labirinto*, draws attention to the plurality of languages that necessarily architecture and criticism have to deal with. The languages of the project, technologies, institutions and history, cannot be related through a universal hermeneutics; they remain fundamentally apart, are essentially untranslatable, and its plurality is irreducible. This means that it is impossible for architectural criticism to establish a direct link with architectural practice. The two disciplines act within different linguistic systems and their objectives are not reconcilable. This does not mean, however, that criticism is a neutral or arbitrary work: the goal remains to produce an analysis "able to constantly question the historical legitimacy of the capitalist division of labor" (Tafuri apud Heynen, 1999: 146). For that reason, history should be seen as a draft; a project aimed at subjecting all reality to crisis: "The real problem is how to define a criticism able to constantly put it in crisis by putting the reality in crisis. The real and not just its individual fragments" (Tafuri apud Heynen, 1999: 146). In the wake of Cacciari, for whom the submission of reality to the crisis is the driving force behind the capitalist development, Tafuri's notion of history finds a language for the more radical implications of negative thinking:

> The Venice authors—writes Heynen—assume that their analyses represent the only tenable position for a critical intellectual, even though they might not be capable of directly influencing the course of social development. (…) They have no illusions about the actual impact their work may have. They still maintain, however, that it is necessary to continue this labour of Sisyphus and to do everything within their power to subject the standard narratives in the realm of history to a crisis. This stance originates in a combination of a "total disillusionment about the age and nevertheless an unreserved profession of loyalty to it" (Heynen, 1999: 147).

The conclusion of Heynen that Tafuri lives disenchanted with his age while declaring it an unreserved fidelity relates to the last topic of Jameson's essay. After scrutinizing the dialectic historiography, Jameson takes another consequence of the neo-Marxist theory of Late Capitalism: that this period of social and economic history corresponds to the new cultural dynamics of postmodernism, that in the United States was related to the criticism of modernism.

Despite postmodernism cannot be confused with post-structuralism, the ambivalence of *Architettura contemporanea*, pointed out by Jameson and Heynen, and the criticism of high modernism tested in *Progetto e utopia* refers Tafuri's work to a "poststructuralist" if not postmodern dimension. This despite Tafuri always rejected the "weak" way the term was appropriated by architecture, preferring instead the term hyper-modernism, to analyse a production that, in his view, is characterized by a constant fall back of reality.

The question, however, is not limited to a mere reaction to the American context. The problem of rejection of postmodernity by Tafuri lies in its assumption that the "social reproduction" in Late Capitalism takes the same form it had in the previous period of modernism, and that those who are called "post-modernists", both in Europe and in the United States, simply replicate the old modernist solutions at lower levels of intensity and originality, not even reaching the tragic tension of the utopias of modernist artists (Jameson, 2000: 459). If the last show to exhaust the impossibility of architecture, its exhaustion as emancipatory tool, carrying an idea of progress, personalities such as James Stirling, Aldo Rossi, Vittorio Gregotti, John Hejduk and Peter Eisenman "end up accentuating by individualist approaches (private languages in Tafuri lexicon) the residual nature of an activity and a discipline totally useless." In the end, pessimistic and disillusioned "they don't do more than to anticipate, with their lucid experience, the emptiness of a speech totally oblivious to reality" (Solà-Morales, 2003: 249).

Jameson notes a certain inconsistency or contradiction in Tafuri's position face to postmodernity. This inconsistency lies in "the deep historical affinity" between modernism analysis promoted by Tafuri and the postmodernist attacks on this same modernism, including Robert Venturi, whom the Italian historian, however, distances itself from criticizing him violently in *Teorie e storia*. In support of this idea, Jameson recalls that in *Learning From Las Vegas*, Venturi focuses "specifically in dialectics (and contradiction) between building and city, between architecture and urbanism", i.e. in recurring themes in the historiography of Tafuri (Jameson, 2000: 459).

Postmodernism tries to do something quite different from modernism. No longer embodies the utopian ideology of high modernism, and in this sense, is void of any utopian or proto political impulse, although, as the suspect prefix "post" suggests, remains in a certain parasitic relationship with the high modernism that it repudiates. With postmodernism emerges a completely new aesthetic, significantly different from the previous one, which is sought to distance:

> It will no doubt be observed that the symbolic act of high modernism, which seeks to resolve contradiction by stylistic fiat (even though its resolution may remain merely symbolic), is of a very different order and quality from that of a postmodernism that simply ratifies the contradictions and the fragmented chaos all around it by way of an intensified perception of, a mesmerized and well-nigh hallucinogen fascination with, those very contradictions themselves (contenting itself with eliminating the affective charge of pathos, of the tragic, or of anxiety, which characterized the modern movement) (Jameson, 2000: 460).

Regardless some ideas of Tafuri can be read in the light of postmodernity, in particular its analysis of the ideology of modern architecture; we cannot forget that the censorship of modernity is part of modernity itself, as shown by the critical thinking in which Tafuri's thought necessarily fits.

Nevertheless, between postmodernism and Tafuri's pessimism, Jameson finds similarities and a kind of dialectic game, in the sense that both share and are determined by a common problem. Among the cultural pessimism of Tafuri, with its entire rigor and ideological asceticism, and the free and complacent game of a postmodernism that entertains shuffling the cards of contemporary social reality, without committing to any radical change, there are not, in the end, many differences:

> It is possible that these two positions are in fact the same, and that as different as they may at first seem, both rest on the conviction that nothing new can be done, no fundamental changes can be made within the massive being of late capitalism? What is different is that Tafuri's thought lives this situation in a rigorous and self-conscious stoicism, whereas the practitioners and ideologues of postmodernism relax within it, inventing modes of perception in order to "be at home" in the same impossible extremity: changes of valence, the substitution of a plus sign for a minus on the same equation (Jameson, 2000: 460–61).

Three years after *Architecture and the Critique of Ideology* Jameson returns to Tafuri with another influential essay, *The Politics of Theory: Ideological Positions in the Postmodern Debate*. This time presents Tafuri simultaneously as "anti-modernist" and "anti-postmodernist" in relation to several other schools of thought.

Tafuri would certainly have been hostile to the "highly simplified" schemes of Jameson, as notes Joan Ockman. Not only because he refused all "easy" ratings of his own cultural affiliations, but also because he refused all the conventions and categories of historiography (Ockman, 1995: 67). Patrizia Lombardo also criticizes Jameson's reading. According to this author, the use of labels such as "post-Marxism" and "postmodernism" is a reckless attitude by those who want to provide an ideologically correct framework of Tafuri's text. These terms may even be clarifiers, but as Lombardo argues, "they imprison an intellectual effort in settings that paradoxically reinforce a linear, historicist approach, that the very Jameson does not stand to call 'history of Marxism Contemporary'." In this sense this author argues that Tafuri, but also Cacciari should be framed in the "Benjamin's project", or the "conceptual tension of metropolis" (Lombardo, 1993: xxxvii).

Regardless the criticism of Joan Ockman and Patrizia Lombardo, these works of Jameson, especially the first essay, have the merit to question the operability of Tafuri's method, calling into question its systematic pessimism often associated with the death of architecture. In our case, it also serves to show how the most controversial dimension of the thought of the Italian historian is central to the postmodernism debate underway in America in the 1970s.

Josep Maria Montaner also drew attention to a number of contradictions in Tafuri's thought. Most of these contradictions have already been pointed out. However, there is an assertion of this author that deserves attention: the statement that the ultimate objective of Italian historian's research is never the architectural work, but the overcoming of ideological barriers that hide the power strategies and the unmasking of modes of production and power systems (Montaner, 2007: 81–84).

This criticism, enunciated in 1999 schematically and apparently without considering all the ambivalence of Tafuri's thought noted by Jameson and Heynen, is developed in more depth by another Spanish author, the architect and historian Ignasi de Solà-Morales, in an essay published in the following year. In this work, Solà-Morales criticize Tafuri inability to incorporate conventional architectural attributes, i.e. all perceptual and/or emotional values that have no relation with the "critical history" historical needs. A structuralist unacknowledged prejudice seems to remain in Tafuri's interpretive system, leading him to reject sensitive values and any experience of architecture that cannot be reduced to grammar (Solà-Morales, 2003: 252). Thus, the whole experience whose object is the sensible pleasure is disqualified for its critical intellectualism, as if this purpose, which Solà-Morales understands to belong to the art of all time, was the "quintessence of a hedonism denounced with the most negative guilt", as can be read in *Le Ceneri di Jefferson*, the last chapter of *La Sfera e il labirinto*:

The "pleasure" that we withdraw from reading the works of Hedjuk, Eisenman or Venturi is totally intellectual. Satisfaction with the subtle mind games that hold the absolute forms (designed or constructed: in this case, it is indifferent): there is no "social" value on it, is clear. By chance this pleasure it is not totally selfish and private? Easily we conclude that these architectures perpetuate a "treason" against the ideals of the Modern Movement. (Tafuri, 1984: 536).

Solà-Morales concludes that after thirty years of the founding gesture of *Per una critica dell'ideologia architettonica*, rereading the historiographical and critical work of Manfredo Tafuri shows "with a certain cruelty" his own dependencies as well as an obvious ideological agenda easy to identify with currents of thought dominant in the troubled years of the post-May 68. In that sense, the work of Tafuri is outdated from the historiographical point of view and today does not make sense to develop a new radical critique from a new unmasking speech of the ideology.

Solà-Morales focus its criticism on the problem of the "experience" of the subject with the architectural object or the city, relying on interpretive resources of phenomenology. Jameson as good post-structuralist remembers, however, that any relationship of this type is necessarily a socially constructed one. Arguments based on human body conceptions, especially in phenomenology, tendentiously are ahistorical, and involve assumptions about a certain eternal "human nature" that supposedly is always hidden behind the data of physiological analysis apparently "verifiable" and scientific.

Against this case, Jameson argues that the body is in fact a social body, and that there is not a human body in advance, but a whole historical set of social experiences of the body, a range of body standards designed by a number of historical "modes of production" and social formations. In this sense, he states that the "return" to a natural "vision" of the body in space proposed by phenomenology also stands as an ideological, if not nostalgic, opinion contrary to what Solà-Morales intimates (Jameson, 2000: 442–43).

Regardless of the judgment we may have about the "monolithic" materialistic and negative view of Tafuri, the truth is, beyond the implementation and expansion of digital resources, with implications in all areas of the discipline—in the design, construction, representation, in languages and forms—architecture has not changed radically since Tafuri ended the study of contemporary. The biomorphic experiments of authors such as Greg Lynn that pointed to a "brave new world" supported by the new digital tools were attenuated in recent years in favour of more conventional approaches and forms, often oriented towards a return to the modernist tradition.

Today also, there is no effort, as it has existed since the mid-1960s to mid-1990s, to transform the architecture, through a practice rooted in ideologies that seek to question, clarify, and improve the world we live in. A critical view of architecture is now changing, if not in crisis, with the last decade witnessing several attacks on the ideas developed by the authors of the Frankfurt School and the French post-structuralist philosophers, considered by many authors as being exhausted. This post-critical position is not only an abandonment of a socio-political conception of architecture in favour of the reality of professional practice and construction, but above all, the adherence to the realistic and pragmatic ideology of neo-liberalism that has become dominant in the West after the events of 1989. In addition, the economic crisis that devastates the world since 2008 and Europe in particular, shows how the "perpetual crisis" scrutinized in *Progetto e utopia* still retains all its relevance.

BIBLIOGRAPHY

Cacciari, Massimo. *Architecture and Nihilism: On the Philosophy of Modern Architecture.* (Translated by Sartarelli, Stephen). New Haven and London: Yale University Press. 1993. ISBN: 0-300-06304-0.

Ciucci, Giorgio. *Gli anni della formazione.* Casabella. 619–620, Gennaio-Febbraio 1995. Milão. p. 12–25.

Cohen, Jean-Louis. *Ceci n'est pas une histoire.* Casabella. 619–620, Gennaio-Febbraio 1995. Milão. p. 48–53.

Hays, K. Michael (ed). *Architecture Theory since 1968.* 1.st ed. reimp. Cambridge, Massachusetts: The MIT Press, 2000. ISBN: 0-262-58188-4.

Heynen, Hilde. *Architecture and Modernity: A Critique.* Cambridge, Massachusetts: The MIT Press, 1999. ISBN: 978-262-58189-9.

Jameson, Fredric. *Architecture and the Critique of Ideology.* in Hays, K. Michael (ed). *Architecture Theory Since 1968.* 1.st ed. reimp. Cambridge, Massachusetts: The MIT Press, 2000, p. 440–461.

Jameson, Fredric. *Postmodernism or the cultural logic of late capitalism.* London: Verso, reprint 2008. ISBN: 978-0-86091-537-9.

Lombardo, Patrizia. *Introduction.* In Cacciari, Massimo. *Architecture and Nihilism: On the Philosophy of Modern Architecture.* (Translated by Sartarelli, Stephen). New Haven and London: Yale University Press. 1993. p. ix–lviii.

Montaner, Josep Maria. *Arquitectura y Crítica.* 2.ª ed. Barcelona: Editorial Gustavo Gili, 2007, GG Básicos. ISBN: 978-84-252-1768-5.

Ockman, Joan. *Venezia e New York.* Casabella. 619–620, Gennaio-Febbraio 1995. Milão. p. 56–71.

Rosa, Alberto Asor. *Critica dell'ideologia ed esercizio storico.* Casabella. 619–620, Gennaio-Febbraio 1995. Milão. p. 28–33.

Solà-Morales, Ignasi de. *Más allá de la crítica radical. Manfredo Tafuri y la arquitectura contemporánea.* in Solà-Morales, Ignasi de. *Inscripciones.* Barcelona: Editorial Gustavo Gili, 2003. ISBN: 84-252-1913-2. p. 243–252.

Tafuri, Manfredo. *Teorias e História da Arquitectura.* 2nd ed. Lisboa: Editorial Presença, 1988. Translation of *Teorie e Storia dell'Architettura.*

Tafuri, Manfredo. *Toward a Critique of Architectural Ideology.* (Translated by Sartarelli, Stephen) in Hays, K. Michael (ed). *Architecture Theory since 1968.* 1st ed. reimp. Cambridge, Massachusetts: The MIT Press, 2000, p. 6–35. Translation of *per una Critica Dell'ideologia Architettonica.*

Tafuri, Manfredo. *Projecto e Utopia: Arquitectura e Desenvolvimento do Capitalismo.* Lisboa: Editorial Presença, 1985. Translation of *Progetto e Utopia: Architettura e Sviluppo Capitalistico.*

Tafuri, Manfredo; Dal Co, Francesco. *Architettura Contemporanea.* Milano: Electa, reimp.1992. ISBN: 88-435-2463-1.

Tafuri, Manfredo. La Esfera y el Laberinto: Vanguardias y Arquitectura de Piranesi aos Anos Setenta. Barcelona: Editorial Gustavo Gili, 1984. Translation of La Sfera e il Labirinto: Avanguardie e Architettura da Piranesi agli anni '70. ISBN: 84-252-1171-9.

House of utopia: An interface for creativity

José Silveira Dias
*Faculty of Architecture, Research Centre for Architecture, Urbanism and Design,
University of Lisbon, Lisbon, Portugal*

ABSTRACT: In the present context of design research, it is proposed the reflection on the utopia (non-place) as a design methodology for the transformation of thinking and the development of ideas as ideal projection of full freedom of thought. Through an intervention in the entrance area of the Architecture Faculty of the University of Lisbon, this reflection aims to present the Pop Up as an ideal artefact that sees itself, stimulating creativity, revaluating and requalifying the role of participants, approaching utopia and perfection. Thus, it is aspire that the entrance (available place) of the FA-ULisbon develops on it an interface in the search for authenticity and feasibility of creative ideal, fertile for social experimentation with people, materials, and processes. In this scope, creativity is presented as a solution method of collective needs through continuous transformation, based on a dynamic network of relations that establishes a new order in opposition to chaos.

The possibility of a place that can stimulate creativity it is given by the use of no-place (Pop Up), which emancipates the emotional state of the individual through the artefact, in its always-amazing action, promoting behaviours, developing identities and expanding memory. It is proposed therefore to systematize a project that, through its utopian manifesto, defends this interface as a place of creativity stimulus.

Keywords: design, utopia, creativity, pop up

1 INTRODUCTION

The design should serve the needs and desires of all of us. By moving away from its traditional areas, developed on the materiality of the artefact, the design approaches the dominant immateriality of human experience, which is reflected in areas such as learning, creating, caring, living, working, playing, and buying.

The evolution of design in terms of human experience could be in the critical and permanent evaluation of its results, with the objective of individual growth in its relationship in the community and the territory. This evolution does not seem to be indifferent to the fact that the design has learned to use its own creativity to stimulate the creativity of most people. In this context, design is the driving force and generates tools and/or models on which it can express creativity, in particular through the creation of productive systems of intangible objects, such as education, health, politics and knowledge.

It seems important to also underline that experience enables participants to focus in a world of wills and own senses. As stated by Flusser (2010, p. 14), all depends on the intention (design), and both are the result of particular views, as the achievement of their goals can enrich projectable attitude. Until very recently, t is recalled that industrial tools deny such possibilities, precisely because they allowed only their creators to determine the meaning and the expectations of others.

In the present context of research design, which presents the Pop Up[1] as an event design strategy, we propose a reflection on the utopia as a project methodology that legitimizes and guides the development of ideas and the transformation of thought, emancipating, ideally, the projection and granting complete freedom to thinking (More, 2004).

2 NON-PLACE (UTOPIA)

Utopia is commonly understood as the motivation behind the design to perfection, establishing itself as thought and process and not as a result. Consequently, the utopia drives the creative development, sustained in the relationship between a fantasist/sensitive thought of emancipation from reality and a technical/practical thinking response to the same reality. As a non-place, it allows creativity; imagination and fantasy intersect with intangibles such as the beautiful, sensitive, and human, a mixed game of identity and relationship (Augé, 2012, 70).

Utopia, as a way of thought, focuses on ways of thinking about reality, in order to reach eventually a situation of social and urban perfection, based on our needs and concerns, drawing on fantasy, but assuming a lower or higher feasibility.

One can therefore engage in an infeasible utopia (intangible assets), in literary form or graphical thought, however inspiring, and might encourage research, or, on the other hand, a feasible utopia as Pop Up strategy, of an ideal object and in an exercise of approximation of utopia to perfection. The ephemeral nature of the project through its constant transformation gives greater emphasis to the process—through which it seeks perfection—than its own perfection. The Pop Up is justified as an incessant building model in search of the path to perfection. This principle also reveals the conviction of the dynamic and changing nature of the being of Taoist philosophy[2]. The possibility of a place to stimulate creativity can involve the use of a non-place, the Pop Up that in its amazing action promotes the relationship. Each location requires a suitability, which contributes to a stimulating environment. This stimulation, to be achieved through this artefact boost the action of the individual in the place, revealing his sense of belonging and responsibility by appealing to act ethically. The contexts differ and architecture, along with the design, conceives one scenography that seems more appropriate to each place. The value of the place is the result of the meaning of each element (person, product, process) and their respective stories, building memories and identity. Indeed, utopia is based on society as a whole and does justice to the interaction between people, products and processes, as well as the interrelation between functions, institutions and goals (More, 2004). In its practice, the utopian method imagines life as a whole made of interrelations, in the union of the parties, as an organic structure (network) with increasing capacity for organization, whose balance must be preserved to promote the development and transcendence (Mumford, 2007, 13). In the strive to find a place for action, challenge, opposition and confrontation, utopia can also establish a new order in opposition to chaos, because chaos always exists where the creative spirit ceases to exercise control of imagination.

3 A PLACE FOR NON-PLACES

This reflection on the utopia has as intervention area the entrance of the Faculty of Architecture of the University of Lisbon. On the need to better define and characterize the entry of FA-ULisbon, it is proposed here an interface to the experience of the material/tangible with the immaterial/intangible. The aim is to stimulate the creativity of the participants, but also dignify the institution. In addition, it is desired that this interface will be a platform for a set of tactics, strategically conceived with the aim of manoeuvres (like Pop Up) of wonder that we can incorporate into fantasy and seduction worlds in order to emancipate the individual and collective creativity. At the same time, it aims to evaluate, reuse and requalifying the institution itself and consolidate its presence within the area as a home for knowledge. This exercise in the FA-ULisbon entrance is appealing to a feasible utopia, in the form of an artefact (Pop Up), an ideal object. Moreover, with the approximation of utopia to perfection, it is intended to demonstrate its application for the recovery of this place, which is available (Figs. 1 and 2). The spatial vagueness in this place makes us feel the need to delimit the implementation of this interface on the existing reinterpretation. The entrance area of the Faculty of Architecture covers the outer space by the stairs to the west, to the Sá Nogueira Street and the area of access to the building 2, the court of the secretariat, bounded on the north by Technical University Avenue and the south by street internal access to the car park. In a functional perspective, it promotes itself in areas that could integrate versatile uses such as the design, the being, and the sharing. In the absence of a program, it is felt the need to define strategies. For the enjoyment of the place, the tactics to stimulate creativity that can be unleashed are: the expectation; the promise (in addition to specific features of the object) that meets the aspirations and moods that encourage

Figure 1. The interface of Faculty of Architecture of University of Lisbon. Image of Google Maps. Accessed 14 February 2016.

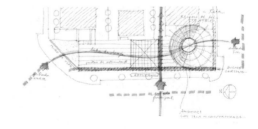

Figure 2. An interpretation of the place, 2016, author's drawing.

and sustain the imagination; by seduction, experiencing the constant call to conversion for innovation; by affection, on relation to the object and the next one. The emotional side, captivated by the repository of memories and stories, triggers the desire to return and thus confirm the promise to expand the experience and intensify the relationship. In an interpretation of the place, three zones are defined: the facade, to call on the promise of an expectation, the route, to the awareness of an experience, the stays, inviting to stay and contemplate, in the confirmation of the relationship (Fig. 2).

To define this interface we developed a facade plan, building a "skin" defined by a reticulated structure (which could be a scaffolding system) representing the relational network. This facade separates the interior from the outside and is covered by a transparent screen. It defines the exterior with a "front" always under construction, a place for dynamic (light, sound, and action) and support for the identification and communication of the institution. For the interior, it surrounds and conforms the territory of the house. This giant and transparent "screen" would allow the identification of the outside of all the "layers" scenographically framing the institution, as a showcase of experiences (Figs. 3, 4, 5 and 6).

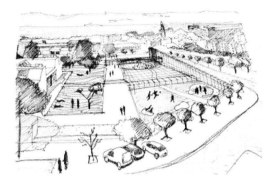

Figures 3, 4, 5, 6.   The House of Utopia, 2016, author's drawings.

4   THE HOUSE

The house is the place where we are most true, authentic, and unique. The house as a general and abstract representation is a "unit" of knowledge that is associated with communication and celebration, the language of objects, colours, and materials, in the will to show who we are and to know ourselves better. We need a house felt physically and psychologically to compensate our vulnerability to the world around us. The spaces where we live are the refuge for our convictions are where our desires take shape and our dreams are preserved. According to Botton (2006, 121), the house is any space able to make more consistent the truths that the world generally ignores or that our heart has difficulty to keep. To feel the house as ours, we must recognize harmony with our own essence, because that is where we meet with ourselves, develop knowledge to interact with the world, and is the place where we lodge references, memories, and stories.

Being a system of references and a perceived image, the house is the abode of authenticity in a physical and emotional interface, where we create the expectation of a promise, practice the experience, and build the relationship (Silveira Dias, 2013).

The entrance (of the house) has the ambition to attract the eye, to seduce and invite into it. The entry should impress as result of the confidence that inspires the house, the prestige assigned to it, and the degree of identification that it transmits. Therefore, we work for the impression of the institution in the territory.

For this impression Zumthor (2006, 11) points out that when we see a person for the first time, we have a first impression. Something similar happens when you enter a building; you see a space and feel immediately an atmosphere. The "first impression" is defined as a promise and its meaning. It creates a resulting expectation of the messages received by the different touchpoints[3] by which the house com-

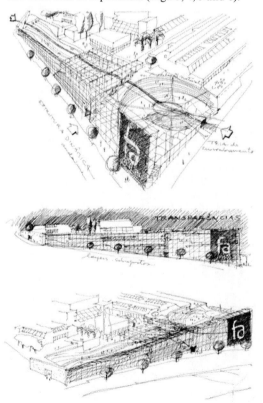

Figures 3, 4, 5, 6.   (*Continued*)

municates: more tangible ones, such as the shape, scale, structure, finishes, light and sound and other more intangible as meaning, identification, movement, anatomy, and atmosphere. At first glance, these are some of the forms of communication that the house has with the individual. Thus, based on the expected, we made the decision to enter and confirm if the house corresponds to the previously formulated promise, longing to see and live scenarios that present themselves as closer with what we dreamed (Loução, 2013, 94).

The entry defines the territory of belonging. We are part of an area in which we had no part before we transpose it. We belong to a place where previously we did not belonged (Bachelard, 1978, 335). It is about that feeling of tension between inside and outside, be in and out, which Zumthor (2006, 46) tells us about the match between the individual and the crowd, between the private and the public, between our deprivation and public display. The facade of a building speaks and a present publicly, give us references and communicates with the outside. It reveals but does not show everything. What is hidden belongs only to those who enter and want to see what goes in there. Even then, not everything is revealed or discovered immediately.

In the invitation to come in, we are encouraged to participate in the experiment that develops during the course: the experience of enjoyment and participation of our consciousness in motion, over the time we remain in this place, inside of it. We stop being visitors to be participants. Each object, surface, texture, light, colour, sound, smell, motion, should continuously strengthen the initial expectation and consolidate our relationship with the house. So, the project in advance, from a cognitive point of view, should try to understand how the participant will "feel" the house and how will you learn, remember and think about this unity of thought. Indeed, strategic thinking in the project is crucial in building a script: the analysis of contextual conditions that determine the materiality or the intangible character of the artefacts in the built environment; the reflection on language, processes, and design, with an emphasis on creative, methodological, and technological issues, in order to encourage innovative and experimental approaches.

In the globalization age, in the context of reformulation of values and identities in which we live, strategic thinking allows structural changes in the issue of addressing the complex systems that form and function perform. In this context, may develop greater permeability tactics in the construction of experience, which is intended to be dynamic and evolving over time. Heraclitus of Ephesus[4] argued that nothing in this world is permanent except change and transformation. Zumthor (2006) promotes a transition architecture, passion for shapes, the effective achievement

of their objectives, which, with the connivance of other arts, architecture refers to the ability to act and link relationships and ability to survive.

5 THE RELATIONAL NETWORK

The definition of the entrance of FA-ULisbon (a house for the knowledge) as an interface for Pop Up events will allow the ephemeral nature to retrieve the place of tradition insofar as it develops a place for the transmission of practice and values between generations and different experiences. In this perspective, it constitutes a specific area of strategic intervention with polyvalent character, reflecting approaches that adequately respond to various scenarios, crossing imaginary, concepts, ideas, techniques, technologies materials and exploiting the complementarities among the various areas of professional interventions (Duarte, 2007, 24).

The institution has the possibility of increasing the level of supply, creating relevant actions, and partnerships. This strategic alignment offers references to students and community, ratings and recommendations, adding value to the experience of knowledge and enhancing the relationship with the institution. The institution, in turn, expands it visibility and recognition with the community in the territory. Partnerships of various areas of knowledge of the institution (architecture, urban planning, design), with the sponsorship of material brands, and individuals invited in demonstrations associated with a unique character communication, will enable the creation of a more holistic view of the perception about the institution. Still, the additional introduction of support for actions services in this interface can provide parallel experiences, providing additional reasons for students to interact more with the institution and feel that they are part of it and develop an ethical responsibility, with the awareness that they contribute to knowledge.

Thus, it is established a network of logic in a system of relations with the actions described above.

Figure 7. The relational network. http://www.takea charcoal.com/string-by-numenfor-use/. Accessed 14 February 2016.

The network morphology seems to be well adapted to the increasing complexity of interaction and to unpredictable models of secondary development of the creative power of such interaction (Kelly, 1995).

6 A PROJECT AS MANIFEST OF UTOPIA

As a reflection exercise developed from the intervention and rehabilitation of the entrance area of the FA-ULisbon, it is proposed an interventional reflection as a new way of thinking, assuming utopia as methodology. On the other hand, facing and appropriating this space as an entry in the Faculty's architectural set (house), is an architectural design challenge on a pilot space where we might want to enable creative responses depending on assumptions and functional arguments associated with large utility spaces as social/meeting/ludic spaces— the fertile interface for social experimentation with people (Bourriaud, 2008, 08). This is therefore a possible definition of constructive design, using sustainable building premises and optimizing recycling, reuse and redevelopment resources.

6.1 *Knowledge areas*

In this particular, we propose the application of the following knowledge areas that through interdisciplinary and their reciprocal action, may decide the project: as main areas, Architecture, Design, and Urbanism; Lighting, Acoustics, Thermal, Multimedia, Inclusiveness, and Sustainability, as secondary areas; Marketing, Cognitive Science, Philosophy, Ecology, Economics, Sociology, and Anthropology as complementary areas.

6.2 *Knowledge areas*

Through the interaction of people, processes and products, is intended to create a synergy between the teachers and the students of this institution, the teachers and students from other national and international institutions (exchange), as well as suppliers and producers of materials, services and technology, brands represented by the respective materials, services and technologies, and invited experts, external to the institution, and the community in general.

6.3 *Intention*

The implementation of this project aims to build an ideal and practical interface, to increase the capacity to reflect, understand and perform actions through the processes, in order to consolidate a strategy of action for, inter alia: to potentiate and to affirm the perception and value of the institution in the knowledge universe; to continuously assess its position in the community; to reclassify the space of the entrance zone as an interface for creativity; to revitalize the image of the institution for recognition, adding value and promoting its growth; to expand the current cultural, technical and artistic spheres, so that it is part of a broader universe; to appropriate areas less consolidated in its structure, such as recreational, social, economic and even political; to meet an economy of means and enhance the "profit" of participants; to stimulate the interaction of participants through support, collaborations and sponsorships; to promote interdisciplinarity between the institution and the community; to observe the behaviour of actors and outline standards of relationship based on ethical responsibility; to promote entrepreneurship activities or the creation of new methods, in order to develop and stimulate services, or any organization of activities and administration; to modify the place through inclusiveness inherent in processes and mind-sets.

6.4 *Methodology*

The methodology proposal is based on an interface for utopia as a way to find results through: anticipation of evolution, growth and any changes through the developed models with visionary sense; monitoring from the start (idea) to the implementation of the project (work), a global, constant and ongoing management among participants, product and processes; continuous evaluation of the behaviour and performance of people face the processes and materials (studied, developed and transformed) and the assessment of their life cycle; observation of the enjoyment of the place (behaviour) through the way it appropriates, adapts and expands people in the place, in relation to the materials and others; constant dynamic construction and guarantee of a file/memory display; reflection of the place for tradition, rapprochement of generations with the transmission of values and practices that contribute to the expansion of knowledge; identification of the life cycle of the interface itself, the ability to anticipate its end, if justified by failure of the overall project, or even through the succession of other more effective. The same applies to the parts (Pop Up) which form the all, if they are exceeded by more effective and specific models.

7 CONCLUSION

In this Project Design is established an ambition for a feasible utopia, validating the Pop Up as ideal object approach to perfection.

Reply and validate through the values of Pop Up (transgression, contamination, portability, revelation, concealment, mishap and manoeuvre) is also using it as a dynamic model of creative strategies with strong economic and social performance, with a view to functional adequacy and enhancement of community life.

This design project aims to answer how the ephemeral sense of Pop Up can help develop behaviour through design and hence consolidate its humanizing capacity (relationship and proximity) in the following possible fields of application: the recovery of the sense of community and new economic and cultural dynamics in the community; in favour of recovery and revitalization of the tangible and intangible heritage; the dialectics with the inside, the outside, the buildings and the mental; the reinvention of application materials (new and used) and their transformation; the reinterpretation of function and form (ready-made) of materials, products and behaviours; the promise, experience and relationship; in the design process as contamination and the definition of its life cycle, for understanding the material ephemerality and the immaterial non ephemerality; the imagination of stimulus globally considered and carried out through inter-relations networks, with increasing capacity for organization, whose balance must be preserved to promote the development and transcendence; and finally the implementation of the non-place for self-consciousness in the appropriation of an affective place.

NOTES

[1] The Pop Up concept is identifiable in the following realities: in a Pop-up book that, when opened, emerges and develops three-dimensional constructions, revealing dynamic and colourful worlds that arise from within the book; in the extra "window" that opens in a web page, whose sudden appearance interrupts the user's routine and sends it to parallel worlds, extra information or advertising; in Retail Design, as a definition of shopping categories. Pop Up Store is distinguished from others by its dynamics in time and space. They are temporary stores, which survive for a short period in the transportation of a brand. They have installation character and derive from guerrilla marketing strategy that uses less conventional and more warlike methods and actions. A Pop Up Store can be mobile or simply settle in an unusual place. It has always meant to be an unforgettable event and in the middle of the crowd. This concept serves to raise the status of the brand and allows consumers access to exclusive products only find in these specific places, or else can also provide consumers to participate in an interactive action and with no product. If we can recognize this category of store used in Retail Design for its event potential (cause/effect), for cell (mobile/ portable) and operation (mishap/ manifest), its evaluation and validation could be a model applicable to other areas beyond retail.
[2] Taoism is a religion arising from the second-century China (Han Dynasty) and originates from an Eastern philosophy, highly valued by Confucianism, known as Tao (way). The origin of the philosophical component of Taoism is attributed to the Chinese philosopher Lao Tse, who lived in the sixth century B.C.
[3] According to marketing strategy, consumer perception about a brand is based on the assumption that the message communicated is always the same in all touch points. Name, logo, product, packaging, service, advertising, environment, building, are some of the contact points—or forms of communication—that a brand has with its audience.
[4] Heraclitus of Ephesus (sixth century B.C.) is considered one of the most important pre-Socratic philosophers, for formulating strongly the problem of permanent unity of being on the plurality and mutability of transitory things.

BIBLIOGRAPHY

Augé, Marc. *Não-Lugares. Introdução a uma Antropologia da Sobremodernidade*.1ª ed. Lisboa: Letra Livre, 2012. 106 p. ISBN: 978-989-8268-14-3.

Bachelard, Gaston. *A Poética do Espaço*. São Paulo: Coleção, Os Pensadores, Abril Cultural, 1978. p. 182–354.

Botton, Alain. *The Architecture of Hapiness*.1st ed. London: Penguin Books, 2006, 290 p. ISBN: 978v0-241-14248-6.

Bourriaud, Nicolas. *Estética Relacional*. 2ª ed. Córdoba: Adriana Hidalgo ed., 2008, 144 p. ISBN: 978-987-1156-56-6.

Duarte, Rui. Imaginários de Futuros Efémeros. ArtiTextos 05, Dezembro de 2007. https://www.repository.utl.pt/bitstream/10400.5/1792/1/FAUTL_13_D_RDuarte.pdf—accessed 14 February 2016.

Flusser, Vilém. *Uma Filosofia do Design. A Forma das Coisas*. Lisboa: Relógio d´Água, 2010. 144 p. ISBN: 978-989-641-036-0.

Kelly, Kevin. Out of Control: The New Biology of Machines, Social Systems, and the Economic World. New York: Basic Books, 1995. 423 p. ISBN: 978-0201483406.

Loução, Dulce. *Paisagens Interiores Para um Projeto em Arquitetura*. 1ª ed. Lisboa: Caleidoscópio, 2013. 128 p. ISBN: 978-989-658-226-5.

More, Thomas. *Utopia*. Brasília: Editora Universidade de Brasília, 2004. ISBN: 85-230-0783-0.

Mumford, Lewis. *A História das Utopias*. Lisboa: Antígona, 2007. 268 p. ISBN: 978-972-608-190-6.

Munari, Bruno. *Fantasia*. Lisboa: Edições 70, 2007. 222 p. ISBN: 978-972-441-357-0.

Silveira Dias, José. *A Casa da Marca—Promessa, Experiência e Relação na A*rquitetura. UTL Repository, Faculdade de Arquitectura, 2013. 326 p. http://www.repository.utl.pt//handle/10400.5/6662.

Silveira Dias, José/Loução, Dulce. *Pop Up the Ephemerality Looking for its Design*. International E-Journal of Advances in Social Sciences, 2015, vol. 1, no. 3, p. 322–330. E-ISSN: 2411–183X.

Zumthor, Peter. *Atmósferas: entornos arquitectónicos, las cosas a mi alrededor*. Barcelona: Gustavo Gili, 2006. 76 p. ISBN: 978-842-522-117-0.

Part III   *Arts*

Considerations on colour techniques in Italian Renaissance painting

Maria João Durão

CIAUD—The Research Centre for Architecture, Urban Planning and Design, Faculty of Architecture, University of Lisbon, Lisbon, Portugal
LabCor—Colour Laboratory of FAUL, Lisbon, Portugal

ABSTRACT: In the Renaissance, with the growth of humanism, painters dealt with classical themes to fulfil commissions for the decoration of the homes of patrons such as the case of Botticelli's 'Birth of Venus' or allegorical content for civic commissions. Other sources of content were found in the writings of humanists or in religious themes, of which Michelangelo's Ceiling of the Sistine Chapel is an outstanding example. The pictorial representation underlying painterly practices were being transformed through the observation of nature, the study of anatomy, the establishment of perspective as a system, as well as the study of colour and light.

This paper emphasizes the use of colour and light in Italy, from Giotto's innovations to the Early Renaissance and how it developed into High Renaissance by invoking the most outstanding technical and artistic contributions of Michelangelo, Rafael and Leonardo da Vinci and their colour systems. The Treatises by Cennini and by Alberti are integrated for their contribution to this innovative era in Florence.

Venice is not contemplated within this paper, not for the sake of its irrelevance, but rather, due to the enormous relevance of the Venetian 'colorito', with its unique and distinct direction and development.

Keywords: Renaissance Painting, Alberti and Cennini, *Cangiante, Unione, Sfumato*

1 INTRODUCTION—THE APPEARANCE OF A PAINTING

Painting is a small part of the whole subject of Medieval and Renaissance art, even though extremely complex. Its complexity involves the origins of a particular technique, periods of its use in different regions, stylistic and technical developments, inventions, circulation of literature and documents regarding materials and technical procedures, some of which considered secret. Thompson's "Materials and Techniques of Medieval Painting" (1956) is an extraordinary account of carriers and grounds; binding media, pigments and metals with descriptions of technical aspects, clarification of effects and functions, variation of methods, craftsmanship, particularities of media and characteristics of the composition of materials, their applications, with and extensive chapter on pigments[1].

One of the variables that determine the appearance of a given painting is the medium used. Besides holding the pigment in place, the medium alters the pigment. The mixing of media with pigment is called 'tempering' and a *tempera* can be any binding agent, even an oil. This is the term's general meaning, although it is associated with a particular technique of mixing egg with pigment

for gessoed panels in medieval painting. Later this technique was less used than oil, and combinations of oil with resin replaced tempera in general in sixteenth century Europe. However, egg tempera was revitalised in early nineteenth century when Mary Philadelphia Marrifield published Cennino Cennini's book "Libro dell'Arte" in its translation to English in 1844. Egg tempera paintings change much less than oil paintings—they dry rapidly and the second coat does not pick up the first as happens in watercolour, and the mixtures in terms of viscosity vary with the amount of water used.

Other factors include opaqueness, transparency and colour changes that some pigments undergo when mixed with some media, as well as the proportion of pigment to medium in a given mixture or even optical effects and appearances that varnishes, waxes, glairs, and gums produced on the resulting work over time. For example, blue pigments are very often more affected by media. Two accounts are given by Thompson (1956: 56–57) that exemplify the influence of media on the resulting colour appearance. The first is "Livro de como se fazen as côres", written in Portuguese, but in Hebrew characters in the fifteenth century or even earlier says to temper dark blues with gum; but for light blues to add white lead and temper with glair. Mixtures of gums and glair were often

Figure 1. Santa Maria del Fiore and Brunelleschi Dome, Florence. Photo credit M. J. Durão, Accademia Gallery, Florence.

concocted and partook of the advantages of each component. The second appears in Ambrogio di Ser Pietro of Sienna writings in 1463:

> If a blue is full coloured, the tempera wants to be weaker in gum; and if the blue is weak, the tempera wants to be stronger in gum.

2 COLOUR IN THE TRECENTO: "IL LIBRO DELL'ARTE"

Medieval colours used in the Renaissance included bone black, lamp black, vine black; white lead, gypsum, kaolin, orpiment, massicot, yellow, burnt ochre, raw umber, burnt umber, raw sienna, burnt sienna, red lead, red bole, vermillion, cinnabar, brazil wood lake, madder lake, sap green, green earth, verdigris, malachite, green, azurite, flower of woad, ultramarine.

"Il libro dell'Arte" written by Cennino d'Andrea Cennini (1933) describes in detail the very methods used in the fourteenth century—the very methods that he used himself. It contains practical advice on "how to paint" almost everything that a late medieval painter would need for painting in egg tempera and in fresco. It was not published before the nineteenth century, but must have circulated among painting workshops at the time and thereafter. To illustrate Cennini's descriptions, Bomford and Roy (1994) chose Nardo di Cione's altarpiece "Saint John the Baptist with Saint John the Evangelist and Saint James", **painted** for the church of S. Giovanni Battista della Calza in Florence ca. 1365. Effectively, the panel is gilded, the haloes punched and worked upon as the artisan procedure of the Trecento. The gold brocade carpet depicted by Nardo di Cione in this altarpiece corresponds to the painting technique used by Cennini for making a cloth of gold:

> If you want to make the cloth red, lay in with vermilion over this burnished gold. If you need to put

any dark on it, put it on with lac; if you need to put on lights, put them on with red lead; all tempered with yolk of egg (...) and in the same way, if you want to make them green, black, or any way you like. (Cennini 1933: 86–87)

The correspondence between Cennini's recommendations for painting in panel with egg tempera and Nardo di Cione's altarpiece extends to the treatment of the draperies:

> Always start by doing draperies with lac ... leave the first value in its own colour; and take the two parts of lac colour, the third of white lead; and when this is tempered, step up three values from it, which vary slightly from each other: tempered well, and always made lighter with white lead worked up.

A very significant part of Cennini's instructions deals with the method he used to create volume with colour in the folds of the drapery (Cennini 1933: 92):

> Start to apply the dark colour, shaping up the folds where the dark part of the figure is to come. (...) take the middle colour and lay in the backs and the reliefs of the dark folds, and begin with this colour to shape up the folds of the relief, and around toward the light part of the figure, then take the light colour and lay in the reliefs and the backs of the light part of the figure.

The system of the three colours is stressed in the final explanation of the procedure:

> When you have got it well laid in and these three colours blended, make another lighter one out of the lightest ... and out of this lighter one make another lighter still; and have the variations among them very slight. Then touch in with pure white lead the strongest reliefs. And make the darks, gradually in the same way, until you finally touch in the strongest darks with pure lac.

Cennini describes colour combinations for the painting of fabrics, none of which correspond to Nardo's, except that he uses 'cangiante' or 'shot' draperies. One saint wears a green 'cangiante': the middle tone is an almond green shade, the shadows are blue, and the highlights are pale green-yellow.

By the end of the 1400s, oil painting was used all over Europe but that did not prevent the development of fresco painting in major works made by Rafael, Michelangelo, and Carracci for example. Although "Il Libro Dell'Arte" is mainly about painting frescoes and egg tempera, a few chapters touch upon the use of oil on a wall, on panel, or on iron. In his account, there is an allusion to Northern Europe´s use of oil: "I want to teach you to work with oil on wall or panel as the Germans are much given to do". (Cennini 1933: 57)

Later on, he recommends tempering with oil when you need to paint velvet cut threads: "make the cut threads as the velvet requires with a minever brush, in a colour tempered with oil..." (Cennini 1933: 89–90).

The second section of "Il Libro dell'Arte" addresses colours specifically in terms of their "character" (the coarsest and the most fastidious; the colours that need to be worked up or ground; the colours that call for tempera or another medium. The approach he uses is overall instructive and informative. As an example, you may find such titles as "How to make a green with white lead and terre-verte; or lime white" or "On the character of the Red colour called sinoper"; "On the character of the Red called vermilion, and how it should be worked up"; "How to make the Red called cinabrese for doing flesh on the wall"; "How to make a green with orpiment and indigo." This section also includes Cennini's naming of different colours; black; red sinoper, cinabrese, vermilion, red lead, hematite, dragonsblood, lac; yellow; light ochre, dark ochre; giallorino; orpiment, realgar, saffron, arzica; green, terre verte; malachite (blue-

green); verdigris; white lead, lime white; azurite and ultramarine. Aligning with tradition, Cennini claims "seven" natural colours, four of which mineral: black, red, yellow and green; three are natural, but need to be helped artificially, as lime white, giallorino, azurite, and ultramarine.

3 TRANSITIONING TO THE RENAISSANCE

From 1434, the Medici dominated Florence and with their wealth supported all the expressions of arts and humanities, gathering around them a group of poets, musicians, architects, sculptors, painters and scholars. A family of tradesmen, the Medici reached all the major centres in Europe and their agent Tommaso Portinari commissioned one of the most famous masterpieces of Northern Renaissance art, the "Portinari Altarpiece" by Hugo van der Goes (ca. 1476). This painting used translucent oil glazes, a technique that produced brilliant colour interactions.

Oil paint was used in glass and stone as early as the eighth century but during the early fifteenth century, Jan van Eyck, Hubert van Eyck, and Roger van der Weyden notably perfected the technique of oil on panel by using the medium to model space, objects and figures. The fact that oil takes longer to dry and that it blends easily on painting surfaces allow for layers of paint to be added without the layers painted underneath. On the other hand, colour modelling may be achieved by addition of colour glazes. The transparency afforded by glazes creates both refined detail and depth of space in the depiction of architectural space—a vehicle for creating the third dimension. Antonello da Messina is often credited with introducing oil painting into Italy and Masaccio as the founder of Renaissance painting with a succeeding generation of painters like Piero della Francesca, Pollaivolo and Andrea del Verrochio.

The technique made possible with the use of oil liberated the painter from the limitations of painting with egg tempera alone. For example, in "Il Libro dell'Arte" volume was modelled by using pure colours for the shadows and by adding white to pure pigment to achieve mid-tones and more white for the highlights. A variation of this technique using a completely different colour for the shadow is called 'cangiantismo' (*cangiante*). In this case, the shadow hue is still unmixed-except with white-, but never with black, to preserve its brilliance. Giotto was credited for the introduction of this technique, and it is therefore not strange that Cennini would pay a tribute to him in various passages while establishing a lineage of painters: Giotto had Taddeo Gaddi of Florence as his

Figure 2. Pacino di Bonaguida, tempera and gold on panel (1320–1329). Photo credit M. J. Durão, Accademia Gallery, Florence.

pupil for 24 years, Taddeo had Agnolo, his son, and Agnolo had Cennino for 12 years as a pupil (Cennini 1933: 46).

In "The Modes of Coloring in the Cinquecento", Hall (1994) assigns the technique of 'cangiantismo' to Michelangelo. She argues that in the High Renaissance there evolved four systems of colouring, each demonstrated by a painter: 'cangiantismo' by Michelangelo; *sfumato* by Leonardo; chiaroscuro by late Rafael and Sebastiano del Piombo; and *unione* by Rafael.

Michelangelo's painting of the Sistine Chapel in 1508 revived *cangiante* changing hue when the initial one cannot be made light enough. The inverse is also *cangiante*: the change to a darker hue when the initial hue cannot be made dark enough. One of the most outstanding practitioners of the technique was Michelangelo in the ceiling of the Sistine Chapel. For example, the use of *cangiante* can be seen in the transition from green to yellow in Prophet Daniel's robes.

A technique not in use at a time that other modelling techniques were experimented, such as *sfumato*, chiaroscuro, and *unione*. Michelangelo's *cangiantismo* takes *cangiante* further by using it not only in draperies of Angels, but in entire compositions, pairing colour hues symmetrically, named 'isochromatism'. 'Isochromatism', a term invented by John Shearman (1987) consisted of a system of organizing colours, such as the symmetrical repetition of the colour of the vestment of an angel. The use of these patterns for aesthetic outcomes, the vibrant colours and saturated hues, the gold and silver, the overall gilding and ornamentation, served the purpose of medieval painter in representing the unearthly and transcendent.

The Sistine Chapel's frescoes have a distinct brilliance, since Michelangelo operated on the shifting of hue, rather than the value or tone of colour, and the range of contrast is wide due to the use of white to de-saturate the hue. The depth of composition is achieved through the balance of luminance. Yellow, for example has enough luminance, so it does not need to be de-saturated. Michelangelo revived Giotto's *cangiante* using vivid hues for shadowing and thus contrasting with the de-saturated highlights, instead of following Alberti's advice.

Approximately forty years after Cennini's publication, Leon Battista Alberti wrote "Della Pittura" (On Painting) (Alberti 1991) in 1435–1436, a treatise of Renaissance painting in Italy. "Della Pittura" was not a manual for the painter. Rather, it was a treatise on the theory of art and even though it considers the use of colour, it has a different approach from Cennini's treatment of each individual pigment and colour uses, or their mixtures and combinations. Alberti's treatise focuses on form and its relationships with light-shadow and colour.

> I have taught how the same colour, according to the light that it receives, will alter its appearance. I have said that white and black express to the painter shade and light; all other colours for the painter are matter to which he adds more or less shadow or light. (Alberti 1991: 82)

Alberti's "Della Pittura" provided advice on how to paint in accordance to nature in a more realistic approach where use of light and shadow are used to create realism, relief, volume, depth, distance and the experience of colour.

Shearman (1962; 1992) addresses 'tonal unity', claiming that it superseded *cangiantismo* and all medieval colour techniques, since they became unappealing toward the end of the fifteenth century.

Leonardo found a balance among brightness of hues, by controlling the dimensions of hue and brightness separately, thus achieving *unione*, i.e. equal brightness. The 'tonal unity' that is found in Flemish oil painting is due to 'down modelling'- one of Alberti's aspects of colour shading.

Leonardo used *sfumato* to attenuate colour outlines and chiaroscuro, a procedure for colour modulation with light and dark as did Andrea del Sarto. While chiaroscuro forces high contrast of value, *unione* is softer in the search for the right tonal key. The tonal key is found when there is harmony between light and dark, without the excesses and accentuation of a chiaroscuro composition.

Searching for greater realism in painting, Leonardo's *sfumato* was a quite different solution from Michelangelo's *cangiantismo* and Rafael's *unione*, and later chiaroscuro. In the gradual transitions of *sfumato*, colour shadows soften, even though defined. If we classify in a sequence the quality of gradual and subtle transitions, then *sfumato* would come first, followed by *unione*, then *cangiante* and chiaroscuro, last.

Rafael's 'School of Athens' represents an innovative contribution to the Quattrocento colour naturalism. That was done using 6 techniques (Bell 1991): (1) variety in colour arrangement; (2) *unione* of the whole; (3) *unione* and *dolcezza;* (4) relief; (5) the massing of light and colours; (6) chiaroscuro in perspective.

Rafael did not want to part from the use of brilliant colour effects, but found an equilibrium in the use of *belezza di colore* and *sfumato*, achieving at once soft shadowing and tonal unity. Later works by Rafael combined *sfumato* with chiaroscuro. Rafael was the master of *unione* leading the way to chiaroscuro, that Hall (1994) considers the modes of colour of the High Renaissance in Italy.

Figure 3. Francesco de Rossi, Madonna and Child, the young St. John and an Angel, oil on wood (1543–1548). Photo credit M. J. Durão, Accademia Gallery, Florence.

4 CONCLUSION

The previous disregard for a naturalist representation of depth and the experience of colour conflicted with Brunelleschi's linear perspective that dictated the representation of spatial depth, recession, formal volume, and relief. This required the understanding of light and its impact on the depiction of the colour phenomenon.

Unlike the unnatural medieval painting, Alberti's system aimed to depict a more natural experience of colour. The emergence this new system was made possible with the introduction of oil as a frequently used medium. Oil allowed for a large amount of experimentation with colour to take place due to the inherent qualities of the medium: transparency; slowness to dry benefiting fusion and blending of colours, underpainting, 'up modelling' and 'under modelling'. When 'Della Pittura' was published in 1435–6 both systems co-existed, but it prepared the direction not only for art, but for the artist and the patron as well.

By 1434 in Florence, Brunelleschi was closing the Dome of the Cathedral, Donatello had completed much of the Cathedral's sculpture, Ghiberti's first doors had been concluded; Masaccio was dead but his work was new in Florence. 'Della Pittura', owing to the art of Brunelleschi, Donatello, and Masaccio, was a fundamental source for other painting treatises of the fifteenth century such as those of Filarete, Piero della Francesca, and Leonardo da Vinci, besides painters all over the world, thereafter. This is one of the reasons why Jacob Burkhardt claims that Alberti was the first universal genius (Burkhardt 1951).

NOTE

[1] On materials and painting techniques, see Maria João Durão (2006; 2008; 2015).

BIBLIOGRAPHY

Alberti, Leon Battista. *On Painting (Della Pittura)*. London: Penguin, 1991.

Bell, Janis. "Writing out Colour in Renaissance Theory." *Genders*. 12 (1991): 77–99.

Bomford, D, and Roy, A. *Colour*. London: National Gallery London, 1994.

Burkhardt, Jacob. *The Civilizations of the Renaissance in Italy*. London: Phaidon Press, 1951.

Cennini, Cennino d'Andrea. *The Craftsman's Handbook (Il Libro dell' Arte)*. New York: Dover, 1933.

Durão, Maria João. "Contributions to the Understanding of Colour in Nature and Art: Diogo Carvalho e Sampayo's Tratado das Cores." In *Views on Eighteenth Century Culture: Design, Books and Ideas*. Eds. Ferrão, L. and L.M. Bernardo. Newcastle upon Tyle: Cambridge Scholar Publishing, 2015. 281–301.

———. "Sketching the Ariadne's Thread for Alchemical Linkages to Painting." *Fabrikart-Arte, Tecnología, Industria, Sociedad*. 8 (2008): 106–123.

———. "O Diáfano e o μέλας." Ar—Cadernos da Faculdade de Arquitectura da Universidade Técnica de Lisboa,6 (2006): 144–147.

Hall, M. Colour and Meaning: Practice and Theory in Renaissance Painting. Cambridge: Cambridge University Press, 1994.

Shearman, John. "The Function of Michelangelo's Colour." In *The Sistine Ceiling: A Glorious Restoration*. Ed. Pietrangeli, Carlo New York: Harry N. Abrams, 1992.

———. "Isochromatic Colour Composition." In *Colour and Technique in Renaissance Painting*. Ed. Hall, Marcia. Locust Valley: Augustin, 1987.

———. "Leonard's Use of Colour and Chiaroscuro." *Zeitschrift fur Kunstgeshichte*. 25 (1962): 13–47.

Thompson, Daniel. The Materials and Techniques of Medieval Painting.

Colour harmony: The ideality of pleasurableness

Saadet Akbay Yenigül
Department of Interior Architecture, Faculty of Architecture, Çankaya University, Ankara, Turkey

Maria João Durão
CIAUD—The Research Centre for Architecture, Urban Planning and Design, Faculty of Architecture, University of Lisbon, Lisbon, Portugal
LabCor—Colour Laboratory of FAUL, Lisbon, Portugal

ABSTRACT: The search for the essence of colour harmony has a long tradition that, being a quest for aesthetic values, remains a contemporary question insofar as it addresses the interrelated issues of both beauty and pleasure. Colour harmony has been discussed in terms of two different points of view. As a measurement of aesthetics, the researches of colour harmony are based on the discovery of its systematic rules by identifying the relationship between colours and its aesthetic value in beauty and harmony. The proportional and orderly arrangements of colours and their relations to mathematics are the main concerns of this first approach. As a measurement of emotion, colour harmony is regarded as subject matter of pleasure, subjective feeling which is peculiar to an individual. Relying on the second approach, many studies have been conducted to identify the reasons behind why colour combinations are perceived as beautiful, pleasant, and harmonious. Thus, this paper is a retrospective review of the literature of colour harmony, its theories, and principles considering the two approaches. The assumption is that, in either case, colour harmony is grounded in a search for the ideality of pleasurableness.

Keywords: Colour Harmony, Complementary Colours, Colour Order Systems, Pleasantness, Aesthetics

1 INTRODUCTION

"The enjoyment of colours, individually or in harmony, is experienced by the eye as an organ, and it communicates its pleasure to the rest of the man."
Johann Wolfgang von Goethe

In 'The Principles of Psychology', James states that the primary layer of emotional responses involves subtle feelings like pleasure which is elicited by harmonious combinations of colours, lines and sounds (apud. Cupchik 1994). According to O'Connor (2010), the general acceptance of colour harmony, upon the definition of Burchett (2002: 28) is "Colours seen together to produce a pleasing affective response are said to be in harmony." Depending on the pleasing effects of colours, searching for the essence of colour harmony has a long tradition throughout the history and scientific world by addressing the question, which colours in combination are harmonious, beautiful, and yield pleasure. Some Greek philosophers, according to whom beauty and harmony were inseparable notions that could only be explained by mathematics, held debates on the evaluation of the nature of colour, its beauty, and harmony. The utilisation

of mathematical order to describe the notion of beauty forms the basis of colour harmony theories from ancient periods towards contemporary times (Holtzschue 2011). The thinking of colour in earliest times was largely based on the hypothesis of Aristotle (350 B.C.), who assumed that "Whatever is visible is colour and colour is what lies upon what is in its own nature visible." (Macadam 1970: 2). Aristotle also asserted that, "[...] sunlight always becomes darkened or less intense in its interactions with objects." (Nassau 1998: 4) thus "[...] what is seen in light is always colour." (Macadam 1970: 3). In accordance, he believed that "[...] white and the black could be juxtaposed in quantities so minute that (a particle of) either separately would be visible, though their combination [...] would be visible; and they could thus have the other colours for resultants." (Macadam 1970: 6). In other words, all colours were the mixtures of black and white, or darkness and lightness, and derived from the four elements; i.e. earth, air, fire and water. According to Aristotle, the element fire was considered as white and the earth was black. The elements of air and water had no colours of their own; however, the air had the ability to make the colour whiter and the water made the colour darker. In his work, 'Sense and the Sensibilia' Aristotle asserted that, between black and white, there existed five colours which were red, violet, green, dark blue and grey (for the alternative was yellow) (Kuehni 2003; Sorabji 1972; Zollinger 1999). The total number of seven in Aristotle's colour scale would later influence Newton's adoption of 'seven' perceived colours of the spectrum (Zollinger 1999). Aristotle's ideas were later employed to develop the modern understanding of science of colour and colour harmony theories (Taft 1997).

Colour science is conventionally accepted to begin with the experiment done by Sir Isaac Newton in late 17th century. He was interested in the physics of colour, the relationship between light and colour, and the mixtures of colours (Zollinger 1999). In his experiment, Newton discovered that when the sunlight passed through a triangular glass prism the light was bent and refracted. This refraction of the light resulted in an array of different colours. Newton stated that "[…] Light [white light or sunlight] is a heterogeneous mixture of differently refrangible Rays, and a mixture of colours" (Apud. Nassau 1998: 5). He distinguished red, orange, yellow, green, blue, indigo (blue-violet), and violet in the visible spectrum. Kuehni states that, Newton nominated seven colours of the spectrum not only because he was in agreement with the spacing of musical scale but also since it was a classical number in ancient periods (Kuehni 2003; Kuehni 2005). More importantly, he believed that there was a strong analogy between musical harmony and colour harmony. In his book 'Opticks', Newton mentioned that the two ends of the colour spectrum, i.e. red and violet, had a correspondence with the two ends of the octave in music (Zollinger 1999). Thus, according to him, whenever the notes and colours were in combinations, their relationship could be harmonious (Green-Armytage 1996).

Although the search for relationships between colour and its aesthetic value in beauty and harmony dates back at least since Aristotle, Newton's work not only on the nature of light as well as on the nature of colour, forms a philosophical and scientific understanding of the colour and colour harmony. Beginning in the 18th century, after Newton's 'Opticks', theories on colour harmony developed and have been discussed throughout the years.

This paper, thus, is a retrospective review of literature on colour harmony, its theories, and principles. Colour harmony is considered both measurement of aesthetics and measurement of emotion (O'Connor 2010), and the underlying assumption of this paper is that, in either case, colour harmony is grounded on seeking for the ideality of pleasurableness.

2 HARMONY IN COMPLEMENTARY COLOURS

Green-Armytage states that "In colour theory, according to a long tradition, a combination of complementary colours will be harmonious and, therefore, pleasing." (1996: 205). In the 15th century, Leonardo da Vinci dealt with the idea of colour contrast. According to him, the most beautiful combination of colours was the pair of contrast colours (Xiao-Ping 2007). Da Vinci claimed that (apud. Green-Armytage 1996: 206) "A direct contrary is […] black with white, […] blue with a yellow as gold, green with red […] and yellow with blue." However, writing on colour being rare, da Vinci's idea was not established. As has already been stated, the milestone for the scientific world on colour is considered the results of the experiments of Newton in the 18th century. Newton, however, dealt with more on the additive mixture of light colours. In about 1730, LeBlon suggested the first concept of subtractive mixture of the pigment colours. He discovered that the three primary hues red (magenta), yellow, and blue (cyan) could not be produced using any other colour and all other colours were the result of the mixtures between these three primaries. (Holtzschue 2011; Westland 2007). With his CMY approach, LeBlon provides a basis for mixing pigments of today printing. In the mid-18th century, Harris used LeBlon's approach of the three primary colours in a colour circle which was based on the subtractive mixing of the pigments (Zollinger 1999). According to Harris, the primary colours red, yellow, and blue were so dissimilar that, they should be positioned at a very far apart from each other on the circle. Later, in the early 19th century, Goethe adopted Harris's circular organisation and suggested that completing colours was pleasing (Green-Armytage 1996; Holtzschue 2011).

In his book, 'Zur Farbenlehre' (On the Doctrine of Colours), Goethe claimed that the complementary relationship of colours was based on after-images of the colours. This effect was called successive contrast. Both successive contrast, and the detailed studies on simultaneous contrast, formed his theory of colour harmony (Green-Armytage 1996; Zollinger 1999). Goethe developed his idea with his six-colour circles that comprised of two sides; the positive side (i.e. red, orange, yellow) and the negative side (i.e. green, blue, and violet). According to him, the analogous pairs of colour combinations did not have a character and their combinations were not fully harmonious (Kuehni 2005; O'Connor 2010; Ou 2004). Kuehni explains Goethe's statement as follows;

There is a scale of declining harmony from perfection to character to without character. Lightness and darkness complicate harmonious relationships [...]. Active colours (yellow, orange, red) [positive side of the colour circle] gain energy when combined with black or dark colours, but loose energy when combined with white or light colours. Passive colours (violet, blue, green) [negative side of the colour circle] look dark and foreboding when combined with dark colours, but gain cheerfulness when combined with light colours (apud. Kuehni 2005: 166).

In accordance, Goethe claimed that, when colours were selected from both sides of the colour circle, opposite to each other, their combinations produced a splendid effect, thus, this resulted in full harmony (Ou 2004). Afterwards, in the mid-19th century, Chevreul developed a more systematic theory of colour harmony.

Chevreul's theory of colour harmony consisted of two principles, harmony of analogy and harmony of contrast (Ou 2011). He also accepted the three pigment primary colours, red, yellow, and blue, and developed his theory of colour harmony according to a 72-hue circle (Kuehni 2003). The first principle of the harmony of analogy included the analogous/adjacent/nearby colours of the same hue, saturation, or lightness. The second principle of the harmony of contrast not only involved the opposing hues, but also the contrast of lightness or saturation within the same hue (Kuehni 2005). O'Connor states that "Chevreul championed 'complementary' colours and their contribution to colour harmony and he equated maximal contrast of the complementaries with maximum harmony." (O'Connor 2010: 268). Chevreul believed that, "In the Harmony of Contrast the complementary assortment is superior to every other." (Green-Armytage 1996: 206). In his book, 'The Principles of Harmony and Contrast of Colours and Their Applications to the Arts', Chevreul classified four types of rules for the harmonies of colour contrast which were direct complementary, double complementary, split complementary, and triad schemes (Xiao-Ping 2007). He also suggested that when colour combinations were seen as dark, the use of white helped them to make lighter; when they looked bright black could be used to balance them; and also when two colours were perceived as disharmonious, white or black could be used as a separation colour. Chevreul's principles of colour harmony have been regarded as the most important guidance for colour education in the departments of architecture, design, fine arts, and etc. (Ou 2004).

3 HARMONY IN ORDERLY ARRANGED COLOURS

Other colour theorists (i.e. Munsell, Ostwald, Itten) later followed Goethe and Chevreul's works on harmony of complementary colours in the late 19th and early 20th century. The major scope of these latter researches was to find out the laws of colour harmony by using rules, control, and order. The measurement of aesthetics of colour harmony was the main subject matter of this period which could only be solved by creating colour order systems with the use of mathematical-balanced representations (Holtzschue 2011). Alongside the hue of a colour, the other two attributes, lightness and saturation of that colour were considered and three-dimensional colour solids or models were presented by many of the theorists (Xiao-Ping 2007). Although Runge developed the first colour solid in the late 18th century, Munsell and Ostwald's colour solids are considered the most important colour models.

Munsell claimed that "What we call harmonious colour is really balance." (apud. Holtzschue 2011: 142) and also stated that "Visual comfort is the outcome of balance." (apud.Arnkil 2008). In 1905, he published a book, 'A Color Notation System', to describe his rules of colour harmony by developing a three-dimensional model. Munsell systematically arranged the colours due to their three perceptual attributes, hue (Munsell hue), value (Munsell value), and chroma (Munsell chroma), in equally graded steps (Kuehni 2005). In his system, the geometric arrangements of the three attributes of colour were in a cylindrical form that is called the Munsell colour space. Munsell hue (H) was placed on an equal numerical interval with an equal perceived difference of attribute that was represented along a circle. This circle consisted of five elementary hues, i.e. red (R), yellow (Y), green (G), blue (B), and green (G); and five intermediary hues, i.e. YR, GY, BG, PB, and RP. These ten Munsell hues were again divided into ten to give 100 hues. On the other hand, the scales of Munsell value (V) and Munsell chroma (C) were based on the ratio scales. Munsell value had ten equal steps from zero (absolute black) to nine (absolute white) where the achromatic scale was represented vertically along the axis of the circle. Munsell chroma represented the chromatic intensity/strength/purity of colours that started from the value scale axis in the centre. Then it radiated in equal steps starting from 0 (zero chroma) and reaching at the brightest hues to a maximum of 17 outward to the periphery of the circle. The complete notation of a colour in Munsell system is expressed as H V/C (Choundhury 1996; Mahnke 1996).

Munsell suggested a set of practical principles in order to attain colour harmony. According to him, colours should be found in a specific path in his colour space. These paths should include the following principles;

a. The grey scale;
b. Colours of the same hue and the same chroma;
c. Complementary colours of the same value and the same chroma;
d. Colours of diminishing sequences, in which each colour is, dropped one-step in value as chroma go down one-step;
e. Colours on an elliptical path in the colour space (Ou 2004: 67).

Munsell also recommended the use of colour strength (chroma) in order to balance colour areas "[...] a small area of high colour strength would balance a large area of low colour strength, [...] a strong colour should occupy a smaller space to balance a weak colour." (Westland 2007: 9). Today, the Munsell colour system is widely known as a uniform colour order system. Alternative to the Munsell system, the Natural Colour System (NCS) was developed by Hård, Sivik and Tonnquist in late 20th century (Hård, Sivik and Tonnquist 1996). The NCS was based on Hering's phenomenological opponent-colour theory from 1874. Hård et al. claim that the NCS as a colour system is based on the relationship between the visual properties of colour precepts and the human sense of colour vision (Hård, Sivik and Tonnquist 1996). O'Connor states that

> while the opponent-process theory of human vision involves pairs of complementary colours [...], there is no evidence to suggest that physiological balance in the human visual system is associated with positive esthetic response of colour harmony. (O'Connor 2010: 268).

However, the two systems have been used as a method of universal colour communication and a tool for colour harmony in colour industries, colour researches and colour education (Holtzschue 2011; Westland 2007). Moreover, several perceptive colour order systems have been developed in late 20th century, i.e. the CIE system, the OSA-UCS system, the DIN, the Coloroid and the Colorcurve system and today they are used for different purposes by different professions.

Ostwald stated that "Colours appear to be harmonious or related if their properties are in certain simple relationship." (Kuehni 2005: 167). In 1917, he began to publish several textbooks on colour science, the first of which was called 'Mathetische Farbenlehre' (Theory of Logical Ordering of Colours). In this book, he described a three-dimensional colour solid in a form of double-cone where the achromatic colours were presented along the axis and chromatic colours were placed along the circle. The idea behind his system was based on spinning-disk mixture of full colour, white and black. Oswald's colour circle consisted of four basic hues; i.e. red, yellow, green, and blue. In circle, yellow and blue were located opposite to each other as red and green. The additional hues were found in equal spacing between these major four hues. According to Ostwald system, opposing hues were complementary when the spin-disk mixture of the proper proportions of colours produced a neutral grey (Holtzschue 2011; Kuehni 2005; Ou 2004; Westland 2007; Zollinger 1999). Ostwald developed some colour harmony principles based upon his ring star shaped colour solid. These principles were explained as follows:

1. Colours harmonise if they are located at the equal white and equal black circle in the solid;
2. Colours harmonise if they have equal white content;
3. Colours harmonise if they have equal black content;
4. Colours harmonise if they have equal hue content (Westland 2007: 9).

He believed that order was the basic law of colour harmony. The Ostwald system was accepted and used by several artists and designers. The Ostwald solid was based on equilateral triangles and was seen much simpler that the structure of the Munsell solid (Choundhury 1996).

Itten followed Goethe's principles of colour harmony and developed a contrast-based theory. He stated that "two or more colours are mutually harmonious if their mixture yields a neutral grey." (apud. Ou 2004: 68). Itten's ideas of colour harmony were based on his 12-colour wheel. He took the idea of three-pigment primary colours, red, yellow, and blue and placed all subtractive complements opposite to each other (Westland 2007). According to Itten, colour harmony was equilibrium and he explained it as follows;

> [Colour] harmony in our visual apparatus then would signify a psychophysical state of equilibrium in which dissimilation and assimilation of optic substances are equal. Neutral grey produces this state. I can mix such a grey from black and white; or from two complementary colours and white; or from several colours provided they contain the three primary colours: yellow, red and blue in suitable proportions. (apud. Holtzschue 2011: 268).

In accordance, Itten developed some colour chords in order to show the complementary relationships of the colours. According to him

[...] all complementary colour pairs and all triples in relationship of an equilateral triangle, a square, or a rectangle [or a hexagon] in the twelve-colour circle are harmonious." (apud. Kuehni 2005: 167).

Although his developed chords were mathematically based, Itten also had an attitude towards individual perception. However, Itten asserted that "The concept of colour harmony should be removed from the realm of subjective attitude into that of objective principle." (apud. Holtzschue 2011: 142). He divided his objective principles of colour harmony into seven contrasts; Hue, Value, Saturation, Extension, Warm and Cool, Complementary, and Simultaneous Contrast. Itten was one of the most important teachers of Bauhaus who made influential contributions to colour harmony and colour education. Today, his book 'The Art of Color' is the most widely used books on colour theory in the departments of architecture, design, fine arts and etc. (Green-Armytage 1996; Westland 2007).

The complementary relationship of colours and their mathematical-based aesthetic measures by developing colour order systems were the main concerns of colour theorists from Goethe through Itten. The general acceptance was that balance between complementary colours was the main and the most important principle of colour harmony. However, Albers who was the colleague of Itten in Bauhaus made an end of the traditions of colour order systems (Holtzschue 2011). According to him, the systematic approaches to colour harmony was inappropriate. He further stated that "no mechanical colour system is flexible enough to pre-calculate the manifold changing factors in a single prescribed recipe." (apud. O'Connor 2010: 269). Albers believed that the visual experience was much more significant than theory. In 1963, he published 'The Interaction of Color', and said about his book that it

> [...] does not follow any academic conception of theory and practice. It reverses this order and places practice before theory, which after all, is the conclusion of practice." (apud. Zollinger 1999: 219).

In the late 20th century, after Albers's contributions to colour, the subject of colour studies moves more into psychological effects of colour, considering colour harmony as a measurement of emotion.

4 CONCLUSION

Green-Armytage states that although some of the colour harmony principles have been widely accepted, these principles are based upon the theorists' own peculiar ideas (Green-Armytage 1996). Kuehni claims that "It is quite evident that there are no universal laws of colour harmony." (Kuehni 2005: 145). In accordance, Burchett asserts that there are not specific models accepted to define the concepts of colour harmony (O'Connor 2010). Judd and Wyszecki define colour harmony as "when two or more colours seen in neighbouring areas produce a pleasing effect, they are said to produce a colour harmony." (apud. Ou 2004: 60). Cupchick, in accordance, indicates that (colour) harmony is a reactive level emotional response which is based on a subjective feeling and peculiar to an individual (Cupchik 1994). This peculiarity not only depends on the individual differences but also is influenced by cultural, contextual and perceptual factors (Hård, Sivik and Tonnquist 1996). Relying on the subjectivity in colour harmony, many studies have been conducted in order to find out the underlying reasons in which colour combinations are perceived as pleasant and harmonious.

Guilford (1931) and Lo (1936), in their studies, tried to determine preference values of colour combinations over individuals' preference ratings. The results revealed that pleasurableness was comprised of its components as a preference value. Hogg (1969) classified 40 bipolar adjectives into four factors in his study of colour pairs which were labelled as strength, pleasantness, warmth, and usualness. Sivik (1983), Sivik and Taft (1989), Sivik and Hård (1989), Sivik and Taft (1992) in their studies aimed to find out the colour harmony judgements of individuals in two or more colour combinations by using semantic variables and also to identify the reasons behind why colour combinations were perceived as beautiful or ugly. In their study, Sivik and Taft (1989) conducted a study to search for the variables of meaning for judging colour combinations and extracted five factors, i.e. general evaluation, articulation, brightness, warmth, and originality. On the other hand, Ou et al. (2004) and Ou and Lou (2006) intended to construct a colour emotion model for colour combinations and defined their model with three factors, colour activity, colour weight, and colour heat.

Ou states that, although accompanied by emotions, "[...] the concept and underlying reasons of colour harmony are still far from definite." (2004: 60). Upon past theories and studies, researches and experiments need to be conducted in order to bring a better understanding to the notion of colour harmony with the motive to discover pleasurableness of ideality.

ACKNOWLEDGMENTS

This paper is part of a post-doctoral research project developed at the LabCor/Colour Laboratory of the Faculty of Architecture, University of Lisbon; supported by 2219-International Post-Doctoral Research Fellowship Program of TUBITAK-The Scientific and Technological Research Council of Turkey.

BIBLIOGRAPHY

Arnkil, Harald. "What is colour harmony?" In Proceedings of the Interim Meeting of the International Colour Association, AIC Colour - Effects & Affects. Paper no 097. Stockholm:2008.

Burchett, Kenneth E. "Color harmony." Color Research and Application 27.1 (2002): 28–31.

Choundhury, A.K. Roy. "Colour order systems." Review in Progress of Coloration 26 (1996): 54–62.

Cupchik, Gerald C. "Emotion in aesthetics: Reactive and reflective models." Poetics 23 (1994): 177–188.

Green-Armytage, Paul. "Complementary Colours-Description or Evaluation?" Colour and Psychology 15–18 (1996): 205–208.

Guilford, J.P. "The prediction of affective values." American Journal of Psychology 43 (1931): 469–478.

Hård, Anders, Sivik, Lars, and Tonnquist, Gunnar. "NCS, Natural Color System-from Concept to Research and Applications, Part I and II." Colour Report F49. An offprint from Color Research and Application 21.3 (1996): 180–220.

Hogg, James. "A principal component analysis of semantic differential judgements of single colors and color pairs." Journal of General Psychology 80 (1969): 129–140.

Holtzschue, Linda. Understanding Color: An Introduction for Designers. 4a ed. New Jersey: John Wiley & Sons, 2011. 258 p.

Kuehni, Rolf G. Color: An Introduction to Practice and Principles. 2a ed. New Jersey: John Wiley & Sons, 2005. 199 p.

——. Color Space and Its Divisions: Color Order from Antiquity to the Present. New Jersey: John Wiley and Sons, 2003. 208 p.

Lo, Ch'uan-Fang. "The affective values of color combinations." American Journal of Psychology 48 (1936): 617–624.

Macadam, David L. Sources of Color Science. Massachusetts: The MIT Press, 1970. 282 p.

Mahnke, Frank H. Color, Environment, and Human Response: An Interdisciplinary Understanding of Color and Its Use as a Beneficial Element in the Design of the Architectural Environment. New York: John Wiley & Sons, 1996. 239 p.

Nassau, Kurt. "Fundamentals of Color Science." In Color for Science, Art and Technology. Ed. Nassau, Kurt. Amsterdam: Elsevier Science B. V, 1998. 1–30.

O'Connor, Zena. "Colour harmony revisited." Color Research and Application 35.4 (2010): 267–273.

Ou, Li-Chen. "Quantification of Colour Emotion and Colour Harmony." Ph.D. Dissertation. University of Derby, 2004.

Ou, Li-Chen, and et al. "A study of colour emotion and colour preference. Part II: Colour emotions for two-colour combinations." Color Research & Application 29.4 (2004): 292–298.

Ou, Li-Chen et al. "Additivity of Colour Harmony." Color Research and Application 36.5 (2011): 355–372.

Ou, Li-Chen, and Lou, M. Ronnier. "A study of colour harmony for two-colour combinations." Color Research & Application 31.3 (2006): 191–204.

Sivik, Lars. Evaluation of colour combinations. Stockholm: Scandinavian Color Institute, 1983.

Sivik, Lars, and Hård, Anders. "On studying color combinations: Some reflexions and preliminary experiments." Göteborg Psychological Reports 19.2 (1989).

Sivik, Lars, and Taft, Charles. "Colour combinations and associated meanings-semantic dimensions and colour chords." Göteborg Psychological Reports 22.1 (1992).

——. "Semantic variables for judging color combinations: An analysis of semantic dimensions." Göteborg Psychological Reports 19.5 (1989).

Sorabji, Richard. "Aristotle, mathematics, and colour." The Classical Quarterly New Series 22 (1972): 293–308.

Taft, Charles. Generality Aspects of Color Naming and Color Meaning. Göteborg: Department of Psychology, Göteborg University, 1997. 149 p.

Westland, Stephen et al. "Colour harmony." Colour: Design & Creativity 1.1 (2007): 1–15.

Xiao-Ping, Gao. "A Quantitative Study on Color Harmony." Ph.D. Dissertation. The Hong Kong Polytechnic University, 2007.

Zollinger, Heinrich. Color: A Multidisciplinary Approach. New York: Wiley-VCH, 1999. 258 p.

Colors with imagined memories

Dulce Loução

Departamento de Projecto de Arquitectura, Urbanismo e Design, Faculdade de Arquitectura, Universidade de Lisboa, Lisboa, Portugal

ABSTRACT: Colours have memory, and it is this memory that act on the visible world, giving the variety and diversity that we see them give the reality that imaginary hue, archetypical colour that everyone recognizes, but no one knows for sure if the colour I see, it is the same as the other sees. However, the word unifies and puts the world of understanding, making the vision a valuable instrument of this knowledge.

In space, the colour as the surrounding material component relates to the objects or elements, which compose and shape, so that the reading is always a reading / interpretation of the relationship between the components of the space. There is not, in fact, the colour as a whole.

Keywords: colour, memory, perception, imagination

Colours are a mental construction, a manifestation of the nature of things, in our eyes. Colours exist because some animals, and we, humans say that they are there, in the things that surround us and which constitute the visible world. They are the things, because we see them. They are the things because we assign them a name, because by appointment, we vogue them.

However, colours exist only because things and our eyes are sensitive to light.

The science of light has its beginnings in Newton; in the eighteenth century he developed the theory, defined the principles, formulated the chromatic phenomenon understanding rules, the colour spectrum, which states that light makes real the real itself, through objective, the measured wavelengths.

The prism refracting releases of the incident light, the fundamental colours, and recomposition, sets colour in addition to colour, the universal colour, white.

Colour no longer belongs to painting or literature, but to the physical, the authority of individuality and materiality of colour. A glass prism refracts colour light.

With Goethe, colour is both connected to light and to darkness. The black and white when mixed give grey, which is then the synthesis of all colours. Although distinguished chemical, objective colour, physical colour, both subjective and objective, and psychological, subjective colours, the preference of Goethe seems to be addressed to the latter, also because the subjective colours allow him to elaborate on the concept of successive and simultaneous contrast.

The science of colour sets seven fundamental colours that appear in the optical prisms, which recover into a single colour, white.

However, colours aligned with the principle of composition, were reduced to three: yellow, red and blue, colours from which come all other. Of these three colours, which in themselves contain the principles of complementarity and contrast, are derived by mixing all others. Thus, yellow and violet, red and green, blue and orange, which are complementary colours, are synthesized and merge as inevitability. The ability of two complementary colours, which together with the primary, stem white, is added or evaded, approximate the ideal phenomenon of light.

Goethe with his Theory of Colours is, to contest the primary character of white light, and the secondary nature of the chromatic sensations, a philosophical nature testimony from Gray, and not the fictional unit of White Newtonian. To see different colours, man thinks colour differently.

The white and black, tend to become, in industrial society, the reference colours, precisely because of their lack of colour, creating neutral colours and a warm grey.

Colours take their place in very specific situations: the flag, the uniform, the frame, the garden; the colours of the country, fashion, painting, nature. The industrialized society is a monochrome community.

The classical theories of vision, inspired by Gestalt are developed based on the perception of the "black and white", considering colour as a product of the form (Arnheim, 1974). The colour, the shape attribute, and the schematic principles,

Kandinsky, Klee, Albers, Sonia Robert Delauny, cannot separate the pictorial work of teaching and artisanal approach to the phenomenon of colour.

Thus arose, in fact, proposals for Chevreul, then the systematic catalogues Munsell, which compared nature with printed inks, reproducible, and the studies of Faber Birren, where the subject of colour, uniting historical approaches and psychological,on the principles of composition.

When we look at the world, we see it colourful. When we dream, some say also the colour paint. When we think of space, when we see images, many idealize black and white. Drawing informs about the form and not about the colour of the objects. However, colour is, or is not, of the nature of the objects?

The physical recognizes pigments. The optical recognizes rods and cones. However, I, when I look at the world, I, because I have memory, I evoke images to recognize what I see, as images of real things, colourful images.

When I distinguish a colour, for example, a red, there is an idea of red inside of me that commands my memory, a certain red apple, a flag, and my mother's dress,

I choose colours because I remember seeing them in situations that I want to play like atmospheres.

My white is the lime in Alentejo, a white right beaten by the sun grazing on a facade at sunset August sun. There is no absolute white, additive synthesis of light. There is a white infinity, taking the reflected colour of the colours that are close to it. Thus, we see white, we see it by opposition. White clarified with the other colours. It is not an abstraction, but in poetry. There is no absolute white.

The colour space is a colour of mixing surfaces, subtractive colour, pigment colour, the colour of the light falling on the surface and is reflected then. Colours acquire shades resulting from the mixture of these two sources, are seen in this optical synthesis that unites matter and light.

When I choose a colour, I carry with it a world of evocations. When you pick or choose a colour, because it will inhabit the space with other colours, this association produces the effect of colour in space. Seeking harmony between colours and forms, between forms and textures between shapes and sounds. Seeking a totality that is clear perceptually, resulting in a whole that is effective in the message transmitted to the user, with the purpose of creating living space, i.e. architecture.

The constitution of the objects, their substance contains constitutively colour, which only manifests itself when, in accumulation, two factors occur: the light that makes the visible colour and the man who sees the colour, which makes it real. In addition, it becomes the real colour because we

see it. This is because our visual apparatus informs us of its existence, but more importantly, because you assign it to a meaning. Assigning a name to a colour is to make it recognizable, that is, it means giving it a set of qualities, associate it with shapes, smells, textures, temperatures.

Because colours have a sense, the blind man sees colours through the evocations that the colour name confers.

Light, is matter becoming visible, which means that it is the illuminated material, which gives reality and meaning to light.

In so-called normal conditions, colours seem to be the one that our memory recognizes as the right to things. However, when the light changes, the sunset, for example, installs another "order". Then we realize that the colour of the world is more than mere perception; it is aware insofar as we know, we see the colour, although a coloured light that interferes with the colour of objects, distancing the image that have of them is contaminating this colour. What is the true "reality"? What we see? On the other hand, what do we know?

We know the world and its features also because when we see it, associate it to experiences already lived. It is this phenomenon that presides over the colour constancy, the recognition that, for example, a prohibition signal is red, even when we see in fog.

The experience of colour is a result of vision associated with awareness of colour. The experience involves perception. We see why there is light, because our eye reacts to chromatic stimuli and because the mechanism of perception occurs, that is, we perceive colour of objects as a feature of the same and therefore, we can determine, judge and evaluate objects. The evaluation of the colour of objects is, in addition to a physical and physiological phenomenon, a manifestation of our experience in the world of our memory and our thinking.

I see why I re-know, based on my personal experience, giving thus meaning to what I see. It is though the experience, not though the vision exclusively that I know the colours of things.

The colour vision is not, then only the joint action of light in the eye and brain. Mancke already refers to the existence of the collective unconscious as part of our heritage, the individual adoption of collective experiences, archetypal impressions that define us as belonging to a particular society. These factors are keys to the experience of colour, in addition to cultural and personal factors.

Awareness of colour in space thus depends on several factors such as the individual experience, the effects of colour and symbolism.

The colour effect depends on the hue, value, and intensity of each colour and its relationship with the colours that are next, accounting and meaning as the word or gesture. The meaning of colour is cultural.

In space, colour as the surrounding material component relates to the objects or elements which compose and shape, so that the reading is always a reading / interpretation of the relationship between the components of the space. There is not, in fact, the colour at all.

Understanding colour space is the result of the relationship with the other elements that make up this space; is contrast to talk to define a colour space, the colour of objects in space results from the fact to distinguish between colours, i.e., to identify them in opposition, so that the colours are or are not from the individual experiment as members of a particular cultural context partner.

I see because I know what I see, in this sense, my view is commanded by my knowledge of the world around me; in view of the world process apprehend what my eye and brain transmit to me based on my understanding of what I see, rather, of what I know why.

The colours are the garments of the world, according to Ionesco.

These garments, made visible by the eye, become sensitive matter for the appointment, by the word, I would say.

When I name a colour, when you assign a name, it's all colourful universe that is called for that particular print; is a whole of evocations of the emerging universe that give reality to make concrete, the visual impression, visible, colour; Then, I see.

Moreover, because I only see what I know, perception is preceded, or, I would say, there is only if all the colours of the world I imagine that colour, I see, as an attribute of something I know, to which I attach an identity, a name.

Sometimes the evocation is fleeting and it is the word that leads us to the recognition of a certain colour, a certain hue that inhabits our imagination. That explains that even those who never saw the colour of the sun, may appoint yellow as the colour of the star king, and imagine its brightness, its light.

By the word, imagination draws a colourful world, values, and hues dictated by our memory. It is she who leads the evocations of colours, colourful things of our real world.

It is the imaginary world, a world of images as the name implies, it commands what I see, and I see.

I look without seeing. Truly seeing implies knowledge, and this knowledge is the domain of the mind, not of the optical device. It is mental

fact, to see the colour of the sky, of the sea, of a flower or of architecture. Colour changes with the light, with pollution, with the time of day, and yet I continue to "see" the same colour, because I know that it exists, in the sky, the sea, the flower...

Called Constancy Chromatic, and is nothing more than the primacy of imagination over reality.

Colours have memory, and it is this memory that act in the visible world, giving you the variety and diversity that we see them give the reality that imaginary hue, archetypical colour that everyone recognizes, but no one knows for sure if the thing that I see, it is the same as the other sees. However, the word unifies and puts the world of understanding, making the vision a valuable instrument of this knowledge.

It is with the colours that we find out that only the imagination known as the Purple Room, birthplace of the Roman emperors, which reveals the power of colour as evocative of a place that no one saw, but we can imagine, and to rebuild, only the evocative power of a hue, in that world that no longer exists.

Colours appear on his appointment, in our imagination, pure, untouched, and full. It is to them that the world dresses.

The sense of "no place" also applies to colour; the fact that colour is a mental event, it has no existence, except in our mind, place of thoughts and memories, dreams and nightmares. Our dreams are colourful, and they evoke a world where the order is of a different kind than the one of the ordinary world. Colours are the mysterious vehicles that connect the real, to the imaginary. By naming a colour, we "place" the sensations in order. By naming a colour, all visible worlds come to an order, reminding us that the utopia is achievable even in everyday life. A place where everything is perfect, in its place, with ideals and organization. Such as in the word of colour, in our minds. Colour describes the world, in place. By naming it, we place the world with our thoughts of harmony, and by the power of the word, which does not exist, becomes real, and, finally, the imagination takes place, creating the new and perfect order of things.

BIBLIOGRAPHY

Arnheim, R., 1954. Art and Visual Perception: A Psychology of the Creative Eye. University of California Press, 1974. 442 pp. ISBN: 0-520-24383-8.

Restoration of historic monuments as utopia: The bell tower(s) of Santa Maria de Belém church

Vera Mariz

Faculdade de Letras, ARTIS—Instituto de História da Arte, Universidade de Lisboa, Lisboa, Portugal

ABSTRACT: This essay was written during the ongoing *Conservation and Restoration Plan of the domes of the Santa Maria de Belém church* (Jerónimos monastery and church).

The Jerónimos church, Portugal's most visited monument, is undoubtedly an iconic historic monument that over the centuries has attracted the curiosity and attention of multiple writers, historians, architects, archaeologists, scientists and, naturally, restorers.

During the 19th century the restorers of Portuguese historic monuments, directly influenced or not by the principles of the French architect Eugène Viollet-le-Duc, tried to recover the hypothetical pristine image of these ancient architectonic constructions. In this context, as we aim to demonstrate by addressing the specific case study of the Jerónimos bell tower, some restorers assumed the role of creators of architectonic utopias.

Through this study we will demonstrate that the importance of this monument as a symbol of the Portuguese age of overseas exploration and the principles of the stylistic restoration and European neogothic movements, played fundamental roles in the development of different but equally utopic restoration plans of the Jerónimos church bell tower.

Keywords: Restoration of historic monuments, stylistic restoration, Jerónimos church and monastery, 19th century

1 RESTORATION OF HISTORIC MONUMENTS AS UTOPIA

In the mid-19th century, the young French architect Eugène-Emmanuel Viollet-le-Duc (1814–1879) began to assert himself as one of the most important restorers of historic monuments. Within this time, in addition to the restoration works of the Sainte Marie Madeleine de Vézelay basilica (1839), the Saint-Chapelle (1840) or the Notre Dame of Paris (1845), Viollet-le-Duc began to understand and disseminate the restoration of historic monuments as a discipline independent of the architecture design (Choay 2008; Neto 2001). Later, between 1854 and 1868, the French restorer, *Commission des Monuments Historiques* Architect and *Inspecteur Général des Édifices Diocésains*, wrote the essential *Dictionnaire raisonnée de l'architecture française du XIe au XVIe*, publication that along with the restoration works themselves disseminate throughout the world the principles of Viollet-le-Duc. This theorist was undoubtedly a successor of Ludovic Vitet (1802–1873) and Prosper Mérimée (1803–1870) ideas, to whom we owe the creation of the previously mentioned establishment of the restoration of historic monuments as an autonomous discipline and practice performed by specialized technicians (Choay 2008).

The importance of Viollet-le-Duc, as noted by Françoise Choay (2008), must not be limited to the following sentence, however this definition is very significant within the present study. According to Viollet-le-Duc *Dictionnaire Raisonné* to restore a building "is not to preserve it, to repair it or to rebuild it, it is to restore a state of completion that may never have existed at any given time." (Viollet-Le-Duc 1866). This reflects the "idealistic" thought of Viollet-le-Duc, the intention of value the structural functionalism of the historic monuments and the perfection of the gothic architecture.

To the French restorer the restoration allowed him to recover or to create the hypothetic pristine image of these uncompleted or transformed monuments with the guarantee of authenticity given by the comprehension of a structure as a logic and coherent organism, a characteristic of his beloved gothic architecture. That is, to Viollet-le-Duc to restore a historic monument was to recover or to create a perfect architectonic model, a construction whose form and mechanical performance of materials were perfectly harmonised, even if it was just a fantasy, a utopian architecture.

Despite this tendency towards the materialization of utopian architecture models, it is important to emphasize that through the valorisation of the gothic architecture as a rational and scientific structure and its graphic re-imagination (Vinegar 1998) Viollet-le-Duc broke through the common mystical and religious interpretations of this construction style. Equally important is the fact that, as noted by Aron Vinegar (1998), to the French theorist, as to the comparative anatomist Georges Cuvier (1769–1832), the main form of revealing and, consequently, of knowing the organism, a biological or architectural one, was the dissection. This exercise was fundamental to the restorer, an architect well acquainted with the structure, the anatomy, or the style of the historic monuments under his responsibility. Thus, the utopian aspect of this process began only after the exhaustive gathering of graphic representations, dissection, and study of the different elements. Actually, the utopian aspect of this restoration plans began with the graphic restoration of these historic monuments according to the ideal-type set by Viollet-le-Duc, even though these constructions do not resemble any real structure.

The theory developed by Viollet-le-Duc was widely disseminated around the world mainly through the translation and publication of his *Dictionnaire Raisonné* and other works. However, the interpretation of these principles was not always the most accurate. In a context marked by nationalist and historicist claims throughout all Europe and by the overall state of ruin of the majority of the French monuments, the idea of restore a possible pristine state of completion of a historic monument by conceiving structures and decorative elements and by eliminating the so called "dispositions vicieuses" or later modifications was very appealing (Choay 2008; Neto 2001).

However, in contrast to Viollet-le-Duc's comprehension of the restoration of historic monuments as a rational, critical, and scientific exercise, a rational utopia, due to influence of the ideological currents of the time, the approach of the 19th and 20th century "stylistic restoration" architects was much more creative. Therefore susceptible to commit excesses driven by the will to create a harmonious structure supposedly characteristic of a particular time without the French theorist scientific concerns.

2 RESTORATION PLANS FOR THE CHURCH TOWER(S) OF SANTA MARIA DE BELÉM CHURCH

In the year of 2015 the Directorate General for Cultural Heritage (Direcção-Geral do Património Cultural) and the World Monuments Fund—Portugal organization signed a protocol regarding the phase "Exteriors 1" of the *Conservation and Restoration Plan of the domes of the Santa Maria de Belém church*, a project that is expected to conclude in 2022. During this specific phase, the works (undertaken by the Nova Conservação team of conservators and restorers) covered the church's west façade, including the bell tower, the subject of the present study. These works included the study of the history of the bell tower, its changes over time and naturally the much-needed treatments of the deteriorated areas. Regarding this works, specifically the historic and artistic studies that we have been developing within this multidisciplinary project, we must emphasize the unique opportunity provided by the scaffolding assembled on the site. The previously mentioned multidisciplinary aspect of this project is equally important and noteworthy, as it is a highly recommended principle that is not always followed in conservation and restoration projects of historic monuments. In such cases as this, the art historian is responsible for the historic and artistic study of the monument, for the identification and analysis of the past conservation and restoration campaign. In this particular project, it was needed to address some issues regarding the Jerónimos church bell tower, such as: the total comprehensiveness of the 19th century campaign regarding this structure, the age of some decorative components, such as the armillary sphere or the medallions with crosses, or the reasons beyond the choice of this type of dome. These studies were always developed in full cooperation with the other team members.

However, as we have mentioned before, this is far from being the first time that this bell tower is the subject of a project of conservation and restoration.

Figure 1. The Jerónimos church in 1832 with the pyramidal roof of the bell tower. João Camacho Pereira, Colecção de Gravuras Portuguêzas, 2ª série, Lisboa, 1947, estampa nº 43.

The beginning of the laicization process of the 16th century hieronymite monastery of Santa Maria de Belém goes back to the early years of the 19th century, dating from 1833 – therefore before the decree that suppressed all male religious orders in 1834 – the suppression of the monastery. At the time that this deliberation of D. Pedro (1798–1834), Duke of Bragança, was made public, it was also established that from then on the suppressed monastery would be used as accommodation of the Casa Pia of Lisboa that until then was located in the highly inappropriate Our Lady of Exile (Nossa Senhora do Desterro) convent. The church would be use as Belém parish. As has been proved by experts in the field (Neto and Soares 2013; Soares 2005) this choice was primarily due to the intervention of António Maria Couceiro, the orphanage administrator, who understood the importance of establishing the Casa Pia in a construction of overwhelming architectonic and historic significance but also the necessity of an intervention that would invert the decay of the former hieronymite monastery. Although the liberal and patriotic D. Pedro IV has welcomed the project of reuse, conservation and restoration of the monastery presented by Couceiro, his death and the withdraw of the orphanage administrator due to a political power change were eventually determining factors to the non-implementation of the monument restoration plan.

The ambitious restoration plan of a historic monument whose past was closely related to the reign of D. Manuel I (1469–1521) and the glorious era of Portuguese Discoveries but that at the time was being used as headquarters of a charitable institution founded in 1780, was resumed during the second half of the 19th century by José Maria Eugénio de Almeida (1811–1872) with the financial and ideological support of D. Fernando II (1816–1885). These restoration works that were started in 1860 and went on for many decades were previously studied by many authors (Anacleto 1991; Carvalho 1990; Gordalina 1986; Neto and Soares 2013; Soares 2005). However, as we have already mention, we are interested in a particular aspect: the restoration plans for the church tower(s) of Santa Maria de Belém church as utopian visions.

The iconographic documents known to date are undisputable proofs of the existence of a single bell tower in the west façade. This tower, partially replaced by the actual one in the second half of the 19th century, presented an octagonal shape and a pyramidal roof. Note however that according to some chronicles and authors but especially the architecture, the original 16th century plan should include two towers much more magnificent than the one visible in the image above.

Because of the church parish situation, the restoration of this part of the former hieronymite compound was not initially included in the restoration plans projected by the Casa Pia administrators. However, it is known that some of the plans presented by the French architect Jean Colson to José Maria Eugénio de Almeida between 1860 and 1862 already included some works in the church. Regarding the bell tower the architect whose proposals were not accepted, suggested the construction of two bell towers in the west façade with a higher and more refined pyramidal roof decorated by acanthus leaves, an element also applied in the surrounding pinnacles. The window vents of the bell tower presented clear similarities with some of the windows of the south façade. A cross should be placed in the top of both towers.

In 1867 the highly respected architect Joaquim Possidónio Narciso da Silva (1806–1896) presented in the *Exposition Universelle de Paris* the result of eleven month and twenty four days of work (Milheiro 1997): a wooden model of the Santa Maria de Belém church after being restored. The aim of Possidónio da Silva was to restore the primitive parts that were demolished or altered over the years after the death of D. Manuel and to continue the works that were not finished, such as the two bell towers, "the most embarrassing part of this restoration" (Silva 1867). As Jean Colson, Possidónio da Silva also proposed two towers whose design was determined by the primitive style of the monument and by the intention of harmonizing the ancient/primitive and new architectural forms (Silva 1867). However, as we can see in the image bellow, Possidónio da Silva's plan was much more ambitious than the one presented by Colson. It was also the result of a much more mature comprehension of the monument's primitive style, a process clearly influenced by the work of Viollet-le-Duc, architect and theorist that Possidónio da Silva knew and admired (Chagas 2003; Martins 2003; Milheiro 1997).

As Viollet-le-Duc, in his quest to restore a "state of completion that may never have existed at any given time" (Viollet-Le-Duc 1866), Possidónio da Silva developed a utopian version of Santa Maria de Belém church, an idealization of something that never truly existed as presented. However it was a rational utopia, the result of a methodical approach to the restoration of this church developed, at least, since Possidónio da Silva prepared several architectural surveys of the most important Portuguese monuments (1848), including the Santa Maria de Belém church and monastery.

In fact we know that as soon as August 1847 the architect who was awarded with the title of "Architect of the Royal House" by D. Pedro in 1834, requested authorization to enter in the Casa

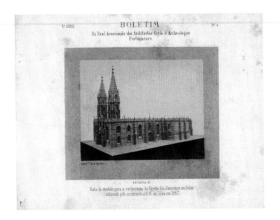

Figure 2. Wooden model of the Santa Maria de Belém church after being restored. Boletim da Real Associação dos Arquitectos Civis e Arqueólogos Portugueses, 2ª série, nº 4, 1875.

Pia domains to draw, to measure and to elaborate the plans of the former church and hieronymite convent (Silva 1847). Later, between 1855 and 1859, Possidónio da Silva would be responsible for the outline and guidance of some masonry and stonework within the restoration of the cloister and south façade of the Jerónimos church (Silva 1860). New drawings and plans, in a total of four, were subsequently made in 1861 within his works on monumental architecture commissioned by the Portuguese government (Silva 1861). It is also important to mention that the knowledge that Possidónio da Silva had of the former convent and church of Santa Maria de Belém was fully recognised by Manuel Raimundo Valladas, the engineer responsible for the conservation and restoration works of the Belém monument from 1884. After all during that year Valladas asked Possidónio da Silva for guidance regarding the completion of the main façade of the church (Possidónio Narciso da Silva 1884).

Thus, we can say that the two bell towers suggested by Possidónio da Silva in the 1867 wooden model are the result of a well-considered analytic process that aimed to decode the characteristics of this unique exemplar of the "manuelina" architecture and to restore the harmony between form and the mechanical performance of the materials. However, it is important to mention that, according to Possidónio da Silva, the utility, the solidity and the proportion of a building should not be the only concerns of an architect whose task was also to introduce some beauty in his work, a condition without which an architect was not worthy (Joaquim Possidónio da Silva 1884). Therefore, we can conclude that despite the scientific method

used by Possidónio da Silva, there is an expression of some creativity of the architect in his utopian vision of Santa Maria de Belém. Finally it is also important to mention the reverence of Possidónio towards History, his understanding of historic monuments as historic documents and of the importance of History to understand these same monuments (Joaquim Possidónio da Silva 1884). This partially explains the interest of the architect in these monument and his attempts of finding the original plan (Silva 1867) and also the rejection of some advices given by the Brazilian historian Francisco Adolfo Varnhagen (1816–1878) (Silva 1868). To resume, how could Possidónio da Silva propose a tower above the transept if there were archaeological and historical evidences of the existence of two towers in the west façade? (Silva 1868).

Nonetheless the serious involvement of Possidónio da Silva in the restoration of Santa Maria de Belém church and his reputation as "Architect of the Royal House", in 1867 the works were commissioned to Giuseppe Cinatti (1808–1879) and Achille Rambois (1810–1882). In fact, the bell tower that we all know is the result of this highly criticized restoration campaign. Among other alterations intended to endow the church of more magnificence, Cinatti and Rambois presented in 1873 a plan of the south façade that included a single tower in the west façade with all the characteristics that can be seen nowadays.

Despite the demolition of the previously mentioned pyramidal roof of the ancient bell tower in 1870 (Soares 2005), when in the interior of this structure or at the same height as the window vents it is perfectly possible to see that the former construction was preserved until the base of the removed roof. In summary, a dome replaced the pyramidal roof and some decorative elements and the lacy flying buttress were introduced in the showy structure. However, as we have mentioned before, the public opinion was extremely critical of this result "highly discordant and profoundly disharmonious" (1879), devoid of "elegance not to say proportions" (1872), "an exceedingly poor domed top" (Watson 1908). In 1895 a public tender launched in order to conclude the works predicted the demolition of the Cinatti and Rambois bell tower and the conclusion of the same two towers proposed by Colson and Possidónio da Silva (1895).

Despite the controversy over the lack of understanding of the distinctive characteristics of the "manuelina" grammar and the public tender of 1895, the bell tower designed by the two scenic designers did not suffer any major changes over the years, remaining as a tangible achievement, the triumph of a fantasy that defeated, among others, the rational utopia proposed by Possidónio

Figure 3. The Cinatti and Rambois dome after the 2015–2016 conservation works.

da Silva within the development of his restoration plans. In the end, the restoration as an act of random creativity and 100% pure invention defeated the restoration as a rational, critical, and scientific exercise, but still a rational utopia.

BIBLIOGRAPHY

"Diário do Governo." 1895 of *20 de Dezembro O Occidente*.26 (1879): 11.

Guia do Viajante em Belém. 1ª ed. Lisboa: Rolland & Semiond, 1872. 108.

Anacleto, Regina. *Alguns aspectos da intervenção oitocentesca do mosteiro dos Jerónimos*. 1925–2015. Vol. vol. LXVII. Coimbra: Faculdade de Letras da Universidade e Coimbra, 1991. ISSN: 0870-4112.

Carvalho, Artur Marques. *Do Mosteiro dos Jerónimos: de Belém, Termo de Lisboa*. 1ª ed. Lisboa: Imprensa Nacional-Casa da Moeda, 1990. ISBN: 9789722704076.

Chagas, José António Amaral Trindade. "Joaquim Possidónio Narciso da Silva (1806–1896). Contributos para a Salvaguarda do Património Monumental Português." PhD. Universidade de Évora, 2003.

Choay, Françoise. *Alegoria do Património*. 3ª ed. Lisboa: Edições 70, 2008. ISBN: 978-972-44-1274-0.

Gordalina, Maria do Rosário. "As obras revivalistas do século XX no mosteiro de Santa Maria de Belém." In *Da mentalidade à criação artística*. Ed. Sintra, Instituto de. 1ª ed ed. Sintra: Instituto de Sintra, 1986. 247–291.

Martins, Ana Cristina. *Possidónio da Silva e o elogio da memória 1806-1896*. 1ª ed. Lisboa: Associação dos Arqueológos Portugueses, 2003. ISBN: 972-9451-45-1.

Milheiro, Ana Vaz. "O Gótico e os Sistemas de Desenho presentes na Arquitectura Oitocentista. Produções teóricas europeias e a recensão portuguesa na obra escrita de Possidónio da Silva." MD. Universidade Técnica de Lisboa, 1997.

Neto, Maria João Pereira. *Memória, propaganda e poder. O restauro dos monumentos nacionais (1929-1969)*. 1ª ed. Porto: Faculdade de Arquitectura da Universidade do Porto, 2001. ISBN: 972-9483-50-7.

Neto, Maria João Pereira, and SOARES, Clara Moura. *O Mosteiro dos Jerónimos*. 1ª ed. Lisboa: Caleidoscópio, 2013. ISBN: 978-989-658-243-2.

Silva, Joaquim Possidónio da. "Boletim de architectura e de archeologia da Real Associação dos Architectos Civis e Archeologos Portuguezes." 7 (1884).

——. *Mémoire Descriptive du project d'une restauration pour l'église monumentale de Belem à Lisbonne. Bâtie en 1500 en souvenir de la découverte de l'Inde par les navigateurs portugais. Modéle fait pour l'exposition de Paris*. 1ª ed. Lisboa: s.n, 1867.

Silva, Possidónio Narciso da. *Fundo Joaquim Possidónio Narciso da Silva*. Vol. Tomo XV. Lisboa: Arquivo Nacional da Torre do Tombo, 1884. 8.

——. *Fundo Joaquim Possidónio Narciso da Silva*. doc. 364. Vol. Tomo III. Lisboa: Arquivo Nacional da Torre do Tombo, 1868. 8.

——. *Fundo Joaquim Possidónio Narciso da Silva*. doc. 2204. Vol. Tomo I. Lisboa: Arquivo Nacional da Torre do Tombo, 1861. 4.

——. *Fundo Joaquim Possidónio Narciso da Silva*. doc. s/n. Vol. Tomo I. Lisboa: Arquivo Nacional da Torre do Tombo, 1860. 4.

——. *Fundo Joaquim Possidónio Narciso da Silva*. doc. 3851. Vol. Tomo V. Lisboa: Arquivo Nacional da Torre do Tombo, 1847. 5.

Soares, Clara Moura. "As intervenções oitocentistas do mosteiro de Santa Maria de Belém: o sítio, a história e a prática arquitectónica." PhD. Universidade de Lisboa, 2005.

Vinegar, Aron S. "Memory as Construction in Viollet-le-Duc's Architectural Imagination." *Paroles Gelées* 16 2 (1998) ISSN: 1094-7264.

Viollet-Le-Duc, Eugène. "Restauration." In *Dictionnaire raisonnée de l'architecture française du XIe au XVIe*. Ed. Viollet-Le-Duc, Eugène. Vol. 8. Paris: A. Morel, 1866. 14 p.

Watson, Walter. *Portuguese Architecture*. 1ª ed. London: Archibald Constable and Company, 1908. 185.

Worlds and real and imaginary borders: Mediation stages or creation as a will to shine

Ana Santos Guerreiro

Faculty of Architecture, University of Lisbon, Lisbon, Portugal

ABSTRACT: The idea of mediation inherent to the process of artistic creation, establishes a relation scope: promotes the existence of multiple meanings, allowing the connection between two realities: being one, the imagination and the world domain and other, the consubstantiation of the work in streams and the internal disruptions to its emergence. It´s domain is therefore not static but dynamic, tilting between these two instances. Artistic creation interprets the world under this tilt (and this action is potentially inquirer, and defining the continuous necessity of art. Mediation we shall call here, and in the sense that matters to art, is proposed as a way of establish relation between the world and the Other, which covers what is human construction in overcoming nature. This coverage is likely to be embraced by the metaphor of translucency: the origin, *mediatio*, from Latin, means divine intervention. To connect or merge takes place through mediation a medium term or medium that unites what appears as opposition, and seems to be excluded. Thus, mediation requires the filing of a medium. In addition, however, to the literal meaning 'of what is between', mediation implies congregate, compose, assimilate the diverse and the multiple, as well as simultaneously requires the establishment of a distance.

Keywords: Artistic creation, Mediation, Work and absence, Interstitial, Symbolic

1

In a certain sense, the idea mediation performs a cut in the nature flow. Man, being a living creature in an environment (that of its reality) acts reflectively over his closest horizons. Mediation can be understood as a facet of this action, with regard to a departure from their primary impulses, which provide interaction with its surroundings by assigning a meaning; it is in this building of sense that reality acquires a human dimension. While looking for ways of mediation, man interrupts the flow of nature, by seeking a departure to what is presented as unavoidable; a first step in relation to nature can be understood as the immediacy of the relationship between individual humans and world objects, the given objects. However, through work, man exceeds the distance between the objective and the subjective world, doing his performance, creating his world.

The search for understanding between the world of the living and the world of the dead has always tried to build a place of intermediation. The heterogeneity of space in the religious experience is a fundamental one: the consciousness of this heterogeneity, according to Mircea Eliade, is the relationship that man establishes the recognition that there is a sacred space and not sacred spaces (Eliade 1989; 1993).

Recognition comparable to 'a world source', the myth foundation, attached to a fundamental religious experience and prior to any discussion on the origins of the world. The understanding of qualitative differences is for the religious man proof that the sacred space is matrix for the real and all the rest is therefore exterior and inform extension building out the surroundings. The recognition that space is not continuous is what enables the condition to think about the foundation of the world. The sacred, in manifestation, leads to a break towards the world`s homogeneity and at the same time it imposes and reveals the existence of an absolute reality, opposed to the surplus—that is the huge extension of the surrounding world. The sacred thus implies the need of men to support the world ontologically, establishing an initial centre, a fixed point of reference in such infinite extent.

The relationship between the two instances, the sacred and profane, is a communication mechanism between the two worlds, and establishes a mediation. The need for this dynamic relates to the dialectical presence-absence, and sets up by the possibility of a dissonant element, a structure based on a binary principle (Burgaleta Mezo 1993: 273–274). The communication, while production of meaning manifests itself because, according to Umberto Eco, the man does not dominate the whole, having need of communication as an alternative among the things known and not known, in a movement between the finite and the infinite, an approach in gradual rapprochement of reality (Eco 1972: 433).

The awareness of this instance as a space of mediation interface is already envisioned by Plato in *Timaeus* 67d (apud. Vasiliu 1999: 15–18), and is the *Khora*, a kind of transparent space, place of sensitive receptacle placed by a Demiurge leaving the world of ideas. This place of sensitive acceptability takes part of the intelligible of an obscure way by an unclear and chaotic movement.

2

Thus, we can characterize the analogy between the communication and the relationship between the worlds of hierophany. The need for a dialogue between the superior world and the ground, or between Heaven and Earth, is long ago linked to the development of human existence and to their need for representation. Accordingly, the achievement of man—and all art—is understood to be representation. The idea of temple, as well as of city, was places of mediation between the sacred and the profane and implied a careful location, with the establishment of the 'Etruscan' rite. In the establishment of constructions, *mundus* and *pomoerium* were intermediate places to relationship and to mediation between different worlds, between the intelligible and sensible. Thus, *Mundus* would be a well dug in the village centre connecting the three worlds; the *pomoerium*, where the wall was built, was related to the rites of passage between worlds, marking the sacred 'land of nowhere'.

There are countless examples in all cultures of mediation places, of which the temples, in all cultures, are the best known, by the appropriation of Christian culture. The gradual loss of these rites, translates in a way, in Western culture, the secularization of society and the implementation of the profane in various ways with a leading role of the rationalism way and its consequences in modernity and in contemporary times.

Back to the idea *of khôra*, it turns out that in *Timaeus*, Plato relates the creation, the Demiurge's work, with the conduct of the ideas of Good, being the Good to communicate. The material and sensitive compliance of the created world reflect also, according to Heidegger's interpretation (Heidegger 1989), the Platonic cosmogony as support for the idea of communication, in the genesis of language needs: Being is only then achievable by the dimension of language, setting up in this mediation the relationship between Man and Being.

3

The residue idea which assists the creation of products, namely artistic (Burgaleta Mezo 1993: 10–12) relates to a whole. P. Burgaleta even alludes to the operating value of syntactic trace elements as hermeneutical keys, sounding like a residual net integrating all operations. It can be followed as a clarifying way of the work, elements net that act as demonstrating entry doors in space, and this space is precisely the hermeneutical space of *khôra*. In this sense, every work of art is a surplus reality of a process, and each work of art can be understood as an attempt to close the gap, the fissure between the two worlds, in an exercise of continuity desire. The work, as residue, is symptomatic of the need for a return in face to an original start, where communication was not necessary because the Sense, the original sense, was inherent to that world.

However, this activity introduces a reflective component on acting way of man, by introducing a 'recognition of self in intuition', performing the assignment of meaning to his action through the symbolic mediation.

It is this attitude, characteristically cultural and symbolic, which performs the specificity and universality of human action in relation to what could be considered purely instinctive domain.

The symbolization process, denying the immediate, installs in mediating the assignment of a meaning to the world, drawing a benchmark, for universality. This process—trying to realize a unit in the world—relates the need to overcome the division between human consciousness and what appears to be immediate given in the world, from an externality in face of this awareness.

The work (Arendt 2007) thus constitutes a fundamental moment at which men begin to transform nature, transforming also themselves. Man opposes to the immediate the built and this reveals the outcome of the mediation. In this way, mediation implies the demand for inter-subjective understanding of the place occupied by Man in the world, and that understanding integrates interpretation, transformation and subversion of reality within which the man has been moving. In this search for understanding of man's place in the world, it is necessary to understand a framework that goes beyond the mere objectification of the world, in a vision of Man only rational but also overtaking image of pure expression of the inner world. Consequently, the mediation process projects and demands the possibility of a constant reflexivity that will imply a certain distance by installing an intermediate space, with porosity capable of apprehending otherness.

4

The need for communication then installs a differential space, characteristic of epistemological research—in which art has an extremely significant domain. Thus, to the residual character of art, aporia inscribes in the work a kind of absence. The absence that produces the symbol and this relates to the metaphysical territory. Interpretation processes of a work reveal, through the ages, the need to put us in face to decipherment and experimentation of Being.

Seen this way, the work, beyond being a residue is also occultation, and installs the space of the difference. The process of symbolization inherent to the work and its interpretive process is the manifestation of a hermeneutical space—place where the Being lives—the space between worlds.

Testimony of an original fissure presence, the meaning process reverberates Western human experience of Being. In this, the modernness of presence has reverberates itself as a place of difference and exclusion (which is inseparable from the question of temporality).

In the sense that, their expression is, also, at the same time, while presence, a failure. It installs in this original sharing of presence and absence, the reveal and veil, a sense that the ancient Greeks associated with intuition idea of truth as revelation and unveiling. In addition, in this survey process to the fissure, as the rift between presence and absence, that Man needs philosophy: it is through this fracture that its metaphysical anguish is installed.

Due to being divided that this presence, reality becomes capable of being symbolized. You can symbolize, trying to upgrade a sin-chronic process an original reality (Trías 2006) that is accessed or understood in diachronic. In the West, the art has over the years considered to be started from the subsistent, the Being or the Idea, which devalued the Time as pure becoming. The historicist scheme that Hegel pointed (one view of time becoming aware of itself through art forms, in which, through the symbolic, classical or romantic stages, art would 'accomplish' the Idea, which thus becomes self-aware) would end in the 'death of art' swinging between the reflective romanticism and 'overcome' of the religious. These issues will be critically enhanced as the 'anti-art' of the twentieth century (Danto 1996).

5

It appears that due to the design of presence not be one—that is, be divided and separate—feels the need to signify and symbolize. The awareness of this

separation installs the need for significance, because the whole is not embraceable. (It is) because there is a deferral at the origin of presence that it makes a puzzle that the need to philosophize appears (Agamben 2006: 229). This differs from the source installs in several antagonies: It is the opposition between the being and his appearance, as harmonious relationship between opposites, or as opposing relationship as the ontological difference that Eugenio Trías established between the being and the Being (Trías s.d.) here creating place-address to Ontology.

Consciousness and the denunciation of the fissure in the symbolic process have been veiled, in a sense, until the 19th century in the West. The issues concerning it in the artistic plane and gnoseological are the example: it`s occultation concerns with metaphysical scope of interpretations related to the truth, such as in paradigm issues, likelihood and copying, among others. It is no accident that the advent of photography in 19th century, will catapult these matters in stark re-examination throughout the twentieth century.

The process of language (linguistic) holds itself, as philological and philosophical problem, the location where it alludes to the representation of the original fissure. The invention of language reflected the need to find the other side of tesserae. In this sense, the sign notion emblemizes the duality of signifier and signified expressed in a unit in nostalgia (melancholy) of the initial Uno. Significance seeks to fill the rift of the presence, establishing relations between signifier and signified where the commitment is the manifestation but also the concealment.

From the symbolic perspective, where part of the meaning is veiled, we realize only a partial compliance. Agamben (2006: 230) refers to the pragmatic necessity that Hegel stipulated in an attempt to bridge this residue: in "Aesthetics Lessons" sought to reify the work of art a model for overcoming symbol which notes the anguish for total adherence between the two parties.

> The symbolic in the sense that we understand it ceases indeed there where the free individuality, rather than indeterminate, general, and abstract representations, constitutes the form and content of figuration. Significance and sensitive representation, internal and external thing and image are no longer so distinct from each other and present no longer as a whole, where the appearance does not already own another essence or the essence another look outside themselves or by their side (Hegel, apud. Agamben 2006: 230)

The original difference in presence is separated in the evidence of a significant convergence built between form and content, internal and external, manifestation and latency.

Time, in its occasions flows and becoming, contaminates and undermines the metaphysics of Western time, so often dedicated to Being and the idea. The otherness of the presence derives from time, from the ephemeral condition that not more immortalizes and installs his variation. The ephemeral captures from time the imperceptible flows. Starting from the consideration of E. Trías on the interstitial as gap or empty space of the shaft still (space) and movement (time), you can understand this interstice as something between the environment and the world, and request the understanding to the ephemeral (Buci-Gluksmann 2006; Byung-Chul 2014; Vattimo 1992). In a way, is out of this exploitation of the interstitial (which paradigmatically has an essential image in the Grand-Verre (1915–23), Marcel Duchamp, with the problematic of *inframince*) and the use of the hybridism and the unveiling of space and time, that many aesthetic of modernity and art forms of the third millennium contemporary are shown.

BIBLIOGRAPHY

Agamben, Giorgio. *Estancias. La palavra y el Fantasma en la Cultura Ocidental*. Trad. Tomás Segovia. Valencia: Pré-Textos, 2006.

Arendt, Hannah. *A Condição Humana*. Tradução de Roberto Raposo; Pósfácio de Celso Lafer. Rio de Janeiro: Forence Universidade, 2007.

Buci-Gluksmann, Christine. *Estética de lo Efímero*. Trad. Santiago Espinosa. Madrid: Arena Libros, 2006.

Burgaleta Mezo, Pedro. "Residua. El valor de lo Margina." PhD. Escuela Técnica Superior de Arquitectura, 1993.

Byung-Chul, Han *A Sociedade da Transparência*. Trad. Miguel Serras Pereira. Lisboa: Relógio D'Agua, 2014.

Danto, Artur. *Après la fin de l'art*. Trad. Claude Hary —Schaeffer. Paris: Seuil, 1996.

Eco, Umberto. *La Estructura Ausente*. Barcelona: Ed. Lumen, 1972.

Eliade, Mircea. *O Mito do Eterno Retorno*. Trad. Manuela Torres. Lisboa: Edições 70, 1993.

——. *Aspectos do mito*. Trad. de Manuela Torres. Lisboa: Edições 70, 1989. 170, 175 p.

Heidegger, Martin. *El Ser Y el Tiempo*. 7ª ed. Trad. L. Gaos. Mexico: F. Cultura Economica, 1989.

Trías, Eugénio. *Lógica del Limite*. Barcelona: Ediciones Destino, s.d.

——. *Primer Libro: El Símbolo y lo Sagrado*. La Idad del Espíritu. Barcelona: Debolsillo, 2006.

Vasiliu, Anca. "Transparent, Le Diaphane et l'Image." In *Transparences*. Ed. Dubus, Pascale. Paris: Les Editions de la Passion, 1999.

Vattimo, Gianni. *A Sociedade Transparente*. Trad. Shooja De Hossein e Isabel Santos. Lisboa: Relógio d' Água, 1992.

Expressionist utopia and dystopia (architecture, literature, film)

Hanna Grzeszczuk-Brende
Faculty of Architecture, Poznan University of Technology, Poznan, Poland

ABSTRACT: In the early twentieth century, utopia and dystopia in expressionism were associated with the metropolis, which brought about the ambivalent feelings of anxiety and disgust but also of fascination and hope for the future. Images and designs of the cities were also a metaphor of man, the present and future society, the relationship with nature and cosmic space as well as the relation between the past, present, and future. Comparing expressionist painting, architecture, and film we could talk about the two scenarios of the future, of a utopia and a dystopia; the vision of the ultimate collapse or restoration of the harmony of the universe. In creating a future world, art was to play an important role, architecture in particular, as the artists attributed to it the ability to transform the human environment and to transform man and society. The cited examples of the works of Paul Scheerbart, Stefan Żeromski, Bruno Taut and Lang's *Metropolis* allow us to show the visions of restoration of the harmony of nature and technology, of art creating the future world unity of man with the cosmic space but also of the fear of progress that is not accompanied by spiritual values and balance at all levels of existence.

Keywords: Glasarchitektur, glass houses, Metropolis

1 EXPRESSIONIST UTOPIA AND DYSTOPIA (ARCHITECTURE, LITERATURE, FILM)

The vision of the world in the expressionist art of the early twentieth century stems from the tensions between the diagnosis of the present and a vision of the future, between damnation and salvation. In the centre of this vision is the city—as a theme, a place, a social and political metaphor. In this context, Ewa Rewers is right when she writes that the images of the metropolis are not just "about capturing some form of urbanity, but they are about the future of the world, the city as a metaphor and metonymy of human destiny" (Rewers 2010: 8). Ambivalent feelings associated with the city, feelings of anxiety and disgust, rebellion, and fascination are combined with hope and future plans and designs. They do not only affect the shape of the city, but the shape of the human fate.

Images of an expressionist monster city continue to support the more and more popular voices of late nineteenth century, when the disease of the city was identified with the disease of the society, of the people who in contact with the metropolis had transformed from

> [the] noble stream of millions into countless thousands of unfortunate people each working in their dens or cells, hating each other. (Howard 2009: 146)

These images could be seen through the layer of the big city glitz exposing what one tried to hide from the view of an average city dweller, ripping off the mask of official culture and revealing its margins, the phenomena undergoing under surface, inseparably fused with the external mask. The other side of the city, its underworld immorality, poverty, and crime was perceived primarily as a threat to the souls of those lost in the solitude of an anonymous crowd. Expressionist paintings wanted to make this cry of the soiled soul heard, while the subsequent architectural projects sprang up from the hope of recovering the spirituality in a world designed by visionary artists, the precursors of a new society. Comparing expressionist painting, architecture, and film we could talk about the two scenarios of the future, of a utopia and a dystopia; the vision of the ultimate collapse or restoration of the harmony of the universe.

The twentieth century is still an era of faith in ideologies, which shall bring redemption to humanity; hence, the disclosure of misery lurking beneath the surface glitz triggers the expressionist search for a better future. Turning to the spiritual values of the past, the expressionists see their fulfilment in the future combined with the technical progress of modernity.

Assuming that—as Paul Scheerbart wrote—"our culture is to some extent a product of our architecture" (Scherbart 1914), artists, not only

213

those representing the expressionism, took upon themselves the task of creating a better world. Within that scope, they referred to the tradition of Renaissance utopia of the ideal city as a model of an ideal society, composed of sinful and weak individuals—of which the Christian tradition constantly reminds us—but they also referred to the romantic vision of the demiurge-artists, who through the power of their art are capable of "turning ordinary people into angels"(Juliusz Słowacki, *Testament mój*/My testament, 1839–1840). Expressionism thus converges with various political, artistic, and social programs in which the desire for a change was coupled with the fear of a violent and destructive force of the revolution. Yet, what distinguishes the expressionist ideas is the concept of a social change towards the reconstruction of spiritual transformation of man and nature—with the participation of the artist or architect as the creator of the future world, whose works are regarded as the proof of the victory of spirit over matter.

In the expressionist architecture, glass became the dematerialized matter. Scheerbart said:

> The earth would really be transformed if brick architecture were replaced by glass architecture everywhere. It would appear as if the earth were clothed in diamonds and enamel decorations. [...] There would be a paradise on earth and we would not have to long for a paradise in heaven. (Osman 2001)

Glass would allow people to break free from the ugliness and misery of the existing world, to regain contact with nature thanks to the new technological possibilities, to reconcile progress with man's original purity. Therefore, in *Glasarchitektur* he thoroughly considers practical issues arguing that a coloured glass reality is at our fingertips and we can create not only glasshouses, but also glass cars and trains or glass furniture. Writing about the coloured glass, Scheerbart regularly refers to the past, mainly to the ancient cultures of the Middle East and to the Gothic era, suggesting in this context the possibility of re-spiritualising human environment. Deliberately, he does not speak about the social forms of the future, and he only refers to certain aspects of the impact of glass architecture on man because life in a glass house, open to the sky and to gardens, would automatically release the dormant [better] spiritual forces.

In 1914, Scheerbart further developed his vision of a paradise on earth, which was first published in 1893, and dedicated it to Bruno Taut, showing the glass pavilion at the Werkbund exhibition in Cologne as the first implementation of a common idea. Taut's exhibition building was an attempt to combine concrete, iron and glass; it was not so much meant to advertise the glass industry, but rather an attempt to "construct the garments for the soul" made of coloured glass mosaics, glowing in the night like a shining jewel (Osman 2001). Surely, Taut's Glass Pavilion with its dome resembling a diamond cut and the walls of coloured glass was the most complete and unprecedented, in the history of architecture, embodiment of the practical and mystical properties of this material, which Scheerbart expressed in one of his aphorisms:

> Das Licht will durch das ganze All—und ist lebendig im Kristall [light penetrating the universe is alive in the crystal].

Perception of crystal and glass as precious materials integrally endowed with spiritual properties has had a long tradition in both art and mystical religious theories, which were invoked within expressionism. Natural crystal structure, the sharpness of its shapes, its transparency and cleanliness constituted a model for new glass architecture and the features of the new man: that was to be righteous like clear crystal and to live in houses that do not isolate, but are closer to nature. The luminosity of the crystal, in turn, was associated with the splendour of God, or rather of the divinity understood not only in terms of religion.

How much this alluring glass architecture turned out to be seen as a picture of a better future is evidenced not only in the fact that the idea was taken up by a circle of German expressionists, but also by its impact on the vision of a new civilization in the novels of Stefan Żeromski. His vision, which almost literally coincides with descriptions of *Glasarchitektur*. Published in 1912, the novel *Uroda życia* [Beauty of life] includes the following considerations by the main character, Piotr Rozłucki:

> Sand has become the most precious treasure, providing glass, which you can build everything with: beautiful glasshouses of farmers with glass furniture and appliances, colourful habitat of a new glass-based art—the source of health [...]. Roads made of glass panes, paved highways reaching far under the linden tree avenues. They smile to the sun, those wonderful houses among the fields. Some are blue, others—pink, white or multi-coloured ... (Żeromski).

Glass houses are also a key theme in Żeromski's next novel *Przedwiośnie* [Early Spring], written in the years 1921–1924, interpreted as a vision of a perfect Poland, which was wasted after regaining independence in 1918. Judging by the title, Żeromski does not give up his faith in the coming of a true spring or the future transformation of Poland into a perfect modern society. It was to be fulfilled by building glasshouses, described in the vision of the dying father of the main character:

> The whole one-storey house [...]. The hot water goes around the walls, (...) Cold water flows through the same inner tubes in the summer, circulating around every room. (...) The same water is used to wash constantly the glass floors, walls and ceilings, spreading coolness and cleanliness. (...) Artists, great artists, (...) wise, useful people, design these houses, they are aware and inspired creators (...). The houses are colourful, with shades depending on the nature of the area, on the inspiration of the artist and on the tastes of the house dwellers. Against the background of the forest the houses are snow white, they are pink in the plains, light green with a hint of violet or nasturtium colour in the hills. The house decorations follow the most sophisticated, the richest and most fantastic indications of the artist and the tastes of buyers because the beam walls can be coloured in its liquid state in any way you like. (Żeromski 1972: 84–86)

Both descriptions are so similar that Żeromski must have known Scheerbart's early writings, both writers, however, attach a different role to the motif of glasshouses. Żeromski strongly associated glass architecture with the social change; the perfect environment, which will allow changing "the man of brick" into the man of crystal, glass architecture, was to endow its users with spiritual properties that for centuries had symbolically been associated with glass.

In Scheerbart works glass defined the characteristics of the future society and its environment, however, the motif of the harmony of the universe and culture regained thanks to the technical possibilities of glass seems to prevail over the motif of social harmony. It was to create a new quality of nature itself combined with art into nature-art or *Kunstnatur*, as defined by Regine Prange (1991). An even clearer idea of the harmony of nature and the human activity transforming it was expressed by Taut in the concept of "Alpine Architecture" / *Alpine Architektur* of (1919), which can be interpreted as a vision of a cosmic *Gesamtkunstwerk* and as a development and visual embodiment of Scheerbart's Kunstnatur vision. Tino Werner writes:

> In this symbiosis of nature and artistic creation, there is a vision of the transformation of people into cosmic beings through light, colour, and form. (Schirren 2004)

The transformation of the Alps into a work of art being a synthesis of architecture and landscape would allow achieving "an eternal harmony of the world" (Schirren 2004).

Glas- und Alpine Architektur are not only a vision of the future society but a broader, actually, an eschatological utopia. Referring to the title of Scheerbart's novel "Paradise. The Homeland of Art"/*Das Paradies. Heimat der Kunst* (1889), we can say that the goal is a utopian unity of man and the world. The visions of the two artists suggest that ideal human relationships and the ideal man will emerge from a transformed world, the man whose creative power will be the human spirit that is capable of defeating, subjugating, and shaping the matter. The organically growing crystal structure with its geometrical clarity and gleam, hardness and brittleness was a harbinger of *coincidentia oppositorum*. Mutually supportive interaction of nature, the creative will of man and the technology endowed with spirituality would allow us to form an individual who, in cooperation with others, would develop ideal forms of social life.

The fantasy of *Alpine Architektur* was regarded by Taut as a starting point for a more "specific" idea of City Crown or *Stadtkrone* that was completed as early as 1917 but was not published until 1919. The common element was the belief in the role of the architect as a "spiritual creator" who is able to discover and express the redeeming desires. While *Alpine Architektur* was a free creation, a vision, which was not constrained with the prospects of its execution, Stadtkrone, was a utopian city of the future, derived from the historical examples and enclosed with the calculation of the necessary costs. Yet, despite the rational reasoning, it remains a kind of spiritual beacon. Scheerbart's text on *Stadtkrone* proclaimed that squandering this opportunity would lead to an apocalypse, plunging the humanity into a sepulchral silence of the night.

Coming up with *Stadtkrone* Taut proposes a kind of inversion, just like in the previous visions: if the city is a clear reflection of the inner structure of man and his thinking, the implementation of the desired shape of the city, expressing the spiritual aspirations, will trigger a feedback reaction affecting the social system and the people. Yet, as he writes, healthy homes, gardens, parks, factories and shops will suffice if this structure is headless. The building complex dominating over the city would consist of social, cultural, and artistic institutions and would simultaneously express the connection with the cosmic order. There were to be four high towers oriented according to the sun, which were to form the basis for the highest edifice towering over the entire composition as pure architecture. The crystal structure illuminated by sunshine would dominate as a shiny diamond, as a sign of supreme peace and serenity of the soul and it would reveal the essence of the matter with its gleam, transparency, and reflectivity. *Stadtkrone* as evidence of individual spiritual harmony, with himself, with other people and the cosmic space would dominate over the city. There one would be able to live not only in health and safety, but also in happiness, in accordance with the recommendations of Aristotle—if happiness is understood as awareness of sense and gratification of higher existential needs.

The *Stadtkrone* concept brings together all the threads, revealed in Scheerbart's and Taut's earlier works and gives them a "practical" dimension, inter alia by reference to other ideas, treated as a partial prefiguring of the future city, e.g., as envisaged by Howard or Sitte. Taut also tried to start the journey towards the city of the future in his housing projects by introducing colour to their architecture, e.g., in the Falkenberg housing estate (1912–1916), which seems to have little in common with *Glasarchitektur*. Yet, looking for signs indicating the revival of the spiritual needs in the society, Taut wrote:

Finally, the colour, colourful architecture is blooming again [...] the entire spectrum of pure colours is poured again onto our homes...

Introducing colour to brick houses would be the first step towards the architecture of coloured glass.

The utopian character of the idea of Taut and Scheerbart lies not only in the fantasies of glass architecture or in ideas of the feedback influence of architecture on the human psyche.

Considered from the perspective of postmodernism, it manifests itself particularly when we realize that we are dealing with an overall vision which is meant to stop the tottering foundations of the European Logos and its "supreme ideas: faith, god, religion", a vision of social cooperation and harmony with nature.

The disappointment felt after World War I, when after the disaster the inevitable rebirth of the world failed to come, as had been expected, was expressed by Żeromski as prior quoted. The fiasco of the glass houses depicted in his novels not only concerned Poland but also reflected the condition of whole European civilization, which once again, had revealed its dark side. Lang's Metropolis of 1927, which shows the effects of the separation of matter and technology from the spirit, diagnosed by Scheerbart as "the hell, the homeland of lust for power" can also be classified as meeting the atmosphere of disappointment.

It is not by chance that Metropolis combines various incarnations of architecture reaching the sky, from the Tower of Babel all the way to the skyscrapers of New York that were Lang's direct inspiration for the film after his travels in 1924. In Europe, the skyscrapers epitomized progress and modernity, particularly after the contest for the Chicago Tribune building design in 1922. Previously, the fantasy drawing of *King's Dream of New York* from 1908–1909 with skyscrapers and airships docking to the buildings' upper floors had become the basis for Antonio Sant'Elia's future city projects. Yet, the Babel Tower was a symbol of pride, while the American cities were also perceived the way Siegfried Kracauer saw them when he wrote for the Frankfurter Zeitung:

Everyone knows the ugliness of New York. Tower monsters that owe their existence to the unbridled will of powerful rogue entrepreneurs stand there wildly and indiscriminately side by side, decorated on the outside and inside with apparent glamour, which in no way befits their goals.

In *Metropolis,* the city with the airplanes flying between the illuminated skyscrapers, with multilevel arteries congested with vehicles, seems to embody the dreams of progress, it appears a visualization built upon Sant'Elia's futuristic fantasy. The meaning of the film, however, emerges from the comparison of the two aspects of technical progress as the skyscrapers reaching the sky are founded on the unified work of the human masses. The impressive machinery of modernity is created in the underground passages of the city by the exploitation of machinelike human masses that would then be replaced through further inventions of mad scientists and unscrupulous entrepreneurs. The dehumanized world of a nineteenth-century metropolis inhabited by the philistine and his victims depicted in expressionist paintings would only be a harbinger of further exploitation if the development of technology and the usurpation of reason were not accompanied with the comeback of the spiritual values.

On the outskirts of the dichotomous and simultaneously integrated world of a gigantic Metropolis is the scientist Rotwang's hut, which seems to be a relic survivor, overwhelmed by the dynamics, momentum, and size of the city. Yet, it is in that modest hut where the new inventions are worked out, including a kind of a golem, a mechanical woman with the appearance stolen from Mary, a positive figure. Despite the appearances, the hut is not an incarnation of a rural idyll, but a dummy, like the humanoid created in its interior. It is a symbol of an inventor unable to create the real life creations and able only to produce destructive substitutes thereof. It appears that when Lang created his film warning against progress, he also warned against a mirage of a return to the past, to nature, and above all to a naive negation of the gains of the progress, as in the case of the rebellion of the workers devastating the machines and at the same time destroying their world.

The message of the film is clearly expressed at its very beginning and its end: a mediator between the reason and the hands must be the heart. If one identifies the heart not only with emotions but also with spirituality, then the film warns against excessive cult of the progress that lures with new opportunities but also destroys the people if they let themselves be overwhelmed by the solutions of their own civilization. After the disaster in the depths of the monster city, the final scene is as follows: in the background of a symbolic portal of the cathedral, a worker and the owner embrace each other, mediated by Mary and the converted Freder, son of the owner. This grip combines the heart, the hand, and the mind.

Stadtkrone and *Glasarchitektur* propose a vision of a new civilization based on the unity of the heart and the mind, the spirit and the matter, progress and nature, and their purity is the result of the exclusion of doubt or uncertainty. In *Metropolis,* the fascination with the future world was turned into a captivating but disturbing scenery of dystopia. Both utopia and dystopia stem from the same source, a pursuit of a better world, a desire for progress. However, it is not the progress itself that is bad, but misuse of its achievements, which is depicted in many juxtaposed planes making up the bases of *Metropolis*. The city above the ground is opposed to the underground city, exploitation to servitude, Mary the prophetess (standing for truth) to Mary the robot (standing for falsehood). Based on these juxtapositions the final gesture of reconciliation gains strength by appealing for existential fullness formed out of harmony and cooperation between all the components of the world.

Confronting *Glasarchitektur* and *Stadtkrone* with *Metropolis* makes us aware that the difference between a utopia and an anti-utopia falls from the adherence of man to the dark or bright side. Lang thus reaches beyond not so much utopia, but rather beyond the belief in the possibility of creating it in a world of shaken Logos, where values are subordinated to ruthless technology.

BIBLIOGRAPHY

Howard, Ebenezer. "Miasta ogrody przyszłości (Garden-cities of Tomorrow)." In *Trzewia Lewiatana. Miasta-ogrody i narodziny przedmieścia kulturalnego.* Ed. Czyżewski, Adam. (Garden-cities and the birth of the cultural suburb). Warszawie: Państwowe Muzeum Etnograficzne w Warszawie, 2009. ISBN: 9788388654817.

Osman, Silke. "Besondere Herrlichkeit. Glasarchitektur bei Bruno Taut und Paul Scheerbart." *Das Ostpreußenblatt* (2001).

Prange, Regine. Das Kristalline als Kunstsymbol, Bruno Taut und Paul Klee, Zur Reflektion des Abstrakten in Kunst und Kunsttheorie der Moderne. Studien zur Kunstgeschichte. Bd.63. Hildesheim: Georg Olms Verlag, 1991. ISBN: 3-440-06874-9.

Rewers, Ewa. "Introduction." In *Miasto w sztuce - sztuka miasta.* (The city in art, art in the city). Kraków: Universitas Kraków, 2010. ISBN: 97883-242-1306-1.

Scherbart, Paul. *Glasarchitektur.* Berlin:1914.

Schirren, Matthias. *Bruno Taut—Alpine Architektur. Eine Utopie.* München: Prestel Verlag, 2004.

Taut, Bruno. *Alpine Architektur.* Hagen Folkwang-Verlag G.m.b.H, 1919.

Żeromski, Stefan. *Przedwiośnie.* 1ª edition 1924. Warszawa: Czytelnik Warszawa, 1972.

——."Uroda życia."https://www.scribd.com/doc/43081039/Stefan-%C5%BBeromski-Uroda-%C5%BBycia. accessed on 17/05/2014.

Nothing but something else. Displaying social dystopia

Marta Germano Marques

Mestrado Integrado em Arquitectura, Faculdade de Arquitectura, Universidade de Lisboa, Lisboa, Portugal

ABSTRACT: Displaying social dystopia is an artificial creation that has the purpose to contrapose a social reality with an extreme hypothesis, one with little if no concern for moralism or economic production, community rules or expected behaviours. The utopian dystopia generates a place where individual minds work independently but with the goal of endlessly improving each other to the limit of acceptance, people objectifying each other with the purpose of leading others and themselves to the closest thing to absolute knowledge, being this knowledge not exactly what is defined by the current circumstances but a knowledge based on corporeal principals and metaphysic experiences, the only necessity of organisms that would eventually destroy each other.

This is a reaction towards the inexplicable lightness that people display in contemporary days, in contrast with more reactive times, apparently closed in the past, times that seem to be forgotten or chosen to be remembered only and stories but not as our common history.

This is a personal recognition of the unbearable truth that human individuals have turned society into, but also the acceptance that the opposite is not practically acceptable. There are no determined conclusions, only the constant idea that what remains in thought has often more power than what is said or done.

Keywords: imagination, lightness, damages, madness, fantasy

1 IDEALLY, THE REALITY

A world connected in every possible way, by cables, waves, ideas, fashion and tides, contacts, calls, messages, conversations, looks, decrees, laws and documents, alliances, unions, hatred and passion, orders and opinions, photography, food, drinks, bridges, highways and roads, by wheels and rotations, news and preoccupations, fear, nostalgia, the History and the stories, wars and conflicts, bombs and explosions, airplanes and trains, cars, travels, escapes and prisons, hidings, tunnels, routes and connections.

Looking for information on different subjects, I often notice that, most of the times, the available information is nothing more but comments about comments, interpretations, and reinterpretations. Moreover, it is rare to find lines written for the first time, or sentences never spoken before. Everything is a constant repetition of the past and the present itself, nothing has the right to a premiere anymore. Because the same things happen so often that keep on overlapping, and running over and torturing each other for the chance of moving forward, not having the awareness that the result will necessarily be the same, repeatedly. The same or the closest thing to that, which means similar or the exact same consequences, so similar that guar- antee no place where to be described. Nothing is new, nothing is regulated by a different set of rules, and everything plays the same inevitable game. Exceptions no longer exist, for even those have already happen too many times to earn the name of exception; they have actually created their own rule of existence—the rule of exceptions that validate every related idea, which paradoxically liberates them from their original character. Therefore, what can someone continue on believing anymore? If you search and seek about the past, about what is recorded in papers and tapes, you find yourself reading or listening about something intimately perverted in its origin: words have already been spoken and written above others and about others; comments and modifications have reshaped the primary facts, burning them and deeply redefining them in content and visual appearance.

The human being is equally capable of infinite production of knowledge and infinite destruction of that same knowledge: *cumulative knowledge* is what could be called to the hypothesis of simply accumulating new skills and learning without ever cleaning the pre-existing ones. However, the problem is that new facts always influence the way we maintain what has already been kept for a while; and the same happens for old ideologies compromising our acceptance of novelty. Why does innovation tend to substitute the pre-existing? Why cannot they coexist without confrontation? Sometimes, it happens that an annulled or vanished object or circumstance loses the possibility to return. And yet, every place you go, people are the same, conversations are the same and sound like each other and like the same nonsense as always: even not understanding the words or trying to do so, it is possible to recognize the same type of smiles, equal, shallow, credulous, always full of that instant joy that seems to completely fill their minds during an exceptional period of time.

How can apparently meaningless moments be meaningful enough to represent everything makes people believe they are happy when in reality they are not, they will never be able to be, because being happy is not a status or an objective condition. It only happens occasionally, in a matter of brief moments and, can only be recognized after a lifetime. It is as fugacious as a shiver, as subtle as feeling cold or a wave of hotness when a door opens for a different environment.

Well, after all, it might be happiness what provokes empty smiles, I will never know for sure. What I know is that those are at least honest, against the ones coloured in yellow and built by

Figure 1.   Pomar, Júlio. Verbena (São João do Porto), 1955. Linocut in two colours. 30,8 × 22,5 cm. Artist's edition.

lips that melt in the corners of the mouth, smiles of pure concern and accommodation to poverty. In fact, as said, instant joy is the closest thing to happiness, even if it is triggered by pure shallowness.

However, what do you think or say when everything seems unequal, when everyone you meet is "happily accommodated" to life in general, when everyone believes to be in the right place. Do you tell them the truth? Or you don't tell them a word because you have the intimate idea that they might well be actually happy, in the most complete sense of the word, and you're the only one you doesn't have the sensibility to acknowledge that. Most people are probably happy, quietly happy, besides all practical issues.

Then again: how can happiness be quiet? Happiness is madness, it is an unbridled monster, and a monster that shows itself for only a few seconds at a time, very occasionally, for it is too dangerous that people believe in its existence. If they knew it existed, they would be obliged by themselves to question the actual concept of what being happy is about and in that moment, they would find nothing but an unworthy void represented by a monster from which they can only see small parts, such as the ears and the tail, sliding behind a piece of furniture once in a while. If that happened, people would not let him go anymore; they would torture him pitilessly, because men cannot deal with strangeness or uniqueness. Men are naturally dirty creatures, for keeping this monster permanently hidden among themselves; not giving him food or water; allowing its anger to grow and turning into inexistence. Men do this continually, not noticing they are contributing to the destruction of something essential to their own maintenance, something they will never be ready to understand or care for in the humane way. Men close their eyes and keep on behaving like insects, incoherent, eternally wild.

> ... o mesmo se passa com o amor, para o qual somos preparados de um modo absolutamente excessivo. (Musil 2008)
> [... the same is applied to love, for which we are prepared in an absolute excessive way. (free translation)]

Love is corrupted by rehearsed words, programmed conversations, gestures that are directed by guides and manuals, social magazines and every kind of trivial comments.

We are slaves of words distorted by socially controlled power, managed by invisible institutions.

Love is distributed in rice packets in humanitarian actions, in delivered vaccines, tears shed for strangers, standardized meals, used cloths and shoes, broken toys, heavenly gifts. Love is a count-

Figure 2. Dacosta, António. Untitled, undated. Chinese ink with gouache. 21,5 × 24,8 cm. Colecção CAM—Fundação Calouste Gulbenkian. Lisboa, Inv. DP1069.

able noun; you can actually pay to give it and pay to receive it. You can acquire it in hour or minute packs, with or without extra accessories and fantasies.

Like happiness, love is nothing of what is believed to be, it just can´t be that way. It is impossible to teach or even to learn how to recognize such things, such translation of words into actions. It can only be taught or learned how to deal with false manifestations of each of those things, monstrous versions, *objectified* versions, possible to describe deviations.

People speak of love as if talking about potatoes or cigarette butts; it is talked about with unconcern, lightness, relaxation.

Relaxation belongs to people that levitate between conversations and chitchats, between coffee shops and rooms, between schedules. Relaxation and lightness belong to people that knows as much from joy as from sadness, belong to the ones that go shopping and look for sales, to the ones who drink beer with friends, to the ones that listen to publicity ads. It belongs to the ones that dedicate their time to things and people they like or want, with careful moderation. Everything they do is inevitably balanced, calculated in a way that makes it not excessive to the body or the mind and neither only little distraction.

The world is measured by the places where normal people move around, the size of their feet and the clothes they dress. Nothing is projected for the ones who live out of the regulated scale, those are always the minority: the *too much something and the something less*, the ones who want everything or nothing at all. Not even these two extremes com-

bined represent the smallest sample, facing all the others, the ones that know exactly in what measure one should want or do something. Because today, people are measured and judged by what they do, no matter what they want.

Imagination should count over practical actions, imagination is everything. It is both fantastic and terrible, driving even the maddest ones to unrecognizable madness: more than what even they can bare, but that they actually do: for the better or the worst, with fewer or bigger damages, internal, external, or collateral.

Teria a vida em geral chegado a um ponto morto? Não, a sua força tinha aumentado ainda mais! Haveria mais contradições paralisantes do que noutros tempos? Dificilmente as poderia haver. No passado não se cometeram erros? Houve erros aos montes! Aqui entre nós: o que acontecia era que se apoiavam as causas dos fracos. E dos fortes ninguém queria saber; acontecia que os estúpidos assumiam papéis dominantes, enquanto grandes talentos assumiam papéis de excêntricos. (Musil 2008)

[Could life had reached a dead end? No, its strength had risen even more! Were there even more paralysing conditions than in previous times? It was unlikely. Were not there mistakes also committed in the past? There were tons of them! Just between us: what used to happen is that people would support the causes of the week. Moreover, no one was interested on the strong ones; it happened that stupid people had dominant roles, while the talented ones were seen as eccentric. (free translation)]

To live out of things or entirely inside them? Ideal shapes or the idealism of living out of them? What is that? If the world had the shape of the strong ones, there would be no room for almost the entire population. Would there be a place for the ones that are not always as smart, not always as brilliant? None of that is real; it is not about that, not at all. It is about something bigger than that, bigger than being a man, bigger than being a person. It is about living each and everything intensively, until exhaustion. Always in one extreme or the other. Always painfully.

Is there any justice in that? Where does that idea fit? Does it fit in being able to know what justice means or in having right to that? Where is the modesty or the shame that keeps us from asking for something that we know to be inexistent, unreachable. I don't actually know it... or I do and because of that I know I could never reach it. I only know it isn't something practical, something possible to get hotter or colder. It is not an animal that can go through metamorphoses or a plant that can blossom to the world. It just either is or it is not, and generally, it is not. On the other hand, maybe, I have never been in the place where it had the chance to be.

Looking to a room filled with chairs, all empty, all the same. Are the people who sit on them, the same? Alternatively, do they transform into the same because they are sitting on the same chair? Well, maybe. They will probably listen to the same, will they wright similar things about it (?) will they talk about the same (?) will they discuss the same (?) they will probably interest for the same or for nothing at all. Those never know what unfairness is about and because of that, nothing will ever seem fair to them (will they say with great lightness), but it can look like it. Is this Utopia? Horror Utopia at least. Unreasonable Dystopia.

O efeito narcotizante daquela música paralisou-lhe a medula e aliviou-lhe o destino. (Musil 2008)

[The narcotic effect of that music paralysed her marrow and relieved her destiny. (free translation)]

2 IN REALITY, THE IDEAL

If a man sees, (wants to see) not everything that proposes himself to, he will probably be able to see any of that, at least in the way he wants to. That inevitable failure comes from the already known impossibility of looking at everything with equal intensity... or maximum intensity, even if there is a great intention being devoted. In that way, even if the one that proposes himself to see and look for everything is inevitably a failed man, than the world he lives in might be nothing more than an illusion, a never accomplished desire. One thing is for sure, it is nobody else's world for no one else will perceive it the same way he does but also, it is not the world he tries to create at every cost, it is something completely different, mutable, animal, independent from its creator. The world that he lives in, is probably the world he generates in is unconscious, more complete that anything he can ever be aware of. That world, being strongest but hidden, is a servant of the world that the man lives consciously, feeds it, keeping it alive. That world, the more complex, and complete, the one that exists in the shadows of the mind is the very essence of the other. It is intensively lived with no intention of that, it is always out of control, highly combustible.

A sound exists because someone listens, a smell torments only because someone smells it. The unconscious world exists because there is melancholia, curiosity, despair, desire, whimper, and dissatisfaction.

If we question the hypothesis of proceeding through this route of eatable stones and delusion and we might find the only possible way of dealing with whatever is happening in any place on the face of the earth; looking for everything till eternity, reaching the out of control status and waiting for balance to come back again... losing the ability of being predictable and assuming as the only possible rule in life, to always remain unsatisfied. To resign, exclusively to live in a continuous *meantime* among things. The *meantime that is everything.*

Is that the possibly innocuous premise of an artist's mind (?): to continuously question and explore, comfortably accepting everything that is purely incomprehensible and that should remain exactly as it is for in reality nothing will turn into the opposite. Is it possible to accept the impossibility of understanding anything and everything in its all and still trying to peel every single layer, knowing that the end will never come...?

Here in limbo, or this world, between the known and the unknown, the rationality and the irrationality; all the extremes collide with each other on a daily basis. Everything exists in direct proportion to the piece of imagination that is devoted to it. What really matters is what goes in your mind at every second, what, in the real world, would make you stumble and fall on the sidewalk. What matters is what we know to be impossible to accomplish but that we still imagine as if it was.

How can someone try to learn about heaven or hell, losing time to question its existence if everything that exists and that is palpable is already here, massively, waiting for someone to notice, everywhere, in the city—that is a sortilege (*a cidade que é um sortilégio*) (Khadra 2010). Why are people afraid of dying but not afraid of living? Why don´t we talk about life the way we talk about death or talk about death the way we talk about life? Or the way we talk about life... There must be life in that death, just as much as there is death in this life. Here, people will start being afraid of simply being and they must learn how to deal with that.

People will live each other the way cities are lived: contributing to its construction, travelling and wandering through it, deforming its sensitive and helpful shapes, infecting its corners and streets with venom, terror, fantasy, and wonder. To leave it absolutely exhausted, destroyed, raped throughout its surface, internally and externally, slaughtering its skin, muscles and bones, an eccentric circus; without apparent reason or easy understanding, without contradicting when people rise at the sight of the elephant of undeniable exuberance.

Figure 3. Dacosta, António. *A morte e o jovem (Death and a young man)*, 1985. Acrylic on cardboard. 31,4 × 19,8 cm. Colecção Fundação Cupertino de Miranda. Vila Nova de Famalicão.

Acham que a felicidade está em protestar. Mas não é verdade. A felicidade, Júnior, é uma pessoa saber calar-se quando as ondas se divertem. (Khadra 2010)

[They think happiness comes from protesting. But that is not the truth. Happiness, Júnior, happens when someone knows how to be quiet at the sight of waves enjoying themselves. (free translation)]

However, is necessary to know how to recognize exuberance, otherwise no one will rise at the right time?

It is necessary to learn what is worth, when everything becomes silent and there is nothing pass-ing through except images, with no subtitles for the represented pictograms. Because living in the foam (Vian 2013) is one of the available options. Living in the foam is also not wanting to leave that place, it's correcting mistakes with spit, mending trousers with nails, it is *dancing at the sound of sunbeams chocking against taps* [free translation, *dançar ao som dos choques dos raios de sol com as torneiras*] (Vian 2013). There is living in the foam and actually knowing the essence of foam, but that is too scary to imagine, too compelling, to serious and dark. Everything that is highly composed by a great percentage of fantasy is fatally destined to equal amount of horror.

To live with no barriers would mean daily ecstasy; with no individual precautions, available to each and every torment acquired as a gift, lived as a wonderful opportunity.

It is for another day, for another life, for another world.

BIBLIOGRAPHY

Brandão, Raul. *Húmus*. Lisboa: Bertrand Editora, Lda, 2011. ISBN: 978-972-25-2302-8.

Huxley, Aldous. *As portas da percepção, céu e inferno.* 1ª ed. Translated by Paulo Faria; Translation of *The Doors of Perception / Heaven and Hell*. Lisboa: Antígona Editores Refractários, 2013. ISBN: 978-972-608-243-9.

Khadra, Yasmina. *O Olimpo dos Desventurados*. 1ª ed. Translated by Maria Carvalho; Translation of *L'Olympe des Infortunes*. Lisboa: Editorial Bizâncio, 2010. ISBN: 978-972-53-0460-0.

Musil, Robert. *O Homem sem qualidades*. 1ª edição. Translated by João Barrento; Trad de *Der Mann ohne Eigenschaften*. Lisboa: Publicações Dom Quixote, 2008. ISBN: 978-972-20-3007-6.

Pomar, Atelier-Museu Júlio. *Júlio Pomar, Edição e Utopia*. Lisboa: Cadernos do Atelier-Museu Júlio Pomar, 2015. ISBN: 978-989-8618-68-9.

Saraiva, Tânia. *António Dacosta, pintores portugueses*. Lisboa: Quidnovi Edição e Conteúdos, S.A, 2010. ISBN: 978-989-554-699-2.

Vian, Boris. *A Espuma dos Dias*. Translated by Aníbal Fernandes; Translation of *L'Écume des Jours*. Lisboa: Relógio D'Água Editores, 2013. ISBN: 978-972-708-644-3.

Notes on the drawing [sketch] as the architect's imaginary

Artur Renato Ortega & Silvana Weihermann
Department of Architecture and Urbanism, UFPR, Universidade Federal do Paraná, Curitiba, Brazil

ABSTRACT: This text analyses the sketch as the architect's imaginary. It theorizes the drawing and explains the action of drawing, mainly in the scope of the architecture project activity. It connects the action of drawing with the visual and spatial perception, with memory and imagination. It considers the architect's drawing as the transformation of mathematical space (rational knowledge—geometry—perspective) into the real, concrete, and experiential space. It is based on the ideas of some authors, researchers in the areas of perception, creativity, drawing, and architectural project in order to understand the relation between the real and the imaginary in the creation process of the architect's[1] of the objects that make up the world surrounding us, or represent the objects that permeate our thoughts, the so-called imaginary[2]. The drawing is, additionally, an object in itself, a product, in other words, a work of art created from lines, dots, or strokes. The drawing also lends, so to speak, its name to the discipline where one learns the occupation of drawing: Observational Drawing, Architectural Drawing, Artistic Drawing, etc. may even be understood as outlines/sketches, studies of a painting, towards the creation and, thus, approach the idea of intention, purpose, something that is planned, a plan, a meaning.

There are, then, several concepts that get involved with the term 'drawing', and each all of them are interconnected and impossible to separate. However, despite the various conceptualizations about drawing, it can be stated that all of them complement and supplement each other in a context of feelings that become a man-impression relation over sketch. I concluded that the drawing consists in "producing" clues towards the project's solution, according to the interpretative and critical sensitivity, as well as organizing the imaginary thinking on what is produced in the mind of the architect.

Keywords: Sketch, Imaginary, Reality

1 THE ARCHITECT'S DRAWING

What is a Drawing? Or more precisely, what is the Drawing to the architect?

Replying to this is not an easy task, as the drawing is enveloped in several characteristics: it can either mean the representation a flat surface.

The drawing comes, this way, from a gesture, a peculiar body mobility strategy that leaves marks from its steps for a support, thus creating, as suggests Copón (2002: 525), a *strange ontological position* in the action as conceived. Since:

> Drawing, the act of making a drawing, implies the direct record of hand movements, meaning a relation between who does it and the results of this doing. It mobilizes the body of the subject who draws, in a unique relation between the hand and the brain, the brain and the eye, imprinting into the results of this action—the drawing—a unique character, directly related to the gesture of doing; an extrinsic harmony resulting from a personal rhythm and gesture. (Rodrigues 2000: 18)

To draw is the drawing, as confirms Bismark (2001). Agreeing with this author, the drawing is a verb. Since the drawing is understood as a process, as an action. The action of processing information from thoughts, and making its representation. This action is directly linked to the spatial and visual perception and, regardless of the type of drawing—to observation, memory or of creation—drawing relies on what has been observed, understood, and internalized. The observational is, at the same time, creation, and memory, as perception takes place in a differentiated form for each individual and is related to the memories and the meaning of what is being observed. A memory and creation drawing, in turn, depends on what has already been observed in order to make sense.

There is no way of establishing, then, the limit for each type of drawing. The words from Arnheim (1996: XIII) are then merged: "all perception is also thought, all reasoning is also intuition, all observation is also invention", and he considers the act of seeing as the one that imposes on reality, in an entirely subjective fashion, form and meaning, as with the ones from Puig (1979), on agreeing that perception is borrowing meaningful structures and not only a record of elements.

Thus, to draw is to perceive, and perceptive activities, according to Piaget (2001), are progressively developed. In the case of the perception of reality, in order to make a drawing, the logic-mathematical notions act as abstract operations, not simply of perceived objects, but also of those actions performed on them. The individual's perception, under this aspect, is conditional to the medium. Piaget (2001) considers as crucial the role of perception in intellectual evolution, which depends on the figurative aspect of the knowledge about the real, whereas actions transform what is real, deriving operations in the course of interiorization and structuration.

For an architect, drawing is no different. To draw is to bring to reality something that is conceived in the mind, by means of the perception of the world where he lives. Something—spatial configurations—that is conceived in a unique, imaginary, world. It is the relationship between the real and the imaginary. It is clear that the architect's drawing has a specific purpose: a document for the realization/construction of a building. It, the drawing, unravels the entire process of an architectural project, from the design sketches; the presentation drawing to the strictly technical drawings of the executive project. In this process, the drawings have origin in generalities—or greater imprecision/doubts -, to definition—the precision/certainties.

Thus, the architect's drawing is the transformation of the mathematical, abstract space, measurable in its three dimensions, within the fundamental concept of the Euclidean geometry notions of the perspective (rational knowledge), in the real, experiential space, that one when architecture is achieved.

What matters here, however, are the drawings from the first phase, in other words, the sketches[3]. As for the "first phase", it is worth mentioning that drawing the sketch is a strong instrument for imagination and creativity, intrinsic to the various moments of the projecture process, but especially at the beginning of the process: as source of inspiration for the creation of the concepts needed for the development of the ideas.

2 SKETCH; IMAGINARY AND REALITY

The image of the "isolated" architect, leaning over the paper implies a unique moment, one of intimacy with the object of study:

This intimacy between the architect and his drawing favours the creative dynamic between designing and representation; it is a conflictive interdependence that the brain that projects must control [...]. (Henri Ciriani, apud. Lapuerta 1997: 95)

The sketch appears as the main instrument to organize the imaginary world of the architect, putting order in the project and moving the process forward. Daher (1984: s.p.) describes these feelings in the architect's personal moment with his ideas: "The initial shapes take form, the outline becomes stronger when the architect writes small numbers, imprints indications about the scale on the drawing, the proportions of those mysterious scribbles. Correction: mysterious for the layman, since for the author they are signs that were only asleep, waiting for their moment. So they appear when the architect is able to awaken them from their silence in either a peaceful, suffered, or pleasurable process".

There is, therefore, a close relationship of feelings between architects and their drawings/sketches. American architect, Stanley Tigerman, comments:

In each project, I experience two moments of pleasure: first when I draw the first sketch and then when the work truly completed. (Lapuerta 1997: 32)

The joy of the first drawing is associated to the spark of creation. The moment that the idea appears in the mind of the architect and he records it on paper. That is not easy, as it is a state of ideas construction. The architect Le Corbusier exemplifies well such a situation by stating that:

[...] When they order a work, I have the habit of including it in my memory, that is, I do not make any sketch for months. The human head [mind] is wired in such a way that it has certain independence: it is a box that we can disorderly fill with the elements of a problem. We allow it, then, to 'float', 'scheme', 'ferment' and, suddenly one day, spontaneously, in its interior, something is triggered; we take a pencil, a crayon, a colouring pencil [...] and we give life to the paper: the idea arrives at the world, it is born. (apud. Lapuerta 1997: 87)

If we could observe the architect, at this very moment, we would see something as Athavankar describes:

[...] a typical scene: cycles of intense introspective 'pondering' followed by 'flashes' of scribbles. Not infrequently, they speak and gesture by themselves, and even change the place they occupy in the environment. The process is similar to the projection of what takes place inside the mental space". (apud. Medeiros 2004: 49)

At any rate, the important thing is that the sketches are and should be used to develop the uncertainties present at the beginning of the elaboration of a project. As sustains Lapuerta (1997, p. 91)

The graphical experimentation, the improvement, the rigor, the constructive dimensions... are, also, things that will complement each other with new drawings.

To draw, therefore, consists in "producing" clues towards the solution of the project, as well as giving order to the architect's imaginary thinking, since:

"Abundant material is produced, and from this, meaningful clues may be selected and used to make concepts emerge. On selecting these clues, mental images are used to interpret, in the draft, more information than it was invested in its lines. When we draw a line, new combinations may appear among these elements, without us having planned this". (Medeiros, 2004, p. 58)

Dorfman (2002) reinforces these values on analysing the drawings from Portuguese architect Álvaro Siza, relating to the project from the Iberê Camargo Museum, in the city of Porto Alegre, Brazil:

The ambiguity that naturally occurs in free drawings favours the onset of new solutions and stimulates the conscious development of intuitive skills and creativity. Through the drawing, the thought follows its course, without the need for giving names or labels to things before they are defined. The drawings offered constitute a communication of the architect with himself, while he tests alternatives, developing his thoughts and registering incomprehensible scrawls. Only accompanied by the explanations and indications made by the architect are the drawings useful for the understanding by part of the public.

Architect Eduardo Mondolfo wrote an article, a while ago, in which he ponders this dialog between the architect and his imaginary drawing. In a moment of "rest" from work (as he was waiting for some customer to call or to arrive), in his office, he saw an image:

[...] a very solid and thick wall, coated in white and grey ceramic (and probably 45×45 cm): thus making a checker pattern. Initially, this wall appeared to be loose to me—but it could very well be a facade of a building (or a home). Thus—the checker pattern of this element, by itself, by the obvious rhythm it presented -, also had no reason for being. None whatsoever... It was when I saw that this orthodoxy was pure appearance. In one of the sides—I say—ends, a row of the ceramic mesh went beyond the wall itself, and advanced into the air, almost suspended, if not for the rounded column, apparently made out of concrete, leaving the ceramics gap in full view: almost 2 and a half meters [...]. (Mondolfo 1986 77)

Mondolfo continues the description of his unique dialog with his drawing, inserting new elements, no more from his mind—as a vision -, but from a debate about several architecture questions, in search of words that would validate the idea (the initial image of the wall ends up being incorporated, in the end, into a drawing of a beautiful home).

However, he finishes the text—and it is what matters—by justifying that he desired to expose, from this evolution of an idea, that there is one

Figure 1. Iberê Camargo Foundation Headquarters. Draft sketches. Architect Álvaro Siza, 2008. Source: Vitruvius Arquitextos: www.vitruvius.com.br/revistas/read/projetos/08.093/2924.

Figure 2. Iberê Camargo Foundation Headquarters, Porto Alegre, Brazil. Architect Álvaro Siza. Source: editora.cosacnaify.com.br/ObraSinopse/11234/Fundação-Iberê-Camargo---Álvaro Siza.

method of projecting by using the drawing as "interpretative and critical sensitivity" on what is produced in the imaginary of our minds. He means, in and by the drawing, or better yet, by drawing, that it is possible to see, review, err, to be right, refuse, accept, destroy, recompose, doubt, clarify, give up, proceed... and to draw more and more—"in each architect, there is a desire of never finishing the project [drawing]" (Vieira 2001: 139).

In short, it is in the practice of sketches that the architect (re)invents shapes, aligns lines and volumes, stipulates empties and fills, defines colours and materials and reinforces the values of his creation. The sketch presents itself, therefore, as a means of personal search for the architect. The drawing and its action: to draw is, therefore, a perceptual activity, something that is not fully complete, as it is open to new visions. It invites us, suggests questions, evokes..., and, thus, exposes our imaginary world for the construction (in theory) of a better world.

NOTES

[1] "The word representation is used here as a way to evoke a situation in his absence. Because, as based on Pistone (2011: 1): "From the sociocultural point of view, the Representation is the system or set of systems through which a society (by means of images, objects and symbols it creates) presupposes a particular view of the world linked to cultural forces that gave rise to it.

[2] The imaginary understood here, in a simple way, as the production of images, ideas, concepts, visions of an individual or a group to express its relationship with the world.

[3] In 1681, Baldinucci quoted in his "Vocabulario Toscano dell'Arte del disegno" the word sketch as: "what painters denominate to these extremely loose brush or pencil strokes that use to parade their concepts without carefully elaborating them; this is what they call to do a sketch".

BIBLIOGRAPHY

Arnheim, Rudolf. *Arte e percepção visual, uma psicologia da visão criadora*. São Paulo: Pioneira: editora da Universidade de São Paulo, 1996.

Bismark. "Desenhar é o Desenho." In Actas do Seminário. Os desenhos do Desenho, nas Novas Perspectivas sobre o Ensino Artístico. Porto: Universidade do Porto, 2001. 55–58.

Copón, Miguel. " El dibujo como máquina conceptual." In *Máquinas y herramientas de dibujo*. Ed. G., Molina; Juan J. Madrid: Cátedra, 2002. 525–577.

Daher, Luís Carlos. *Sobre o desejo – digo o desenho do arquiteto*. São Paulo: Museu Lasar Segall, 1984.

Dorfman, Beatriz Regina. "O ensino do desenho no curso de arquitetura: a construção do pensamento visual." In *As relações arquitetônicas do Rio Grande do Sul com os países do Prata*. Eds. Miranda, M.M and N.F.D Brum. Santa Maria: Pallotti, 2002.

Lapuerta, Jose Maria de. *El croquis, Proyecto Y Arquitectura [scintilla divinitatis]*. Madrid: Celeste Ediciones, 1997.

Medeiros, Ligia M.S. *Desenhística: a ciência da arte de projetar desenhando*. Santa Maria: sCHDs Editora, 2004.

Mondolfo, Eduardo. "Esboço de uma gênese." *Revista Projeto* 88 (1986): 77–80.

Piaget, Jean. *A psicologia da criança*. 17ª Ee. Trad. Octávio Mendes Cajado. Rio de Janeiro: Bertrand Brasil, 2001.

Pistone, Santiago L. Dimensiones de la Gráfica Arquitectónica. Material del Taller de Expresión Gráfica. Rosário: Cátedra arq. Pistone, 2011.

Puig, Arnau. *Sociología de las formas*. Barcelona: Editorial Gustavo Gilli, 1979.

Rodrigues, Ana Leonor M.M. *O Desenho*. Lisboa: Editorial Estampa, 2000.

Vieira, Álvaro Siza. "A Construção de um projecto." In *Actas do Seminário. Os desenhos do Desenho, nas Novas Perspectivas sobre o Ensino Artístico*. Porto: Universidade do Porto, 2001. 136–149.

Sources and references in the utopian work of Bartolomeu Cid dos Santos

António Canau
Department of Arts, Humanities and Social Sciences, Researcher CIAUD, Architecture Faculty, Lisbon University, Lisbon, Portugal

ABSTRACT: Bartolomeu Cid dos Santos was an artist that found in printmaking a privileged means of expression. His work took shape by the combination of drawing and the expressive particularities of printmaking, learned and developed at the Slade School of Fine Art, as well as the artist's strong cultural background, being his work sources and references of scholarly nature, mainly based in the areas of literature and cinema. "I am two" the artist use to say. One that looked inside and another one that looked outside. The part of his work in which he looked inside, reaches themes of metaphysical character. The part in which he looked outside approached reality, with works of political and and social intervention nature. His work had as recurring themes, Space, Time, and Memory. In the works of metaphysical nature we found thematic of a utopian character that morphologically take the form of pyramids, labyrinths, and architectural interior spaces of monumental scale. In this paper, we will discuss the sources and references of the thematic of utopian nature in the work of Bartolomeu Cid dos Santos, trying to be as comprehensive as possible.

Keywords: Bartolomeu, Printmaking, Utopia, Influences, Labyrinths

1 INTRODUCTION

Bartolomeu Cid dos Santos was born in Lisbon at midnight and ten minutes of the 24th of August 1931, son of the doctor João Afonso Cid dos Santos and Nazaré Vilhena dos Santos. According to his sister Alexandra, his parents decided to call him Bartolomeu because he was born on the day of St. Bartolomeu (Sampayo 2010).

In 1951, he enrolled in the painting course of E.S.B.A.L. attending classes until the year of 1956, in this year he enrolled as a free student at the Slade School of Fine Arts—University. College.—London, attending the course as a student between 1956 and 1958 (Gulbenkian 1959: 2). Between 1958–1961, he attends the course of painting of the Lisbon Fine Arts School, not completing the course. He taught printmaking at the Slade School between 1961 and 1996, succeeding Antony Gross after his retirement at the leading of the Printmaking Department at Slade School of Fine Art in 1976.

Bartolomeu Cid dos Santos was an artist who found in printmaking a privileged means of expression. His work took shape by the combination of drawing and the expressive features of printmaking, learned and developed at the Slade School of Fine Art, with its strong cultural background, having his sources scholarly character references, especially to the level of literature and cinema.

The artist said he was two. One who looked inside and another that looked outside. The one that looked inside addressed metaphysical nature themes. The one that looked outside approached reality with works of interventionist nature at political and social level.

His work of great size and diversity, had has recurring themes, Space, Time and Memory, having Bartolomeu recurred on an original way to experimentation, and also on the way he mainly used acids, transversely in different formats and scales, from printmaking to the public art, for obtaining the visual record.

Bartolomeu in the initial phase of its work until 1956 works in drawing painting and ceramics, from 1956 until 1996 he works on printmaking and public art, since 1996, he combined printmaking and public art, with painting, assemblages, sculpture, and installation. Having developed his artistic activity in London between 1956 and 1958 and 1961 and 1996, the Portuguese art panorama was not a reference to Bartolomeu, despite the artist exhibited regularly in Portugal in that period and exclusivity with the 111 Gallery from 1969.

Specifically regarding to printmaking, Bartolomeu at the end of 50 and 60 was the reference to

Portuguese artists. However, because the Cooperative GRAVURA was bringing foreign printmakers to give courses, and the Portuguese artists started to integrate the circuits of international print biennials, his reference is no longer exclusive. Bartolomeu contributed however at the Slade, to the training in this area of some Portuguese artists in particular Fernando Calhau, Victor Fortes, and Guilherme Parente.

Concerning its public art works in etched stone, Bartolomeu pioneered using acids, away from the traditional engraving in stone, or graffito, examples of Almada, with its work in graffito at the University of Lisbon.

2 SOURCES AND REFERENCES ON THE UTOPIAN WORK OF BARTOLOMEU

In this paper, we will discuss the sources and references of the works of utopian thematic nature in the work of Bartolomeu Cid dos Santos.

Bartolomeu sources and references vary depending on the thematic and on the various phases of his work. He takes his influences as something natural and inherent to the creative process, in terms of the history of art in general and in their particular case, stating:

We do not start alone. All creators seek inspiration in others. I for example, initially drank in Giorgio de Chirico metaphysical painting, later on the music of Shubert or on Pessoa's poetry (especially Álvaro de Campos). In addition, these artists drank on others who have been drinking on others. This almost endless chain dates back perhaps to the Altamira cave.

Art is always constructed from the existing art. Even when it is to break up with her. (...) Of course, we have to know very well what existed before. Who is considered self-sufficient is just embarking on the most stupid of arrogance. We do, in fact, part of a series of links that are lost in time ... we never start from scratch. (Silva 2001: 14–19)

As noted above, Bartolomeu's Work takes on a dual character, according to the artist:

I am two people, I am a person, but I am ... I am two and I am the one that is also me. One looks inside, are the dreams. And the other looks outside in a critical way are the realities. (Melo s. d.)

One looks out to the world around us and reflects or is critical of it. The other, looks inside, and relates to memories and the passage of time. If you want, you can call it a metaphysical attitude towards life. (Santos 2007)

These two facets, as different from each other, having lyricism as opposed to a gloomy side, lethal,

ironic, political, scathing, are revealing of what structure this author. About the lyrical component in which it is inserted the part of the work in which the artist looks inside, Bartolomeu says:

As for the other aspect, the looking inside, apart obviously of course being part of my persona has many sources of information many derived from literature, music, cinema, as well as works of other artists. (Santos 2007)

In the thematic line, consisting of the works of metaphysical nature, a reflection of the inner world of the artist, we have a whole imaginary, whose strong artist's culture is an inexhaustible source of funds for the creation of his works in these thematic in which are included the works of utopian nature of which we name some of the prints that are part of them: *Landscapes*: Prints, *The Explorers,* 1974, etching/aquatint. *Labyrinths*: *The Visitor,* 1971, etching/aquatint. *Atlantis,* 1972, etching/aquatint. *The Secret Place,* 1974, etching/ aquatint.

The works of Bartolomeu incorporating the theme of utopia are mainly related to its series of labyrinths, with references to the lost continent of Atlantis, and some works with closed architectural interior spaces of monumental scale. In the first works of the series of the labyrinths that Bartolomeu developed between 1970 and 1979, there is only the labyrinth door, case of *Untitled* 1970, aquatint, and on the last work, *False Landscape* 1979 etching / aquatint only appears between mountains a piece of wall with the typical bricks used by Bartolomeu in their labyrinths. This series of works, according to Bartolomeu came following his reading of the work of Jorge Luis Borges in 1969, advised by Paula Rego. (Melo s. d.)

As mentioned, the series of labyrinths was started by representations of a door (Figure 1) as noted in the first prints and Bartolomeu himself says:

This is I looking for a proof that was one of the first labyrinths I did. Had not yet the labyrinth, only had the door. (Melo s. d.)

Little by little, Bartolomeu inserts the labyrinth and the sphere in his compositions, in which the architectural elements with sphere are influenced by the Newton's Cenotaph of the architect Boullée and its utopian architecture and by the series of prisons of Piranesi (Fig. 2). According to Bartolomeu the prints of this series:

They are metaphysical (...) they raise questions that can be philosophical, have to do with power, with force, with fear. (Melo s. d.)

Figure 1. Ascending Sphere, 1972 água-forte/água-tinta.

Figure 2. Atlantis, 1972 etching/aquatint.

The sources and references of Bartolomeu's plastic work are mostly of scholar character and come from various areas of knowledge and of the most varied artistic expressions, yet being their main influences, literary, musical, and cinematographic.

According to Bartolomeu all that is creator has references, regardless of the creative area (Santos 2002). In his case, the artist says that when works he usually is not really thinking about the references that he is using at that time in his work, and he only exercised retrospectively and more methodically in 1996, a reasoning in order to discover and understand these same sources and references of his work. The reason Bartolomeu did not carry out this exercise before, we find it in his own words:

> You will note that from 69 spheres, circular shapes, or even circular prints are part of my form of expression. I have been asked many times why I like to use such elements, and if you ask me frankly, I do not know or seek to know it; it seems to me that on the day that an artist knows to the smallest detail the reasons and motivations that lead him to create a particular work, it risks to lose the sense of adventure that the unknown generates, and that artistic creation requires. (Sarre 1988)

We enumerate bellow the sources and references of the utopian character of works of Bartolomeu Cid dos Santos:

2.1 *References in painting*

One of the references in painting, was a small painting in the possession of Bartolomeu's family, from a disciple of Bosch with 22 cm in diameter (Fig. 3) and according to his sister Alexandra (Sampayo 2010) belonged to his grandfather Reynaldo dos Santos and was inherited first by João Afonso Cid dos Santos and then by Bartolomeu. This painting

Figure 3. Painting of a Bosch disciple from the Bartolomeu's collection.

Figure 4.   *The End of Atlantis*, 1973, etching/aquatint.

is the genesis of the form of all its round prints, as Bartolomeu told us in a telephone conversation.

In addition to the circular shape itself according to Bartolomeu (Santos 2002), the frame composition with a set of figures on the left, and another one on the right in the foreground, with an open space in the middle, also influenced the composition of some of his prints, namely *Atlantis revisited*, and *The End of Atlantis* (Fig. 4).

Max Ernst with his works *Forêt-arêtes* oil on canvas 1926 and *La ville entière, oil on canvas 1935/37* (Bishoff 1992: 41, 58,59) is reference to some Bartolomeu's landscapes, namely *The explorers,* and his collages *Une semaine de Bonté* (Cam 1993) for some works, like *the abandoned cannon.* Piranesi with his *Carceri d'Invenzione* series (Ficacci 2006) is a reference for some of the prints of the series of labyrinths, with closed interiors of monumental scale.

2.2   *References on literature and cinema*

The series of labyrinths comes from Borges and of the reading of his complete works bought in Madison in 1969. They are labyrinths where according to the artist there is no exit or entry.

> Well, I discovered Borges in 1969. It was the end of a cycle on my work and I went to America to teach at the University of Wisconsin. It was in the year of release of the "2001 Space Odyssey". Two or three days before leaving, I saw the film in London and was very impressed. Speaking with Paula Rego on Kubrick, she told me, "Look, you know Borges? So you read it, you will like it. "I came to America and saw a bookstore showroom in Madison, full of

Borges. I bought right there the "Aleph" and other books. When I returned, I already was someone else. (Sarre 1988)

> On my return to London, the combination of Borges with Kubrick, both playing with ideas and concepts, with mysteries spaces and questions, completely changed the character and the direction of my work." (Sarre 1988)

2.3   *References in architecture*

In the print *Atlantis*, the labyrinth ends up on a centre but has a wall behind and a watchtower. Bartolomeu says that without realizing used the architecture of Boullée and its Cenotaph of Newton (Fig. 5) and opened the sphere like him without realizing. The architecture of Boullée (Le Comte 1969) is a constant reference in the buildings with spheres of the labyrinth series.

Boullée tried to create an architecture for a new world, for a future society, tried to create monumental buildings that constitute the expression of a revolutionary ideology and constituted a starting point for a new world, revealing thus his disposition to a utopia and renewal of society.

Through the work of Boullée, the formal vocabulary of Bartolomeu acquires geometric solids, geometrical pure shapes, spheres, pyramids, and his imagination combines them with other elements, turning them into intriguing and poetic compositions by the nature of its formal purity in transfiguring contexts (Fig. 6).

The University College Library staircase giving visual access to 3 floors (Fig. 7) influenced the print *The Secret Place*, 1974 (Fig. 8) with the sphere on top of a foreground wall with two lower floors behind (Santos 2002). The door that we see in this photography with the arc top may also have constitute reference to the doors of the series of labyrinths.

Figure 5.   Cenotaph of Newton of Etiene Boullée.

Figure 6.   *Atlantis Revisited* 1973, etching/aquatint.

Figure 8.   The secret Place, 1973, água-forte/água-tinta.

doors that he rented according to his sister Alexandra by 1970, they are also an influence, once that they are morphologically close to the doors used in its labyrinths.

2.4   Other influences

The influence that triggered the appearance of the sphere in his prints of labyrinths, the utopian architecture of Boullée, is a reference to which we arrive by morphological comparison. The same is no longer true about the other influence that Bartolomeu said in a letter to us (Santos 2007), the sparkling sphere that appeared in the television series *The Fugitive*. An unlikely influence, if we think in terms of references as those originating primarily from the universe of visual and related arts, and therefore only possible to identify by the author's own testimony.

Another difficult reference to spot, which we only reached also through the testimony the artist (Santos 2002) is the space of the UCL library access staircase in which we see three floors, which Bartolomeu incorporated in some prints of the series of the labyrinths, namely, *The Secret Place*,

Figure 7.   University College Library staircase photo taken by Bartolomeu.

In addressing the photographs taken at the home of Bartolomeu's sister, Alexandra Sampayo Mello, we realized that the example of what Bartolomeu states about the influence of this staircase, It may also be applied to the the shape of his Sintra house

1973 etching/aquatint, in which the viewer is placed in an observation point that allows him to see one or more plans below is observation point.

After the knowledge of this fact, we were more attentive to this kind of situations, and wen photographing Bartolomeu's family albums (Sampayo 2010) an interior photograph of Bartolomeu's Sintra house caught our attention. This photo shows a door, which apart from one or another detail, assumes quite relevant similarities with the doors used by Bartolomeu in their series of labyrinths.

The purchase of the house by Bartolomeu around 1969 (Sampayo 2010) coincides with the beginning of the series of the labyrinths.

These are all the sources and references of the utopian work of Bartolomeu Cid dos Santos that was possible to us to identify during the investigation for the preparation of this paper. If others exist, they are not obvious and therefore, only the artist himself would be able to say it.

BIBLIOGRAPHY

Bishoff, Ulrich. *Max Ernst 1891–1976 Oltre la pittura*. Koln: Benedikt Taschen, 1992.

Cam. *Max Ernst: obra gráfica e livros*. Colecção Lufthansa. Lisboa: Fundação Calouste Gulbenkian 1993.

Ficacci, Luigi. *Giovani Battista Piranesi*. Roma: Taschen, 2006.

Gulbenkian, Fundação Calouste. *Boletim de Bolseiro da FCG Lisboa*. Lisboa: Fundação Calouste Gulbenkian, 1959.

Le Comte, Daniel Boullée Ledoux Lequeu, Les architectes Révolucionaires. Paris Éditions du Sénéve, 1969.

Melo, Jorge Silva. "Bartolomeu Cid dos Santos." *Por Terras Devastadas*: RTP 2, s. d.

Sampayo, Alexandra Mello. "Entrevista pelo autor. Porto, 30 de Junho." 2010.

Santos, Bartolomeu Cid dos. "Assunto: Information for PhD." Ed. a.canau@sapo.pt, António Canau 2007.

———. Conferência de Bartolomeu no ISPA sobre fontes e referências na sua obra. [Recurso electrónico]. Lisboa: ISPA, 2002.

Sarre, Inês. "Entrevista ao Professor Bartolomeu Cid dos Santos." *Peregrinação*. 20 (1988): 84–86.

Silva, José Mário. "Bartolomeu dos Santos." *Diário de Notícias: Suplemento* 10 de Mar. de 2001.

Contemporary utopias in art and architecture: Searching for spiritual significance in the work of John Pawson and Anselm Kiefer

Sarah Frances Dias

CIAUD—Centro de Investigação em Arquitetura, Urbanismo e Design, Faculdade de Arquitetura, Universidade de Lisboa, Lisboa, Portugal
Associação Portuguesa da Cor, Lisboa, Portugal

ABSTRACT: Utopias create an imagined world of possibilities and allow for the visualization of impossibilities; like a dream, they allow for the triggering of the imagination, opening spaces for new possibilities to emerge. At the heart of modernism, was a utopian desire to create a better world, and the belief that design, art, and architecture could transform society and move towards better ideals. In contemporary art and architecture, this utopian desire still defines creation, however in a new and distinct way. Two case studies are presented exemplifying two types of contemporary utopia in art and in architecture. In architecture, with John Pawson, the research clarifies a type of utopian spiritual depuration through minimalism, creating a space that allows for a contemplative way of life. In art, with Anselm Kiefer, the work explores a metaphysical and cosmological utopian type of existence, allowing for a deeper connection with the primitive and natural order of the universe. The research shows that not only is utopia part of contemporary art and architecture, but that those utopian visions do in fact, to some extent, come alive in reality. In doing so, we become closer to More's utopian dream, where men are led by the soul and by spirituality.

Keywords: Utopia, Architecture, Art, Significance, Meaning

1 UTOPIA: AN IMAGINED PERFECTION

Utopia can be understood as an 'imagined perfect place'. The word was coined by Thomas More (1478–1535) in *Utopia* (1516) and was derived from two Greek words: '*u*' (no) and '*topos*' (place), which literally translates to 'no place'; however More suggests that the word *eutopia*, which means 'good place' "(...) is a better descriptor." (Sargent, 2005); which means that utopia can either be a place, which has no real existence or can be an imaginary 'good' place. This duality clarifies two contradictory conditions that are innate to the concept: that it is both an impossibility (no place) and simultaneously an optimal 'real' condition (good place). The word became part of common language even prior to the various translations of the book. However, multiple 'utopian societies' had been imagined prior to his literary work; all civilizations have had a certain form of utopia fuel their existence: for example, the utopian society of Magnesia, imagined by Plato in Ancient Greece described in *Laws*, one of Plato's last dialogues, whose aim was "(...) to produce citizens who possess complete virtue." (Bobonich and Meadows, 2013). Contemporary uses of the word and concept always encompass a form of imagining a 'perfect place' or a 'perfect society', which defines a human aspiration towards a form of perfection that is always unreachable and in some way impossible. Utopias create an imagined world of possibilities and permit a visualiza-

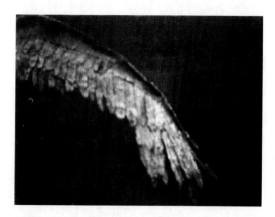

Figure 1. Artistic utopia: an allusion to the condition of freedom and flight. Detail of Anselm Kiefer's Buch mit Flügeln; Source: https://www.flickr.com/photos/futurowoman/1435307819/.

tion of impossibilities. Like a dream, it allows for the triggering of the imagination, opening spaces for new possibilities to emerge. This type of imagining a new world, a new vision, a new way of life or even a new type of dream, has always been part of what art and architecture has aimed to do. More so, it defines art and architecture's essence.

1.1 *The legacy of More's utopia: Spirituality and imaginative possibilities*

One of the main points brought forth by Thomas More in his book Utopia (1516), of relevance for the present research paper, is the merging of the real and the imagined world that creates a new imagined possibility, as clarified by author Dominic Baker-Smith (2014), when he explains that Utopia emphasizes "(...) the relationship between imagination and experience." (Baker-Smith, 2014), were the real and the imagined merge becoming one single dimension. Another focus point of the work is the spiritual way of life led by the Utopian's, the people that composed the utopian society. They possessed a unique mind-set and way of life that was based in the belief that true happiness could be found in two things: firstly the cultivation of the mind, and secondly, (the most important ingredient for true happiness), the cultivation of spiritual pleasure. As such: "(...) the key to their thinking lies in the soul and its destiny (...)." (Baker-Smith, 2014). So what happens when, like the Utopians, we place the soul in the centre of our lives and our creations?

2 MODERNISM'S DREAM FOR A BETTER WORLD

At the heart of modernism, was a utopian desire to create a better world, and the belief that design, art, and architecture could indeed transform society and move towards better ideals. Richard Rinehart (Director of the Samek Art Museum at Bucknell University) in his research paper 'Utopia Now and Then' (2014) explains that Modernism, is characterized by a 'utopian impulse', and that there were multiple modernist utopias. An example of this was the Arts and Crafts Movement, that with William Morris aimed to re-join people to their labour and to their land, "(...) through use of natural forms and materials and a return to village/communal living." (Rinehart, 2014, p. 4). In Italy, the Futurist Movement, with the publication of the *Manifeste du Futurisme* (1909) by Italian poet and painter Filippo Marinetti, proposed a society run by the aesthetics of the machine and industry. In American Art, it is possible also to trace this utopian idea in Abstract Expressionism with; for

example, Jackson Pollock and Minimal Art with Donald Judd, that aimed to create a new type of art that (...) points to the perfection of the void (Rinehart, 2014, p. 7), in a spiritual way. In both art and architecture, modernism aimed to create a new form of expression that would reflect and cater to the human intellect, spirit and soul, at all levels of existence, and believed that architecture and art, in a new way, could shape the minds and souls of all humanity.

3 CONTEMPORARY UTOPIAS IN ART AND ARCHITECTURE

Author Rinehart (2014) continues explaining that utopia implies a break with the past, and that this break also defines a new type of contemporary utopia that lives in the present time. This contemporary utopia in both art and architecture, he claims, is an attempt "(...) to reconcile the universal with difference". (Rinehart, 2014, p. 22). However, the main aspect of contemporary utopia proposed by the author is a new type of utopia, which lives in the lives of the observers instead of existing in a distance place. In his words, he claims:

> So utopia may not be some far off place, but an impulse that exists here and now whose realization is created contingently, relationally, and immediately in the space between utopia and the subject. (Rinehart, 2014, p. 26)

Stephen Duncombe (2010), professor at Gallatin School, New York University, in his conference talk in the Walker Art Centre, entitled 'Utopia Is No Place: The Art and Politics of Impossible Futures' also explains that a growing number of artists are abandoning truth-telling political art for a boldly imaginative utopian practice. He explains that this type of art, allows for a more positive change in society: because by dreaming the impossible we create a sort of imaginative positive change in society, allowing for a bridge between what is real and what could be, by opening up a space for participation and engagement from the audience. He claims: "Everything has to be an engagement with the idea of truth." (Walker Art Centre, 2010). This idea of truth, in art, is presented in a more fluid and dynamic way, allowing for dreams and imagination to define the reality. Because truth is ever changing and evolving (we are still producing truth), he explains, that we, as humans, need a dream to move us forward and reach towards.

It is this dream, this 'utopia', this possibility of a better life associated to an idea of truth, that contemporary art and architecture is presenting to the world. In this sense, as the author Duncombe

(2010) explained, art has stopped critiquing what is, and has begun looking at what might be and at what could be, trying to materialize those impossibilities. Although Utopia is the sense of something that is always unreachable, these art and architectural forms, give us a compass to move towards, a possibility in which our minds and souls can inhabit and live in, in present... even if always a little unreachable.

4 TWO CASE STUDIES JOHN PAWSON'S SPIRITUAL DEPURATION AND ANSELM KIEFER'S METAPHYSICAL CONNECTIONS

The two examples that follow exemplify two types of contemporary utopia in art and in architecture. In architecture, with the work of John Pawson, the research clarifies a type of Utopian Spiritual Depuration through Minimalism, which embodies simplicity and elegance in order to create a space that allows for a meditative and contemplative way of life. In art, with the work of Anselm Kiefer, the work explores another type of Utopia; one that is focused on creating a metaphysical and cosmological type of existence, allowing the viewer to understand a deeper connection with the primitive and natural universal order of the universe.

4.1 *John Pawson's utopia: Spaces for spiritual silence*

John Pawson was born in 1949 in Halifax, Yorkshire, United Kingdom, where he spent his childhood years. Educated with a Methodist Religion background he was sent to Eton College Boarding School, at the age of thirteen. In his early twenties, he worked for a couple of years at his family's textile industry, and then left for Japan to become a monk, but ended up teaching English at the Business University of Nagoya for some years. He moved to Tokyo during the final year of his stay, where he visited the studio of Japanese architect and designer Shiro Kuramata. Following his return to England, he enrolled at the Architecture Association in London where he studied for three years before dropping out. At the age of thirty, he decided to establish his own architecture practice, in 1981. Pawson has established himself ever since as a leading exponent of the minimalist aesthetics. He has received multiple architecture awards such as the Blueprint Architect of the Year (2005), the RIBA National Award (2008), and the RIBA Arts & Leisure Regional Award (2008). Although private houses define the core of his work, his projects have spanned a wide range of scales and building typologies, ranging from designing objects, bridges

(Sackler Crossing across the lake at the Royal Botanic Gardens), shops (his most prominent being the Calvin Klein store), monasteries (Cistercian Monastery in Bohemia) and even designing objects in multiple sites (Yorkshire, Okinawa, the Alentejo, California, the Veneto, Manhattan or the Belgian North Sea).

4.1.1 *Characterization*

Pawson is known for his minimalistic approach on architecture and he is generally considered the master of minimalism. Minimalism as an art movement began in the 60 s and 70 s; it is derived from the depuration and reductive aspects of Modernism, and can be seen as the opposing art movement of abstract expressionism, in content, form, and significance. The term is generally and loosely used in architecture and design to describe the aesthetic sense of reduction to the bare minimums or essentials. Its aim is to create an atmosphere of absolute simplicity and subtle elegance or beauty.

In most of his architectural work there is a predominance of white and light, defining an atmosphere that is clean and pure. He eliminates superfluous features and reduces design and architecture to its bare minimum, in order to create the essence of the space. The elements that are added to the space, such as furniture or flooring, use natural materials, mainly wood, that give the space a sense of a natural environment and connection to nature. Natural materials allow for the echo of time to manifest themselves. Another aspect that is important to refer in Pawson's spaces is the sense of proportion and 'human' scale, accentuated by the sharp sense of geometry, and the use of pure forms that echo throughout all the elements of each space. John Pawson in an interview by Sherryl Garratt (2010) for the Telegraph explains a spatial installation at an exhibition at the Design Museum:

> It will be a room in which nothing happens at all. But you'll hopefully get a feel of the things we do. Some sense of precision, light and proportion; of scale, and sensual material." (Garratt and Pawson, 2010, p. 1)

These four elements (light, proportion, scale, and sensual material) define Pawson's architectural considerations; they are the concepts upon which his spaces are developed. Overall, his spaces combine those elements to create spaces and atmospheres that act on the senses and on the sensitivity of man. Moore (2010) emphasizes this when he claims: "A Pawson space offers a higher form of hedonism which, like snow in a sauna, heightens the senses." (Moore, 2010, p. 1). He also claims that the spaces allow man to become immersed in

Figure 2.   A John Pawson space detail. Source: https://www.flickr.com/photos/ndecam/6390303221/.

another sort of dimension, where simplicity of life and sensuality of material live combined in one single existence.

4.1.2   *Utopian minimalism: Visual clarity, simplicity, elegance, spirituality, and grace*

The idea of simplicity, John Pawson explains in his book Minimum (1999), is an ideal shared by many cultures and religions, and has its roots deeper than just aesthetics ones: from the Japanese Zen, to Thoreau's quest for simplicity, from the Quakers to the Buddhists, the simple way of life has always been an aspiration and has always granted a sense of liberation. Simplicity, then, possesses a moral dimension and as such implies the notions of 'selflessness' and 'unworldliness'. This type of minimalism is a form of 'utopia' creating an intangible space. Attached to simplicity and to this simple way of life, is the idea that things have to be reduced to their minimum conditions. Minimum, for Pawson can be defined as "(…) the perfection that an artefact achieves when it is no longer possible to improve it by subtraction". (Pawson, 1998, p. 7); he continues explaining that it is the quality that something has or possesses, when every component, every detail, and every function has been 'reduced to its essentials'. He summarizes as such; "It is the result of the omission of the inessentials." (Pawson, 1998, p. 7). And is the underlying idea that defines simplicity in its core. Overall, the minimum as a way of life that Pawson identifies and describes in all spheres of existence is as much an aesthetic quality as it is a philosophical and a religious one: it is close to a spiritual type of depuration, a cleansing of the superfluous and a focus on what really indeed matters and is meaningful.

Either as a physical experience, a merely visual one or a philosophical one, Pawson's spaces echo simplicity, restraint, minimalism and harmony, portraying a world that seems to be at a perfect balance within itself and with the world around it; they act psychologically, emotionally and spiritually, on the viewer transmitting those specific emotions and meanings that only the 'minimum' and overall 'simplicity' can portray. They form a type of 'utopia', where life is clean, where meditation, calm, and tranquillity are part of the everyday way of life and where harmony exudes from every surface and from every material. Devoid of clutter, things, and noise, these spaces of silence elevate human life, allowing man to become in tune not only with his surroundings but also with his self (his mind, his body and his spirit).

4.2   *Anselm Kiefer's utopia: Metaphysical and poetical places*

Anselm Kiefer was born on March 8, 1945, in Donaueschingen, Baden-Württemburg, Germany, a few months before the end of the war. Raised near the region of the Black Forest as a catholic, in a small village in Germany, he began studying law; however, he abandoned his studies in 1966 to study in an Art Academy in Dusseldorf, where he studied under Peter Dreher (a realist and figurative artist). Kiefer went on to study with Joseph Beuys during the late 60 s and 70 s, an artist that deeply influenced his views on art. Besides paintings and sculptures, Kiefer also creates watercolours, woodcuts, and works with photographs. He has had numerous solo exhibitions in various prestigious art galleries and museums: at MoMA, New York (1987); Neue Nationalgalerie, Berlin (1991); The Metropolitan Museum, New York (1998); Royal Academy, London (2001); to name a few. In 2007 Kiefer became the first artist to be commissioned to install a permanent work at the Louvre. In 2014, Kiefer was given the first retrospective exhibition in the Royal Academy of London. Many art historians and critiques consider Anselm Kiefer as "(…) our greatest living artist" (Pritchard, 2014).

4.2.1   *Characterization*

Kiefer's art is generally characterized by an individual and unique artistic language, physical materiality, monumental scale, visual complexity and the weight of history and time; through his art he balances powerful imagery, 'critical analysis' and an understanding of the poetical and mythical being in the world. Prodger (2014) explains, that to describe Kiefer "(…) perhaps the only word that makes some sort of sense is "shaman"." (Prodger, 2014) and emphasizes the importance, that "(…) for the artist, there is something meaningful going on, however knotted or nebulous that meaning might be." (Prodger, 2014). He is varied in his fields of artistic expression ranging between sculpture, architectural objects, spaces, paintings, altered photographs, and some of his creations blend the lines between the fields of sculpture, architecture, installation, and painting. Using either a monumental scale or a smaller one, his work is mostly characterized by layers of multiple meanings (histories, mediums, techniques, influences, and themes). In order to do so Kiefer explores multiple concepts, themes, influences and subjects: the idea of civilization, the concept of time, the opposition between creation and destruction, the power of myth and legends, the weight of history, literature, philosophy, poetry (Goethe's poetry), architecture, music, archaeology, astronomy, alchemy and other references. Ian Alteveer (2000), in an article for the The Metropolitan Museum of Art, about Kiefer's work summarizes:

> Their encrusted surfaces and thick layers of impasto are physical evocations of the sediments of time and meaning they convey." (Alteveer, 2000)

4.2.2 Utopian connections: Metaphysical, cosmological, existential, poetical, and spiritual

Kiefer's art can be generally categorized as 'romantic anti-capitalism', aspiring to promote a romantic sense of existence and an integrated kind of life with the natural world and the universe. The term was first coined by Georg Lukács in his studies on Dostoyevsky, explains Rampley (2000), and can be seen in various fields of creation, portraying a sense of a utopian vision. This type of 'romantic anti-capitalism', the author continues:

> (…) can be traced back to early Romantics such as Friedrich Schlegel and Hölderlin, who sought to overcome the subject-object dichotomy in the name of a pre-conceptual intuitive immediacy. (Rampley, 2000, p. 7)

Rampley (2000) further clarifies, that what these movements and schools of thought have in common is the focus on a 'pre-reflective experience' of the world, prior to modernization (technological and economical), characterized by 'purity' and 'openness'. Kiefer's work, besides this characterization as a 'neo-romantic' or 'anti-capitalist', is also linked to what is described as New Symbolism, a movement that sustains the same core principle: the return to a primal world, where technology and economy has not drowned out the mysticism and purity of the world.

Kiefer, in an interview for Tim Marlow (2014) speaks of his art as a journey of self-understanding claiming that with each work, the 'self' becomes clearer: through these new connections and unions that derive from this inherent questioning and longing; as such, it is through the establishment

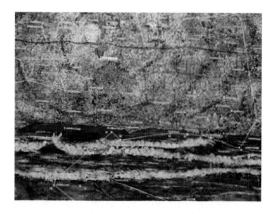

Figure 3. Anselm Kieder at the Grand Palais Exhibition, paintinf Andromeda, la voie lactée, 2001, detail. Source: https://www.flickr.com/photos/dalbera/549340096/.

of this 'another world' and establishing new connections that he expands his own existence and his own consciousness and can convey that to others. In this regard, Kiefer further explains that every work of art he creates is a kind of self-projection or self-aspiration of a continuation of the self; in Kiefer's words:

> A 'me' projected into a geological time [...] kind of like a 'self aspiration', to be continued into the geological time, and then I have more time... you know?" (Imagine, 2014)

Alan Yentob later clarifies:

> Kiefer's fascination with time seems to be getting deeper and wider. Geological time is the oldest time on earth, stretching back millions of years. A time scale in which the human race, and everything that is achieved is little more than a blip..." (Imagine, 2014)

This fact and fascination with time establishes another utopian connection in Kiefer's work: the utopian desire of a 'living forever' in the fabric of time itself. Through mythology, through art as a sort of religion, thought the self and the projection of the self, Kiefer finds in his artworks a way of creating a unification of the world, and an imprint in the world that seems to propel throughout the ages. In creating 'another world': a 'utopian world' that is in tune with nature, with the senses, with matter and with the universe, using poetry and metaphysical aspects at its core, Kiefer creates the possibility of the understanding of the self in a universal field of existence; a self that is aware, conscious, spiritual and metaphysical, and belongs to the fabric of space and time itself. This 'another

Figure 4. Von den Verlorenen gerührt, by Anselm Kieffer, 2004.; Source: https://commons.wikimedia.org/wiki/File:Anselm_Kieffer__Von_den_Verlorenen_gerührt,_die_der_Glaube_nicht_trug...JPG.

Figure 5. Harmony, nature and the place of man in the universe: Contemporary utopias; Source: https://www.flickr.com/photos/31246066@N04/4167653247/.

world' creates another understanding of existence for the viewers, where life is whole, complete and true, and where the universe's song is heard, understood and lived.

5 ENDING REMARKS

The research has shown, that these utopian visions when materialized into artistic and architectural creations (as described above with the two contemporary examples: in architecture with the work of John Pawson and in art with the work of Anselm Kiefer) can indeed become a reality. Both exam-

ples use different means to execute their utopian visions, however what both have in common is the profound desire to communicate that 'utopian' dream and make it a reality for themselves and for the viewers, allowing for a more in tune balanced and spiritual existence to emerge. In this case, not only does utopia command creation, but also it becomes alive in the hearts and in the souls of man. In doing so, it allows for the people that live and experience these works, to be a little bit more similar to More's Utopians, and to become more in tune with their souls and their destinies.

BIBLIOGRAPHY

Alteveer, Ian. *Anselm Kiefer (born 1945)*. In Heilbrunn Timeline of Art History. New York: The Metropolitan Museum of Art, 2000. [WWW Document]. URL http://www.metmuseum.org/toah/hd/kief/hd_kief.htm (accessed 4.26.15).

Baker-Smith, Dominic. *Thomas More*, in: Zalta, E.N. (Ed.). The Stanford Encyclopedia of Philosophy, 2014. URL. http://plato.stanford.edu/archives/spr2014/entrieshomas-more/.

Bobonich, Chris, Meadows, Katherine. *Plato on utopia*, in: Zalta, E.N. (Ed.), The Stanford Encyclopedia of Philosophy, 2013. URL. http://plato.stanford.edu/archives/sum2013/entries/plato-utopia/.

Garrat, S., Pawson, J. *John Pawson interview: Plain speaking*, The Telegraph, United Kingdom, 2014.

Imagine Anselm Kiefer Remembering the Future (2014), 2014. URL. https://www.youtube.com/watch?v = CK 9 N2 NbSehE&feature = youtube_gdata_player.

Moore, Rowan. *John Pawson: Plain Space*. The Guardian, United Kingdom, 2010.

Pawson, John. *Minimum*, mini format edition. ed. Phaidon Press, London, 1998.

Pritchard, Claudia. *Anselm Kiefer*, Royal Academy, preview: Is he our greatest living artist? The Independent, 2014.

Prodger, Michael. *Inside Anselm Kiefer's astonishing 200-acre art studio*. the Guardian, 2014. URL. http://www.theguardian.com/artanddesign/2014/sep/12/anselm-kiefer-royal-academy-retrospective-german-painter-sculptor.

Rampley, Mathew. *Anselm Kiefer and the Ambivalence of Modernism*. Oxford Art Journal 23, 2000, 79–96.

Rinehard, Richard. *Utopia Now and Then*, 2014. URL. http://www.coyoteyip.com/rinehart/papers_files/utopia_now_and_then-rinehart.pdf.

Sargent, Lyman. *Utopia*. New Dictionary of the History of Ideas, 2005.

Walker Art Center, Utopia Is No Place: The Art and Politics of Impossible Futures. United States of America, 2010.

The ambiguity of micro-utopias

Gizela Horváth

Department of Fine Arts, Partium Christian University, Oradea, Romania

ABSTRACT: "Micro-utopia" is one of the central concepts of relational aesthetics. Nicolas Bourriaud promotes such relational art projects that construct "hands-on Utopias", "micro-utopias".

The present paper considers the concept of "micro-utopia" ambiguous and analyses this ambiguity by reconstructing the conceptual framework of relational aesthetics, based on three notions: the concept of "society of the spectacle" (Debord), the concept of artistic critique (Boltanski and Chiapello), and the concept of "pragmatic utopia" (Vieira).

I argue that the concept of micro-utopia as tool of interpretation does not function without any problem. It can be proved on some of the works of Rirkrit Tiravanija, the famous artist of relational art that the ambiguity of micro-utopia leads to conceptual problems within the framework of relational aesthetics. There is concern that Tiravanija's cooking-performances are utopias only for few, because these events strengthen the exclusivist harmony among the natives of the art world. Then I turn to such artworks that thematise the labour process. In these cases works imagined as alternatives against the current capitalist system illustrate or reproduce the very functioning of capitalism.

The difficulties of the concept of micro-utopia suggest that nowadays we cannot think coherently about the utopia, not even in the world of art that is the land of freedom.

Keywords: micro-utopia, relational art, contemporary art, Nicolas Bourriaud, Rirkrit Tiravanija

1 CONCEPTUAL FRAMEWORK OF MICRO-UTOPIA

Micro-utopia is one of the key concepts of relational aesthetics, the theoretical framework used by Nicolas Bourriaud in order to interpret a significant group of art phenomena that proliferated in the nineties. These artists did not create objects of art, but "a specific sociability", "the set of 'states of encounter'" (Bourriaud 2002, 6), a framework for the relations between people.

According to the author, the relational art is needed because we live in the society of spectacle where human relations are not authentic anymore; they are not direct, but transmitted by representations that dominate every aspect of social life. Bourriaud's favourite artists create relations between persons—they do not merely represent or thematise them, but create them, so these works "outline so many hands-on Utopias" (Bourriaud 2002, 3). These are not utopias that redeem the entire society because the time of such utopia is over—these are only micro-utopias. Bourriaud stresses:

> It seems more pressing to invent possible relations with our neighbours in the present than to bet on happier tomorrows. (Bourriaud 2002, 20).

Despite his undeniable success, the "micro-utopia" concept of relational aesthetics seems to be ambiguous: on the one hand, its coherent conceptualization is difficult, on the other hand, it seems that is does not function flawlessly as a tool of interpretation, not even in the case of those artists that gave the impulse to develop Bourriaud's theory.

According to my hypothesis this ambiguity is rooted in the ambiguous position of relational aesthetics concerning three references: one historical reference (The Situationist International), and two theoretical references (the concept of utopia, respectively that of criticism). In what follows I will present the ambiguity of micro-utopia along the analysis of these references, then I will investigate some of Rirkrit Tiravanija's works from the perspective of the concept of micro-utopia.

1.1 *Relational arts and the Situationists International*

One of Bourriaud's explicit theoretical resources is Guy Debord's text entitled *The Society of the Spectacle* that was published in 1967 and influenced the student uprising from '68 in Paris (the students wrote quotations on banners from this text). Debord describes the society of his era as "spectacle", as a state where the place of the relations between people is taken by representations and dialog is replaced by passive contemplation. This alienation is dangerous because it presents the *status quo* as unquestionable: "THE SPECTACLE MANIFESTS itself as an enormous positivity, out of reach and beyond dispute. All it says is: "Everything that appears is good; whatever is good will appear"" (Debord 1995, 15). Debord saw the possibility of resistance in art and for precisely this reason, he played a major role in The Situationist International (SI) avant-garde art movement. The SI refused the concept of the autonomous art and used art as tool for regaining life. SI tried to fight against alienation by creating "situations" that were aimed to recall authentic impressions, desires, and experiences. In this sense, the situations that were created follow the way of a utopia that in the future would re-establish the authentic relation between humans, based on dialogue. Situations that were created spring out from the happenings of everyday life, so we speak of situations that although are "created", do not fall far from life.

The similarity between "created situations" and the "states of encounter" created by relational arts is outstanding. Both are born in opposition to the "society of spectacle", both aim to re-establish the authenticity of everyday relations between people. Still, Bourriaud tries to highlight the distance from SI and relational arts: while situationists saw art situations as a stage of the life liberating revolutionary activity, Bourriaud considers that nobody wants revolution anymore, we just simply want to "inhabit" the world. Bourriaud mentions another formal difference: for relational aesthetics, dialogue and meeting are substantial, while in SI situations the dialog was not an important aspect. It seems that there is a misunderstanding or misinterpretation. According to Jennifer Stob, Bourriaud's text deals unjustly with SI, offering a "mischaracterization" of it:

> His texts service this post-war avant-garde group in the same way one might have a vehicle serviced (replacing older parts under the hood with newer ones) or a building restored (gutting the building's interior while maintaining the historical façade). (Stob 2014, 26).

Bourriaud's relation to the SI is ambiguous because, on the one hand he accepts the premises of the SI as starting point (the spectacle-society hypothesis and its description, the thought that it is possible to react to it with art tools, that these art answers do not need to be objects, but situations), but at the same time he considers relational art as being conformist, and not revolutionary. This is problematic because he thinks that relations created through relational art are an alternative in comparison to the relations alienated by representa-

tations—but art itself also belongs to the field of representation. It is questionable how art can be an alternative, in spite of adapting to circumstances. It is a great danger that situations of relational art do not stop spectacle, but they continue it.

1.2 *Relational art and criticism*

A few years after the events from '68, SI split up and the critical mood started to disappear from the art scene.

It is not by chance that the student uprisings from '68 were so closely interlaced with art movements, art activities, and ideas. In a highly instructive and serious study that was dedicated to the new spirit of capitalism, Luc Boltanski and Ève Chiapello (2005) enumerate those "sources of indignation" that fed the different forms of criticism against capitalism:

– Capitalism as the spring of inauthenticity of objects, persons and feelings
– Capitalism as the spring of oppression—that manifests itself in paid work (wage-labour), and the opposite of which are freedom and autonomy
– Capitalism as the spring of poverty and inequality
– Capitalism as the spring of opportunism and egoism that eliminates social bonds.

Furthermore, the authors differentiate between the two forms of criticism: an artistic critique and a social critique.

The resources of artistic critique are the first two "sources of indignation" and it highlights the loss of meaning that may be deducted from standardizing production mode and commercial system. The artistic critique considers wage-labour a pointless slavery that only strengthens the rule of profit. In opposition to wage-labour is set the artistic life, the freedom, and autonomy of the artist.

The social critique aims firstly the "immorality" of capitalism, inequality, and poverty.

The authors note that the events from '68 meant a kind of climax in critique, and in the following years critique faded, even more, it almost ended.

If we analyse the movement from '68, then it becomes clear that it embodied artistic critique: we witness not the revolution of labourers, but (first of all) of the educated young people, of the future „cadres" (managers), who did not get out on the streets because of subsistence, but for freedom and authenticity. This is precisely the reason why the critique later faded: the capitalist system incorporated the demands of '68 – the greater flexibility in the shaping of work programme, in the execution of tasks, the division of responsibility, the avoidance of mechanical work execution.

The relation of relational aesthetics to both forms of critique is ambiguous. On the one hand, it refers to the "society of the spectacle" and it represents the position of art critique: it presents the society of the nineties as „the final stage in the transformation to the "Society of the Spectacle" as described by Guy Debord" (Bourriaud 2002, 3), and it complains that

> We feel meagre and helpless when faced with the electronic media, theme parks, user-friendly places, and the spread of compatible forms of sociability, like the laboratory rat doomed to an inexorable itinerary in its cage, littered with chunks of cheese. (Bourriaud 2002, 2)

Therefore, similarly to the revolutionaries of the '68, it ascertains the alienation, the lack of autonomy and misses the free activity and the authentic relations. On this level, Bourriaud also joins the artistic critique. On the other hand, he suggests a reconciliation with the given situation that totally takes the sharpness of the critique, and gives place to a kind of cheerful fatalism. The present art practice "appears these days to be a rich loam for social experiments" (Bourriaud 2002, 3), but it turns out that these attempts do not intend to change the world, but they want to inhabit it, the relational artist aims "to turn the setting of his life (his links with the physical and conceptual world) into a lasting world" (Bourriaud 2002, 5). Bourriaud's statements raise uncertainty: on the one hand because it paints a dark image of the present and indicates the ability to resist as the stake of art that is based on the idea that the artwork stands up to the mill of the "Society of the Spectacle" and „the invention of individual and collective vanishing lines" (Bourriaud 2002, 13) is the task of art. On the other hand, he states, „any stance that is "directly" critical of society is futile" (Bourriaud 2002, 13), even more, it seems that relational art does not do/may not do anything else than to perpetuate the situation that was already declared unsustainable. We have to agree with Ildikó Szántó, who arrives to the conclusion that what Bourriaud's and Parreno's artistic approaches fail to deliver is exactly what Boltanski and Chiapello call "social critique" (Szántó 2015). At the same time, it seems that Bourriaud does not clearly undertake even art critique. This way it is hard to understand why he calls the situations created by relational artists micro-utopia or hands-on utopias, considering that the necessary condition of utopia is the strong dissatisfaction with the present situation.

1.3 *Utopia and micro-utopias*

Utopia always presents two sides. On the one hand, we can see the positive side: the state of Paradise on the Earth, which is projected into an unrealizable future. On the other hand, this bright notion is always fed from the dissatisfaction with the present, so directly or indirectly utopia is the criticism of the present, as well.

Utopia never refers to the isolated subject, but to the society. According to an idea formulated in the Enlightenment, people do not only suffer the (social) circumstances, but also shape them.

After the historical traumas of the twentieth century (the two world wars, the totalitarian regimes, the Holocaust, socialism), nineteenth-century optimism was replaced by a more sceptical vision.

Nowadays, utopia does not seem to have a place in our world: neither in space, nor in time. In the present globalised world, it is hard to imagine an isolated island that corresponds to the original concept of "utopia" formulated by Thomas Morus (the meaning of which is: no space). Today there is hardly any isolated place in the world; we live in a global village.

We also cannot handle the second meaning of utopia, which does not connect utopia to a non-existent place, but to a time that does not exist now. I am talking about the meaning that Fátima Vieira calls "euchronia" and that "is the product of the new logic of the Enlightenment" (Vieira 2010, 9) and it describes a future earthly paradise to humanity. The optimism of the enlightenment lost its power; no more big visions about the perfect society are born. Maybe we have to agree with Terry Eagleton, according to whom "utopian thought is hardly in fashion in these sceptical, politically downbeat days" (Eagleton 1999, 36).

The death of utopia has two scripts, an optimistic one, and a pessimistic one. According to the optimistic script, utopia is not needed anymore because capitalism fulfilled the desires (of welfare, of freedom) of people, so utopia is here and now with us. According to the pessimistic script, in the global capitalist system the possibility to resist has disappeared, and:

> the best we can do is decorate our cells, rearrange the deck chairs on the sinking liner, prise open the odd fissure in this otherwise seamless monolith into which a stray beam of freedom or enlightenment or gratification may infiltrate. (Eagleton 1999, 37)

Fátima Vieira, who dedicated a serious study to the concept of utopia, although deals with the hypothesis of the death of utopia, in the end she rejects it. She considers that nowadays utopia did not disappear; it only altered its form. Since the anthropological base of the utopia—the wishing nature of man—does not end, so utopia may not end either. The author also highlights the ability of utopia to alter and argues that nowadays it takes new shapes that were unknown before.

Let us spend some time upon this new form that the author calls "pragmatic utopianism". In this form utopia is not a proposal for a completely new social structure, but a kind of gradual reform:

> Utopia is now asserted as a process, and is incorporated in the daily construction of life in society, and is focused now on operating at a micro-level [...] indicating a direction for man to follow, but never a point to be reached." (Vieira 2010, 22).

Pragmatic utopianism is a strange concept. This kind of utopia loses the essential characteristic of the utopia: the unfeasibility. Utopia is a place that exists nowhere, because it cannot be realized—or at least during the time when it is elaborated, it seems impossible to realize. The inexistence of the utopian state is not contingent, but it is necessary. Such a pragmatic utopia, that is actually with us, that is it actually has space and time—is not utopia. It is more a kind of an optimizing policy, that— since we implement it into practice, we apply it here and now—would not be called utopia, but a plan.

At the same time, the pragmatic utopianism described by Fátima Vieira that functions on the "micro-level" is remarkably close to the "micro-utopia" term used by Nicolas Bourriaud. At the end of the nineties, it might seem that this form of utopia has central place in the contemporary art.

Bourriaud's point of view concerning utopia is not unequivocal. On the one hand, from his writings it turns out that the time of utopia has ended: "art, likewise, is no longer seeking to represent Utopias" (Bourriaud 2002, 21) and:

> the role of artworks is no longer to form imaginary and Utopian realties, but to actually be ways of living and models of action within the existing real, whatever the scale chosen by the artist" (Bourriaud 2002, 5).

The author cannot sufficiently emphasize that "the radical and universalist Utopia" is not anymore on the agenda.

On the other hand, as a result of the relational art the author mentions precisely the creation of micro-utopias: "social Utopias and revolutionary hopes have given way to everyday micro-utopias" (Bourriaud 2002, 13), and:

> These days, Utopia is being lived on a subjective, everyday basis, in the real time of concrete and intentionally fragmentary experiments. (Bourriaud 2002, 20)

We find here the same ambiguous attitude as in the case of the criticism: on the one hand, in post-modern style Bourriaud refuses the outdated utopian ideas, because he finds that nowadays nobody wants to change the world, but he holds on to the idea that the relational art would create a kind of a paradise-like state in small, in micro communities, during an art exhibition. Artists that he calls the engineers of inter-subjectivity, create functional models (interstices):

> These interstices work like relational programmes: world economies where there is a reversal in the relationships between work and leisure (Parreno's exhibition *Made on the 1st of May*, Cologne, May 1995), where everyone had a chance to come into contact with everybody else (Douglas Gordon), where people once again learnt what conviviality and sharing mean (Tiravanija's itinerant cafeterias), where professional relationships are treated like a festive celebration (the Hotel occidental video by Henry Bond, 1993), where people are in permanent contact with the image of their work (Huyghe). (Bourriaud 2002, 32).

Against Bourriaud's micro-utopia, the same objection can be formulated as against Vieira's pragmatic utopianism: both of them lack that what makes utopia utopic—the promise of the unrealizable paradise on the earth. The inventor of utopia is a dreamer, not an engineer.

We can conclude that on theoretical level the concept of micro-utopia is unclear, it emphasizes what it rejects (the global redemption of the world), rather than what it affirms. It states "precarious constructions", which comply with the current world's "precarious" state and that is what they lengthen (Bourriaud 2009).

It is not excluded that a concept that cannot be unanimously defined still functions, as a fulcrum of the interpretation of art. That is why we should analyse one of Bourriaud's favourite examples, some micro-utopia of Rirkrit Tiravanija from close point of view.

2   RIRKRIT TIRAVANIJA'S MICRO-UTOPIAS

Rirkrit Tiravanija is one of the stars of the relational aesthetics, who surprised the art world when instead of preparing objects, he created meeting possibilities: since 1990, he cooked and served Thai food in galleries to visitors on more occasions. This type of work of art is the perfect example of the relational art: the artist brings people together, creates the opportunity for them to be together and talk to each other. In the society of the spectacle, this would be the way of escaping from control, a micro-utopia at hand. According to Bourriaud, we can learn here again what the friendship relation and sharing means.

It is characteristic of this type of works that they do not criticize the social, economic, and political status—everything happens on a „micro" level, the actions that get place here recall the personal living spaces: eating, games, rest. That is why these are neutral spaces, they do not pronounce judgment above anything, they do not want to change anything, and they "only" intend to bring people together. Probably, sharing the meal always "brought people together." However, why is such an event so special that it can be designated as utopia (even if "micro")? We can read the description of art critique Jerry Saltz that relates his experiences at Tiravanija's "exhibition" that took place in 1992 at the 303 Gallery in New York (Saltz 1996). From the description it turns out that he returned 11 times to the gallery and it happened that he set down to eat next to strangers, but most of the times he ate with acquaintances, artists, critiques, people from the gallery, art dealers. They had fun and they shared the gossips of the art world.

Obviously, such a time spent together can be pleasant, it can almost achieve the limits of utopia—but how else would it be if people who have common topic, common interests would be brought together? Such a utopia does not refer by any means to those crowds cursed to passivity, but only to the natives of the art world. Claire Bishop also emphasizes exactly the lack of social dimension:

> Tiravanija's micro-utopia gives up on the idea of transformation in public culture and reduces its scope to the pleasures of a private group who identify with one another as gallery-goers. (Bishop 2004, 69).

Rirkrit Tiravanija has works that are more bound politically than the mentioned cooking-eating platforms that bring the community together. I find especially interesting those works that were presented at the Venice Biennale in 2015 and that were connected to topics that traditionally appeared in the criticism of capitalism as labour or protests. In these works, the critical view returns so one of the conditions that underpins utopia is present. The question is whether hands-on utopias can be found in these works or not?

The first work is the *Untitled 2015 (14,086 unfired)* which created a brick factory in the exhibition space, where three workers produced relatively big unfired bricks on the spot. 14,086 is the number of the bricks of which a smaller family home can be built. On the bricks, a motto of the '68 uprising is engraved in Chinese: *Ne travaillez jamais!* The bricks were on sale on the spot for 10 euros per piece, the amount that was collected went to an organization that protects workers' rights.

"Ne travalliez jamais" is direct reference to the SI—utopian—aims that wanted to change the alienating wage-labour to such extent that work may not be differentiated from not work. Moreover, we can go as far to interpret bricks as the steps made towards the construction of a new society—although the 14,086 number warns us that it can be only about the micro-utopia of the personal, familial space. This interpretation is obliterated by the fact that here real workers produce real bricks and they are paid for their work. So: Rirkrit Tiravanija's bricks, while they create a new world, are still another brick in the wall of capitalism.

The title of another work of Tiravanija presented at the biennale is: *Demonstration Drawings (2006–2015)*. It is about two hundred drawings that plant such photos into the traditional medium of drawing that were made at the International Herald Tribune about the uprisings, at mass demonstrations, strikes, protests. The work is special because not one of the drawings is Tiravanija's work, although the whole work is his: he asked art students, whom he did not mention by name, to make drawings based on the photos chosen by him. Although the press dealt more with the brick factory, I found the later work more interesting. On the one hand, since with these numerous drawings in uneven style and quality it directs attention to social dissatisfaction. On the other hand, in this case he does not ask workers to produce an object (brick), but pays artists to use their abilities in a project invented by him. On the one hand, the hired artists do—as paid work—what in principle an artist does freely and driven by inner motivation, that being precisely the reason for which the artistic life is the symbol and flag of the so-much-desired freedom. On the other hand, the creativity of the artists that participate is limited: the project was not their idea, they do not choose the photos that will be drawn, they do not give title to the drawings, and they do not choose the medium. They are executants with extraordinary abilities, also another brick in the wall. Moreover, although they drew the 200 drawings their names appears nowhere, even their numbers is unknown to us. Similarly, to the capitalist factories, we do not know the names of the workers producing objects. Also, most probably, they were paid for their work—but not as the star artist is gratified by the art world. David Cohen's note in connection to the 2008 exhibition of work is entirely adequate:

> A sense of exploitation is palpable in a work where the sum is exponentially greater than the individual parts, and where the originating and organizing agent, Mr. Tiravanija, reaps infinitely greater reward—of attention, thought, and obviously financial, too—than the individual scrawlers. (Cohen, 2016)

The situation of the Thai artists is an interesting objectivation of the "ne travaillez jamais" utopia: on the one hand the work of art, the drawing is not alienated work, but self-expression, on the other hand, in this concrete situation it is still paid and controlled from the outside (wage-labour).

Although the *Demonstration Drawings* uses classic medium—the drawing—this is aesthetically almost irrelevant. Maybe not even because of its evident message: directing attention upon the problems of the current society and the forms of resistance. What is most interesting is the paradox situation in which the artist who creates the micro-utopias gets: he formulates the criticism of the current society, he propose a utopian transgression—with the tools of art, within the institutional framework of the art world. The relational artist gets into a strongly ambiguous situation, being given that he himself is the privileged of such a system that he wants to set in opposition with his own micro-utopias. As Claire Bishop notes:

It is arguable that in the context of today's dominant economic model of globalization, Tiravanija's itinerant ubiquity does not self-reflexively question this logic, but merely reproduces it. (Bishop 2004, 58)

It seems that on the art's own territory, on the nowhere land of creativity, autonomy and freedom, some of Tiravanija's projects function like any other production process in the capitalist system.

3 THE SADNESS OF MICRO-UTOPIA

The analysis of Bourriaud's concept of micro-utopia and that of some artistic manifestations that served as reference, presented a constitutive ambiguity.

On the one hand, if we apply the concept of micro-utopia to those art events, which Bourriaud prefers to refer to in the *Relational Aesthetics*, then we find harmonious spaces, situations on micro level. However, besides lacking the ungraspable character of utopia (because they are at hand), these events are hard to differentiate from non-utopian everyday events: eating together in a fast food, drinking coffee together, spending pleasant time with others.

If we apply micro-utopia to those relational art events where the criticism against the present can be found we will encounter other difficulties: if art does not only represent its subject, but also it creates and operates it, it becomes part of that reality to which he wants to set his own micro-utopia in opposition.

The sadness of micro-utopia springs from its impossibility: but in opposition to utopia that is impossible to realize, it seems that micro-utopia is impossible to conceive coherently.

BIBLIOGRAPHY

Bishop, Claire. Antagonism and Relational Aesthetics. *October Magazine* 110, Fall 2004, 51–79.

Boltanski, Luc and Chiapello, Ève. *The New Spirit of Capitalism*. New York, London: Verso, 2005.

Bourriaud, Nicolas. Precarious Constructions. Answer to Jacques Rancière on Art and Politics. Essay—November 1, 2009. In *Open* 17, 20–40.

Bourriaud, Nicolas. *Relational Aesthetics*. Dijon: Les Presses du Reel, 2002.

Cohen, David. Rirkrit Tiravanija: Demonstration Drawings at the Drawing Center. In *Artcritical* 2008.09.29. http://www.artcritical.com/2008/09/29/rirkrit-tiravanija-demonstration-drawings-at-the-drawing-center/#sthash.KXshEzrP.dpuf, last download: 2016.01.13.

Debord, Guy. *The Society of Spectacle*. New York: Zone Books, 1995.

Eagleton, Terry. Utopia and Its Opposites. In Panitch, Leo and Leys, Colin. *Necessary and Unnecessary Utopias, Socialist Register 2000*. Woodbridge: Merlin Press, 1999, 31–40.

Saltz, Jerry. A Short History of Rirkrit Tiravanija. *Art in America* 82/2, 1996, 82–107.

Stob, Jennifer. The Paradigms of Nicolas Bourriaud: Situationists as Vanishing Point. *Evental Aesthetics* 2, no. 4 (2014), p. 23–54.

Szántó, Ildikó. Die Fabrik im Museum. Perspektiven auf den Fordismus im Kunstfeld seit 1990. In *Kritische berichte Heft 3* Jahrgang 43 (2015), 31–42.

Vieira, Fátima. The Concept of Utopia. In Claeys, Gregory (ed.). *The Cambridge Companion to Utopian Literature*. Cambridge University Press, 2010, 3–27.

Tirana: Colour and art on the way to utopia

Verónica Conte
CIAUD—Faculty of Architecture, University of Lisbon, Lisbon, Portugal

ABSTRACT: "Greening and Painting" was presented by the creator Edi Rama as a political intervention, involving the painting of several façades of the Albanian capital, among other actions in the public space. This innovative *Colour Manifesto* announced the end of the old oppressive times, an affirmation of a new political regime, and a change from grey to vibrant colours.

Today, paintings on façades have become a widespread practice all over urban public spaces, giving exposure to many renowned artists, as well as social art and design activist projects, some of which also *aspiring* to promote social change. Thus, the study of this forerunner project, due to its complex process and context, grounded in a visionary ideal of recovery for the city, can not only provide inspiration, but can also generate insights, questions, and improve critical reflection on processes and achievements.

After a bibliographic review, field research and an interview with Edi Rama in 2010, throughout the present paper, we will revisit the three main phases of the city transformation by colours: firstly, by Edi Rama's project creation (2000–2005); secondly, by invited artists from the Architecture Biennial (2003 and 2009); and finally, by municipality architects' drawings. Its imagery and results will be presented and discussed, as well as some consequences on the public space. By doing so, we will[1], began a huge transformation in the city in 2000. The aim was to recover, or rescue the capital, on one hand, from the hermetic dictatorship of Enver Hoxa (that ended the Communist regime in the late 90 s), and, on the other hand, from the subsequent illegal construction, that invaded the former public spaces, enabled through periods of "freedom" and "anarchy". When Rama was elected, there were neither paths analyse, in the author's words, a path to achieve a Utopia: "Tirana: not a place of fate but of choice" (Sala, 2003)

Keywords: Tirana, façade painting, city renewal, social change, aspiration

1 TIRANA

Edi Rama, the artist and Mayor of Tirana nor trees on many streets, and many buildings were also left unpainted. Tirana "was not a city, with the problems of a city, it was a "dead city", said Edi Rama in the given interview, "It was necessary to reinvent it". From the beginning "Greening and Painting"[2], along with other structural work, brought bright colours to facades—bright red, blue, lavender, or gold—and black, with unexpected painting interventions. Thus, the colour on the exterior walls was the first sign of change.

The paintings were developed in three phases. The first group occurred in 2000, on facades of state buildings at the city core, by Edi Rama as its colour creator, (Fig. 2). The mode adopted was "free-style" in geometric shapes, completely autonomous and detached from the original colours and forms of the architecture. It should also be noted that in some buildings the paintings only occurred on the façade of the main street, while the inside of the blocks was left unpainted, strengthening the contrast between "before" and "after." By doing so,

citizens themselves would come together to paint the interior of the blocks, (something that had not happened yet by 2010). Deciding how to paint the inside of the block could be done according to each individual's intention, a gesture framed as a sublimation of dictatorial anodization, a chance to participate in the collective construction of the city, in a freer way.

In the second edition of the Biennial of Tirana, 2003, the curators Hans Ulrich Obrist and Anri Sala, created the "Façade Project" passing the authorship of the drawings into the hands of international artists like Rirkrit Tiravanija, Dominique Gonzalez Foerster, Liam Gillick and Olafur Eliasson, (Fig. 3). Later on, in the 2009' Biennial, six new artists were invited to make drawings for another six buildings, Franz Ackermann, Tomma Abts, and Adrian Paci among others[3]. These were more conceptual proposals and intended to reflect on topics such as indwell and urban environment.

In the last phase of the project, after 2009, and according to the given interview, the design of the paintings was delegated to the Tirana's municipality architects.

Figure 1. Tirana: Painted facade. With cordial words "ju lutem" (please) and "me fal" (I am sorry), (photo by the author, 2010).

Figure 2. Painted facade: Intervention from 2000–2003, (photo by the author, 2010).

Figure 3. Painted facade by Olafur Eliasson, 2003, (photo by the author, 2010).

By the time of the field visit (2010), the existing paintings, were located mostly in buildings from the dictatorship period, especially in a circular line around the city centre.

Initially, the process was not participatory. As explained by Edi Rama:

After the elections where I had promised very different things, imagine that I began to ask: What colour do you want to paint your house? It was a policy measure in which people had to accept the paintings

Later, after some controversy, the population was consulted.

This happened quickly. The change was sudden and public. It became necessary to discuss colours, assess their significance, their usefulness. The mayor organized a survey. He realized that 60% of the population were in favour of colour and 40% were opposed. A second question: Do you want this to continue? 90% answered «yes». It seemed as if the colour had the power to overcome the taste and hostile negativity." (Nesbit, 2004, p. 167)

Edi Rama had created "the impossible", had generated a change in the poorest capital in Europe through something as unusual and seemingly fragile as a make-up painting, but that was simultaneously its motor.

Colour has an accelerating impact on the rhythm of breathing. It allows the dusty veil to be lifted and creates a new era for the city. It is a paradox because it is the poorest country in Europe, riddled with problems, and I do not think you can find another one in Europe, even among the richest where they also discussed colour so passionately

248

and collectively. The biggest debate, from my point of view, in cafes and on the street was: What will come along with the colours? (Sala, 2003)

"Greening and Painting" received a strong international recognition. In 2003, Anri Sala took images of its documentary film "Dammi i Colori", to Poughkeepsie, in preparation time of the "Utopia Station" project for Venice Biennale. This way, Sala brought a discussion around the subject—colour in the city—to an elite audience, unfortunately, far from Tirana's inhabitants:

Why should the colour used by an artist necessarily produce art? Why should colour be in a certain shape? Why couldn't it get into the trappings of city life and accomplish the task of generating hope in the vision of the future? Why couldn´t pain be dismissed with colour? May the shape blend in so well in reality that aesthetic issues become political again? (Sala, 2004, p. 7)

Tirana became a different city, with a new freedom, materialised and symbolized by colour. However, even if we can see a few buildings painted with bright colours, as inhabitants' spontaneous way to join the colour movement, should not we ask to which extent, didn't inhabitants become alienated from what was happening on the streets? Was the reason for the change always understood, considering that many painted walls were still missing sheetrock? At this point, I remember Constanzo & Constanzo (2004) pointing out that rather than following chromatic plans made by experts, in reality, what matters is not when people are committed to an urban "facelift", but when actually already know what they want for their space. This ideal calls for a long path. In Tirana, the notion of public space had been lost during anarchic times, where the common good had no recognition or protection. The surprise brought with colour was a way to restart the discussion about the concept of public space and its value, a value that belongs everyone. The colour was a promise, a way of generating hope, and a trigger for a subsequent transformation.

2 COLOUR AND ART ON THE WAY TO UTOPIA

Since the last decade, the number of artists and collectives of artists whose work is dedicated to covering the cities walls with unexpected paintings has been increasing. Among others, some of the most renowned are the works by the Dutch architects Jeroen Koolhaas and Dre Urhahn, "Favela Painting" in Santa Marta, Rio de Janeiro, Brazil (2010)[4]; by the French photographer JR, "Art" Kibera in Nairobi, Kenya (2008) (Rao, 2010), and by the collective originary from Madrid "Boa Mistura", in

particular "Luz nas Vielas" in Sao Paulo, Brazil (2012)[5]; again surprising by their large scale intervention, in contrast to the poverty of the slums.

In addition, the conversion of most deprived areas of cities throughout the facade paintings by Municipal initiatives (or supported by them), can be found in different places from Latin America to Asia. "Las Pamlitas" in Pachuca, Mexico, (2015)[6] and "Villa Carlos Mugica", Buenos Aires, Argentina (2011) (Fernandes, 2013), just to name a few examples subsequent to the preceding.

Conceptually we can relate these interventions with the Utopian construction by Ernst Bock:

In his point of view, the question is not about jumping into the abstract construction of perfect societies; instead, it´s about involving what is hidden, buried—aspirations to another life, the desire of other relations with the world beyond utilitarianism, the extension of the self through "us". (...) The material for Utopia goes through time, is made up of religious themes, legendary tales, but also of poetry, music, of those things that transport you out of yourself. (Riot-sarcey, & al., 2008, p. 43)

In all examples mentioned above, this "transport out of oneself", seems to be achieved by colour and painting, by contrasting landscapes from before and after, by intervention in places never seen or promoted before.

Along with the visual transformation, some interventions have parallel goals such as to integrate marginal places, to give more dignity and visibility, and, in some cases, to empower (often through the activity of painting by local-participants) those who live there; to make dreams come true, to believe and to aspire.

However, how many people are actually covered by those invisible and more profound effects, from large-scale painting interventions?

Are residents and their culture, sufficiently involved in the process, to make a solid construction of the "us" possible?

What if paintings disguise a risk of gentrification? On the other hand, in another direction, if paintings just delay the construction of basic infrastructure as electricity or sewers?

Quoting Italo Calvino

With cities, it is as with dreams: everything imaginable can be dreamed, but even the most unexpected dream is a rebus that conceals a desire or, its reverse, a fear. Cities, like dreams, are made of desires and fears, even if the thread of their discourse is secret, their rules are absurd, their perspectives deceitful, and everything conceals something else.

But, if we pay attention to smaller scale interventions, such as "Paredes Pinturas / Wall Paintings", started in 1999 by the Brazilian artist Mónica Nador, (Nador, 1999), "Pinta Bien" in Arturo Illia, Tartagal, Argentina (2008) promoted by the civil association Más Color (Fernandes, 2013); or "Vivercor Corabitando", São Cristóvão, Montemor-o-Novo, Portugal (2012) (Fernandes, 2013)[7], developed during my PhD, we can observe that walls can speak for themselves. In these projects, where creative dialogue is established closely, between promoters of the action and residents-participants, the local traditional culture, poetry, and histories of now and of ancient times were brought to inhabit the present. However, the most cross-bounded paintings, with people and local culture, implies slower processes, they need the time for the establishment of trust and relationships between people, for developing creativity and singularities under a common concept. Quoting the words of the Uruguayan journalist and poet Eduardo Galeano "Utopia is on the horizon. I move two steps closer; it moves two steps further away. I walk another ten steps and the horizon runs ten steps further away. As much as I may walk, I will never reach it. So what is the point of Utopia? The point is this: to keep walking"[8]. Small steps made the road seem longer, but what is the point of rushing if the road is endless? Shall we start our slow walking right now?

The present research made part of my PhD under the theme "Participatory Paintings on Residential Facades—Identity, Culture, and Public Participation" developed at the Faculty of Architecture—University of Lisbon (Portugal) in collaboration with the Faculty of Architecture, Design, and Urbanism—University of Buenos Aires (Argentina).

ACKNOWLEDGMENTS

I particularly want to thanks Edi Rama who kindly received me, enabling the interview at Tirana's City Hall.

I also would like to express my gratitude to my advisers Professor Maria João Durão and Professor Jose Luis Caivano, and to the Fundação para a Ciência e Tecnologia, for giving me a PhD grant. A special thanks also to my friend Siobhan Cronin for her generous editorial input.

NOTES

[1] Edi Rama held the position of Mayor of Tirana from 2000 to 2011. Prior to embracing a political career, he was an artist. In 2004, he won the recognition of "World Mayor".

[2] "Greening and Painting 2000–2005. Reinventing the Wheel. When colours Become politics ...", was the title given to the action dissemination text. Many works of (re)construction of the public space were done, namely paving streets, demolishing many illegal buildings, transformation of the river banks into a green line and the rescuing of the remaining public space into new squares and gardens. All this action became possible also due to EU support, World Bank, and private investment.
[3] http://tica-albania.org/TICAB/Facades/index.html.
[4] http://www.favelapainting.com
[5] http://www.boamistura.com
[6] http://razoesparaacreditar.com/urbanidade/grafiteiros-pintam-comunidade-inteira-no-mexico-e-violencia-na-regiao-diminui/
[7] http://www.vivercor.com
[8] http://www.azquotes.com/quote/439781

BIBLIOGRAPHY

Appadurai, Arjun. The Capacity to Aspire: Culture and the Terms of Recognition. In: *Culture and Public Action*. Palo Alto, California: Stanford University Press, 2004.

Barriendos, Joaquín. Las utopias delarte (y vice versa). Variaciones sobre un tema, In *Arte y activismo. Miradas cruzadas Europa/Argentina*. Ramona 55. Revista de artes visuales. Octubre 2005.

Constanzo, Alexandre. & Constanzo, Daniel. La Politique des Couleurs. In Sala, Anri. *Entre chien et loup. When the night call it a day*. Paris: Musée d'art moderne de la Ville de Paris, 2004.

Calvino, Italo. *Invisible Cities*. USA: Harcourt, inc. 1974.

Fernandes, Maria Verónica. *Co-design em acções de pintura em fachadas residênciais. Expressão das identidades individuais e colectivas na construção do espaço público*. Lisboa: FA, 2013. Tese de Doutoramento. Available at http://www.repository.utl.pt/handle/10400.5/8706.

Nador, Mônica. *Paredes Pinturas*. (Tese de Mestrado). São Paulo, Departamento de Artes Plásticas da escola de Comunicações e Artes da Universidade de São Paulo, 1999.

Nesbit, Molly. Les Choix qui S'offrent a Nous. In SALA, Anri. *Entre chien et loup. When the night call it a day*. Paris: Musée d'art moderne de la Ville de Paris, 2004.

Riot-Sarcey, Michèle; Bouchet, Thomas; Picon, Antoine. *Dicionário das Utopias*. Lisboa: Edições Texto & Grafia, 2008.

Rao, Vyjayanthi. Slum Theory. *Learning from Favelas*. Lótus Internacional, 14 Agosto 2010, Volumen 143.

Rueda, Maria. *Arte & Utopia, La Ciudad desde las Artes Visuales*. Buenos Aires: Assunto Impresso Ediciones, 2003.

Sala, Anri. *Entre chien et loup. When the night call it a day*. Paris: Musée d'Art Moderne de la Ville de Paris, 2004.

Sala, Anri. *Dammi i colori* (film), 2003.

Part IV *Humanities*

The ideal city for Leonardo Bruni: An analysis of *Laudatio florentinae urbis*

Fabrina Magalhães Pinto
History Department, Universidade Federal Fluminense (UFF), Campos dos Goytacazes, Brasil

ABSTRACT: Leonardo Bruni, chancellor and historian of Florence, and among the first men to translate Plato and Aristotle in the 15th century, is also the author of one of the main praises to the city of Florence in that period. By writing the *Laudatio florentinae Urbis*, in 1404, the humanist shows Florence and its government as the ideal to a fair city, organised, beautiful, and where freedom and civic engagement against all tyranny reigns. This article aims to understand the *Laudatio* not only as a political brochure and rhetorical work, defending the republican ideals on a key moment of the city history, but also as one of the earliest texts to see Florence as a model of the ideal city; the city built according to reason and perfectly suited to the new political man and the city-state which then flourished in the Italian Renaissance.

Keywords: Leonardo Bruni, Italian republicanism, Renaissance

1 INTRODUCTION

The *Laudatio Florentinae Urbis (Panegyric to Florence*) was written in 1404, by the, then, humanist and historian, and later Florentine chancellor Leonardo Bruni. Conceived shortly after the death of the Duke of Milan, Gian Galeazzo Visconti, in 1402, and the consequent threats that Milan tyranny represented to the survival of Florentine city, *Laudatio* has been interpreted over the last decades basically in two ways: first as the founder text of the civil humanism, as proposed by Hans Baron (1968), and, secondly, as an exclusively rhetorical work, as postulate his opponents (Hankins, 1995, pp. 309–338). Based on these two opposing currents, several interpretations of this controversial text have been discussed. This short article aims to develop a hypothesis so far scarcely approached by researchers: the association between the *Laudatio* and the projects of ideal cities in the early Renaissance.

The discussion surrounding the construction of an ideal city in Florence has existed at least since the 13th century, when became evident the concern with the architectural and urban aspects, such as the Santa Maria del Fiore cathedral expansion project, and the public buildings construction. In 1293, during the term of the Florentine Republic, the notary Mino Cantoribus suggested replacing the Santa Reparata church by an even larger and more magnificent cathedral, so that "the industry and power of man could not invent or even try anything larger or more beautiful". When Arnolfo di Cambio, architect of the cathedral, was put in charge of the project in 1294, he was asked to build "the largest and most beautiful church ever built". Therefore, he did: with 153 meters long, 90 meters wide at its broadest point, and 86.7 meters high. His project also included the construction of a large dome that would become the new symbol of Florence; The dome should be so grand (41 meters) that the solution to their coverage only would become reality a century later, in 1418, with Filippo Brunelleschi. About the Florentine buildings, Bruni says in the *Laudatio*:

> What on Earth can be more splendid or so magnificent to be compared with its buildings? I am ashamed of all other cities every time such comparison comes to mind. This is because they usually have only one or two well-constructed roads, and in every other the absence of any beautiful constitution is such that they greatly blush, embarrassed when viewed by foreigners. (Bruni, *Laudatio*, 2008, ch. 10)
>
> Am I a fool by enumerating all this? If I had a hundred tongues, a hundred mouths and an iron voice, I would still not be able to show all the magnificence, the beautiful buildings, the treasures, the luxury, and the luminance of Florence. (ibid, ch. 12)

The *Duomo*, and its grand architectural design, the Baptistery and Giotto's bell tower, the Hospital of the Innocents, the *Palazzo Vecchio*, and several other public buildings already pointed to the construction of an ideal city, and a free city.

In this sense, as the Republic advances also did the desire to create an urban and architectural model for the city, which should be able to supersede all other rivals in opulence and magnificence. These buildings would be directly associated with the political, republican and expansionist ambitions, constituting an ideal image of a city: where justice, rationality, and freedom should reign. Such an elevated model could only be as thought and built in a republic like the Florentine, heir of Rome and of the principles of virtue of the ancient. Thus, if we look closely, we can mention the Panegyric to Florence (*Laudatio Florentinae Urbis*) as part of the earliest writings on ideal cities; this would cause an advance of at least half a century in the dating made by analysts that are more traditional.

Some of the major works about the ideal cities concept were written after the mid-fifteenth century: as *L'architetura* (*De Re Aedificatoria*), by Leon Battista Alberti, published in 1486, the *Trattato di Architettura*, by Filaret, (written between 1461 and 1464), and the *Codex Atlanticus*, by Leonardo (written 1478–1519). It is sufficient to recall Leonardo da Vinci's representation of the city of Milan. In his *Codex Atlanticus*, advises Ludovico il Moro to beautify the city: worrying about hygiene, water supply, population distribution and creating neighbourhoods that are more organized. Therefore, we agree with Eugenio Garin (1963, p. 15) when he says that Leonardo's sketch, far from being a purely imaginary plan, is already connected to the real aspirations of the Italian city-states, and to the desire that Milan could also become an urban model in Italy. The ideal cities are connected to a reform project of the real cities. Their constructive space is not equidistant from the real city. As for the theories on utopian cities—such as the Utopia of Thomas More, for example—they are even posterior, dealing with the rational construction of a non-place, or a place in the future.

For the purpose of conceptual distinctions, we classify the ideal cities relating them to existing cities. However, both ideal cities and utopian ones are projects, surpass the reality limits, and therefore are likely to be described in fiction form, but not logically opposing to any of the real aspects of the city, merely placing them in more evidence and modifying their uses, adapting them to the criticism of the real situation. (Rodrigues, 2000, p. 136).

Beyond the writings of Alberti, Filarete and Leonardo, it is also possible to see the concern with the city representations in art, as we have already, since the 14th century, several images on ideal villages, as seen on Simone Martini (Figure 1), Botticelli (Figure 2), Piero della Francesca, Perugino and Vittore Carpaccio. Below, we analyse these images.

Figure 1.   Simone Martini. Equestrian Portrait of Giudoriccio da Fogliano, 1328. Palazzo Pubblico, Siena.

Figure 2.   Sandro Botticelli, the Tragedy of Lucretia (1498). Isabella Stewart Gardner, Museum of Boston.

2   REPRESENTATIONS OF THE IDEAL CITY

This fresco by Simone Martini is perhaps the most interesting and emblematic example of the exaltation of republican victories. Commissioned between October and December 1328, it is part of the cycle of "Castles conquered by the Republic of Siena", which used to occupy the two largest walls at the the Siena Public Palace room. The whole cycle started in 1314, and have continued on this work by Simoni Martini, and finished by himself in 1331, with the representation of the Arcidosso and Casteldelpiano Castles (unfortunately, both are now lost).

The Republic of Siena was one of the most prosperous cities in non-pontifical Italy. The Sienese people had just promoted a true political revolution: in 1262, the *popolani*—the new people coming from the effervescent trade but with no political representation—established a people Council, presided by the *Capitano del Popolo* (People's Captain), a directly elected leader. (Skinner, 1996, 45). The conquest and annexation of the latter castle—the Montemassi castle—was part of a new campaign to obtain the control of Maremma region. Montemassi was conquered by the Sienese after a long and costly siege, which took place between January and August 1328.

In the painting landscape, on the background, it is possible to see represented an archetypal Tuscan town, with a fortified castle and a large church. The wall gate is open, such as the castle, creating a sense of vulnerability that surely refers to the surrender of the place. The centre of the fresco represents the character of Guidoriccio da Fogliano (*condottieri*, conqueror of Montemassi and winner of the battle against Castruccio Castracani), goes far beyond a simple equestrian representation. It was, on the contrary, a propaganda image of Siena and the Government of the Nine's desire to "expand, enlarge and conserve the city" (Castris, 2007, p. 265). Similarly, this politic and civic painting could be seen as a form of celebration and permanent remembrance of the achievements of Siena Republic against the Guelph military forces; a kind of monument to its resistance against the papal claims to predominate in the city. Therefore, it is already possible to see here, which interests us more specifically, the affirmation of a republican ideology, of an ideal government, fairer than the others.

In Botticelli, in his The Tragedy of Lucretia (1498), we can already perceive the presence of classical elements (arches, columns and temples) in the representation of the city; in addition, of course, of elements such as perspective, mathematical precision, and perfection of the lines.

The key issue presented in this work is the honour violation. Lucretia, Collatino's wife, after suffering sexual violence from Sixth, son of Tarquinius Superbus, commits suicide, causing the revolt that led to the overthrow of the monarchy in Rome. On the background, a great architecture is assembled, richly decorated, as in Calumny of Apelles (1494), with a big triumphal arch inspired by the Roman Forum arches, decorated with bas-reliefs and golden statues.

Lucretia, who suffered a violent death at a young age, was the protagonist of a tragedy motivated by desire, and by the need to save his reputation. However, the moral drama and the celebration of Lucretia's chastity also take on a political significance. Livy had already emphasized how the Roman heroine was a victim of the tyrant's abuse, and how his death incited the rebellion against an unfair government. This combination is significantly present, both on Roman fountains and in those from the 14th and 15th centuries, as well as in painting; particularly in this Botticelli composition. We can assume that the painting also deals with the political question of the legality of the uprising against the dictatorship, connected to the Florence's situation at those years, with Pedro de Medici's expelling and the Savonarola's Republic establishment.

In the city of ideal forms of Botticelli, the tyranny imposed by a religious leader against the republic should be broken. As Lucretia, Roman heroine, had been abused by tyrants, also the Republic of Florence was victimized. Moreover, the same way her death caused a rebellion against unfair government, the fall of Medici's Republic should generate the conflict against the Dominican friar's religious radicalism.

We agree with Eugenio Garin when stating that such an attitude of mind would have begun to manifest at least since the 14th century, as during the expansion of the Italian city-states would have concomitantly been developed both the desire for a new political organization, as the search for a better adapted architectural reconstruction to an economic growth situation (Garin, 1963, p. 16). For this reason, we cannot neglect the development of the republican cities—which had pursued, since the beginning of the 12th century, freedom and self-government, be it against the interference of the Holy Roman Empire, or against the Church—are disconnected from urban issues and the building of a model: one of free cities, where civic participation and political autonomy should reign. Likewise, we must not forget that this choice of a political model different of feudal monarchies enabled an intense evolution of the republican cities, especially Florence. Solving the problems of certain popular conglomerates, distribute them more rationally and take care of general aspects such as public health, epidemics contention, popular disorders control, defence against external attacks, as well as the building of imposing and representative public works to show the strength of the republic became, from that moment, a constant in the ideal cities imaginary creation.

Thus, says Garin (1963, p. 17), the political treaties have become treatises on architecture and urbanism, raising the demand for the city rationalization, be it at administrative or architectural level.

It is not difficult to find, in the the the late 14th century and early 15th Florentine texts, a parallel between the *res publica* institutions and their buildings, and that is precisely what makes Florence an ideal type of city. The *Palazzo della Signoria* and the *Duomo* became more than republic symbols, but expressions of Florence's strength and wealth. The city should transform itself at the exact pace that men politically evolve. In this sense, Florence sought to harmonize the political advances with the urban concerns. Men so brave—defenders of the republic against the cruellest opponents, and legitimate defenders of freedom, not just for the city but the whole peninsula—should live on the most complete of cities. Leonardo Bruni says:

> Just as we see many children have so much resemblance to their parents that it is possible to recognize the affiliation most clearly in their own faces, also this most noble and celebrated city is in such harmony with its citizens that it seems that they, by a high reason, have not been able to dwell in any other city, nor this city have been able to have any other citizens. This is because these same citizens are better than all other men in genius, pru-

dence, elegance, and magnificence, and so is the city, located in the most prudent way, surpasses all others in splendour, architecture, and cleanliness. (Bruni, *Laudatio*, 2008, ch. 4)

Similarly, it is no less important to note that the humanist writings recognize as ideal the form of political organization in the city-state, in open opposition to the large empires of ancient and medieval world, especially the Holy Roman Empire and the Church. It is clear, therefore, that the city defence, as an ideal of political organization, is closely related to the struggles for the conquest of autonomy, and to escape the interventions of the empire and the papacy. After these brief considerations, we can point Bruni's *Panegyric to Florence* as one of the representative texts of this tradition: that associates politics, rhetoric, and idealization of the city. Let us move to the analysis of this work.

3   THE *LAUDATIO FLORENTINAE URBIS*

Leonardo Bruni, historian, humanist, philosopher, and chancellor of the Florence Republic between the years of 1427 and 1444, has published numerous speeches in which the republican city is systematically praised. The *Laudatio* is one of his earliest texts and an instrument to expose their republican convictions. Written shortly after the death of Duke Giangaleazzo Visconti, in 1402, and of the threats that the Milan tyranny represented to the survival of the Florentine city, the paper presents a set of rhetorical arguments defending the republican cities, in opposition to the attacks from their Empire connected enemies. At this time, armed and ideological conflicts walked alongside, and therefore the eloquence was a clear instrument of political persuasion. However, there is no need to follow only this route: namely to consider Bruni's text as a rhetorical exercise of incitement to Florentine citizens to defend the republican system. We can look at this text also on the other bias, analysing it as praise, seeking to turn Florence in to a model to be followed by other cities: whether in its architectural, artistic, political, and social aspects, and even an example in care of health related issues.

It is not hard to find among the classical texts, the inspirations, and models followed by the early Renaissance politicians and architects for the creation of the ideal cities. According to Hans Baron, one of the main inspirations of Bruni would be the Greek author Aelius Aristides, with its *Panatenaico* (mid-second century). Although we agree with the thesis pointed out by Baron, of the influence of *Panatenaico* in the *Laudatio*, we do not share his view that this would be the only source that Bruni would have used as a basis to write his work.

We believe that other sources, as Coluccio Salutati (also chancellor of the Republic of Florence and Bruni's personal friend), Menander, and Plato himself may have had great influence on Bruni. Let us move on to the specifics of the text.

The text of the *Laudatio* can be divided into three parts: 1) Site and situation of Florence and its territory (chapters 4–29); 2) Origins of the Florentine people (chapters 30–47) and 3) Florence's actions and *virtùs* (chapters 48–91). Bruni divides this third part in two distinct chapters, initially describing Florence's actions and *virtùs* during war. After detailing Florence's foreign policy, Bruni examines their domestic institutions.

Soon after the exordium, Bruni starts describing the city's site and situation, and its territory, accordingly to Aristides' model. The Greek orator uses the following description order: Attica's situation by the perspective of the sea (Pan 9–13.), the land (Pan 13–16.), the climate (Pan 18–19.), and the territory nature, on the topographic matter (Pan. 20–23.). What draws our attention here is that Aristides only analyses the territory of Attica as a whole, and just briefly refers to the city of Athens (Pan. 16: 6–9).

The humanist, in turn, begins, in his first part, describing the city, rather than its territory. Great importance is given to Florence's health, as well as its beauty, thanks to its monuments and splendid buildings.

> Many are the filthy cities, which overnight produce all kinds of manure, that every morning is exposed to everybody's eyes and feet by the streets, the most fetid thing imaginable. (…) As a deformed body, even having many perfect parts, there can be no happiness also in the cities, if they are filthy, even if everything else is present, there cannot be beauty at all. (Bruni, *Laudatio*, 2008, ch. 7)
> [...] Florence is, in fact, as far as we can discern, to such point healthy and clean that one cannot find any other that is brighter. This city is unique and, all around the globe, only in it nothing offends the eyes, nothing is unpleasant to the nose, and nothing is unclean to the feet. Given all the supreme diligence of the inhabitants, as well as their precautions and care, all filth are moved away, so one can only find in the way joy and sense gratification. (Idem, ch. 8)

Therefore, by reversing Aristides' plan, Bruni has the freedom to address modern issues relating to the city planning and hygiene issues, that emerge due to the economic growth of the city and the political strengthening of the Italian city-states. Bruni's ideal city must have the necessary conditions to assure harmonious citizens experiences. Regarding the Florentine buildings, the author says: "what, all around the globe, could be more splendid or so magnificent that can be compared to its buildings?" (Bruni, 2018, ch. 10). Thus, by the magnificence of its buildings and sacred temples, as well as their care with hygiene and urbanization, it exceeds all existing cities, having "nothing to shock the eye, no uncomfortable odour" and nothing that is repugnant to visitors (Ibid, p. 211–213). As for the strength of the city, it is revealed by its military power and their victory over an extremely dangerous enemy.

However, why does Bruni reverse Aristides' plan, first writing about the city, and only later about its territory? We may suspect that Bruni puts the initial focus on what is most important to him: the city of Florence (with its men, its army's bravery, its architecture's beauty and its streets' healthiness), and only then he turns himself to the issues which he attached less importance to: the climate, the topography, the territory and the distance between Florence and the sea (that protected it from floods, disease and unhealthy conditions of coastal areas) (Bruni, 2008, ch. 25, p. 233).

As stated by Stauble, Bruni made an idealized description of the city, not taking much into account some aspects that medieval sources would consider to be more relevant: such as the city situation, all of its buildings, its people and its different social groups, its political organization, its everyday life (Stauble, 1976, pp. 157–64). Such a difference in the approach of the themes signals to a difference in design.

At the core of the work, we can see the subject of freedom, only achievable by the preserving of civic freedom provided by the republican system. Bruni writes *Laudatio* shortly after the conflict with Jean Galeazzo Visconti, whose threatens aspired to turn Italy into a major force, under Milan's domain. Against a unification that would put a tyrant in the Italian cities' political control, Florence throws itself in defence of the republican freedoms. Similarly, resuming an old theme, Bruni said that the city should be free and fair:

> First of all, therefore, one has to take care that justice remains sacred in this city, for without justice, there can be no city, nor could Florence be called a city. Next, there must be freedom, for without freedom these people would never consider worth living. (Bruni, 2008, ch. 77)

This idealized socio-political structure (of an equity-seeking repairing state) is also related, according to Bruni, to the architectural structure of the city, as in the *Palazzo della Signoria*, the *Duomo*, rationally arranged houses, large squares, and sanitary precautions to prevent further incursions of the plague. This is the reason why Bruni's praise is important, precisely because his ideal city, so filled with ancient references, is not a fantasy born from out of reality, but it is instead identified with an existing city, which needs both praise and to intensify the rationality of its existing features. Florence is the one that, by its history—and by the strength of its Roman heritage—seems destined to be displayed as the ideal for human coexistence, and for the achievement of a common life based on rational grounds.

The ideal city, as pointed out by Eugenio Garin, perfect in its buildings and political institutions is not from out of this world, from heaven or from the country in Morus' Utopia: it is present in an exemplary city, even if not perfect: Florence, that can only be an ideal model thanks to its option for justice, freedom and *civitas*. Due to this reason, unlike Dante's treatise *De Monarchia*, for example, the ideal model is the republic. Moreover, the Florentines are those who can best imitate this model, since they are heirs of the republican Rome and, therefore, has always defended the political ideals against tyranny.

> Therefore, it was: this splendid Roman colony was established in the exact same period that the Roman people domain greatly flourished, and when mighty kings and warrior nations were being conquered by the armies of Rome and its virtue. (...) Due to this reason, I believe that something was true and is still true in this city more than it is to any other: the people of Florence love freedom above all things, and they are great enemies to any tyrant. Therefore, I believe that from its foundation to this day, Florence has developed an immense hatred for the invaders of the Roman Empire, and for the destroyers of the Roman Republic. If any trace of even the names of the corrupters of Rome has survived to the present, they are hated and scorned in Florence. Political diligence is not new to the Florentines, nor did it begin a short while ago, as some think: its origin lies in ancient disputes. It began long ago when nefarious men committed the greatest of all crimes: the destruction of freedom, splendour, and dignity of the Roman people. (Bruni, 2008, ch. 35)

BIBLIOGRAPHY

Baron, H. *From Petrarch to Leonardo Bruni: Studies in Humanistic and Political Literature*. Chicago: University of Chicago Press, 1968.

Bruni, Leonardo. "Laudatio florentinae Urbis" In: *Histoire, éloquence et poésie à Florence au début du Quattrocento*. Textes choisis, édites et traduits par Laurence Bernard-Pradelle. Paris: Honoré Champion, 2008, pp. 205–301.

——. *Opere letterarie e politiche*, a cura de Paolo Viti. Torino, 1996.

——. *Lettres Familières*. Tome II. Édition, traduction et notes de Laurence Pradelle. Presses Universitaires de la Méditerranée, 2014.

Cambiano, Giuseppe. *"L'Atene nascosta di Leonardo Bruni" in: Rinascimento. Rivista dell'Istituto Nazionale di Studi sull Rinascimento*, volume XXXVIII. Firenze: Leo S. Olschhki, 1998.

Chastel, Andre. "O temple" in: Arte e Humanismo em Florença. Cosac Naify, 2012.)

Clastres, Pierluigi Leone de. *Simone Martini*. Milano: Federico Motta Editore, 2007.

Elio, Aristides. "Panatenaico" In: *Discursos I*. Introducción, traducción y notas de Fernando Gascó y Antonio Ramirez de Verger. Madrid: Editorial Credos, 1987, pp. 113–254.

Garin, Eugênio. "La citè idéale de la Renaissance Italienne" In: *Les utopies à la Renaissance*. Presses Universitaires de Bruxelles, 1963.

——. *Storia dell'utopia*. Donzelli, 2008.

Pernot, Laurent. *La Rhétorique d'éloge dans le monde gréco-romain*. Tome I et II. Paris: Éditions Brepolis, 1993.

Pradelle, Laurence Bernard. "L'Influence de la Seconde Sophistique sur la *Laudatio Florentinae urbis de Leonardo Bruni*" In: *Rhetorica: A Journal of the History of Rhetoric*, Vol. 18, No. 4, 2000, pp. 355–387.

Rodrigues, Antônio Edmilson Martins. "Das possibilidades de cidades utópicas" In: *Revista MORUS* - Utopia e Renascimento, n. 6, 2009.

——. "Cidades ideais e cidades utópicas" In: *Tempos Modernos: ensaios de História Cultural*. RJ: Civilização Brasileira, 2000.

Salvador González, José María (org.). "La Imagen del Poder. Epifanías de la Potestas" In: *Mirabilia* Ars 3 (2015/2), ISSN 1676–5818 126.

Skinner, Quentin. *As Fundações do pensamento político moderno*. São Paulo: Companhia das Letras, 1996.

Viti, Paolo. *Opere letterarie e politiche di Leonardo Bruni*. Torino, 1996.

Creating settings: Thomas More and the imagination of places

Natacha Crocoll

Faculté des Lettres, Université de Genève, Geneva, Switzerland

ABSTRACT: Despite the scholars' interest for the topography of the island of Utopia, place descriptions are often overlooked in favour of narration. Nevertheless, the study of these secondary topographies enables a deeper understanding of Thomas More's major work.

In this paper, we will first analyse how the author constructs such places and highlight their functions in the general structure of *Utopia*, building a bridge between reality and fiction. We will do so in the light of Thomas More's formation as a humanist, which made him aware of literary traditions that take roots in classical literature and develop through the Middle Ages into the 16th century.

Thus, the Polylerites and Zapoletes' homelands, as well as the desert surrounding the equator, correspond to traditional commonplace scenery whose symbolism has to be pointed out in order to understand fully their importance. We believe that the lack of interest in these fragments is due, at least partly, to this affiliation, which could be considered a lack of originality, a question we aim to discuss in the last section.

Keywords: topography, space, literary traditions, *topoi*

1 IMAGINING NEW HORIZONS

Underlying the concept of *Utopia* is the idea of new horizons. Though the social and political aspects of Thomas More's masterpiece are not to be overlooked, topography deserves special attention and treatment.

Indeed, More lived and wrote during the golden age of Renaissance exploration, right after all the known world's frontiers had been expanded, giving birth to a completely new concept of other-worldliness. Scholars who had always considered "the other" within the limits of Europe, northern Africa, and Asia had then to add an entire continent to their conception of the world[1].

Leaving aside the theological and scientific implications of such a change, the impact on men's imagination was colossal. At the beginning of the 16th century, the New World was far from being fully explored and thus became a source of inspiration for many writers, an example of which being the innumerable travel narratives appearing at that time.

More never clearly states that the island of Utopia can be found in America. However, most editors of the text have explained that, through Raphael Hythloday's story, we can deduce that Utopia is situated close to the newly discovered continent (More 1983: 11)[2].

Thus, though one may consider More to be constructing the imagery of his fictional land on a blank page, he nevertheless proves, on several occasions, to be heir to literary tradition. The purpose of this article is to analyse how places are constructed in Thomas More's narrative. We will give special attention to the link between the text and the images this evokes in the reader's mind.

1.1 *Places in* Utopia

First, we have to admit that the study of *loci* in Thomas More's text could be regarded as highly ironical, considering the meaning of the word "utopia" (Vieira 2010: 4). Nevertheless, the mere fact that it is a non-place allows More total freedom when creating the setting of the island and the places related to it.

Before going any further, it is essential to remember that the famous description of the island is not the only mention of places in the text. Actually, in the first book, the author indicates that the Polylerites live far from the sea and surrounded by mountains, as do the Zapoletes in the second book. These are briefly mentioned, barely touched on, but they do matter. There is also this wild desert environing the equator, which does not seem to have attracted the scholars' interest.

The next two sections will be dedicated first to these brief mentions of surroundings and their link to medieval tradition and, second, to the second longest description of the text, that is the equator, which appears in the first book. Since a serious study has already been made by Brian Goodey on the details of the mapping of the island of Utopia (Goodey 1970), we will recommend his analysis and will direct our study towards the function of this extract from the second book. Moreover, the topography of the island obeys to the genre of travel narratives and thus is not related to our topic[3].

2 FOLLOWING MEDIEVAL LITERARY TRADITIONS

Thomas More was a man of his time. Born at the end of the Middle Ages, his years of studies turned him into an exemplary humanist connoisseur of the *studia humanitatis*[4].

As such, More's admiration and debt to classic thinkers—be they Greek or Latin—is huge and well documented. In our opinion, More is also strongly connected to the medieval tradition of place description. Alternatively, as Fátima Vieira rightly says:

> More did not work on a *tabula rasa*, but on a tradition of thought that goes back to ancient Greece [Plato] and [...] traverses the Middle Ages." (Vieira 2010: 6).

Even if the subject of landscapes and topography in medieval literature is too complex to be exposed in a few pages, some aspects have to be outlined before proceeding to the analysis of More's extracts. For the majority of this long period, nature and characters' environment are mostly skimmed over, bar a few select cases. Indeed, authors and copyists are accustomed to resorting to *topoi* and using nature description mostly as a symbol or allegory underlining of the events they are recording. However, even if until Petrarch landscapes and nature weren't considered to be the protagonists of a work of art from a purely aesthetic point of view, this by no means implies that medieval scholars were indifferent to aesthetics when describing a place or that their topographies completely lacked sensibility (Orozco Diaz 1968: 26–27).

2.1 *Polylerites and Zapoletes isolated by mountains*

Back to *Utopia*, this approach to topography can be clearly identified already in the first book with the description of the Polylerites' land. Raphael Hythloday mentions this fictional place situated close to Persia—to whose shah they owe a tribute–, as exemplary, referring to the judging and punishing of thieves. Though the political and social structure of this land is clearly stated, its situation remains unclear, Raphael only briefly

mentioning that it lays surrounded by mountains and far from the sea (More 1989: 23). Of course, any precise location would be impossible since Polylerites only exist in the author's mind and would thus jeopardise the whole process of hypotyposis[5]. Nevertheless, the mention of mountains and the distance to the sea is more significant than it first seems.

It is not necessary to go back to the Middle Ages to affirm that mountains are synonymous with isolation[6]. Actually, high mountains were not really explored until well into the 18th century, leaving this place mysterious and remote. Thus, mountains were considered throughout the medieval period to be a non-place, along with deserts and forests. This means that they symbolized "new horizons", marking a steep difference between the public—and the author's—reality and the figures that were described in the text (Zumthor 1993: 63). When it comes to the Polylerites, this symbolic tradition of mountains allows the author to justify, on one hand, why nobody besides Raphael knows of this place and, on the other hand, why this land has maintained completely different customs.

However, mountains are not only a natural defence protecting autarky. Since they are mysterious, they could also be dangerous, in which case they are synonymous with savagery and deceit. We could of course mention the Moors' attack against Charlemagne's forces in the Pyrenees or the adventures of Juan Ruiz with the *serranas* in the Spanish *Libro de Buen Amor*, but More's description of the Zapoletes is just as accurate. In the second book, the reader discovers these mercenaries occasionally hired by Utopians when they need to go to war. Just like the Polylerites, the accent is placed on their characterisation and habits rather than on their homeland. In this case, the only mention that we get is also contained in one sentence:

> The forests and mountains where they are bred are the kind of country they like: tough and rugged. (More 1989: 91)

As aforementioned, traditionally, forests and mountains are similar to deserts in that they are opposed to civilisation, indicating the opposition between savage and belligerent tribes on one side, and utopians on the other.

Thus, in both examples, More used literary symbolism to back up his narrative and reinforce the characterisations of the different peoples. Ultimately, he does not offer a detailed and specific description of the places he presents, but uses traditional patterns. Of course, More does not just follow medieval motives; he also plays with them, integrating them into his game between reality and fiction.

Just as in *Utopia*—and especially in the first book—the narrator is playing with the reader, mixing real people and locations with invented characters and places, this ambiguity appears also in the description of both the Polylerites and the Zapoletes' homeland. For the first one, as we have seen, the satiric etymology of the name indicates to the reader that such a population is fictional, and yet the mention of Persia and the isolated location add some verisimilitude[7]. For the latter, the name shows that this is a literary construction too, but are the Zapoletes and their surroundings a pure product of More's imagination? No. As several editors of the text have pointed out, the Zapoletes are to be identified with Swiss mercenaries and, given that these are renowned for living in the Alps, the location amongst forests and mountains is not simply symbolic. Thus, More is transferring, or hiding, real references into fictional characters.

2.2 *The equator:* A locus agrestis

The longest description of the text, after the one of Utopia's island, takes place at the very beginning of the first book, when More first meets Raphael Hythloday and the latter describes his travels after Americo Vespucci's departure. With his men, Raphael leaves the fortress and begins his journey to Ceylon by crossing the arid land that supposedly surrounds the equator (More 1989: 11).

Just as More had referred to traditional literary motives for the homeland of Polylerites and Zapoletes, this description brings to mind the *topos* of the *locus agrestis*, i. e. the 'savage place'. Its characteristics can vary according to the authors, but the main components are always the following: heat, a desolate land, extreme survival conditions, total absence of water and vegetation. Ramón Pérez Parejo traces its origin back to the Bible and then adds that throughout the Middle Ages, this *topos* tends to appear in narrative situations where all seems desperate, or as an initiatory obstacle for the protagonist (Perez Parejo 2004: 265–267). As for María Teresa Rodríguez Bote:

> Más que una imagen en sí, casi podríamos afirmar que se trata más bien de una actitud hacia el paisaje, o, mejor dicho, un temor hacia lo incógnito de la naturaleza y los peligros que por ende presentaba cualquier desplazamiento". (Rodriguez Bote 2014: 384)

In the case of the equator, it seems that this description fulfils another goal. Of course, it is a motive that evokes of the dangers of travelling at that time, reminiscent of the travel narratives that were so successful back then; yet it is also a double place of transition. First, it could mean that Rap-

hael and his men actually had to cross savage lands before encountering new civilisations. Secondly, and this is more probable, the *locus agrestis* could be seen as a metaphor used to symbolize the transition from reality—Brazil and Vespucci—to More's fictional world—Utopia.

The author could also have chosen other images for that purpose, like a voyage through the sea, but, as we will see thereafter, the components of this horrible desert are interesting for the narrative, offering at the same time the opportunity for a symmetrical construction and a contrast for the upcoming description of Utopia.

The first thing that transpires from the fragment is the hyperbolic enumeration of words with a negative connotation. The vastness of the desert is emphasized by its emptiness due to the lack of vegetation and the perpetual heat. Of course, this sterile area offers a strong contrast with the fertile ground, which allows utopian autarky.

Furthermore, the savage aspect of this *locus agrestis* is composed of beasts like lions—that are not here reminders of nobility and dignity but rather of the Roman executioner of the Christian martyrs–, snakes whose symbolism is quite clear, and human beings who are closer to beasts than to men. Again, this offers a welcome contrast to the utopian island presented as a territory. In ancient Rome, the term *territorium* was used to designate the farming land attached to a *urbs*. Leaving aside the popular evolution of this word, it is interesting to note that in the 13th century, this exact term is found again in some documents to describe the space controlled by a jurisdiction (Zumthor 1993: 80). Thus, after explaining the etymology of the medieval conception of "territory", Paul Zumthor defines it as follows:

> L'union de l'homme et de l'espace fonde le "territoire", espace civilisé de qui, par son travail, se l'est approprié et y a créé un droit". (1993: 78)

Of course, this cannot be said of the tribes that live in the desert, since their bestiality proves them unable to create a "civilised space". Consequently, the absence of society in the desert offers another contrast to the perfectly ruled society from Utopia that reflects the Renaissance ideal of order.

A last word should be said about the symmetrical construction that was mentioned. In an editorial note, Marie Delcourt affirms that More projects onto the cities across the desert the characteristics of northern Europe towns (More 1983: 12). Should this be true, this would establish a clear parallel in the first book between real cities and fictive ones, which reinforces the social and political criticism towards actual customs and institutions.

3 CONCLUSION

In her introduction to utopian literature, Micheline Hugues writes that:

Le rêveur, le voyageur utopique, prennent leur essor vers d'autres lieux, vers d'autres temps, où tout semble possible, où il dépend de l'imagination de créer une réalité plus conforme au désir. L'utopiste a pour modèle le Dieu créateur [...]. Mais il n'est pas Dieu. Il lui faut donc se contenter de donner au monde qu'il crée une existence virtuelle. Cependant, dans son geste créateur, rien ne semble pouvoir entraver la liberté de son imagination" (Hughues 1999: 87)

Yet, as we have seen, Thomas More does not just create out of his imagination, but relies on literary traditions that he projects onto a new world. Does that mean that he lacks creativity?

The reader should never forget that *Utopia* is not about a dreamy paradisiacal land but that the fiction is here as a pretext for a political treaty—a term which also appears in the original title. As such, descriptions are here to serve the narration.

Yet, the fact that descriptions are just a frame does not imply that they are unimportant. They matter because they support the narration and allow either to enlighten certain aspects of the states the author is presenting or to structure the text, as in the case of the *locus agrestis*. Indeed, once the symbolism of all these *topoi* has been pointed out, it becomes obvious that the hierarchy between places reflects the hierarchy between men. Utopians and Polylerites benefit after all from urbanised structure and abundant soil that is not dissimilar from the famous *Pays de Caucagne*. On the contrary, Zapoletes and the bestial tribes from the equator see fit to live in inhospitable lands.

Finally, mountains and snakes are not here for an aesthetic purpose. They appear because Thomas More plays with traditional literary motives and uses them to indicate when his narration transitions from reality to fiction.

NOTES

[1] On the idealisation of the East as a place of wonder, see (Freedman 2012)
[2] The mention to Brazil is in a note made by the editor to the text.
[3] There is also an analysis on the city of Amaurotum in Hughues (1999: 40–42).
[4] For a detailed presentation of More's formation, see George Logan, and Robert Adams "Introduction"(More 1989: xiii-xv).
[5] Philippe Hamon explains that hypotiposis is a form of description (not necessarily of places, but also of customs, people, etc.) that is realistic enough to create in the reader's mind a triple judgement of existence, aesthetic and conformity (Hamon 1991: 18).
[6] Marie Delcourt, editor of one of the French versions of *Utopia*, writes in a note that this fragment is a reminding of Epicure's principle of living hidden to live happily (More 1983: 32).
[7] Louis Marin has studied the distance established between the Polylerites and reality (1973: 205–210).

BIBLIOGRAPHY

Freedman, Paul. "Locating the exotic." In *Locating the Middle Ages: The spaces and places of Medieval Culture*. Eds. Weiss, Julian and Sarah Salih. London: King's College London, 2012. 30–33. ISBN: 9780953983872.

Goodey, Brian R. "Mapping "Utopia": A Comment on the Geography of Sir Thomas More." *Geographical Review* vol. 60.1 (1970): 15–30. ISSN: 00167428.

Hamon, Philippe. La description littéraire. De l'Antiquté à Roland Barthes: une anthologie. Paris: Macula, 1991. ISBN: 9782865890316.

Hughues, Micheline. *L'utopie*. Paris: Nathan Université, 1999. ISBN: 9782091910451.

Marin, Louis. *Utopiques: jeux d'espaces*. Paris: Les Editions de Minuit, 1973. ISBN: 270730400X.

More, Thomas. *Utopia*. Eds. Logan, George M. and Robert Martin Adams. Translation and edition by George Logan and Robert Adams. Cambridge: Cambridge University Press, 1989.

———. *L'utopie ou le traité de la meilleure forme de gouvernement*. Genève. Ed. Delcourt, Marie. Translation, edition and notes by Marie Delcourt. Librairie Droz. 1983.

Orozco Diaz, Emilio. Paisaje y sentimiento de la naturaleza en la poesía española. Madrid: Prensa Española, 1968.

Perez Parejo, Ramón. "Simbolismo, ideología y desvío ficcional en los escenarios y paisajes literarios: el caso especial del Renacimiento." *Anuario de Estudios Filológicos* XXVII (2004): 259–274. ISSN: 0210–8178.

Rodriguez Bote, María Teresa. "La visión estética del paisaje en la Baja Edad Media." *Medievalismo*.24 (2014): 371–397. ISSN: 1131–8155.

Vieira, Fátima. "The concept of Utopia." In *The Cambridge Companion to Utopian Literature*. Ed. Claeys, Gregory. Cambridge: Cambridge University Press, 2010. 3–27. ISBN: 9781139798839.

Zumthor, Paul. *La mesure du monde*. Paris: Seuil, 1993. ISBN: 9782021180824.

Translating the "figures of sound" of More's *Utopia* (book I) into Brazilian Portuguese

Ana Cláudia Romano Ribeiro
Departamento de Letras, Universidade Federal de São Paulo, São Paulo, Brasil
Escola de Filosofia, Letras e Ciências Humanas, Guarulhos, Brasil

ABSTRACT: The Latin text of *Utopia* is full of sound and semantic resources, few of which have actually been taken into account by its translations. Those resources are important—even if they have been scarcely noted by translators to date—because, as pointed out by Edward Surtz (1967, p. 108), if the manner in which a writer expresses himself is molded by ideas, the ideas are also molded by the tools of expression. The fact that they are inseparable was already noticed by Juan Vives in the sixteenth century, when he mentioned two reasons for reading *Utopia*, its language and its subject (*apud* Surtz, 1967, p. 109). Nevertheless, one of the least studied aspects of the *libellus aureus* is precisely the specificity of the language in which it was written, its style, and its particularities. Among them are its musicality or rather the "physical aspects" of More's language—"rhymes and rhythms" (cf. André Prévost, 1978). This paper aims at appreciating some "figures of sound" (in Surtz's words) of *Utopia*'s book I. For that purpose, we will compare the Latin passages with two Brazilian translations, the most recent Portuguese version, as well as my own.

Keywords: *Utopia*, translation, figures of sound, style, Latin

I had never been interested in the Latin text of Thomas More's *Utopia* published in 1516 and had always read it in translations. In 2002, I took up the study of French utopias of the sixteenth and seventeenth centuries and became curious about the text, with which all French utopias seemed to dialogue, at a time when Latin was still the *lingua franca* of the literate circles. This eventually led me to translate the Latin version of *Utopia* into Brazilian Portuguese. What was my surprise when I found a "musical" text full of sound features that were missing in the translations, which seem to emphasize the spirit, rather than the letter! That separation between form and meaning, however, is not consistent with *Utopia*'s actual wording, in which, as noted by Edward Surtz (and, as it happens with all great literary works), "The expression is molded by the ideas, and the ideas seem to be reciprocally influenced by the expression" (1967, p. 108). It should also be remembered that Juan Luis Vives, in 1523, recommended reading *Utopia* for both the use of language and for its subject.

In fact, one of the least studied aspects of *Utopia* is precisely the specificity of the Latin in which it was written, its style, its peculiarities. This gap, already highlighted by Surtz in 1967 (p. 107), prevails, as observed by Clarence Miller in 2011 (p. 71). I understand style here, in quite a general way, as described by Surtz in his article from 1967: "the selection and the arrangement of words and phrases." (p. 93).

I aim to present here some of the "figures of sound" (Surtz's expression) of Book I of *Utopia*, compare them to the published translations (two Brazilian ones and the latest Portuguese translation), and then suggest my own translation. I assume that the form and especially the sound mechanisms of the Latin text result from the idea that wants to be communicated and, at the same time, reinforced it (as observed by Surtz). Therefore, it is important to try to translate them, so as not to neglect this aspect of More's wit.

The "figures of sound" mentioned by Surtz are figures of repetition, i.e., they are "the repeated placement, in the discourse, of a phrasal part that already had been employed". These repeated figures "hold the flow of information and give the reader more time to emotionally 'enjoy' the information presented as important" (Lausberg, 2004, p. 166). In Book I of *Utopia*, they are usually inserted within short periods, typical of *Utopia*, which, as you may remember, was written in dialogic form. Virtually all of it is made up by Rafael Hythlodeu's speeches, who only twice utters long sentences in Book I (quite long, in fact, and called "marathon sentences" by C. Miller, 2011, p.73). We agree with Surtz when he states that "The dearth of long, complex, formal periodic sentences corresponds to the conversational and familiar nature of the discourse." (1967, p. 94)—, a feature More shares with Seneca (an author held in high regard by the Utopians, together with Cicero).

Translators of *Utopia*, such as Marie Delcourt, André Prévost, George Logan, and Clarence Miller point out the "varied language" used by More. Delcourt even hypothesizes that it is a text of oral nature, dictated by More to his secretary, John Clement, a hypothesis rejected by Aires Nascimento, although he agrees that "the pace of the sentence, in its variety, corresponds to the practice of the court bar, which More [who was a lawyer] used to perfection." (Morus, 2009, p. 150). The figures of sound certainly contribute to the varied character of *Utopia*'s language and their presence, according to Prévost, is not surprising "for an author particularly sensitive to the virtues of music." (1978, p. 249).

This colloquial style of the utopian text and even More's judgment about it, in his letter to Pieter Gillis, in which he evokes the "negligent simplicity" (*neglectam simplicitatem*) of his "little book" (*libellum*), should not lead us to believe that it is a simple text, as Elizabeth McCutcheon notes. On the contrary, the impression of natural and effortless writing is an effect More creates to achieve rhetorical sophistication, one aspect of which McCutcheon studied accurately, i.e., the use of litotes, in her article of 1971 (republished in *Essential articles for the study of Thomas More*, 1977, p. 263–274). Prévost notes that simplified syntactic structures, as well as modes and tenses that don't follow the uses of the *consecutio temporum* of classical Latin (and seem to reproduce the spontaneity of spoken language) are used side by side with formally more elaborate structures "and, to meet the 'research' of his stylistics, he goes to the limits of preciousness and to what is later called, in England, *euphuisme*" (1978, p. 243).

Let us now analyze some excerpts of Book I—that feature some patent sound-semantic resources—according to the translations of:

– Luiz de Andrade (Moore (sic), 1937; abbreviated LA—it is not known if his translation was based directly on the Latin text); this is *Utopia*'s most reproduced translation in Brazil, by several publishers; the Portuguese spelling of that edition was preserved here;
– Camargo and Cipolla (More, 2009; CC—who translated George Logan and Robert Adams' English version, 2002); this is *Utopia*'s second most widespread translation; although it has admittedly been translated from one of the English translations, I decided to examine it;

– Aires Nascimento (Morus, 2009; AN—translator of the Latin text); this is the latest and most complete edition of *Utopia* available in Portugal;
– And, my own translation, Romano Ribeiro (RR), in progress.

The number in parentheses that follows the translated excerpts refers to the page of the above-mentioned editions.

Let us analyze some examples of the figures of sound. Among these, we focus on those relating to the repetitions of word parts.

I won't fully comment on the translational options, but will only briefly mention some plays of sounds that caught my attention while I was translating *Utopia*'s Book I.

Excerpt 1 was taken from the dialogue between the characters Thomas More, his friend Pieter Gillis and Raphael Hythloday, the Portuguese sailor-philosopher. Pieter, after hearing Raphael tell about the customs and institutions of several peoples he came across while traveling the world, suggests that such amount of experience would be of high value to a king, who would benefit from his advice, which would result in improved life conditions for Hythloday and his relatives. Raphael replies that his family should not expect to improve its life at the price of him becoming a servant of kings; after all, he already shared his possessions with them.

1. – [Hythloday:] [...] *quos debere puto hac mea esse benignitate contentos, neque id exigere atque expectare praeterea, ut memet eorum causa regibus in* seruitium *dedam.* [Pieter:] *Bona* uerba *inquit Petrus, mihi* uisum *est non ut* seruias *regibus, sed* inseruias. (*CW* 54/27–28)[1].

LA: Êles não se queixarão, espero, do meu egoismo; não exigirão que, para cumulá-los de ouro, eu me faça escravo de um rei. [...] entre estas duas palavras latinas, *servire* e *inservire*, [os príncipes] vêm apenas uma sílaba a mais, ou a menos. (20)

CC: Creio que isso os deixou satisfeitos, e não vão agora insistir, e muito menos exigir, que para agradá-los ainda mais eu tenha de me tornar um escravo de reis.

"Compreendo", respondeu Peter, "e concordo; mas a sugestão que fiz foi de *serviços*, não de *servidão*." (24)

AN: Julgo que eles devem ter ficado satisfeitos com a minha liberalidade e por isso não esperarão que eu, por sua causa, me ponha ao serviço de reis.

– Palavras certas, diz Pedro; a mim querer-me-ia parecer não que deverias servir às suas ordens, mas secundá-las. (239)

RR: [...] penso, devem estar contentes com esta minha benevolência, e não exigir mais coisas nem esperar que eu mesmo, por causa deles, seja um serviçal de reis.

– Boas palavras, disse Pieter. O que me parece, em relação aos reis, é que não seria <u>servidão</u>, mas <u>serviço</u>.

The word play between *seruitium* (neutral noun, "servitude"), *seruias* (2nd person of the present subjunctive of *servio*, "serve") and *inseruias* (2nd person of the present subjunctive of *inservuio*, "serve the interests of") is based on a paronomastic organic change that confronts three words with the same root, similar in sound, and whose difference in meaning is emphasized by Pieter's answer.

Lausberg classifies as organic the changes based on the linguistic system, so words are etymologically connected. He defines paronomasia as "a word play based on the meaning of words, which arises due to the change of a part of the word body, a process which often corresponds to an almost imperceptible change of the word body, a surprising ("that causes strangeness"), "paradoxical" change of the word meaning." (2004, p. 179)

LA chose an explanatory translation, which explains the play of similar sounds, transformed into the Latin infinitives *servire* and *inservire*. The syntagma *bona uerba* was eliminated, thus suppressing Hythloday's judgment about Pieter words, but *uerba* is resumed in "two Latin words", highlighting the infinitives, calling the reader's attention to the fact that they are Latin words.

CC's translation maintains the paronomasia "<u>serviços</u>"/"<u>servidão</u>" [<u>services</u>/<u>servitude</u>] and creates another one, "<u>compreendo</u>"/"<u>concordo</u>" [I understand/I agree] (inorganic or pseudo-etymological), which is not found in the Latin text.

In AN's version, the trio "<u>serviço</u>"/"<u>servir</u>"/"<u>secundá-las</u>" (<u>service</u>/to <u>serve</u>/to <u>second</u> them) seems to be an attempt to translate, to some extent, the paronomasia, by repeating the first letters: "se-".

In my translation, I adopted CC's solution, but maintained the singular and the word order of More's text (first, the negative form, "would not be servitude", then the affirmative clause, "but service"). I also maintained the original spelling of Pieter's name to reveal his origin (Antwerp, in the region of Flanders).

The alliteration of *uerba*/*uisum* was only maintained by AN and RR in the pair "<u>palavras</u>"/"<u>parecer</u>" [words/to seem].

Excerpt 2 was extracted from a speech by Hythloday, in which he reproduces the dialogue between him, a lawyer and Cardinal Morton. Prior to excerpt 2, we find Hythloday's analysis of the reasons that lead people to steal, the main one being the breeding of sheep that become devourers of men (the picture of sheep that devour men is perhaps Book I's most famous one). Landowners, no longer satisfied with their income, turned all their land into sheep pasture, stopped cultivating it and expelled the peasants. They are the subject of the verbs in the following section.

2. – [Hythloday:] [...] ***nihil** in publicum <u>prosint</u>, **nisi** etiam <u>obsint</u>,* [...] (66/6)

LA: [...] <u>às expensas</u> do público e <u>sem proveito</u> para o Estado [...] (29)

CC: [...] que <u>em nada contribui para o bem</u> da sociedade—precisam, agora, <u>fazer-lhe positivamente o mal</u> [...] (35)

AN: [...] já **não** <u>se incomodam</u> com a utilidade pública, a **não** ser que os <u>prejudique</u>. [...] (248)

RR: [...] **coisa alguma** fazem de <u>favorável</u> ao bem comum, **coisas** <u>desfavoráveis</u> **somente** [...]

The pair *prosint*/*obsint* (two verbs in the third person plural of the present subjunctive) plays with organic changes, the prefixes pro- and ob- resulting in opposite meanings of a verb formed by the same radical (*sum*). The first three translations don't maintain the paronomasia, which, through repetition of form, emphasizes the antonymy.

Still in this excerpt, observe the repetition of the first syllable in *nihil* (noun)/*nisi* (conjunction), an inorganic change in a paronomasia that emphasizes the difference in meaning. This iteration is maintained in the translations by AN ("já não"/"a não ser") (no longer/unless) and by RR ("coisa alguma"/"coisas") (nothing/things).

Excerpts 3 and 4 belong to the same reply made by Hythloday that contains excerpt 2. They discuss the financial aspects of economic disorders arising from extensive sheep breeding: the concentration of the wool trade in the hands of a few, which keeps prices high, regardless of the number of sheep.

3. – [Hythloday:] *Quod si maxime **increscat** ouium numerus, precio nihil **decrescit** tamen (CW 68/4–5)*

LA: É verdade que o número de carneiros <u>cresce</u> rapidamente todos os dias; mas nem por isso o preço <u>baixou</u>; [...] (30)

CC: Não que os preços <u>caíssem</u>, por mais carneiros que houvesse, [...] (37)

AN: Assim, mesmo quando o número de ovelhas <u>atinge uma cifra elevada</u>, nem por isso <u>baixa</u> o preço. (250)

RR: Mas, ainda que o número de carneiros <u>acrescesse</u>, o preço em nada <u>decresceria</u>.

In this excerpt, *increscat* and *decrescit*, compound terms similar in form, not only have different prefixes (in-/de-), but also differ in tense (third person singular present subjunctive/third person singular present indicative). None of the first three translations preserves the semantic tension of that paronomasia created by the sound play of the iteration of the primitive verb "grow" (*cresco*).

4. – [Hythloday:] *quod earum, si <u>monopolium</u> appellari **non potest** quod non unus uendit, certe <u>oligopolium</u> **est**. (CW 68/6–7)*

265

LA: [...] porque si (sic) o comércio das lãs **não** é um monopólio legal, **está**, na realidade, concentrado nas mãos de alguns ricos açambarcadores [...] (30)

CC: [...] pois o mercado da lã tornou-se, **se não** estritamente um monopólio, pois isto implicaria a existência de um só vendedor, pelo menos um oligopólio. (37)

AN: Quanto a isso, se **não se pode** falar em monopólio das ovelhas, já que não é apenas um a vender, certamente **existe** um oligopólio. (250)

RR: Isso porque dos carneiros, se **é impossível** falar em monopólio (pois não é apenas um proprietário que vende), de oligopólio certamente **é possível**.

Two parallelisms are found, here. First, the sound match between the neutral nouns *monopolium/oligopolium*, a paronomasia maintained in CC's and NA's translations. The second parallelism, formed by the litotes *non potest* and by repeating the primitive verb in *potest*, sounding like *est* (*possum/sum*), was maintained by LA and AN. I tried to preserve both using "impossível [...] monopólio"/"oligopólio [...] possível" (impossible [...] monopoly/oligopoly [...] possible). Here, too, it can be said that the meaning is emphasized by means of repetition, by means of the figures of sound.

Excerpt 5, as excerpts 2, 3 and 4, is also part of Hythloday's long speech in which he recounts the dialogue that took place at a dinner given at Cardinal Morton's home. Taken from a part called comic *intermezzo* by Prévost, the sentence below involves two characters present at that dinner, a parasite mimicking a jester and a mendicant friar. It's part of Hythloday's description of that jester, who, when he realizes his success in annoying the friar by comparing him to vagabond, keeps teasing him. In the end, the fool proves wise and the friar a fool.

5. – [Hythloday:] *I am* **scurra** *serio* **scurrari** *coepit.* (*CW* 82/28)

LA: Então o nosso bufão gracejou com seriedade (43)

CC: Foi então que o bobo começou a esmerar-se em suas ironias [...] (51)

AN: O bufão começou a representar a sério [...] (263)

RR: Logo o bufão começou a bufonear seriamente [...]

The repetition *scurra* (masculine nominative)/ *scurrari* (present infinitive) is an etymological figure that produces semantic intensification, throwing light on the character and the actions of the jester. The etymological figure consists, according to Lausberg, of repeating the radical and "is used to intensify the semantic force", as in *Hml.* 3,2,1 *Speak the speech*; *vivre la vie*; *sein Leben leben* (2004, p. 181). It's a type of paronomasia.

In LA's, AN's, and RR's translations the opposition madness/seriousness (*serio*) that runs through the entire *Utopia* is maintained and produces ironic effects, since Hythloday's comparison between European countries and others, non-European ones, shows the reader that what some consider wise may be madness, and vice versa. However, the etymological figure that emphasizes that contrast was not maintained in LA's, CC's, and AN's translations.

Excerpt 6 is also part of Hythloday's long speech. It refers to the lesson of the Azorean, a people located in the south-east of Utopia, in which Hythloday describes an exemplary aspect that contrasts with the French politics of that time: after experiencing the evils arising from a territorial expansion policy, the king of the Azorean decides to content himself with his current territory. At the end of that excerpt, the Portuguese sailor advises the French king to follow the example of the Azorean who loves his people and who is loved by it.

6. – [Hythloday:] *Amet suos & ametur a suis* [...] (*CW* 90/18–19)

LA: [...] amai vossos súditos, e que o amor dêles faça a vossa alegria [...] (49)

CC: [...] amar seus súditos e fazer-se amar por eles [...] (60)

AN: [...] ter amor pelos seus e ser por estes retribuído [...] (271)

RR: Que ame os seus e seja amado pelos seus [...]

There are two inflectional changes of the word body (polyptotons), here: *amet/ametur* (the verb *amare* in the third person present subjunctive, in the active voice/same verb, in the same person and mode, but in the passive voice) and *suos/a suis* (plural masculine possessive pronoun in the accusative/ same pronoun in the ablative). I tried to translate this excerpt maintaining the polyptotons, i.e., preserving the repetition of the verb and the possessive pronoun, whose redundancy I believe to be intentionally emphatic. I also chose not to emphasize the fact that the possessive refers to subjects, a meaning that gets clear if one reads the entire former passage, so as to maintain the ambiguity between *suos/a suis* which, in this excerpt, could refer to family members, but also to his children, thus evoking the metaphor of the king as a father and his subjects as his children.

CC maintains the repetition of the verb and the alternation between the active and the passive form, but explains the referent of the possessive and doesn't repeat it. The choice of non-repetition (*uariatio*) seems to generally characterize the translational choices of the first three translators in the examples analyzed so far.

Before the next excerpt, Hythloday imagines counsellors meeting to find ways to increase a king's wealth. The first strategy is presented in **excerpt 7**.

7. – [Hythloday:] [...] *dum unus intendendam consulit aestimationem monetae, quum ipsi sit ero-*<u>*ganda*</u> *pecunia. deijciendam rursus infra iustum, quum fuerit* <u>*corroganda*</u>. (*CW* 90/25–27)

LA: Éste propõe elevar o valor da moeda quando <u>se trate de reembolsar um empréstimo</u>, e de fazê-lo descer muito abaixo do par quando <u>se trate de tornar a encher o tesouro</u>. (49)

CC: Um deles sugere que se aumente o valor da moeda sempre que o rei <u>tiver de fazer algum pagamento</u> e se reduza esse valor sempre que <u>tiver dinheiro a receber</u>. (60)

AN: Um conselheiro sustenta que se faça a apreciação da moeda, no momento em que o rei vai <u>lançar uma nova</u>, para de seguida a fazer baixar além do valor real logo que precisar <u>recolher</u>. (271–272)

RR: Enquanto um decide pelo aumento do valor da moeda quando se deva <u>desembolsar</u> em dinheiro e em seguida, pela diminuição abaixo do valor real quando se for <u>embolsar</u> [...]

Again, in the first three translations, the paronomasia containing the participles *eroganda/corroganda* in the singular feminine nominative form was not translated. LA and CC maintained the parallelism ("se trate de reembolsar/se trate de tornar a encher" and "tiver de fazer algum pagamento/ tiver dinheiro a receber") (it's about refunding/it's about filling anew and have to make some payment/ have to receive some money). In my translation, I tried to maintain the organic change using the pair "desembolsar"/"embolsar" [to disburse/to pocket].

The excerpt 8 is the conclusion of a part of Hythloday's speech in which he reasons that the impossibility of eradicating all political mistakes at once should not be a reason to turn away from public life, just as a ship shouldn't be abandoned because the winds may not be stopped.

8. – [Hythloday:] *Nam ut* <u>*omnia*</u> *bene* **sint**, *fieri non potest, nisi* <u>*omnes*</u> <u>*boni*</u> *sint* [...] (*CW* 100/2–3)

LA: [...] porque <u>tudo</u> só será <u>bom e perfeito</u>, quando <u>os próprios homens</u> forem <u>bons e perfeitos</u> [...] (56)

CC: Pois <u>as coisas</u> jamais serão <u>perfeitas</u> enquanto <u>os homens</u> não atingirem a <u>perfeição</u>. (69)

AN: De facto, não é possível que <u>tudo</u> dê em <u>bem</u> senão quando <u>todos</u> forem <u>bons</u> [...] (279)

RR: Pois não será possível estar <u>tudo</u> <u>bem</u> a não ser que sejam <u>todos</u> <u>bons</u> [...]

Two polyptotons are found here: *omnia/omnes* (nouns in the nominative and in the plural, the first one neutral, the second one masculine) and *bene/boni* (adverb/adjective in the masculine plural nominative). The regular rhythm of this excerpt is further reinforced by the alliteration in *nam/nisi* and by the repetition of the verb.

LA's translation features an *amplificatio* ("bom e perfeito" [good and perfect] instead of *bene*) following the first English translation of *Utopia* (Ralph Robynson, 1551), while CC translate *bene/ boni* as "perfeitas"/"perfeição" [perfect/perfection], which takes the *Utopia* closer to platonic thought, something suggested by More himself in several passages of his book. The polyptotons of that passage are particularly difficult to maintain during the translation: only LA managed to maintain the same verb, at the price of translating the adverb *bene* by the adjectives "bom e perfeito" ["good and perfect"]. I chose to translate *omnia* by "tudo" ["everything"] (instead of "todas as coisas" [all things], my previous choice), which has the same number of syllables than "todos" [all], just as the pair *omnia/omnes*. Translating *bene* using an adverb forced me to vary *sint/sint*, translating the first occurrence using "estar" [to be]. I also changed the word order to maintain a rhythm (*mutatis mutandis*) that's similar to the original. I keep trying to find a more accurate solution in terms of form and sound.

In AN's translation, the alliteration of *nam/nisi* finds a match in **não**/senão [no/else]. Its phrasing also seems to be closest to the original, of the four ones.

Those examples show us that in the Latin text, the figures of repetition or figures of sound, are built using terms of the same family (*scurra/scurrari*; *bene/boni*), words derived by means of prefixes (*seruias/inseruias*; *prosint/obsint*; *increscat/ decrescit*; *monopolium/oligopolium*; *potest/est*; *eroganda/corroganda*), words with inflectional differences (*amet/ametur*; *suos/a suis*; *omnia/omnes*) or even the consecutive repetition of words (*sint/sint*). Such procedures seem to aim to draw the reader's attention on both what is said and how it is said, indicating that a semantic tension accompanies these figures of repetition. I aim to translate these and other figures following the assumption that, as already shown by scholars of the *libellus aureus*, such as Edward Surtz and Elizabeth McCutcheon, <u>the form means</u>: "The devices of sound, judiciously employed, spring from idea and feeling and, at the same time, reinforce both—as they should" (Surtz, 1967, p. 104). In continuing that work, I intend to dedicate myself to the way form signifies in the narrative and argumentation of More's *Utopia*.

Translated by Christian Greis

NOTE

That reference refers to the edition of the *Utopia* published in the *Complete Works* of Thomas More, followed by the page and line numbers.

ACKNOWLEDGEMENTS

This paper was written as a part of a post-doctorate project in Classical Studies, Department of Linguistics, University of Campinas, under the supervision of Isabella T. Cardoso, whom I thank for her comments. My post-doctorate studies were funded by the CNPq—National Council for Scientific and Technological Development—from 01/03/2012 to 28/02/2013 (process 150068/2013-1). I also would like to thank my colleagues Matheus Clemente De Pietro and Bruna P. Caixeta for their critical reading.

BIBLIOGRAPHY

Lausberg, Heinrich. *Elementos de retórica literária*. Tradução, prefácio e aditamentos de R.M. Rosado Fernandes. 5ª ed. Lisboa: Fundação Calouste Gulbenkian, 2004. ISBN: 972-31-0119-X.

Mccutcheon, Elizabeth. Denying the contrary: More's use of litotes in the *Utopia*. *Moreana*, 31–32, 1971, p. 107–122.

Mccutcheon, Elizabeth. Denying the contrary: More's use of litotes in the *Utopia*. In: *Essential articles for the study of Thomas More*. Edited with an introduction and a bibliography by R.S. Sylvester and G.P. Marc'hadour. Hamden, Connecticut: Archon Books, 1977, p. 263–274.

Miller, Clarence. Style and meaning in More's *Utopia*: Hythloday's sentences and diction. In: *Humanism and style. Essays on Erasmus and More*. With an Introduction by Jerry Harp. Bethlehem: Leigh University Press, 2011.

Moore [sic], Thomas. *A Utopia*. Tradução e prefácio de Luiz de Andrade. Rio de Janeiro: Athena, 1937.

More, Thomas. *Utopia*. In: Surtz, Edward S.J.; Hexter, J.H. (ed.). *The Complete Works of St. Thomas More*, vol. 4. New Haven and London: Yale University Press, 1965.

——. *L'Utopie*. Présentation, texte original, apparat critique, exégèse, traduction et notes d'André Prévost. Paris: Mame, 1978.

——. *Utopia*. [1ª ed. 1993] George M. Logan e Robert M. Adams (orgs). Edição revista e ampliada. Tradução Jefferson L. Camargo e Marcelo B. Cipolla. São Paulo: Martins Fontes, 2009.

——. *Vtopia ou A melhor forma de governo*. Tradução, com prefácio e notas de comentário de Aires A. Nascimento. Estudo introdutório de José V. de Pina Martins. Lisboa: Fundação Calouste Gulbenkian, 2009.

Surtz, Edward L. Aspects of More's latin style in *Utopia*. *Studies in the Renaissance*, 14, 1967, p. 93–109.

Sir Thomas More's *Utopia*—glimpses of a presence in 16th century Portuguese chroniclers

Ana Paula Avelar

Universidade Aberta, Lisboa, Portugal
CHAM, FCSH, Universidade NOVA de Lisboa, Lisboa, Portugal
Universidade dos Açores, Ponta Delgada, Portugal

ABSTRACT: In the 16th century the encounter with new spaces in Africa, Asia, or America, meant for European countries a questioning of their own conventional identities. Portugal assumed then a nuclear role in the way Europe has to know the Other—different places and different peoples. Sir Thomas More's *Utopia* fictionally mirrors the Portuguese role in the unveiling of new worlds, namely through Raphael Hythlodaeus' character, the traveller who tells about his presence in an ideal land. Eventually this paper analyses the dialogue between Portuguese 16th century chroniclers and More's text.

Keywords: chroniclers; Utopia; historiography

1 INTRODUCTION

In the 16th century meeting new spaces in Africa, Asia, or America, meant for European countries a questioning of their own conventional identities. Portugal assumes then a nuclear role in the way that Europe knows the different places and peoples. This becomes clear when we analyse the changes that take place in cartography. Didactic narratives of Christian history and places give way to reality as such. Urgency of information required a rigorous worldwide configuration of oceanic expedition routes, since the sailors should follow their paths far away from the shores. This means that they needed a faithful representation of seas, oceans, and continents; besides the notion of earthly space had changed and its representation had become much more accurate (Albuquerque, 1987, 9–56).

The maps that represented the earthy space transmitted the novelties and followed the information arriving in Europe by the hands of the Portuguese navigators. That evolution is clearly present when we compare, for example, the 15th century Catalan world map (Fig. 1) and the 1502 Cantino's map (Fig. 2).

In the later we can clearly see the impact that the Portuguese navigations had in Europe, namely when we compare the representation of the African coast with the Indian coast.

Figure 1. The Catalan World map. http://www.gettyimages.co.uk/detail/news-photo/cartography-15th-century-catalan-world-map-around-1450-news-photo/122337703.

Figure 2. Cantino's map. (https://pt.wikipedia.org/wiki/Planisf%C3%A9rio_de_Cantino#/media/File:Cantino_planisphere_%281502%29.jpg).

2 THE CURIOSITY AND THE IMPORTANCE OF THE NEW SPACES ACROSS EUROPEAN INTELLECTUAL CIRCLES

They echo in Sir Thomas More's persona Raphael (Fig. 3) "... a stranger, who seemed past the flower of his age; his face was tanned, he had a long beard, and his cloak was hanging carelessly about him, so that, by his looks and habit, I concluded he was a seaman." (More, 2012, Sec 4:29) Besides, he is an educated man; as Sir Thomas More says:

...for he has not sailed as a seaman, but as a traveller, or rather a philosopher. This Raphael, who from his family carries the name of Hythlodaeus, is not ignorant of the Latin tongue, but is eminently learned in the Greek, having applied himself more particularly to that than to the former, because he had given himself much to philosophy. [...] He is a Portuguese by birth, and was so desirous of seeing the world, that he divided his estate among his brothers, ran the same hazard as Americus Vesputius..." (Ibidem, Sec 4:30).

It was not only the Brazilian coast that Raphael had met. He also travelled to Indian shores; he explored Ceylon, and arrived at Calicut, where he got a passage in a Portuguese ship, using the Cape of Good Hope Route, and returning home.

Both the conflict between classics and modern thinkers, and the emerging modern knowledge flow in *Utopia*, (Fig. 4) namely when Raphael refers the new nautical techniques and compares his sailors' skills with the utopians:

"The first vessels that they saw were flat-bottomed, their sails were made of reeds and wicker, woven close together, only some were of leather; but afterwards, they found ships made with round keels and canvas sails, and in all respects like our ships, and the seamen understood both astronomy and navigation. He got wonderfully into their favour by showing them the use of the needle [The magnetic needle of a compass] of which until then they were utterly ignorant. They sailed before with great caution and only in summer time ..." (Ibidem, Sec 4:33)

We must realize that in the 16th century Europeans considered that only the compass and the skill of European navigators and cartographers made possible the oceanic journeys, not only the rounding of Cape of God Hope, but also the novelty of the New World "America". Johannes Stradanus' 1589 (Fig. 5) engraving symbolically shows the importance of the first encounters between a European citizen and a Native American: Vespucci shows an astrolabe while America is rising from sleep.

In the first half of the 16th century, Portugal and Spain control the seas; they do not control the land, the continents. The world Europe knew

Figure 3. Raphael Hythlodaeus. Detail, The Island of Utopia', woodcut by German painter Ambrosius Holbein in 1518 edition (http://4 umi.com/more/utopia/).

Figure 4. The Island of Utopia', woodcut by German painter Ambrosius Holbein in 1518 edition (http://4 umi.com/more/utopia/).

Figure 5. Symbolic representation of Amerigo Vespucci discovering America—Johannes Stradanus' 1589 engraving. (http://www.artchive.com/web_gallery/A/%28after%29-Straet%2C-Jan-van-der-%28Giovanni-Stradano%29/Columbus-Discovering-America,-plate-2-from-Nova-Reperta-New-Discoveries-engraved-by-Theodor-Galle-1571-1633-c.1600-2.html).

turned out to be only a parcel of the world; the desire to rule the world that what was missing would enhance the next move. In early 16th century, Portugal would play a significant part in the rigorous way this new world would be described. Description was then strictly connected with quality, although a new mercantile mentality anchored in emerging techniques would provide a new quantitative approach. Some of the glimpses of this new mentality appear in the narratives of the Portuguese presence in the Eastern world.

There was a growing need of knowing what could be seen in those new spaces and an eager desire to transmit personal experiences. This desire goes hand in hand with the emerging idea of a perfect world. Portuguese authors describing their travels to Asia reveal and mix their personnel experiences with the new cosmos they meet. Both Duarte Barbosa (1996–2000) and Tomé Pires (Loureiro, 1996; Pires, 1978) mention the natives' customs in the distant lands they have reached. Garcia da Orta describes various plants and their medical applications in his *Colóquio dos Simples e das Drogas…* (Orta, 1895).

Novelty was experienced in a different way when Europeans met Western lands. The marvellous becomes a rather important element emerging in Columbus' letters (Columbus, 1992) Vespucci's *Four Voyages*, (Vespucci, 1986) or Martyr's *Decades*.

Recent explorations in the West revealed not just a New World but also a golden one where the natives were supposed to live in a state of inno-

cence. We also find the marvellous in Pero Vaz de Caminha's letter to his King, where he tells of Cabral's first expedition, and of his arrival at Brazilian lands:

> My Lord, this people's innocence, his ignorance of shame matches Adam's. Since they are innocent Salvation may be taught to them." (Caminha, 1974, 81).

As I have mentioned before European experience in the East, in the desired lands of spices, in India, was quite different. When the Portuguese narrated their presence there, they wanted to state the importance of the Portuguese Kingdom in a European political arena. Both works first describing Portuguese encounter with Asian spaces were printed in the second half of the 16th century. Their authors, Fernão Lopes de Castanheda and João de Barros, claimed priority, originality, and… their King's protection. Both actually got it. In 1551 Castanheda's first book was published under the title of *História do Descobrimento e Conquista da Índia pelos Portugueses*. In 1552 João de Barros' *Ásia, Dos feitos que os Portugueses fizeram no descobrimento e conquista dos mares e terras do Oriente*, also known as *Décadas*, was also published. (Avelar, 1997, 83–88).

The writing of History stated that collective memory had played a significant role in the struggle for power: "oblivion and silence" versus "reference and remembering" had so far manipulated collective memories, History. The importance of memory and its social function thus becomes a relevant presence. Either in the Royal Chronicles or in the Expansion Chronicles, writing History meant building and preserving collective memory. The Royal Chronicles of those monarchs who played a significant role in the policies of Portuguese Expansion, namely D. Manuel and D. João III, mirror this strategy (Avelar, 2003, 23–31).

In his Prologue to *Crónica do Rei D. Manuel*, Damião de Góis identifies what he considers to be the right method for those who take themselves as History writers, and denounces those who have not fully related facts as they were (Góis, 1949, I, 1). Writing History meant for him a careful selection of facts, an analysis of each action reported, and of each actor involved; besides it also meant decorum (AVELAR, 2003, 117–119). All this should be present in the Prologues. An ethical issue is raised here: since History always surpasses its chronicler, facts are always more important than writing. Portuguese chroniclers mention state perpetuity since History has a pedagogical function in the education of the Princes.

According to Barros and Castanheda, perpetuity also derives from the vassals' worth. History provides examples either of the analogy between "great leaders and great vassals" - in Rome and Greece -, or of those deeds which have been forgotten because the vassals did not prove to be worth of their leaders – "barbarians, Greeks, and Latin people". Castanheda stated that this would not be a danger since the deeds of the Portuguese—of the vassals—were important enough to perpetuate the Portuguese Empire.

As we have said, recording a fact implies a dialogue with memory. Writing History was underlined then by a common methodology, whose main vectors were classical references—the Past, and facts—the Present (Castanheda, 1979, I, 3–5).

It is in this context that we must evaluate the evocation of Thomas More in Portuguese 16th century texts. Portuguese navigations impact on Sir Thomas More's *Utopia* has already been approached by historians (Morus, 2006, 71–89) as Luis de Matos, "L'Utopia: Realité et Fiction". The English humanist may have read Vespucci's *Quatuor Nauigationes and Itinerarium Portugallensium*, a selection of narratives, first printed in 1508, describing the main Portuguese sea expeditions, namely those to India under king D. Manuel I's initiative. Having these narratives in mind Luis de Matos ponders on the distinction between fact and fiction, between reality and utopia:

> Rien ne prouve que l'entrevue d'Anvers entre Thomas More et Hythlodée, à laquelle certains ont cru, ait eu lieu. Tout ce que l'auteur de l'Vtopia fait dire au marin portugais est le résultat de sa lecture de relations authentiques, recueillies dans les *Quatuor Nauigationes et dans l'Itinerarium Portugallensium*, parus plusieurs années avant l'ouvrage de More. (Matos, 1991, 422)

Luis de Matos provides some information on More's *Utopia* reception in Portugal. We know that in 1640 it was one of the forbidden books by the Inquisition. João de Barros mentions it in his *Espelho de Casados*, printed in 1540, by the priest Heitor Pinto, in his 1563 *Imagem da vida Cristã*. António de Gouveia wrote an epitaph about it and Damião de Góis expressed his admiration in his letters to Erasmus. In a letter, he wrote in Padua, in December 1535, he mentions Cardinal Pole's Italian report on More's death. In his answer, Erasmus says that, although he does not know how to read Italian, he will make an effort in order to translate it. In the early days of 1536, Damião de Góis thanks him the words of praise of his *Ecclesiastes* version. John More, Sir Thomas More's son, translated into English Góis' *Legatio*, a narrative of Prester John's embassy to the king D. Manuel (Góis, 1945).

We must also bear in mind that João de Barros mentions *Utopia* in his Prologue to Ásia Third Book (Década)... (Matos, 422). Both Góis and Barros write about what was then Portuguese recent History: the first a chronicle on king D. Manuel, and the later a chronicle of the Portuguese presence in Eastern lands.

More's *chiaro/scuro* (present days England darkness and conflicts versus Utopian Ovid's *Metamorphoses* golden age of light and harmony) somehow echoes in those Portuguese texts. This strategy of binary opposition, this *chiaro/scuro* also flows in Portuguese 16th century chronicles. In the first half of the 16th century, Gaspar Correia states, in his *Lendas da Índia*, that he writes about the "India golden days" which preceded "present days iron reality". (Correia, 1975, I, 1–3) This metaphor may be seen as a subliminal criticism to the situation in India, which may have been the reason of his book rather late publication.

Present day events have an influence in the writing of History. Barros even supports the idea that the Portuguese narratives about Eastern lands should be didactic. Writing History becomes part of a new modern perspective about what immortality means. As we have said before Barros mentions More's *Utopia* in *Ásia* .... Third Book (*Década*). Portuguese navigations have for him part history, didactic, and moral function. In the First Book (*Década*) Prologue, he explained his discursive strategy; in the Second Book (*Década*) Preface, he developed what he thought to be the building process of Portuguese expansion; in the Third Book Prologue, he stated the main principles of History writing, namely the difference between classic and modern writers. At this moment, he stresses the importance of the fable. Picking up the historical example of classics such as Xenophon and Apuleius, he points the difference of the modern fable.

According to Barros a new paradigm emerges from More's *Utopia*, namely in its didactic function of teaching the English how they should rule themselves. In his evocation of a "dark present" which should be enlightened by the "sweetness of the fable" offering the milk of "moral doctrine" there's a subliminal criticism to his contemporaries; those who write without paying attention to History moral lessons, thus wasting their time, and the time of those who read them.

The modern fable provides the narrative means to reflect on the lessons of History. More, the humanist who revised the fable provides a living example, in the dawn of modern days.

BIBLIOGRAPHY

Albuquerque, Luís. *As navegações e a sua projecção na ciência e na cultura*. Lisboa: Gradiva, 1987.

Avelar, Ana Paula. *Fernão Lopes de Castanheda-Historiador dos Portugueses na Índia ou cronista do governo de Nuno da Cunha*. Lisboa: Edições Cosmos, 1997.

——. *Figurações da alteridade na cronística da Expansão*. Lisboa: Universidade Aberta, 2003.

Barbosa, Duarte. *O livro de Duarte Barbosa: edição crítica e anotada*. Lisboa: Instituto de Investigação Científica Tropical, 1996–2000. 2 vols.

Barros, João, *Ásia…Primeira Década*. Lisboa: Imprensa Nacional-Casa da Moeda, 1988. (http://purl.pt/26841/3/)

——. *Ásia Segunda Década*. Lisboa: Imprensa Nacional-Casa da Moeda, 1988. (http://purl.pt/26841/3/)

——. *Ásia…Terceira Década*. Lisboa: Imprensa Nacional-Casa da Moeda, 1992. (http://purl.pt/26841/3/)

Caminha, Pêro Vaz de. *Carta a el-rei d. Manuel*. Lisboa: Imprensa Nacional-Casa da Moeda, 1974.

Castanheda, Fernão Lopes de, *História do Descobrimento e Conquista da Índia pelos portugueses*. Porto: Lello e Irmão, 1979. 2 vols.

Columbus, Christopher, *The Four voyages*. London: Penguin, 1992.

Correia, Gaspar. *Lendas da Índia*. Lisboa: Typographia da Academia Real das Sciencias, 1858–1866. 8 vols.

——. *Lendas da Índia*. Porto: Lello & Irmão, 1975. 4 vols.

Góis, Damião de. *Crónica do Felicíssimo Rei D. Manuel*. Coimbra: Imprensa da Universidade, 1949. 4 vols.

——. *Opúsculos Históricos*. Porto: Livraria Civilização, 1945.

Loureiro, Rui M. *O manuscrito de Lisboa da "Suma Oriental" de Tomé Pires (Contribuição para uma edição crítica)*. Macau: Instituto Português do Oriente, 1996.

Matos, Luís de. *L'expansion portugaise dans la littérature latine de la Renaissance*. Lisboa: Fundação Calouste Gulbenkian, 1991.

More, Thomas. *Utopia*. New York: Minor Compositions, 2012. (http://theopenutopia.org/wp-content/uploads/2012/09/Open-Utopia-fifth-poofs-facing-amended.pdf (accessed 29/02/2016)

Morus, Thomas. *Utopia, edição crítica, tradução e comentário; estudo introdutório de José V. de Pina Martins*. Lisboa: Fundação Calouste Gulbenkian, 2006.

Pires, Tomé. *A Suma Oriental de Tomé Pires e o Livro de Francisco Rodrigues*. Coimbra: Universidade de Coimbra, 1978.

Orta, Garcia da. *Colóquio dos Simples e das Drogas*. Lisboa: Imprensa Nacional-Casa da Moeda, 1895. 2 vols.

Vespucci, Amerigo. *Cartas de viaje*. Madrid: Alianza Editorial, 1986.

Gil Vicente and Thomas More's construction of a perfect community: "Frágua d'Amor" in the imagination of a new world

Maria Leonor García da Cruz

Centro de História-FLUL, Faculdade de Letras, Universidade de Lisboa, Lisboa, Portugal

ABSTRACT: Two contemporary 16th century authors, close to the ideas of Christian humanism, one Portuguese, Gil Vicente, the other English, Thomas More, write in the first quarter of the 16th century using the created image of a perfect principality (Castle/Kingdom and the Island of Utopia). This supposedly real archetype is revealed following exploratory land and sea voyages, depending on the case. In comparison with societies whose experience moves away from that model (European societies mainly of Portugal and England), they critically and sarcastically analyse social and mental flaws by way of the use of a carnivalesque narrative discourse. At the same time, they reveal virtues desirable in political ethics and put forward practical civic solutions to correct the regime. The comparative analysis of two testimonials from that era, *Utopia* by Thomas More (1516) and *Frágua d'Amor* (Love Forge) by Gil Vicente (c. 1525), of their similarities in terms of objectives and discursive techniques but also of their own definitions of identity, enrich our knowledge of mentalities in the Modern Age.

Keywords: Gil Vicente, Principality, Public decency, Laughter, Utopia

1 AN ARCHETYPE OF THE PERFECT PRINCIPALITY

In Gil Vicente, in *Frágua d'Amor*, two travellers discuss wonders, the first about a perfect Castle (Pilgrim's speech) then about a perfect Prince (Wanderer's speech). In Thomas More, it is a philosophising Traveller, Hythlodaeus, who defines in a narrative style his impressions of his voyage to the island of Utopia.

If the first author posits that, the *voyage* results from a tradition of journeys, especially over land and symbolically religious, the second author introduces the news of journeys and sea discoveries and the experience of a Portuguese navigator who travelled across the seas of America to Asia and contacted different peoples who were unfamiliar to the European. In both cases, the voyage represents discovery and, after all, initiation in a journey in pursuit of an idealised imaginary world, albeit supposedly real.

By comparing both speeches, considerations of a moral nature, enhancement of civic virtues in addition to the images of an ethical-religious nature, in a renaissance environment, connect the works to a time of transition that characterises 16th century Europe, at least in the first half of the century.

In the play by Gil Vicente a Pilgrim is interested in a Castle founded by the mighty that had for long been praised (its fame is universal, spreading through palaces and shantytowns), built on a good spot, somewhere in Castile, turreted and walled, strong and graceful, happy. In the context of the marriage (*desponsório*) of D. João III King of Portugal with the Spanish infant D. Catarina, the castle represents the princess as being of noble and virtuous birth.

Among the virtues that safeguard the fortress, Gil Vicente registers three theological virtues—Faith, Charity, and Hope—presupposing belief and full and free submission to God, trust in eternal life and love for one's fellow man.

Although the inhabitants of More's Utopia do not follow a religion in an institutionalised way as is done in the Christian principality, the similarities to the Portuguese author's thoughts are remarkable: valuing the belief in the immortality of the soul and eternal happiness[1]; total and free submission to a belief and to a complete philosophy of public good that presupposes relations of communion of interests and peace between the inhabitants of the island.

Other virtues referred to by the Portuguese author can also be enhanced with similarities found in More. I am talking about Goodness, the Castle's tallest tower, and Wisdom. It should suffice to recall the belief shared in the writings of the two thinkers in the value of good works to guarantee happiness after death. On the island of Utopia, it becomes the duty of a human to save and comfort his fellow man.

Love (Cupid, principal captain) will have led the King of Portugal, already famous for his victories and the extension of his conquests, for having easily and quickly taken the Castle *whose gates are of honesty / the keys of devotion / accoutrements of reason / weapons of sanctity.*

In Gil Vicente, it is the final definition of the principality / castle itself with Portuguese arms / political regime / kingdom that guides its conduct by reason and by faith. The foundations of its increasingly greater power are its *merits*, i.e., the practical manifestations of a political ethics of a Christian ideal whose virtues and good works / merits are extended to the subjects and provide them with a way of salvation / eternal happiness.

The laws of the island of Utopia also have an objective of happiness that goes beyond everyday practice and the short period of time and seek to guarantee what More's work so often insists on: public safety.

2   THE FORGING OF A NEW MANKIND

Gil Vicente's work uses a time of festivity and renovation to transform positively people according to the virtues of the new queen, the prince, and his politics. Love—Cupid had already renewed bonds with the Emperor[2], altered feelings, and calmed dissensions.

It is now time to change men. How? By way of a huge forge and the intense work of the smiths, in a fitting renaissance environment. In fact, on stage are shown smiths-planets with their hammers, accompanied by young woodland girls with tongs ("gozos d'amor"), captained by Cupid.

The arrival of such an excellent Queen and the start of an era of new governors were sufficient reason for the re-forging of the *Portuguese people* and the creation of a *new world,* in the words of the god of love. The objectives emerge clearly explained: all individuals, in life, may correct themselves.

According to the god of love, Negroes, moors and rustics (villains) may alter their condition and gain happiness. Anybody can refine him or herself and enhance their value similar to transformed gold from the mine. Such aims appear to come close to the pleasure (but good and honest) that in Utopia would be the main objective of life.

It should be noted however that among the characters that in Gil Vicente are submitted to the forge to undergo change, it is especially Fernando, a converted Negro, who comes out unsatisfied. He manages to change his physical appearance, becoming white and resembling a European but not in his speech, stereotyped, that reveals him as an African. In a situation that becomes ironic and causes laughter, the essence of the truth is also

reaffirmed here, the truth that cannot / must not, in the viewpoint of the time, be changed.

Fernando's conversion did not bring him religious culture nor did it educate him in the standards of language. In addition, the Fool, in his parlance, apparently reveals himself highly uneducated.

However, the topic of education is much more complex and to it is connected the usefulness and functionality of that same education in society, in the common good safeguarded by virtuous leaders.

For whom and for what does culture serve and does education as it is oriented in the Portuguese society of the time result in friars with no priestly conscience, noblemen with no honour, corrupted justice? More also highlights the idle education and the contagion of vice in England in his dialogues. Thus, admiration for the regime on the island of Utopia that promotes learning of letters and useful things and guides the education of the inhabitants towards contempt for the ostentatious and superfluous (use of precious metal and jewels, for example).

In close relationship with the usefulness of the individuals themselves and social strata is also their productivity. On the island of Utopia only a very restricted number of intellectuals are not occupied with crafts or farming. Among them are included magistrates, chosen for their integrity and superior intelligence among the more gifted citizens, as well as priests of an exceptional nature and the prince himself. The man of the street in turn who has been trained liberally and in agriculture can only travel when his services are temporarily dispensable. The point of contact with Gil Vicente's thoughts is clear in several of his works[3]: promoting continuity in successive generations; overrating the farmer and the cattle producer; condemning social parasites, and in *Frágua d'Amor* when he creates a monk who was a former carpenter who wishes to become a nobleman and warrior (of a just war) in order to be useful for the State.

3   EMERGENCE OF A NEW WORLD BY CORRECTING A CORRUPTED SOCIETY

Among the standard characters, that Gil Vicente uses in his play is Justice. It is a vehicle used to criticise strongly one of the pillars of royal governance. It is no surprise it is chosen to figure among the facets of society that should be regenerated to build a new world. It is an increasingly hunched representation due to the weight of the gifts it receives. Corruption invalidates its symbols *par excellence*, the stick, and scales, from remaining upright.

Only after being hot forged a number of times can the chickens and partridges received as gifts be extracted as slag and with even greater difficulty large bags of money. It appears that the final desired transformation is obtained when forging a beautiful and upright woman.

Would it cease to want to answer favourably to *Gentlemen's requests*, as before it complained, by begging for smaller hands and less hearing?

The advice of the god of love is that it follows as its model the acts of Emperor Trajan. It is not known whether at any time or place Justice ever listened to and followed such wise advice. If we reflect on the words of Hythlodaeus in More's work, good advisors were few and rarely heeded for a governing practice of public good, given the use of sycophancy in many who prefer self-promotion. In the England of the time a cruel justice with penal codes was practiced that made no distinction between the severity of the punishment and the nature of the crimes.

The advice of the god of love is that it follows as its model the acts of Emperor Trajan. It is not known whether at any time or place Justice ever listened to and followed such wise advice. If we reflect on the words of Hythlodaeus in More's work, good advisors were few and rarely heeded for a governing practice of public good, given the use of sycophancy in many who prefer self-promotion. In the England of the time a cruel justice with penal codes was practiced that made no distinction between the severity of the punishment and the nature of the crimes.

Other characters that Gil Vicente puts on stage are members of the regular church, i.e., monks. One of them, Friar Funil, shows a total absence of awareness of his priestly condition, the rules of his order, and the uncontrollable desire to live a worldly life. In a similar position to his, ready to be submitted to the forge and become laymen would be another seven thousand...

Nevertheless, it is not only the Church and the lack of pastoral vocation of many of its members that Gil Vicente is critical.

The religiosity and Catholicism of the playwright resist his very often critical and even sarcastic spirit with regard to the secular and regular members of the Church of Rome (from the simple parish priest to the monk or the Pope himself) when they remove themselves from the doctrine and example of Christ. He shows an evangelical humanism in many of his works.

Several passages of More's work refer to priests as merciful men and how important they were in forging the character of youths and in correcting the conduct of the inhabitants of the island of Utopia.

Other critical observations are noted in the work of Gil Vicente by way of the creation of another monk, Friar Rodrigo. After several crafts, among

which that of carpenter in the neighbourhood of Lisbon's Ribeira, Rui Pires became a monk. He nevertheless continued to live a restless life outside the monastery. He wanted to become a layman once again. However, what kind of layman? He wished to be a nobleman. He wanted to be a warrior. Moreover, he was passionate, a vassal of the god of love.

The references he makes in his speech to the extraordinary abundance of monks in all Christendom are of great interest, especially as to their uselessness for war, by notably referring to the war against the Moors. He considers that at least a third of monks should be re-forged and transformed into lay warriors[4].

Members of an ultimate social stratum deserve in *Frágua d'Amor* the observation of spectator. I refer to nobles of different categories.

These are not only the warrior-noblemen that Gil Vicente considers of great usefulness in the Portuguese society of the overseas Expansion, through the speech of Friar Rodrigo. It is not the Marques of Vila Real either, to whom he refers respectfully by considering him a perfect gentleman and by fully dispensing him from being submitted to any transformation in the forge.

The absence of noblemen on stage is nevertheless significant. We see pages and servants that bring their messages. Such a situation on its own qualifies another part of society that follows the rules of privilege, which does not beg for a transformation but naturally expects to be served.

The representations are close to the harsh criticisms of parasitism and idleness in English noblemen made in More's dialogues.

The marquis demands that the forge be immediately taken to his residence. Another nobleman, Vasco de Foes, already ridiculed by the stratagems that he uses to appear younger, futilely orders to be transformed into a child. Finally, a third nobleman, the Earl of Marialva enquires of the possibility of being rejuvenated in his body and soul but on the condition that his money is not touched. What he really wanted was to be embellished in the forge in order to make more money without giving any to anyone else.

The latter two cases are negative characters that reveal falseness and futility, greed and lack of goodness. They are close to the representation of the corrupt and rapacious noblemen of the observation in Thomas More's speeches on the England of the time, that just like some of the wealthy and clergy, they are not content with the idle life and life of pleasure that they live nor with the income they receive that, as is stressed, give nothing to the common good, but they still steal arable space, provoke mendacity and vagrancy in the English countryside and cities, despair and violence.

On the island of Utopia, justice is established because it is a true community that had no private property and did not use currency (channelling that obtained to the national treasury).

4 CATHARTIC LAUGHTER AS A CONCLUSION

In the explanation of how the forge works, the *frágua do amor* or *the love forge,* with the ability to fix the Portuguese and make a new world emend, laughter is roused by the way candidates are enticed to change their physical appearance, the colour of their skin, even their age. In fact, what is wanted, instead of superficial and futile alterations, is to improve the conduct and moral values of the Portuguese, rendering them more truthful in their nature and upright in their conduct in accordance with the virtues shown by those governing them.

Those who seek the forge to be transformed are the strata that feel maladjusted, like the monk who wanted to be a layman and get married, or being ambitious wanted to be a nobleman, or the idle and futile nobleman who was only concerned about his appearance or he who wants to improve himself by becoming even more miserly and thieving. The love forge, however, is guided by Christian morals and only changes the condition of the monk with no vocation who wants to become useful in the war against the infidels, and straightens Justice that has become hunched, by ridiculing the idle or vile claims of the remaining characters.

Gil Vicente also uses in a carnivalesque narrative discourse, the Negro in his stereotyped and simple speech, closely connected to ideas of food, drink, and carnal love. He shows to be a slave of Venus and wants to be physically transformed to have white traits and white skin. After having been intensely hammered in the forge, he comes out a white man but instead of speaking proper Portuguese, he expresses himself in a "Negroid" form of speech from "Guinea". Sorry, he imagines the rejection he will be the target of by any woman, black or white, who will not cease to ridicule and despise him. Another carnival-type figure is the Fool, Vasco de Foes's servant, whose dialogue serves to increase the ridicule of the nobleman who wants to become young.

Circumstances are distinguished by laughter and the jester / Negro or Fool is removed when seriousness is lost in more profound matters, as also occurs in More (attitude of the cardinal when the jester mixes up vagrants with mendicant monks). In Gil Vicente, the forge does not alter the social structure (transforming the Negro into a white man) nor does it operate the miracle of youth but it only corrects how each person works for the common good.

Affectation, haughtiness, and luxury are the target of ridicule in Gil Vicente's plays, as they are by the inhabitants of the island of Utopia. This can be seen in the scene of the luxuriously dressed noble ambassadors, regarded as jesters.

Idleness, greed and evil are not rewarded but are rather addressed ironically and condemned, by seeking solutions for mankind by reversing experiences and discourses, in forms that bring the two authors closer. Both wish for public decency and that virtue is esteemed and rewarded.

NOTES

[1] On the island of Utopia atheism is absolutely rejected and Christianity begins its dissemination.
[2] King of Castile and Emperor of Germany, Carlos V, brother of D. Catarina.
[3] Gil Vicente's known published works range from 1502 to 1536.
[4] It should be noted that as far as this is concerned an extended comment could be made regarding the position of the two thinkers on war and its legitimacy.

BIBLIOGRAPHY

Ackroyd, Peter. *Thomas More: Biografia.* Lisbon: Bertrand Editora, 2003.

Afonso, Luís U. As representações da Justiça em Gil Vicente e a relação do dramaturgo com a arte manuelina. In *Teatro do mundo: Teatro e justiça. Afinidades electivas,* 4 (2009), p. 127–50.

Battistini, Matilde. *Symboles et Allégories.* Paris: Hazan, 2004.

Blanchard, Ian. Population Change, Enclosure, and the Early Tudor Economy. *The Economic History Review,* 23, n. 3 (Dec. 1970), p. 427–445.

Cristóvão, Fernando (ed.). *Condicionantes culturais da literatura de viagens: estudos e* bibliografias. Lisbon: Edições Cosmos; Centro de Literaturas de Expressão Portuguesa da Universidade de Lisboa, 1999.

Cruz, Maria Leonor García da. *Gil Vicente e a Sociedade Portuguesa de Quinhentos-Leitura Crítica num Mundo de «Cara Atrás» (As personagens e o palco da sua acção).* Lisbon: Gradiva, 1990.

——. Gil Vicente e o Império. In Medina, João (ed.). *História de Portugal.* Amadora: Ediclube, 1993, IV, p. 333–340.

——. Gil Vicente—jogo de identificações sociais num mundo de ambivalências. In C.P. Cruzeiro and Rui O. Lopes (eds.). *Ciclo de Conferências «Arte & Sociedade / A condição social da e na arte».* Lisbon: Faculdade de Belas Artes—UL, 2011, p. 72–80.

Cruz, Maria Leonor García da, Maria José Teles and Susana Marta Pinheiro. *O discurso carnavalesco em Gil Vicente no âmbito de uma história das mentalidades.* Lisbon: GEC Publicações, 1984.

Ferguson, Arthur B.. Renaissance Realism in the "Commonwealth" Literature of Early Tudor England. *Journal of the History of Ideas,* 16, n. 3 (Jun. 1955), p. 287–305.

Martins, Mário. *O Riso, o Sorriso e a Paródia na Literatura Portuguesa de Quatrocentos*. Instituto de Cultura portuguesa. Lisbon, 1978.

Monleón, José B.. De la utopía política a la utopía ética. In Diago, Nel and José B. Monleón (eds.). *Teatro, utopía y revolución : Acción Teatral de la Valldigna IV*. Universidad de Valencia, 2004, p. 117–136.

More, Thomas. *Thomas Morvs: Vtopia*. (Intr. José V. de Pina Martins; Cr. ed., transl. and notes Aires A. Nascimento). Lisboa: Fundação Calouste Gulbenkian, 2006.

Mumford, Lewis. *História das Utopias*. Lisbon: Antígona, 2007.

Pereira, João Cordeiro. A Estrutura Social e o seu devir. In Dias, João J. Alves (ed.). *Nova História de Portugal: Portugal do Renascimento à Crise Dinástica*. Lisbon: Editorial Presença, 1998, V, p. 277–336.

Riot-Sarcey, Michèle, Thomas Bouchet and Antoine Picon (eds.). *Dictionnaire des Utopies*. Paris: Larousse, 2002.

Souza, Marisa de Assis. Uma Forma Curiosa de falar em Lisboa: dos imigrantes africanos do século XVI e de certos aristocratas do século XIX, retratados por Gil Vicente e Eça de Queirós. In *Anais do IV CLU-ERJ-SG*, Ano 4, n. 3. Rio de Janeiro: UERJ-FFP-DL, 2007, p. 1–15. http://www.filologia.org.br/cluerj-sg/anais/iv/completos/comunicacoes/Marisa%20 de%20 Assis%20Souza.pdf.

Teyssier, Paul. *A língua de Gil Vicente*. Lisbon: Imprensa Nacional-Casa da Moeda, 2005.

Vicente, Gil. *Copilaçam de todalas obras de Gil Vicente* (Pref. Maria Leonor Carvalhão Buescu). Lisbon: Imprensa Nacional-Casa da Moeda, 1984, Vol. 2.

——. *Obras Completas*. (Pref. Marques Braga). Lisbon: Livraria Sá da Costa, 1943, Vol. IV.

Vieira, Fátima. The concept of utopia. In Claeys, Gregory (ed.). *The Cambridge Companion to Utopian Literature*. Cambridge University Press, p. 3–27. http://dx.doi.org/10.1017/CCOL9780521886659.

Gregorio Lopez: An alter(-)native image from a 16th century globalising world

Lia F.A. Nunes

Department of Christianity and History of Ideas, Graduate School of Theology and Religious Studies, Rijksuniversiteit Groningen, The Netherlands

ABSTRACT: Out of a biographical approach to history, I will present us some thoughts related to the ideas of collective identity or sense of belonging bridging the early- and our perhaps-post-modernity. The example of Gregorio Lopez [c. 1542–1596] is somehow paradigmatic of a re-search or attempt of a better understanding of what the historical narrative about modernity entangles, but also to an informed critical position in order to disentangle it. How Gregorio Lopez understood the idea of belonging to a community; how he lived the environmental change of the *Americanisation* of the "western indies"; but also how we can share perspectives now with a '(non) religious' character from the 16th century Atlantic world; these are the challenges for this time and space trip. Living in, what we now see as, the (un) clear border between the medieval and the modern times, Gregorio Lopez shared his perspectives in ways most of his contemporaries could not imagine. Can we?

Keywords: early-modern globalisation, 'alter-native', Gregorio Lopez, 'Americanisation', historical imagination

1 WHO IS THIS PERSON?

Can we continue on the road for Utopia living in-against-and-beyond the Apocalypse? The story of Gregorio Lopez is the one of a walker in Modern World History of/for both non-places: five hundred years ago, Thomas More's book was being printed, and many saw old worlds being destroyed and other/new worlds being revealed. In order to deny a monolithic depicting of Gregorio Lopez—or any other 16th century man—we have to proportionally harmonise the frontiers of the imaginary in those literary *loci* and *topoi*, admitting his lived experience as part of the realities created and enacted through the readings and interpretations of both re-presentations of the future-real.

We know where he died: in Santa Fe of Mexico City, one of *pueblos-hospital* created by Vasco de Quiroga in New Spain, inspired by the reading of Thomas More's *Utopia*. Did he share the Utopia? On the other hand, did he felt living the Apocalypse? In the next few pages, we will dive in his story to question its "modernity". I will focus on the identification processes that were put to work to control people's movements by then, to try to understand what is still present of such processes.

The official/traditional/historical narratives go like this: Gregorio Lopez was born in Madrid (in Spain) or in Linhares da Beira (Portugal), prob-

Figure 1. [Portrait of Gregorio Lopez] Print; 121 × 158 mm. Matías de Irala. Ilustración de Vida del Siervo de Dios Gregorio López...por el P. Francisco Losa. Fourth Edition. Madrid, 1727. Biblioteca Nacional de España, IH/4951/3.

ably in 1542. Supposedly, he crossed the Atlantic when he was around 20 years old; and in what was called New Spain—a "discovered" realm, essential pièce in the game of thrones of the desired Western Indies-, he started enduring a solitary and virtuous life. For three decades of such endurance, notwithstanding the different geopolitical regions where he lived; the diverse people he connected with; his own physical and mental conditions; the Catholic Church and the Hispanic Monarchy decided to endorse his canonization.

The "heroic" virtues for which he was almost sanctified probably do not appeal much to our contemporary taste. His intellectual activities are much more attractive: not only he wrote a commentary about the Apocalypse's book, as we also can appreciate his pharmaceutical compendium—a work where he popularized the recent developments generated by the contact between European and native botanic, by translating the Latin-produced knowledge into Spanish. Not works of genius, but important for the time when knowledge was eager to break through, when press was massifying, when people like Gregorio could produce their own world-maps and earth-globes—as several people affirmed he did.

For centuries, he has been compelling people to read his exemplary life, written by his hagiographer, Francisco Losa; and to use him as a time capsule to the second half of 16th century, and to several places. Although he was inscribed in History as a religious figure (no matter how heterodox, or polemic, or indefinable), Gregorio Lopez is for sure one bridging-worlds man. Why?

1.1 *An alter-native…*

Gregorio Lopez struck me as a fellow compatriot, from Serra da Estrela. Therefore, diving into his historical biography I almost drowned in the matter of his 'nationality' or, better put as Portuguese and Spanish languages did in the 16th century, in the questioning of Gregorio Lopez's *naturality* (*naturalidad*). It could not only be *fait-divers*, after all, he kept the secret of his birthplace and family his entire life. Not even his biographer was able to get this mystery out of him; fuelling multiple possibilities: Castilian, Portuguese, Old Catholic, Converso, or Jew, even son of kings.

I tried to overcome the historicity of the problem: the concept of *nation* and community existed; and to cross the Atlantic in the 16th century (… till now) every passenger had to state it. Officially, by order of the King, all migrants had to prove: where one was from, what his creed was, what her social and marital condition was, what their profession was. At some point, Gregorio either fulfilled the required conditions to pass legally—

and we lost the sources attesting it—or he went to the 'indies' in an irregular situation. Nevertheless, from the moment he set his feet in the 'new world', he could be who ever he decided to be. Apparently, so he did.

From a place where borders were still so fragile, he passes to another where there was not even a clear notion of where the land ended. A place we would call and we turned into America, alluding an idea of unity of many different lands, different people, different geographical structures, different cultural habits, different languages, and polis—all native, natural from those lands.

Gregorio's (hi)story then invites us in that process, but in a slight different manner. In fact, when asked where he was from and who his parents were, he got used to respond: *my father is god, my homeland is heaven*. Obviously, it is a mystical answer; but can there be something more than the theological translation of his enigma?

There is more to it than the obvious religious mind-set in talking through such an epitome. Firstly, and as all the other "European" migrants, Gregorio Lopez was the *other*—at least to all the million people who were *natural* from the territories occupied and conquered by the 'Spanish' (the *Spanish* is somehow also a creation of this time and process). Then, being there as Gregorio did, from the early 60 s, the differences between the natives from Tlaxcala, Mexico, the diverse nations of the Gran Chichimec (Northern Mexico), from the Huaxteca (East and Northeastern Mexico) et al., were differences easily noticeable and practical for anyone paying attention.

The generalization was easy to be made though through asserting the natives an impossible *alterity*: their worldview, their perspective was completely downsized. Their religions but also the absence of a religious type of thought, their paganism, their rituals, their monuments; in a way their natural culture was not only not understood, it was demonized. It was then easy to declare inferiority; and to open the room all sorts of prejudices—even when it was possible to acknowledge some degree of sophistication or to admire the diversity. The greed for power at the expense of exploring the dignity and freedom of the other was there in Iberia, and in New Spain, between all of those nations and groups. Gregorio stepped aside one place and the other, and imagined himself as an alter-native, a native from another place, as heaven could be.

1.2 *…in those places we called America*

Getting back, and getting some other questions aside (because the 'Indians' from the western indies weren't all friends playing around; because not all migrants were living-demons; because there is no such thing as saints) Gregorio Lopez saw many things we can bridge to ours places and times and concerns. He saw ecological devastation provoked by his compatriots' domination: massive mining, enormous deforestation, and cattle-production causing big disequilibria, destruction of centennial sustainable urban and rural practices. Not to mention all the pervasive social and political changes from the first decades of contact to a colonial system eager to sub-and-ex-tract more profit out of such wide new territories.

Unfortunately, we know those stories too well. They became familiar, part of our DNA—not only because we are learning to accept the past; sadly we are also too aware that all of it is still happening! I do not recall who referred peoples from America do not mind being called *Indians* for it actually recalls how the *Europeans* were mistaken. And I also bring to this conversation Eduardo Viveiros de Castro, when he keeps reminding us of taking other perspectives; like the Amerindian one—we all know the Brazilian's anthropologist historical-imagination exercise of trying to picture the end of the world, the apocalypse…"ask the Indians", he says. For now, let us just think how long it took for the New World to stop being called the Western Indies; and try to consider how long it will take before it ceases to be called America.

Anyhow, that land was good, wide and wild enough for Gregorio Lopez to walk continuously through it, learn from it, from those who lived there—who knew he was not from there, but recognized him as one, like them. For instance, and although we are not sure of a change, Gregorio became noticed by his vegetarian diet: he would reject offers of meat and fish, coming from the settlers production or from the natives hunters; but he would gladly accept offers of fruit and exchange his own gifts/competences/skills by a piece of land where to grow his house-garden.

In those places, we called America, in those artificial entities we call, for instance, Mexico, there are too many alternatives. Living them without a special will to explore, evangelize, enslave, mine, took Gregorio to various worlds. The Desert, the Jungle, the Volcano, the Lagoon; the Wind, the Water, the Earth, the Tree, the Colour, the Light, the Sun, the Moon: did they convert him? Was he able to recognize them as such, with their big capitals?

Julio Glockner advanced the possibility that the reason why the Popocatepetl' Volcano is also called Don Goyo—from Gregorio—might be connected with the canonization of the first hermit of the Indies. Just as Iztaccíhuatl called Rosa after Rosa de Lima; and seen that Gregorio lived at least for two years in Atlixco, a town near the Vulcan; it is not such a farfetched hypothesis. And

how amazing it would be! It is still so overwhelming now to walk between the two lover-volcanos; that it is hardly impossible not to understand the power and sacredness asserted by the natives to those places.

Yet, their voices were silenced, for the impossibility to communicate and to understand—Pannikar targets the problem, in my opinion, by understanding colonization not as a moral evil but as an intellectual mistake. Gregorio Lopez was not the only silent voice (perhaps his silence echoing in time like a scream) criticizing the state or status of things by not engaging with it. There were already other utopias in place. How?

1.3 *Religiously online…*

What we know is he never took a Religion: translating from the 16th century meaning, he never took the vow of any religious order. Besides, there are too many uncertainties on what he believed or not—some say he was a crypto-Jew, others say he was a hidden *alumbrado*, the Church tried to make him a saint- and it is time to ask whether we really care. He definitely possessed a strong intellectual character, for no matter his ailments (from depression to typhus, dysentery, and probably inflamed throat for all that he had to swallow), Gregorio was always capable to communicate, to recognize the uniqueness of any interlocutor.

A quick look through the list of processes and accusations present to the Mexican Inquisition, but also a careful reading of multiple sources, take us to picture we are not used imagining. Disguised in the sources, as blasphemy or conversion, there are multiple stories of critical voices and dissatisfied people. Monks that became miners, soldiers that became priests; the roles were not that ecstatic.

It is said that it was God who made him walk; but Gregorio's feet made the steps of the way. We find him in a modern time not only because of the distances he travelled physically. What an adventure must have been for Gregorio Lopez to cross the sea and experience the radical changes in the worldview that proceeded from the *planetarization* of the Earth. It was already mentioned he made his own earth-globes and world-maps. He discussed the position of the North with seamen. And he went to places completely unheard of few decades, even years, earlier.

(T)his life happened in what is commonly called Early-Modernity. Gregorio fully lived a new time for the 'socialization' of knowledge, ideas, ideologies: science and technology, monotheisms, capitalism—all turned read, theorized, practiced, interpreted through easily compiled in small and easily transportable objects, called books! All the worlds he could now visit inside a story, a treaty, a novel, a study, a *vita*, a chronicle.

Gregorio was able to visit even worlds that could only be constructed because they were imagined through a book. Gregorio lived in one of those places, in the last years of his life: the pueblo-hospital of Santa Fe, in Ciudad de Mexico. Long story short, the *pueblo-hospital* was the political/social/economic translation of Vasco de Quiroga's reading of Thomas More's *Utopia*. Vasco de Quiroga, whom Gregorio might have encountered when he first arrived in Ciudad de Mexico, believed it was possible to create the ideal society *with* the Indians: there would be a clergyman supervising the catholic and Christian morality/identity/principles of the community, but for the rest all its citizens were provided work, communal tasks, physical and spiritual assistance. With all its flaws, we can see from now/here; these communities were oasis in a completely different broader panorama—where there was *one* imperial policy that could not work out the same way so many places with complete diverse situations.

Either rejecting or embracing the new, Gregorio Lopez was online with his time. As Hallier Dueñas explained me about the changing time we are living now with the internet, teaching me the obvious—we/I belong to the last generation that grew up without the world wide web -; Gregorio must have felt the other way around. Gregorio was the first generation living in a world with books. We can go even further, if we think of the *nau* and the *galleon* as the early-modern spaceships—what was dreamed of, out of necessity and creativity, became reality.

1.4 *...or potentially exemplary?*

The first time I wrote about Gregorio Lopez, I entitled the text with adjectives in the form of a question: "unadapted or atemporal?" and after years of research, the question starts to fade. An initial rush to identify or classify the man transformed into an immense curiosity on the processes of identification of his time and spaces; moreover, on the way he apparently defied those processes. For now, let me go back to Viveiros de Castro and his appeal for a distinction between what is a model and what is an example.

Avoiding slaves when it was currency; avoiding eating meat when it was proudly introduced to others who barely used them—Gregorio retired from the dealings of the Indies that stayed in the winner's History. For sure, Gregorio would question the search for official recognition of his sanctity. He was no hero; he just knew he could not rest if there was one of his brethren in danger, suffering. Yet, was he not a hero everyday?

Gregorio was also an *Indian*, in the sense of *native*, natural from a place: a place he lost, run away from, missed; we do not know. For sure, he moved, and became an alter-native; in-against-and-beyond any nation and community. Nonetheless, he became a part of those new spaces, and he even claimed belonging to another one—heaven.

Amongst others, this was one of the reasons—I believe—the Catholic Church would not canonize Gregorio Lopez. Turning him into a Saint would make him a model of virtue, and a model we apply, we imitate. For the 17th and 18th century cardinals, Gregorio was a servant of God worthy of admiration but not imitation: it was dangerous to proclaim so independently a solitary and such a close connection with god—and although they could not stop that idea, they obviously could not sponsor it. The paradoxes of a closed process of identification of the modern period: for what was/is the Catholic community if not a nation without borders? A nation in expansion closely sponsored by one side of a divided 'Europe', which only relaxed after starting the secularization process. Alternatively, was Modernity the process of questioning, deconstructing, and disentangling of that Christian identity that, from Western Europe, colonised inner and outer worlds?

Albeit the macro-histories/pictures, could Gregorio be a potential example of life in-between worlds, times, spaces, changes? In his commentary of the Apocalypse, he conveys a trust in the future of the *homo viator*, of the man who walks alone through life. Walking, individually, finding ways through contemplation of the divine—that ultimately is everywhere. Online in the virtual world connected by satellites, with no unbreakable walls. All in god, by oneself; I wonder what he would say of life now... are we living Gregorio's utopia?

Again, what is the role of religion still? Some say it is definitely over; some say it is re-entering our lives more than ever. Does it even have to be assigned a role? However, what is religion, after all? Could your diet become a religiosity in the way we choose to prepare, produce, cultivate, recollect, and eat whatever we eat? Could our work entail a certain mindset or worldview that allows us to identify and avoid people in our own or between others' professional networks? Can politic/social/climate/gender/race-related activism be ascertained some degree of sacredness for the intense solidarity that generates inside the movements, the communities, the fighters?

2 GLOBALISING SOCIAL AND HISTORICAL IMAGINATION

Gregorio Lopez lived the Utopia that was transformed in common sense: ships crossing the seas, the centrality of the Sun, a breaking through worlds and frontiers of the imaginary trying to meet limits or borders in new-old realities. More. He saw the Revelation of the nuclear bomb effects: millions dying of illness, greed, devastation, hunger, violence, traffic. Is there any good left?

The bridges between times, spaces and peoples are what animate me the most when I stop to hear Gregorio Lopez's voice in the echoes of the universal mind, the *cosmoteandric* reality, in the historical-architectures of time. Always getting online in the world wide web of authors and stories through the books he read, Gregorio Lopez was also sharing that knowledge to all sorts of others—and the books had just started, and were just starting to become targets of conservative reactions towards the new and the unstoppable socialization of ideas and knowledge. Laurie Anderson reminded us that the book was as much a technology as a computer is: we make it alive, we use it, and we act upon/through it.

Using is different from collaborating, valorising is not consuming. Santa Fe is a good metaphor of the *longue-durée* of *our* modernity. As said, this was one of the autonomous islands of alternative settling built by Quiroga, where Gregorio spent the last years of his life receiving all kinds of people with all kinds of questions: women in search of help; man looking for answers, even a viceroy looking for a bit of peace (I like to think it is no coincidence that during this period the Chichimec war was put to an end by Luis de Velasco, the Youngest; a man seen talking for hours with Gregorio whom, about the same 'Indians', would only say "let them be"!) Nowadays, Santa Fe is not only one of the more contaminated places of Ciudad de Mexico; it is also the high finance neighbourhood of the whole country. A place from where the best waters would come to the city, a colourful place turned into concrete-money grey. Yet, the ruins of Gregorio's hermit are still there, in a rare oasis of green, trees, spring-water that a part of the community wants to preserve and remember.

Gregorio Lopez lived enough even to see the Utopia vanishing... forbidden books, forbidden people, forbidden. The Time is now, and only in the *now* I can somehow relate to his actions. It was too dangerous to be him, to say what he said, not to do what he did not do during his lifetime. He actually built his own utopia, his non-place without crosses, with earth-globes. In addition, fortunately, he had no martyr vocation, and there I find a way to connect him here.

It is still a hard choice to be-come critical on what we eat, on what we do, on what we are told (by the institutions, by society, by History) that we are. We are all online already, and we are getting ready to colonise Mars, so they say; but we still

Figure 2. Detail of Gregorio Lopez's commemoration mural in the Church of Santa Fe, in Ciudad de Mexico. Photographed by the author.

can turn a sea into a cemetery. We continue to talk "us" and "them", and they are all too confused with Good and Evil. If we are going to space, if we find other beings out-there, will we continue to use terms such as "western", "Christian", "European", "American", "Portuguese", and all that jazz? We might, but probably we will be from the planet Earth first. Therefore, what does it mean still in here, in this globe?

More than an example, Gregorio is an image we can dialogue with to imagine something in-against-and-beyond our own now. (In that new world, who cared if he was from Madrid or Linhares da Beira: "are those cities; where are them? Portugal, where is it, in Spain? You come from a village? Are there places as small and as lost as ours in Europe? Cities, like Mexico? Is it cold, is it hot, is it dry, is it humid?"—Gregorio was from the other side of the sea, where heaven, the sky was the same.) Lennon taught us some verses on imagining; and in a concrete-bridge entering Bremen, we can read: "the only good nation is imagination".

Well, right now, challenging the social and historical ideas of nation/nationality, production, and consumption systems, even the human constructions of religions, societies and Histories; can we be alter-natives? Can our screams of anger and our dreams hope become realities? The Man walked, Gregorio, until the Moon, you know. However, some insist in building and hiding behind the walls. What if we make a step back before we destroy and cross them? What if we contemplate before?

Imagine that!

ACKNOWLEDGEMENT

And so many beautiful conversations, in and out side Academia; too many to mention them all. Impossible though not to name David Areola,

Hallier Dueñas Morales, Thomas Hillerkuss, João Paulo Oliveira e Costa, Mirjam de Baar and Mathilde Van Dijk, Julio Glockner, Edith González, Manuel Rozental, Marian e Marinella, that taught me to re-think the world.

BIBLIOGRAPHY

Archivio Secreto Vaticano. Congre. Riti Processus Mexicana Beatificationis et Canonizationis Gregorii Lopez, Primi Anacoretae Indiis Occidentalibus. Vol. 1704–1736.

"A escravidão venceu no Brasil. Nunca foi abolida". Entrevista de Alexandra Lucas Coelho a Eduardo Viveiros de Castro. *Público*. Published 2014-03-16. https://www.publico.pt/mundo/noticia/a-escravidao-venceu-no-brasil-nunca-foi-abolida-1628151

"A revolução faz o bom tempo" conference by Eduardo Viveiros de Castro, in *Os Mil Nomes de Gaia*, 18-05-2015, accessed in 2016–02–10 in https://www.youtube.com/watch?v = CjbU1jO6rmE.

Biblioteca Nacional de España. Sala de Recoletos [*Información sumaria que se hizo en México de las virtudes y milagros del venerable Gregorio López en el año 1620 y siguientes, a petición de Felipe III para su beatificación*]. http://bdh.bne.es/bnesearch/detalle/bdh0000056044.

Bodian, Miriam. *Dying in the Law of Moses. Crypto-Jewish Martyrdom in the Iberian World*. Bloomington & Indianapolis: Indiana University Press, 2007. IBSN: 978-0-253-34861-6

Cardoso, Jorge. *Agiológio Lusitano*. Lisboa: Oficina de Henrique Valente de Oliveira, 1657. Tomo II.

Derrida, Jacques. Above All, No Journalists!. In Vries, Henk de, Weber, Samuel (eds.). *Religion and Media*. Stanford: Stanford University Press, 2001. P.56–93. ISBN: 9780804734974.

Guerra, Francisco. *El Tesoro de Medicinas de Gregorio Lopez*. Madrid: Ediciones Cultura Hispanica del Instituto de Cooperación Iberoamericana, 1982. IBSN: 9788472322967.

Holloway, John. *Change the world without taking power. The meaning of revolution today*. 3rd ed. New York: Pluto, 2010. ISBN: 978 0 7453 2918 5.

—— *Crack Capitalism*. 1st ed. New York: Pluto, 2010. ISBN: 978 0 7453 3008 2.

Huerga, Alvaro. Edición, Estudio Premilinar y Notas por Alvaro Huerga. In Lopez, Gregorio. *Declaración Del Apocalipsis*. Madrid: Universidad Pontificia de Salamanca, 1999. Coleccíon "Espirituales Españoles", Série A Textos, Tomo 46. IBSN: 84-7392-425-8.

Losa, Francisco. *La Vida que hizo el Siervo de Dios Gregorio López en algunos lugares de esta Nueva España*. México: Juan Ruiz, 1613.

Milhou, Alain. Gregorio López, El Iluminismo Y La Nueva Jerusalem Americana. In Sarabia Viejo, María Justina (coord.). *Europa e Iberoamérica, cinco siglos de intercambios*. Sevilla: Junta de Andalucia, Consejería de Cultura, 1992. Vol. 3, p. 55–84. ISBN: 84-604-1354-3.

"On the dark side of history: Carlo Ginzburg talks to Trygve Riiser Gundersen", *Eurozine*, Published 2003-07-11. http://www.eurozine.com/articles/2003–07–11-ginzburg-en.html_.

Panikar, Raimon. *Ecosofia. La saggezza della Terra*. (translated by Rivarossa, Dario). Milano: Jaca Book, 2015. ISBN: 978-88-16-30546-5.

Ragon, Pierre. *Les saints et les images du Mexique (XVIe-XVIIIe siécles)*. Paris: L'Harmattan, 2003. Coll. Recherches-Amériques latines. IBSN: 2-7475-4941-0.

Rubial Garcia, Antonio. La Hagiografia, su evolución histórica y su recepción historiográfica actual. In De Peralta, Doris Bienko; Bravo Rubio, Berenise (coords.). In *De sendas, brechas y atajos. Contextos y crítica de las fuentes eclesiasticas siglos XVI-XVIII*. Mexico: Inah, Conaculta,Promet, 2008. ISBN: 9789680303267.

Rubial Garcia, Antonio. *Profetisas e Solitarios. Espacios y mensajes de una religión dirigida por ermitaños y beatas laicos en las ciudades de Nueva España*. Mexico: Fondo de Cultura Economica, 2006. IBSN: 9789681679842.

Salma Lavie "West, East and Bureaucratic Culture", Lecture on the 23rd February 2016, organized by the Center for Religion, Conflict and the Public Domain—Faculty of Theology and Religious Studies/RUG.

The City of the Sun: The specular reverse of the counter-reformation

Carlos Eduardo Berriel

Instituto de Estudos da Linguagem, Universidade Estadual de Campinas, São Paulo, Brazil

ABSTRACT: Tommaso Campanella is the author of a utopia that reveals, more than any other one, the complex issues of both the Reformation and the Counter-Reformation, as well as the entire spirit of that time. In *The City of the Sun* (1602), he describes his ideal republic that would, in his view, be the political model that could save the Church from the setbacks of the Reformation and lead it to the top of the kingdom of this world. His idea of salvation of the Church, however, diverges from the predominantly Jesuitical and hegemonic vision produced by the Council of Trent. His theocratic polis is based on Bernardino Telesio's philosophy of nature—the same philosophy that had brought Giordano Bruno to the stake shortly before and would soon bring Galileo before the tribunal of the Inquisition.

Keywords: Campanella, Utopia, Counter-Reformation, Colonial system, Scientific revolution

1 INTRODUCTION

Tommaso Campanella is the author of a utopia that reveals, more than any other one, the complex issues of both the Reformation and the Counter-Reformation, as well as the entire spirit of that time. In *The City of the Sun* (1602), he describes his ideal republic that would be, in his view, the political model that could save the Church from the setbacks of the Reformation and lead it to the top of the kingdom of this world. His idea of salvation of the Church, however, diverges from the predominantly Jesuitical and hegemonic vision produced by the Council of Trent. His theocratic polis is based on Bernardino Telesio's philosophy of nature—the same philosophy that had brought Giordano Bruno to the stake shortly before and would soon bring Galileo before the tribunal of the Inquisition.

The structure of this utopian polis is large, complex, and audacious. To understand it, one needs to relate it to the historic period marked by a religious crisis that includes the reorganization of the Catholic Church after the Council of Trent, the consolidation of the Spanish monarchy, the French wars of religion, the transition from mercantilism to manufacture and, eventually, the scientific revolution. Just as Thomas More shrewdly read the reality of his time, so did Campanella and he portrayed it in *The City of the Sun*.

One of the most traumatic aspects of that era is the hostility of the Church against the spirit of scientific discoveries. Although it had remained objectively neutral to those discoveries in the beginning, as well as to the philosophy that produced them, from the Council of Trent, the Church identifies the scientific revolution as one of the breeding grounds of Protestantism, as its thought emancipated science from religious culture and made man—since Pico della Mirandola and Erasmus—an analogue of God, fully capable of establishing sufficient contact with the Creator of the world and his work without the intervention of an intermediary—the Church.

The scientific revolution in progress was based on the empirical axiom that the truth of science was achieved by means of the human senses, the seat of experience, shifting the criterion from the truth of Revelation to the sensitive material structure of man: the senses, which before had been a source of error and perdition, now host the sound knowledge of the physical world. As a result, science becomes autonomous, breaks away from the theological field, actually disallowing the Church and making it superfluous in this field. This is what Telesio advocates.

In principle, there was no opposition, rather a process of making scientific reason autonomous from faith. The Catholic decision to turn into opposition that which was merely automatization was taken at the Council of Trent and had severe consequences—the most obvious one being the violent persecution of scientists, including the bloody episodes commonly known. After all, this entire process started—or seems to have started—in absolute connection with the Protestant Reformation.

The fact is that the Reformation weakened the power of the Church in a substantial part of Europe; only Portugal, Spain, and Italy remained unaffected. It is also true that those three countries, the major European powers at the outbreak of the

Reformation, would become the poorest and most deficient countries of the continent a few centuries later. The catholic Tridentine complex—Italy, Spain, and the subordinate kingdom of Portugal—refused to adopt the openly bourgeois procedures that, by way of the new scientific reason, would have led their economies to manufacture and the free development of trade. The triad of those Catholic countries closed itself into the circuit established by the Inquisition, resulting in an economic system based on colonial exploitation. Therefore, inquisitorial irrationalism also became an economic and political system.

It can be argued that this decline could not have been suspected. However, Campanella's *The City of the Sun* seems to dispute exactly that way.

To put it briefly, after the Reformation, Europe is primarily divided into two blocks:

1. The catholic nations, which also are the metropoles of their tropical colonies and which are going to freeze in the regime of the colonial pact, and
2. The protestant nations, still devoid of a colonial empire and which, over time, are going to develop their economies towards manufacturing—making them more modern and open to future industrialization.

The Tridentine church is thus the expression of the pact with the Iberians states, which it depends on and is part of—in other words, the Church mingles with the Iberian state. Rome's fate is subsumed in the process of mercantile capitalism, which is its economic expression and, at the same time, its limits—a scenario that will also define the social character of the Iberian colonies. That choice, as we see, entailed the exclusion of manufacturing, which was closely associated with bourgeois rationality, as a hegemonic decision made by the catholic economic circuit. The result is a Jesuitical missionary Church that opposes science and actively engages in the inquisitorial processes, which ultimately decides the fate of the Iberian metropoles and their future aspect. The Church's destiny is thus closely tied to the colonial pact, to this particular form of Iberian mercantile capitalism, leading to the incorporation of new territories occupied by the Jesuit missions. After all, it was not the Church that ran the Inquisition, but rather the Inquisition that ran the Church.

The opposite, which did not take place, would there have been a Church that was opened to the new sciences, that encouraged its modern investigative processes, welcomed unprecedented books, keen on new knowledge. In short, a Church that would tie its fate to the manufacturing bourgeoisie and, as its ally, seized the throne of the World.

What could have taken place, but did not, is described in *The City of the Sun*.

2 THE CIRCUMSTANCES OF A PIECE OF WRITING: THE REVOLT OF CALABRIA

Before *The City of the Sun*, Campanella had prepared a grandiose and bold project. The advent of the Lutheran Reformation had deepened the gap between the temporal and the spiritual powers, which Campanella opposed. In his view, the Christian world should have a single government, exercised by an authority both sovereign and sacerdotal—i.e., the Pope. In practical terms, Campanella argued that a powerful European sovereign should invest his economic, political, and military resources to achieve that universal monarchy and put it at the feet of the Roman pontifical throne. He suggests that mission first to Philip II of Spain and then, 40 years later, to the king of France, Louis XIII, and Cardinal Richelieu.

The City of the Sun was written to demonstrate and prove that the union between faith and reason, between the Church and the new science would rationally lead to caesaropapism and that the path of the Roman symbiosis with the Iberian metropoles, optimized by Jesuit action, had turned faith and reason into antagonistic and irreconcilable axes, resulting in the closing of the Church to the modern world—withdrawn in its particularism, averse to the universality of scientific reason.

The City of the Sun is the description of what the world could have been without the counter-reformation alliance of the Church with Iberia: universally Christian, rational, and unified under the rule of the throne of St. Peter.

In addition to being a philosopher, Campanella supported direct political practice and the revolt of Calabria in 1599, shortly after Philip II's death, provided an exceptional occasion. Campanella threw himself into that visionary operation: by joining the crowd of the deluded, rebels and thugs, he planned to set Calabria free from feudal nobility and the Spanish rule and to establish a communal and theocratic republic by force, free of private property and social classes, of which he himself would be the legislator and the leader, organizing the region according to the principles of a City of God. Campanella was a Neapolitan subject, which means that he was a subject of the Spanish Empire, as the Aragonese ruled over Naples. Between 1503 and 1713, Naples was the second most important city of the Spanish Empire; it owned the largest fleet in Europe and was its most populous city, a fact that contributed to make it also the largest cultural centre of the empire and an outstanding centre of political and economic power.

Campanella openly fomented the rebellion without the support of the Church, which ignored his ecumenical proposals and, supported by his astrological studies, announces the future transformations of the world and predicts its imminent return to a political and religious unity based on the presages and significant wonders for the year 1600, a date on which the Christian republic promised by Santa Catarina and Santa Brigid would come true, as described in *The City of the Sun*.

The revolt fails; two traitors denounce it to the Spanish government. Campanella is arrested, along with hundreds of rioters, and accused of *lèse majesté* and heresy, as he is identified as the leader of the revolt. Savagely tortured, the Spanish keep Campanella in prison in Naples for 27 years, where he writes his utopia, aiming to explain his political project based on his theory, to get possibly his acquittal as a result.

3   THE CITY OF THE SUN

The City of the Sun is a peculiar utopia that departs from the common utopian expressions, since it is not a satire, but rather a neat scientific prediction of a certain and fatal event that would occur in the real world when the astral conjunctions were propitious. It thus blends, with astonishing energy, prophecy, and natural philosophy.

In its original draft, *The City of the Sun* covered the entire earth; it was later reduced to a single city.

The City of the Sun is a dialogue between the Grandmaster of the Knights Hospitaller and a Genoese sea captain who had been Columbus' helmsman and who, as he landed in Taprobana, discovered the City of the Sun in the middle of a wide plain. That philosophical polis reproduces Copernicus' model representing the solar system by seven strongly fortified concentric and circular areas featuring the names of the seven planets. At its centre lay a circular temple, the city's heart and brain.

The political regime expresses the hierocratic system. Above all is the Metaphysical or the Sun, called Hoh, the "ruler of all, at both the spiritual and temporal level", a sacerdotal prince who stands out due to his wide knowledge. In *The City of the Sun*, the power exercised by the sovereign emanates from God and belongs to him, represented on earth by Hoh. This priesthood, invested with political power, owes its prestige to metaphysical knowledge, to historical and technical experience, which is consecrated by a sacramental order and divine investiture.

Hoh is permanently supported in the exercise of power by a supreme council, a triumvirate composed of:

A. Pon, or Power, who has jurisdiction over the activities of war and peace; under his command, men, women and children are continuously prepared for war.

B. Sin, or Wisdom, supervises the production of knowledge and education, which is universal. All seven city walls are frescoed with geographical maps, mathematical figures, personalities, animals, plants, etc., making the city a kind of book in itself. Such radical pedagogy produces wise men, such as Hoh, the Metaphysical, an omniscient creature and, at the same time, a priest, a scientist, and a philosopher, an intermediate between the divine will and human intelligence. The Solarians believe that only the one who masters all the sciences is fit for the science of government, since he is the one who best masters the constitutive knowledge of the social order.

C. Mor, or Love, is in charge of procreation and marriage, closely controlled functions that are not subject to the free choice of individuals; they dutifully deliver their bodies to the State for an essential function, i.e., the reproduction of citizens. Doctors according to physical features defined by the State choose couples and copulation takes place at a time set by astrologers to favour specific features also defined by the rulers. The defining criteria are based on eugenics to improve the race and the production of the type of Solarians required by the state. Not the citizens choose the City; it is rather the City that chooses its citizens. As in Sparta, the children belong to the State.

In principle, no absolute power governs *The City of the Sun*, since the Assembly of the people has the power to dismiss all magistrates, except for the top four ones. Dismissals are based on consensual decisions, since the magistrates are so wise and virtuous that they voluntarily resign from their position when they come across a more suitable and fit citizen able to perform their function.

Fair laws produce social uniformity: there is neither dissidence, nor opposition, nor claims, nor any active minorities, nor political parties. Every citizen is a part of the whole, of a big wheel of which he is just a small cog. His will instinctively follows the rationally fair course of the State and all individuals are permanently subject to the requirements of order and perfect balance.

The city's geometric structure reveals the total control of the State over all matters. Their passion for symmetry shows its love for order, which has reached a mystical level. Its inner workings are perfect as a clock.

The polis is organized like a convent, including perpetual adoration before the altar in the temple, which is internally decorated with a figuration of the Copernican universe. Divine will inspires everything and the City hierarchy ranges from man to

God, by mediation of the priests in the first place. Religious abstraction of deism prevails, supported by rational inquiry: a generic devotion to the Creator, the belief in divine providence, in immortality of the soul, and in reward and damnation hereafter, are the sound foundations of the solar religion. No material creature is worshiped, there is no idolatry, and the sun itself is merely honoured as God's sensitive representation.

In *The City of the Sun*, only a residual form of trade subsists. There is no idleness, no marginality, no corrupt clergy, nor do the strong abuse the weak; actually, there are neither strong, nor weak. There is no poverty or hunger; there is no shortage, disorder, civil disobedience, uprisings, crime, nor any violence against the institutions. A sovereign order reigns that reveres the Sacred, which in turn is investigated by scientists.

Collective life is not subject to fantasy and exceptions, but to the standard of rationality that the City implies. "Natural" development is to be avoided, as it is subject to the disruptive actions of history and common events of real life. Therefore, The City of the Sun does not have any past and it is not the result of historical evolution. The Solarian utopia is. It exists in an immutable present that knows neither past, nor future; perfect as it is, it will never change.

The lives of the Solarians are entirely ruled by reason, i.e., they live according to the dictates of Campanella's metaphysics. Their religion is not yet Christian, but little is missing: the Solarians, "who merely follow the law of nature, are very close to Christianity" and the difference lies in the absence of the Revelation, i.e., the coming of Christ. As a result, they do not believe in the Trinity, but in a single God represented by the life-giving sun. According to Campanella, Christianity "doesn't add anything to the law of nature, except for the sacraments" and therefore "the true law is the Christian one; once its abuses are eliminated, it will be the mistress of the world".

Thus, in *The City of the Sun*, faith is aimed at a creator God who is revealed by the spectacle of nature and understood by reason. The City of the Sun is the "philosophical evidence that proves that the truth of the Gospel is in accordance with nature". The Solarians, who live that way, are "almost catechumens of Christian life" merely by complying with the law of nature.

Campanella depicts a republic that is founded not by God, but by philosophy and human reason, to demonstrate that the truth of the Gospel is consistent with human reason. Discovered by philosophy and founded on reason, the natural religion of the Solarians is based on a standard that allows evaluating the value of historical religions, to choose the true one among them, and to bring it back to its true principle by eliminating deleterious abuses. He thus believed that natural religion, which is innate, is always true, while constructed religion is flawed and may be false. Natural religion is found in all beings and, as they originate in God, they tend to return to Him.

Therefore, a rational polis demonstrates a superior capacity of constituting a power that ensures the Church the throne of the world.

The Iberian colonial system is the reverse of *The City of the Sun*.

4 SCIENCE AND RELIGION IN HARMONY IN THE CITY OF THE SUN

In *The City of the Sun*, autonomous natural science and faith are not mutually exclusive—which contradicts the thesis of the Council of Trent -, but act to complete each other: since nature is the living statue of God, scientists who research the natural laws and phenomena get closer to God, as He is present in them. Scientific research is similar to prayer. Hoh is therefore both the high priest and the most important scientist.

The underlying philosophy of nature is found in a work by Bernardino Telesio (1509–88), *De rerum natura iuxta propria principia* that Campanella had read in his youth and that had influenced him in a permanent and structural way. Telesio made of philosophy of nature what Machiavelli made of politics: a sphere of reflection, emancipated from other spheres of thought, such as Morality and Religion. It primarily aims to recognize objectivity in nature, since the things *per se*, when properly observed, reveal their nature and features. This explanatory principle has universal validity, i.e., nature is one, in all times and places. Man is apt to know nature because he himself is nature; the senses are efficient means of knowledge, and man as nature is sensitivity. Therefore, what nature reveals of itself coincides with what the senses perceive: thus, sensitivity is self-revelation of nature by its part that is man. That principle was adopted by Galileo and taken as a thesis to its ultimate consequences.

Telesio developed a rational method for the appropriation of tangible reality according to which in nature, only natural forces exist, which must be explained through their intrinsic principles only, excluding metaphysics. Man would possess the ability to know the natural entity and would thus have power and dominion over all things that integrate nature. The natural world is thus a vast heritage available to man, a domain that needs to be benefited from, materially and ideally. This philosophical principle comes up at the early stage of manufacturing and the opportunity and the historical logic of these ideas are easily understood.

This philosophy would thus allow the Roman faith to merge with the scientific revolution, since it admitted the existence of the sacred in the world and it simultaneously set the philosophical foundations of modern science. An intimate union between reason and faith becomes feasible—which means to say that the Church could ally with the modern manufacturing system, already under implementation in Northern Europe.

If Campanella's thesis had prevailed, the Church and its allied empires would be able to embrace the new science, could conceive nature as an entity open to the profane mind that, nevertheless, would find God. The sovereign would be not only a scientist, but at the same time the high priest. Investigate nature through rational science and the practice of prayer would be the same kind of activity.

Eventually, that is more efficient to defend the Church's interests than what was practiced by the counter-reformation universe, by the triad of the Catholic states. Campanella's work, entirely coherent and aimed at a single target, intended to restore the power of the Church. He considered the Tridentine reform to be insufficient, not only to lead the world back under the domain of the Church, but unfit for the rehabilitation of Rome as an efficient entity apt for that task.

5 WHAT CAMPANELLA'S PROJECT MEANS

To compose *The City of the Sun*, Campanella based himself on the social imagination typical of utopia immanently to criticize the Tridentine Church, its options, its commitments. At the same time, he calls for a thorough review of the counter-reformation assumptions. Essentially, as we have seen, Campanella wants to reconcile faith and reason by leading science back to the Church—a vital action to save her from becoming irrelevant for the States that were absorbing the scientific revolution and that would soon become the main hegemonic centres and colonial metropoles.

To be more specific, Campanella wanted the expression of faith built by the Counter-reformation give in to the mentality of the scientific revolution so that it could conquer the throne of the world. That would require a radical change in the policy of the Church, which, through the structural alliance with the Iberian metropoles, was establishing the deepest meaning of its new identity. Basically, he aimed to save the Church from itself, correct the course it had adopted.

This empire, to reproduce itself, dismissed the most advanced points of the scientific revolution in progress and chose instead the Jesuitical directive of the *Propaganda Fide*. This is the inquisitorial aspect, which expressed itself by obstruction, often by means of stakes, to prevent the adoption of the new science and of manufacturing, i.e., the Inquisition as an actual economic agent. By burning scientists, the Jesuit Inquisition makes an economic and political choice. The persecution of science is even a constitutive act of America's colonization, an expression of the Iberian power. At this point, a fundamental question comes up: what would be the best way to constitute the policy of the Church so that it may fulfil its purpose as the mistress and saviour of the world?

Campanella does not need to invent a picture to oppose it; it is simply the opposite of *The City of the Sun*.

Campanella's utopia, taken as a blend of scientific reason and faith, is thus a complex system that combines scientific rationality and prophetic irrationality, a modern mentality and religious traditionalism. Campanella is not an adept of the Counter-reformation, but an exotically rational reformist. With its play of mirror and lights, of criticism and defence of the Church, his work might be, paradoxically perhaps, the most baroque utopia of the seventeenth century[1].

NOTE

[1] The Spanish Empire, sustained by the American gold and silver, became an aristocratic and ecclesiastical system, irrational in economic terms, despising work, in which the Church and the nobility were given the bulk of wealth, which immobilized it in sumptuous buildings or governmental obligations and dissipated it in luxury instead of investing it into productive economic activities. In its role as the metropole, Spain was unprepared for the growing demand from the colonies for production materials and equipment and became dependent on countries that were open to manufacture and science—the Protestant countries—for their supply. It thus enriched those who practiced what the Inquisition had crippled and even banned from its domain.

BIBLIOGRAPHY

Campanella, Tommaso. *Apologia di Galileo. Tutte le lettere a Galileo Galilei e altri documenti.* Introduzione e cura di Gino Ditadi. Este (PD): Editore Isonomia, 1992, 284 p.

Campanella, Tommaso. *La Città del Sole: Dialogo Poetico. The City of the Sun: A Poetical Dialogue.* Translation based on the Italian text edited by Luigi Firpo, by Daniel J. Donno. London: University of California Press Ltd.,1981.

Campanella, Tommaso. *Philosophia sensibus demonstrata*. Testo latino a fronte. Editore: La Scuola di Pitagora. Collana: Collezione Vivarium, Gennaio 2013.

Ernst, Germana. *Tommaso Campanella. Il libro e il corpo della natura*. Roma-Bari: Laterza, 2002.

Firpo, Luigi, Per uma definizione de Utopia, in *Utopie per gli anni ottanta*.—Studi interdisciplinari sui temi, la storia, i projeti. A cura di Giuseppa Saccaro Del Buffa e Arthur O. Lewis. Roma: Gangemi Editore, 1986.

Moneti Codignola, Maria. "Campanella, a cidade histo-riada", *Morus – Utopia e Renascimento* 5 (2008), pp. 86–106.

Pico Della Mirandola, Giovanni. *Discorso sulla Dignità dell'Uomo*. A cura di Giuseppe Tognon. Prefazione do Eugenio Garin. Bescia: Editrice La Scuola, 1987.

Skinner, Quentin. *The foundations of modern political thought*. Volume One: The Renaissance. Cambridge: Cambridge University Press, 1978, p. 255–256.

Telesio, Bernardino. *De rerum natura iuxta propria principia, libri IX*. Nápoles: Horatium Salvianum, 1586.

The feast of the family and the fragmentary aspect of the representation of socio-political structures in Bacon's *New Atlantis*

Helvio Moraes

Departamento de Letras, Universidade do Estado de Mato Grosso, Brasil
UNEMAT, Campus Universitário de Pontes e Lacerda, Mato Grosso, Brasil

ABSTRACT: This study aims at analysing the relation between the fragmentary aspect of Bacon's *New Atlantis* and the representation of the socio-political structures of his utopian island by the reading of the narrator's report of the Feast of the Family, a ceremony granted by the State to the man who is able to father thirty children or more. Firstly, I discuss the fragmentary—or deliberately inconclusive—nature of the work, pointing to some divergences among Baconian scholars who have discussed the problem. I will argue that, beyond the author's apparent conservative attitude towards patriarchal rule, his silence is, at the same time, an evidence of: (a) his awareness that a social and political transformation is unavoidable after the establishment of his program of scientific reform; as well as (b) his awareness of being incapable, given the historical circumstances he lived in, to provide a complete description of the world under de rule of a scientific enterprise. Thus, the second part of the article is the reading of the ceremony of the Feast of the Family, not as an idealized description of Bensalemite society, but also as a satiric representation of English political and social institutions in the author's time.

Keywords: English utopia; Francis Bacon; socio-political representation

1   ON THE INCOMPLETENESS OF BACON'S UTOPIA

A problem that Bacon scholars always face while reading *New Atlantis,* is the absence of a detailed description regarding the social and political institutions of the utopian island of Bensalem. Some hints of social relations and, implicitly, political institutions are described in a passage, which precedes the meeting of the foreigners with the Father of Salomon's House: The Feast of the Family. According to the narrator, it was:

> a most natural, pious, and reverend custom [...] granted to any man that shall live to see thirty persons descended of his body alive together, and all above three years old, which is done at the cost of the state. (Bacon, 2002, pp. 472–3)

This man is the *tirsan* (the father of the family) and, in this ceremony, he is greatly honoured.

I share the opinion of those scholars for whom it was above Bacon's capacity to provide a detailed description of the social and political life under the control of scientific enterprising—obviously given the historical circumstances in which he lived in, of which he proved to be fully aware by clearly presenting his personal limits (as in the paragraphs dedicated to the explanation of the fifth part of the plan for his *Great Instauration*; *Works,* IV, pp. 31–2). The passages in *New Atlantis* that refer to the social-political structure of the island are lacunar and complex. Firstly, I believe that those seeking, in these pages, only idealized, albeit fragmented, representations of his political thought, are wrong. Secondly, if we are to give credit to Rawley, Bacon's secretary and chaplain, who in the preface of the work speaks of the absence of "a structure of laws", "or of the best state or mould of a commonwealth", it seems to me that this is due to his clear-sight and confessed limitation, possibly also due to an intentional silence, rather than to an interest diverted towards more urgent affairs.

Above all, I believe that, despite being largely considered a conservative regarding his political vision and certain English customs—in his defence of monarchy and patriarchy, for example—Bacon kept at least some degree of suspicion that it would be impossible, with the advancement of knowledge, to keep the English political and social institutions unchanged.

The representation of the Utopian society and its ruling figure, are ostensibly conventional and generate a glaring contrast to the description of its scientific and technological superiority.

Since, in fact, Salomon's House must be seen as the end of the foreigners' "initiatory journey", it should be noted, however, that on the way, none of the interlocutors provides any clarifications regarding the political organization of the island. Some references are made to a governor, whose function seems to be simply to confirm the decisions made by the *tirsan* (during the Feast of the Family, as it is seen further on), and a mention is made to the delivery, during the ceremony, of a "King's Charter", a document "containing gift of revenue, and many privileges, exemptions, and points of honour, granted to the Father of the Family" (*idem,* p. 474).

However, in addition to having his dignity overshadowed by the *tirsan*, this king never makes himself seen. He does not host the travellers, he does not visit them during their stay in the House of Strangers, he does not, at any moment, participate in the Feast of the Family, and he is not in the procession marking the visit of the Father of Salomon's House. Even more impressive is that the highest honour that a foreigner can have in Bensalem is not to be presented to the king, but to get a hearing with the Priest.

Albanese (1990, p. 115) makes interesting considerations regarding the use of the terms "kingdom" and "state" in the Baconian utopia:

In striking comparison to its humanist model, *New Atlantis* never elucidates its civil hierarchy, never gives articulation to its structure of power. Rather, the text uses the words "state" and "kingdom" interchangeably, although the two words summon up images of government that pull [...] in divergent directions: one is corporate, if faceless, the other incorporated by a titled head, known to all. It is possible to read this silence about the authority at the heart of Bensalem as an affirmation of social and political conservatism, the more likely because the text that contains it is the product of a Jacobean bureaucrat in search of patronage for his philosophical program.

Tradition, according to Weinberger (1976, pp. 866–7), believes that *New Atlantis*:

... is incomplete because it does not contain a teaching about government or political rule", and because "it fails to combine an account of the organization of the project of science with an account of the political rule necessary for the best form of human life.

One of the aspects, which occupies Baconian utopia scholars the most, is precisely its fragmentary character, or, using a more adequate term, its incompleteness. Having this in mind, specific considerations are called for.

Formally, its incomplete condition is doubly emphasized: the story ends asserting "the rest was not perfected". In the preface, *To The Reader*, mentioned above, besides presenting the utopia as an "unfinished work", Rawley alludes to the author's goal and tries to justify the text's lack of completion: the urgency to finish another kind of work which he had already started (which was considered a favourite by the author) is the reason presented by Rawley in order to explain the lack of a detailed description of the island's political and social structures. However, the contradiction is easy to observe, by comparing the preface's final part (where the reason for the work's lack of completion lies) with the beginning, where it is mentioned that Bacon's objective was to provide a description of an instituted school for "the interpretation of Nature". This makes us wonder if the (repeatedly) declared lack of completion of the work must be interpreted in a literal manner.

This issue is considered irrelevant by Berneri (1962, p. 150), because, in her view, even if there was any intention of writing a part in *New Atlantis* dedicated to the political organization, Bacon probably would not change the fundamental character of a republic where science plays the role of protagonist. This because, just like Plato:

... who refers, in detail, to the laws with which the existence of the guardians is regulated, and speaks very little of other social classes, Bacon only pays attention to the institutions and the work of the members of Salomon's House—the guardians of his ideal society [...]. However, what is not said might be just as significant as what is: Plato and Bacon are concerned only by the ruling class because, in their view, this is the one that matters.

In fact, in classic utopia, once the core institution is marked, all others are necessarily subordinate. In the utopian world, the clash between two levels of power would be inadmissible. Obviously, for Bacon, who by now was already politically ruined, despite still expecting to go back to his public life under the king's consent, it would be extremely hard and risky to represent in details the submission of the state's apparatus to his collegiate of scientists, perhaps even impossible, considering that the figurative representation of the monarch exonerates the very idea of such apparatus.

In this sense, venturing an answer to Ablanese, it could be said that Bacon's prudent silence, instead of a conservative attitude, suggests that power, once in the hands of the scientific community, significantly alters the way that society is organized, of which, however, the author can only provide some sketches regarding religion and social life, but none regarding the government since the current form would already be overcome in the ideal world.

Moreover, the idea that the state places itself on an inferior, or even subordinate, power level below Salomon's House is explained by the Father himself:

> We have consultations, which of the inventions and experiences which we have discovered shall be published, and which not: and take all an oath of secrecy, for the concealing of those which we think fit to keep secret: though some of those we do reveal sometimes to the state, and some not" (Bacon, 2002, p. 487).

Thus, I believe that the formal incompleteness of the utopia corresponds to the completeness of the subject. Alternatively, using White's fortunate formulation (1956, pg. 344), it is a work of deliberate incompleteness. Formally, Bacon remains true to his model, Plato's *Critias*. However, in the Platonic fragment, while Zeus is about to speak on the punishment brought upon old Atlantis (although he does not, for the dialogue abruptly ends), in the Baconian utopia, the Father speaks about the preservation and the glory of new Atlantis. To the extent that the content of the speech suggests, the text ends only when the speech is complete (Spitz, 1960, pg. 61). Thus, apart from the silence, which may, however, be partially filled when scrutinizing in between the utopia's lines, Bacon's plan should be seen as accomplished, along with the description of Salomon's House.

So, in what manner could the parts dedicated to some of the Bensalemite customs, such as the Feast of the Family, be read? I believe that through the perspective of satirical representation, as it will be seen.

2   THE FEAST OF THE FAMILY

The foreigners, who have ported in Bensalem after a series of hardships in their journey through the Pacific Ocean, are placed in quarantine in the House of Strangers, and by the end, are gradually being brought into the island's daily life. They exchange "acquaintance with many of the city, not of the meanest quality" (Bacon, 2002, p. 472), which suggests some sort of social stratification in the island, contrary to Morus' communist utopia. The generosity of the people and the continuous discovery of "many things right worthy of observation and relation" (which, however, the narrator does not describe), gradually makes them forget of all the things which are dear to them in their own countries and accept the Bensalemite life style in a much more sympathetic manner. One of these wonderful things seems to be the Feast of the Family, a celebration funded by the state to praise the father figure, which is entitled to " any man that shall live to see thirty persons descended of his body alive together, and all above three years old" (*idem*, pp. 472–3).

Considering that conciseness is one of *New Atlantis'* attributes, it ends up being rather odd to read a passage that is characterized by such a detailed description of a ceremony. In addition, as I said, the representation of the customs and the mention of some particularities regarding social relations, centred on the notion of the patriarchal family, are ostensibly conventional, even more when compared to the final speech of the Father of Salomon's House, which I believe, is intentional. It has a certain aspect of decaying, the weight, and mould of overcome traditions, which make use of a kind of liturgy devoid of sense in its reverence of dying principles. It is no coincidence, therefore, the representation of the householder, or *tirsan*—the "shy", the "fearful"—being supported by two of his sons, to listen from the herald the tribute that the State pays him. Nor is the evident turpitude in the image of a father, alone at the table, even in public, having at his disposal some of his children, who serve him on their knees. In contrast with the affected language and the ceremonial ritual, there is the elegant reception and the urban and illustrated language of the Father of Salomon's House.

Therefore, in my view, the passage cannot be seen as an idealized representation of social relations based on the notion of the nuclear family. I think that, at this point, maybe more than in any other passage, we must discern what is presented from the perspective of the character and, above all, how the plot is woven on an authorial level. I think that it is possible to notice a "Baconian laughter" in certain details which make up the bigger picture, although the philosopher seems to have been a monarchist sympathizer throughout his life and no stranger to a misogynistic point of view. There is some truth in the paradox indicated by Smith (2008, p. 116), for whom "the *New Atlantis* is only funny when we take its apparent absurdities (such as the Feast of the Family) seriously". This is because, through a satirical perspective, Bacon lays bare the mechanism that maintains the English patriarchal monarchy. This is not just a sarcastic laugh.

Obviously, Bacon was aware that the conventions and principles which kept the English state working, as efficient—and, therefore, legitimate –, as they might have been, also had inconsistencies and arbitrariness, not being, however, principles "wherein nature so much preside" (Bacon, 2002, p. 476). On the contrary, the narrator of his utopian state holds to be true in the world he describes.

In the case of the Feast of the Family, even more instigating and suggestive is the fact that it is the only event the narrator describes without having witnessed—he reports a ceremony which two of his companions have been invited to, and it can be assumed that the expressed admiration in the description was intensified specially due to the indirect experience of the listener who exacerbates what is already reported to him with enthusiasm. In other words, the ceremony can be admirable in the eyes of those who have witnessed it, or for the narrator who heard about it, but not necessarily, at least not in all aspects, to Bacon. Therefore, his is a witty laughter, of one who has noticed very early and has lived very closely to the inherent contradictions of political life.

I make a final consideration, summarizing what has been exposed so far in this regard: I believe that the passage can be read at various levels, hence its complexity. You can find amalgamated elements which, in fact, are part of the Baconian notion of the greatness of the State and his general political thought: elements of social criticism, elements of political satire, under the the philosopher's perception that in a world where institutions created through scientific advancements became predominant, many aspects of the world in which he lived would necessarily go through transformations. Since he cannot clearly predict what those transformations would be, he highlights certain issues related to the social political structure, such as they were established in the time he lived while at the same time, paradoxically, addressing them through satire, as Morus also had in the first book of *Utopia*.

Thus, how can the Feast of the Family be read?

Above the father figure as the central element of the family institution—and, by extension, the Bensalemite society—what is exalted in the Feast of the Family is fecundity, the ability to provide the state with numerous offspring. The description of the ceremony is divided into parts that suggest certain characteristics of collective life in *New Atlantis*, in the case the image of the patriarchal family as a microcosm of the state is considered. In the two days before the opening ceremony, the *tirsan*, accompanied by three friends, is given all authority (confirmed by the assistance of the governor of the city where he lives) to, among all his children, who are officially called, advise, resolve disputes, mitigate financial problems, censor vices, guide in issues pertaining to marriage, among other actions. Everything he decides, although always in accordance with the "order of nature" and rarely disobeyed is legitimized with the State's seal. On the day of the Feast, after a religious ceremony, this father, accompanied by all his offspring, heads towards a hall, where, in one of its extremities, there is a platform, and in the middle, a table, and a chair at his disposal. Over the chair:

> is a state, made round or oval, and it is of ivy [...] curiously wrought with silver and silk of divers colours, broiding or binding in the ivy; and is ever of the work of some of the daughters of the family; and veiled over at the top with a fine net of silk and silver. (Bacon, 2002, p. 473)

The mother does not participate in the banquet. If anything, she just watches:

> if there be a mother from whose body the whole lineage is descended, there is a traverse placed in a loft above on the right hand of the chair, with a privy door, and a carved window of glass, leaded with gold and blue; where she sitteth, but is not seen. (*ibidem*)

After everyone is set, the ceremony begins. A herald, accompanied by two young men, read the "King's Charter", a document which grants a money donation, privileges, exemptions and honour titles. While the letter is read, the *tirsan* "standeth up, supported by two of his sons, such as he chooseth" (*Ibidem*). After reading and delivering the letter, an acclamation is made—"Happy are the people of Bensalem" (*idem*, p. 474)—, and the herald finally presents the honoured father with a bunch of grapes ("in number as many as there are descendants of the family") made of gold. The *tirsan*, in turn, passes the gift to one of his sons, which he has chosen to live with him, "who beareth it before his father as an ensign of honour when he goeth in public, ever after; and is thereupon called the Son of the Vine".

After the ceremony, the father withdraws, returning soon after for dinner, in which, alone under the canopy (unless one of his sons, being a member of Salomon's House, follows him), is "served only by his own children, such as are male; who perform unto him all service of the table upon the knee" (*idem*, p. 475). In the hall filled with guests, dinner is also served. After the feast, hymns are sung praising Adam, Noah, and Abraham. The *tirsan*, then, retires again to his prayers and when he returns, gives his blessings to all his descendants, one by one, with the following words:

> Son of Bensalem, (of Daughter of Bensalem,) thy father saith it: the man by whom thou hast breath and life speaketh the word: The blessing of the everlasting Father, the Prince of Peace, and the Holy Dove be upon thee, and make the days of thy pilgrimage good and many. (*Ibidem*)

What is reported in these pages is, therefore, the acts and privileges granted to a man who will one day, or more precisely three days, exercise the duties of a king.

As I mentioned, everything that is said and done in these three days, by his own sovereign will, will have permanent validity, with the approval of the State. In fact, his interests converge with the internal policies of the island, in the guarantee of civil peace, in the resolution of legal force of impasses between citizens, in the condemnation and punishment of disorderly conduct, in the orientation, education and counselling. Not coincidentally, the descendants who, in the beginning of the report are called his offspring, in his final blessing are referred to as "the sons of Bensalem". Under the stratagem of the description of the Feast of the Family, I believe that Bacon, therefore, makes references to the aspects of social organization in Bensalem, founded on the principles of the patriarchal family, but, at the same time, also elaborates a commentary regarding the English monarchic government and the figure of the king as the "father of his people".

Politically ruined, after exhaustive efforts to win James I's sponsorship for running his program of reform, Bacon engenders his revenge in the figure of an old man, "shy" or "fearful" (as the term *tirsan* suggests), who, although "father" of a crowd, is a solitary being, and only remains with the pomp and pageantry of a state liturgy lacking of sense.

In this regard, some information provided by Smith (2008) is indispensable. The author analyses in great detail the description of the Feast of the Family, and observes a syncretism in the ordering of textual citations belonging to different times and cultures, such as references to episodes and names of the Judeo-Christian tradition, the autumnal rites of Bacchic festivals and the influence of Hebrew culture in the development of the image of the *tirsan* as a ruler or province official, whose authority is entrusted by the emperor.

In her reading of the ritual itself, she demonstrates how, implicitly and comically, Bacon refers to specific aspects of the protocol apparatus around James:

The Tirsan's relationship to his children is governed by state-enforced "reverence and obedience". Again, without overdoing the comparison, one may note that [...] the Tirsan is given his privileges through a direct grant from a higher authority. In the case of the Tirsan, that higher authority is the "king of Bensalem". [...] The Tirsan [...] is not described as a subject of the king, but as his "friend and creditor". What could be a more effective parody of the notion promulgated by James I that kings are like gods, and that kings receive their privileges directly from "on high"? James is well-known for having regarded and represented himself as a figure very much like both a god and a father of many; in his 1610 speech to Parliament, he famously announced that "Kings are not onely GODS Lieutenants vpon earth, and sit vpon GODS throne, but euen by God himselfe they are called Gods ... Kings are also compared to Fathers of families: for a King is trewly *Parens patriae*, the politique father of his people." [...] Secondly, one must consider the implications of the setting that Bacon creates for the Feast of the Family. [...] The description of the Tirsan's seating arrangements is directly evocative of the arrangements described in the House of Lords Precedence Act of 1539. [...] The Tirsan's repeated withdrawals for private prayers are directly evocative of the king's withdrawals to his own traverse in recorded ceremonies involving James. Even the fact that the Tirsan stands up supported by two of his sons reminds that King James was "ever leaning on other men's shoulders". Thirdly, the cluster of imagery associated with Dionysus or Bacchus subtly evokes the debauchery and dissipation that were known to have characterized the Jacobean court and the drunkenness that was said to have characterized the Jacobean monarch on more than one occasion. (Smith, 2008, pp. 117–8)

However, as the author herself says, this is just a reading possibility, or one of the levels, that the report gives us, and it would be reductive to stick to just this perspective.

In his tacit comment on patriarchy and the divine right of the king, Bacon indicates that, under the designation pursuant to the "natural order", in fact, what is observed is a political institution whose conventionality, or anti-naturalism, is a striking feature. It helps to ensure civil peace, the good order of the State. However, it is not founded on the natural order.

Based on the true natural order, another, more relevant, institution is built in the island: its college of scientists. This supersedes all the old building structure, whose instability becomes manifest when, finally, it is possible to glimpse the new order established by the progress of science. It is inevitable that part of the old structure falls to the ground with the rise of scientific research. In the eyes of the narrator—and the group of foreigners—it is understandable that the outward signs and "ornaments" the new reality in which they find themselves inserted cause admiration since, essentially, they are not conflicting with the image they can formulate of a well-ordered nation, from the paradigms produced by European civilization. However, all of this happens before they have knowledge of the structure, the functioning, and the wonders produced by Salomon's House.

This idea becomes even clearer if we contrast the description of the Feast of the Family with the speech of the Father of Salomon's House. If the image of the *tirsan* can be metaphorically viewed as that of the king, the priest's seems to be far greater, who, solely by the force of his speech is able to inspire respect and admiration. His superiority is undeniable when we compare the acts emanating from the unquestionable authority of the former and the outstanding achievements of the College he presides, where it is possible to breathe a more democratic atmosphere, as well as to observe a long yearning of Bacon fulfilled: "[...] through my hands to endow the human family with new mercies" (*Works*, IV, p. 20). Besides, as Smith (2008, p. 120) observes:

The Tirsan's personal authority is exhausted in the very exercise of it (although his decrees and orders evidently persist after the Feast is over), while that of the Father of Salomon's House is designed to last. It endures because it is transmittable through the dispersal of knowledge and because of the goal it seeks ("knowledge of Causes and secret motions of things; and the enlarging of the bounds of Human Empire, to the effecting of all things possible") can never be wholly attained.

Finally, there is a great disproportion in the idea of the authority of a man over his family, which, though numerous—as of a king over his people—, in contrast to the image of a House whose power extends over all of humanity.

Salomon's House is the historical overcoming of patriarchal monarchy and the principles that sustain it. If a contrast—or clash—between two spheres of power is displayed in *New Atlantis*, this is not by an inconsistency or internal contradiction to the work, or by an incapacity or lapse of the author in finding a solution to various claims of power.

I believe that Bacon did not advance this issue because, from the start, he already had in mind the idea of working out a fable in which the centre would be:

[the] model or description of a college instituted for the interpreting of nature and the producing of great and marvellous works for the benefit of men.

Therefore, the instance of the power is the scientific community. The contrast that is presented should be read in light of what the literary genre of the fable allows us to perceive, specially while considering the philosopher's commitment to the study of its structure and meaning, besides its ample employment in a considerable part of his work. Read in this light, the Feast of the Family is presented with a plot composed of various levels, as can be observed, and the contrast mentioned, instead of indicating a contradiction regarding the utopian world created by Bacon, serves for a comparison between instances of power that belong to distinct time periods.

ACKNOWLEDGEMENT

This work had the benefits of CAPES support.

BIBLIOGRAPHY

Albanese, Denise. The *New Atlantis* and the Uses of Utopia. *ELH*, 1990, vol. 57, nº 3, pp. 503–528.

Bacon, Francis. *New Atlantis*. In *The Major Works*. Ed. Brian Vickers. New York: Oxford University Press, 2002.

Bacon, Francis. *The Works of Francis Bacon*. Ed. J. Spedding, R. Ellis & D. Heath. 14 vols. London: Longman, 1857–74.

Berneri, Maria Luisa. *Viaje a través de Utopía*. Buenos Aires: Editorial Proyección, 1962.

Smith, Suzanne. The New Atlantis: Francis Bacon's Theological-Political Utopia? *The Harvard Theological Review*, Jan., 2008, vol. 101, nº 1, pp. 97–125.

Weinberger, J. Science and Rule in Bacon's Utopia: An Introduction to the Reading of New Atlantis. *The American Political Science Review*, Set., 1976, vol. 70, nº 3, pp. 865–885.

White, Howard B. Political Faith and Francis bacon. *Social Research*, 1956, vol. 23, nº 3, pp. 343–366.

Utopia in early modern Portugal and Spain: The censorship of a nest of vipers

Hervé Baudry

CHAM, FCSH, Universidade NOVA de Lisboa, Lisboa, Portugal Universidade dos Açores, Portugal

ABSTRACT: This paper revisits the issue of Thomas More's inquisitorial censorship in the Iberian Peninsula. Some historians have investigated the Portuguese case as *Utopia* was first censored by the Portuguese Inquisition in 1581; little critical consideration has been given to the Spanish censorship. An Iberian perspective on this question has never been previously adopted. The premise of my unifying approach is not historical, according to the Double Monarchy, or the traditional relationship within this geographical area. Both censorships appear at the same historical moment but submit the narrative and the other texts often published alongside it to different measures, legally binding through the Indexes, which suggest different receptions of More's writings in a similar context. Describing these differences leads us to distinguish two forms, or levels, of censorship, micro and macro-censorship, or, expurgation and prohibition. Both exist in this instance, proving the Portu-guese reaction to the texts to be the most repressive. A second aspect is the question of the effectiveness of controlling access to books A difficult problem when it deals with simple prohibition, the methodology of micro-censorship studies helps evaluating to some extent what was achieved by the censors. The present analysis, limited to the Portuguese data, tends to invalidate previous conclusions on a wide and lasting reception of *Utopia* in Portugal and to confirm the efficacy of the local controlling system.

Keywords: Thomas More, Erasmus, Censorship, Portugal, Spain

1   PRELIMINARY NOTES

In early modern European history, *Utopia* was subjected to measures of censorship that appear to be different from the local systems of book controlling. For instance, in France, the first translators of Latin texts "censured attacks against religion and politics" (Pierrot, abstract). The present study deals with the Inquisitorial postpress censorship. Its most visible face is reflected by sets of measures against books we conventionally describe using the expression "put in the Index". These Indexes were printed catalogs of names and works elaborated throughout much of Catholic Europe. The best known of these were Roman and entitled *Index Librorum Prohibitorum* (Index of forbidden books). The Inquisitorial destiny of Thomas More's works is interesting because it reflects the variety of censorship policy within that world. In Portugal, censor-ship was complete, that is to say, *Utopia* was prohibited: legally, no one could possess, sell or read it. In Spain, the censorship was partial: owning, selling or reading were authorized once the texts, and not only *Utopia*, had been expurgated. This kind of censorship can be described as microcensorship. The Inquisitions of both countries published specialized Indexes for this purpose, commonly referred to as *Index Librorum Expurgatorum*, beyond the Indexes of prohibition.

The present investigation lies within the framework of both kinds of censorship, prohibitive and expurgatory. The issues of microcensorship, whatever the field, are commonly approached in a national per-spective that is from the point of view of local Indexes. Fernando Moser and Pina Martins paid attention to More's censorship in Portugal (Moser, 1983); in the field of history of science censorship in Spain, José Pardo Tomás published a book reference in 1989. Moser and Martins used exclusively the Portuguese Indexes of 1581 and 1624 in the first case; Tomás, exclusively the Spanish ones (1584, 1612, 1632 and 1640). This study requires a more multilateral approach.

Microcensorship studies are intrinsically transnational and transtextual. Partially exploring the history of books, partially the history of texts, they are straightly dependent on the history of the leading texts of both histories, the Indexes. Moreover, the genealogy of this literary genre proves they all are strongly intertextual. Let us say that, methodologically, none of them can be studied in isolation.

In More's case, the close dependence of both Iberian countries seems even stronger in the perspective of material bibliography. As we shall see, one censored copy possessed by the Portuguese National Library cannot be explained but with the use of a Spanish edition of the Index.

What is the relevance of these methodological and historical points? Microcensorship studies have a great deal to say about books and the reception of ideas. The present paper first describes the variety of variety of ways More's book was received by the censors in the early modern period and attempts to evaluate the effectiveness of book control. It is important to note that the short title, *Utopia*, refers to More's narrative and to the successive editions (Gibson, n°. 1–37: *Utopia*; 74–77: *Collected Works*). Since 1516, but mostly from the Basilean edition of 1518 by Froben, to Raphael Hythlodaeus's travel in the Utopian Republic were joined various texts by More and other authors, including Erasmus of Rotterdam, a heretic. According to the conception of the printed book as a metonymy and metaphor of the author (Betteridge, 298), the different receptions of *Utopia*'s editions in circulation in the Peninsula reflect the censor's hesitations between saintliness (More) and heresy (Erasmus). As to evaluating the effectiveness of censorship, the present study will only deal with the Portuguese side.

2 *UTOPIA* IN THE INDEXES

Here follow, chronologically organized, the texts of the Portuguese and Spanish Indexes in which appear Thomas More's *Utopia,* from 1581 to 1624.

A few words about the first entry. The same year, in 1581, Antonio Ribeiro published two Indexes, the Roman one in Latin, and a Portuguese, entitled *Catalogo dos livros que se prohibem nestes Regnos* (= Catalog of the books that are forbidden in these kingdoms; the plural is due to the royal title, King of Portugal and the Algarves). It is divided in two sections: a local list of prohibited books in Latin and in Portuguese («en Lingoajem»); a list of works to be corrected («para se emmendarem»). *Utopia* is listed in the first section under letter V; in the second section, one More's epigram must be erased from the anthology by Leodagarius a Quercu (*Flores Epigrammatum*, 1560, vol. 1).

2.1 (Portugal) *Catalogo dos livros que se prohibem nestes Regnos.* Lisboa: Antonio Ribeiro, 1581, in-4

* fl. 16v°: Utopia Thome Mori.
* fl. Q v°: Ex Thoma Moro, de Nautis ejicientibus monachum. fol. 230.

2.2 (Spain) *Index librorum expurgato-rum,* Madriti: Apud Alfonsum Gomezium, 1584, in-8°, p. 193

Thomas Morus.

Ex Thomae Mori, viri clarissimi, scriptis, in impres.[sione] Basileae apud Episcopium, anno 1563.

1. *In Epistola Guillielmi Budaei ad Lupsetum, de Thoma Mori Utopia, fol. 3. epistolae, lin.[ea] ult.[ima] dele ab illis verb.[is]* Quo certe instituo Christus, *usque ad,* ac fata nostra regere.
2. *Lib.[er] 1 Utopiae, pag. 31. lin. 7 deleat.* Non Hercule magis, quam si essem sacerdos.
3. Lin. 20 eiusdem folii, deleatur ab illis verb. *Nam Cardinalis,* usque ad, *hoc quoque dictum.*
4. *Lib. 2 Utopiae, ubi agit de religionibus Utopiensium, pag. 146 deleatur in marg.[ine]* O Sacerdotes nostris longe sanctiores.
5. *Pag. 261 ex epigrammate de novo testamento verso ab Erasmo, deleatur ab illis verbis,* Lex nova nam veteri, *usque ad,* Christi lex nova luce nitet
6. Pag. 524 linea 22 epistola de morte Thomae Mori, deleatur, *Multo magis licuisset hic esse tacitum.*
7. Lin. 27 eiusdem paginae, deleatur, *Simplici, synceraque conscientia errasse.*
8. Et pag. 530 lin. 6 deleatur, *Forte fefellit eum persuasio*
9. Deleatur etiam tota Apologia pro Moria Erasmi ad Martinum Dorpium.

2.3 (Spain) *Index librorum prohibitorum et expurgatorum,* Madriti: apud Ludovicum Sanchez typographum regium, 1612, in-fol., p. 722–723

Thomae Mori Lucubrationibus adiunctae Epistolae.

Ex Thomae Mori Angliae Ornamenti eximii, lucubrationibus, etc. quibus additae sunt duae aliorum Epistolae de vita, moribus, et morte Mori, Basileae apud Episcopium, 1568 [*i.e.* 1563]

1. In Epistola Guilielmi Budaei ad Thomam Lupsetum Anglum, cuius initium, *Gratiam sane,* pagin. 6 in fine, post illa verba, *communionis legem,* dele usque ad illa, *Utopia vero,* exclus. [ive].
2. Et pagin. 11 post illa verba, *quod Erasmi,* dele, *amicus est,* exclusive.
3. In Epist. G. Courini [Nucerini] ad Phil.[ippo] Mont.[ano] cuius initium, *Quoniam iuxta,* pag. 524 ad finem, post illud, *malevolentia peccasse,* dele usque ad, *Hoc sibi penitus,* exclus.
4. Et pag. 530 sub initium post illud, *quam mortem oppetere,* dele usque ad, *At demiror,* exclus.[ive]

2.4 (Portugal) *Index Auctorum dam-natæ memoriæ,* Ulyssiponæ: ex off. Petri Cræsbeck, 1624, in-fol.

*p. 180 (*Index Lusitanus*),T, 2ᵈ class:
*p. 180 (*Index Lusitanus*),T, 2ᵈ class:
Thomae Mori, viri alias sanctissimi, seu (ut habetur in editione Lovaniensi 1566 Raphaelis Hythlodaei Utopia omnino prohibetur. At vero Epistolae lucubrationibus eiusdem Thomae adiunctae, donec emendentur iuxta praescripta in Expurgatorio.
*p. 186:
Utopia Opus sic inscriptum De optimo Reip. statu, quod etiam Italico sermone circunferur in opere Francisci Sansovini De Gubernatione Regnorum et Reip. impresso Venetiis 1561.
*p. 1030 *(Index Expurgatorius)*:
Thomas Morus.
Ex Thomae Mori Angliae Ornamenti eximii, lucubrationibus, etc. quibus additae sunt duae aliorum Epistolae de vita, moribus, et morte Mori, Basileae apud Episcopium, 1568 [*i.e.* 1563] sequentia expungantur.
[1 to 4] [reproduces 2.3 n°. 1–4]

5. A primo fol. usque ad 18 continetur Utopia, sive sermo Raphaelis cuiusdam de optimo Reipublicae statu, incipit, *Cum non exigui momenti, etc.* qui totus praecidatur cum multa in eo commendentur a Christianae Reipub. statu abhorrentia.
6. Eadem Utopia inserta est operi de optimo Reipub. statu edito Italico idiomate a Francisco Sansovino Venetiis anno 1561 estque totus liber 18 sive ultimus,
7. atque adeo inde quoque praecidentus fol. 59 praefatio, sive Epistola Lutheri *Gratia et pax, etc.* tota auferatur usque ad titulum, *Respondetur ad epistolam, etc.* exclusiv.

All the following Indexes are Spanish (1632, 1640, 1707, 1747, 1790) and reproduce the text of 1612.

3 *UTOPIA*'S IBERIAN CENSORSHIP

The main difference between the reactions was noted above: Portuguese macrocensorship and Spanish microcensorship. Comparing A and D, the Portuguese censors reinforced the traditional prohibition by adding the Spanish expurgation of Budé's and Courinus's (*i.e.* Erasmus, according to Mesnard, 371) letters.
From 1584 onwards, all Indexes precise the edition(s) involved, which probably signifies that the censors had it/them in hand. The error in 1612 ("1568") does not exclude a direct consultingconsultation of the text. It is reproduced in all later

entries and may be due to an incorrect reading or confusion with Thomas More's letter to Johannes Pomeranus, the only work of the English published in 1568.
Here are some remarks on the passages concerned by *Utopia*'s microcensorship (texts 2.2, 2.3 and 2.4).
In two cases, the textual difference only corresponds to how passages are located in the book, not to the contents to delete (in this case, expurgations are only suppressions; elsewhere, microcensorship instructions can specify additions or substitutions of words): inclusive style in 2.2 (indicates the first and the last words of the passage): corrections n°. 1 and 7; exclusive in 2.3 (indicates the last words before and the first after the passage): corr. n°. 1 and 3. In corr. n°. 8 (2.2), only four words should be erased; corr. n°. 4 (2.3) indicates the same passage but exclusively.
The extent of passages to expurgate is varied: from four words (2.2, n°. 8) to 63 pages (2.2, n°. 9). Their location reflects the variety of the book: three corrections into More's narrative (2.2, n°. 2–4), Guillaume Budé's (a French humanist also indexed at first in the 1581 *Catalogo*) letter (2.3, n°. 1–2), two More's epigrams (2.1; 2.2, n°. 5), Courinus's letter (2.2, n°. 6–8; 2.3, n°. 3–4), More's letter to Dorpius on Erasmus's *Moriae Encomium* (2.2, n°. 9) and Luther's letter to Sebastian Schlik (2.4, n°. 5). In 1624, the Portuguese Index adds to the corrections of 1612 a letter by the heresiarch Martin Luther which can only be found in the 1566 edition (printed in Louvain).
These texts were subjected to typical deletions, following the rules that were established in Latin (from the Tridentine and Roman Indexes) and vernacular at the beginning of the Indexes. All censorship inter-ventions were made to defend the faith and good customs.
The suppressive instructions of 1584 and 1612 apply to propositions dealing with religion (2.1, 2.2, n°. 1–3) and the writings of Erasmus (2.2, n°. 5, 9; 2.3, n°. 2):
2.2, n°. 1: Pythagorean communalism was laid down by Jesus Christ, who put an end to the quibbles of civil and canon law which remain predominant. It is a typical humanist attack based on faith against medieval law practices; n°. 2: a joke against priests; n°. 3: an attack against mendicant orders; n°. 5: six verses of an epigram in praise of Erasmus's translation of the New Testament; n°. 9: the apology of the *Praise of folly*, one of Erasmus's prohibited works (Bataillon, 743–780). In 1612, censors followed only correction n°. 1 and suppressed laudatory words about Erasmus (2.2, n°. 2); More's narrative remains untouched.
Beyond More's writings, two texts of the collected works had to be expurgated. The *Epistola*

de morte Mori, entitled *Expositio fidelis* in 1535, bears three (2.2, n°. 6–8) then two corrections (2.3, n°. 3–4). Their extent is limited to a few words of moral and psychological significance: on the idea that silence is much more licit than abjuration (2.2, n°. 6); that we can sin although our conscience is simple and pure (2.2, n°. 7; 2.3, n°. 3); and that More's belief might have weakened (2.2, n°. 8; 2.3, n°. 4).

To conclude briefly on this point, some aspects can be highlighted concerning the intellectual context within which the censors (late) received *Utopia*: this follows the anti-Erasmian trend of the Counter-Reform and reflects the founding of the *persona* of Thomas More as a hero of Catholic faith. Budé, Erasmus, Luther: are all names of strongly censored humanists and the Latin editions of *Utopia* might have seemed highly problematic to the censors. But all of them did not take identical measures, showing that their perception of the danger was different. For the first censors of More's narrative, Rafael Hytlodaeus's discourse was perceived as a text of "high toxicity", an expression of mine in accordance with the medical metaphors commonly used in the war against heterodoxy. We do not know what edition the men of 1581 had in hand, but it was certainly a Latin text. The text of 1624 proves the initial prohibition was not a provisional decision, which happens when a text is forbidden until its expurgation ("prohibitur donec expurgatur"). The spirit of this full prohibition follows the radicalism of the Portuguese *Rol dos Livros defesos nestes Reinos* of 1561 (a4 v°): though Catholic, some authors ought to be prohibited because they were not aware of the perilous times they lived in and then did not see that their words might unintentionally damn.

4 THE QUESTION OF EFFECTIVENESS

It may seem inappropriate to study More's microcensorship, or expurgation, in a country where macrocensorship, or prohibition, was the measure taken against the book. But it is extremely difficult to attempt an evaluation of the actual censoring of books in the field of full prohibition. On the contrary, a decisive step of the microcensorship approach consists in the inventory of existing copies and the analysis of quantitative and qualitative information they may deliver, at first, the degree of achievement of the instructions provided by the Indexes, but also any institutional or individual marks of ownership and readership.

At the present time (February 2016), eleven XXIst century Portuguese libraries have been consulted. Six of them have no copy printed in the

sixteenth and seventeenth centuries. In five libraries from Lisbon, Coimbra and Oporto, twenty-three copies including More's works have been located, twelve of which include the narrative of *Utopia* (editions from 1518 to 1689), with eight in Latin and four in Italian. The rest is composed of editions of Lucian of Samosata by More and Erasmus (eight copies), More's correspondence (two) and Epigrams (one).

Out of these statistics, one copy (Basel, 1563; Gibson, 74) was located in the old catalog (XVIIIth century) of the library of the Santa Cruz monastery (Coimbra), which is unlikely to be the one actually possessed by the City Library of Oporto.

Six copies, that is 27% of the total, show evidence of microcensorship, among which only one copy of *Utopia* (that is 7% of the total) and five of Lucian. In this case, only Erasmus's texts and name have been censored.

The unique copy expurgated (deleted passages) of More's narrative is an edition of 1566. The passages deleted prove the censor used the Spanish Index of 1612: only two passages are involved (2.2, n°. 1 and 2) because the rest of the pieces are not included in this edition (Gibson n°. 75b). No annotation indicates the place of expurgation but it is probable that it was achieved in Spain as it can be observed through the inventory of expurgated medical books (Baudry). We do not know when it has entered the country.

Do these elements help to evaluate the effectiveness of prohibition in Portugal? Let us recall at this point that, formally, the list of 1624 remains the reference Index in Portugal at least until 1768, when the Inquisition was extinguished. In principle, during this time, no book entered the country or remained in circulation. So, one would be tempted to assert that the number of copies is an evidence of reception, hence that the book controlling was "lax", as Patricia Manning notes about Spain (Manning, 73). But our issue deals with the long term, from the sixteenth century until the present day. Methodologically, from the material presence of an early modern print in a library we may not deduce its presence, circulation or even readership in the past. More indications are needed.

The analysis of the thirteen copies of More's writings (one *Utopia* is expurgated, the seven Lucians do not form part of our present consideration since, in these editions commented by both humanists, only Erasmus is censored, not More) provides some interesting information about their traceability and possible readership. Only one copy has evidence of an ownership in the sixteenth century (edition of Louvain, 1566): it belonged in 1587 to Baudouin de Glen, the abbey of Hénin-Liétard

(Belgium). Seven copies were possessed in the eighteenth and the nineteenth centuries, one of which, for instance (the first translation in Italian, 1548, at the National Library of Portugal) by Cypriano Ribeiro Freire, the first ambassador of Portugal in the United States of America (1794–1799); more interestingly, three copies of seventeenth century editions were owned by the Colegio de Santa Rita (Coimbra), founded in 1750. Of the two oldest editions of *Utopia* located (Basel, 1518, at the National Library), one belonged to Paulo de Carvalho Mendonça (1702–1770), the other, to an unidentified cleric of Isny (Switzerland), Paulus Fredericus Renzius.

None of these copies gives evidence of ownership or readership in early modern Portugal. Of course, it does not prove nobody read or possessed More's works but it suggests that we would better think in terms of a late, rather than an early, reception, at least, from the second half of the seventeenth century.

Fernando Moser and Pina Martins agree on the wide influence in early modern Portugal of Thomas More and his best-known text. But the reduced number of short references, which total fewer words than the instructions of the Indexes, say little in favor of a wide and continuous influence. Interestingly, Moser states that the printing of More's narrative in Portugal was not "essential" at the time (Moser, 20). This remark brings to the forefront the Peninsular dimension of our issue and that of the book market. Moser's argument is based on the fact that many prints from Europe and "even from Spain" were in circulation. What does this mean in practical terms?

By mid-1500, the *Utopia* was translated in vernacular (in a chronological order: German, Italian, French, English and Dutch). The majority of the non-cultivated Europeans could read or hear it in the common language. In the Renaissance, the Spanish works were dedicated to the life and martyr of Thomas More (Pedro de Ribadeneyra, 1588; Fernando de Herrera, 1592, 1617). As to a Spanish version of the *Utopia*, it was not printed until 1637, at Cordoba (and the first Portuguese translation appeared exactly three centuries later, in 1937). The translator, Geronimo Antonio de Medinilla i Porres probably followed Sanso-vino's *Del governo de regni et delle republiche antiche et moderne* (1561) who, contrasting with the full version of 1548, only translated the second book. A long note on chapter 9, "On religion", seemingly written by him, reminds the necessary Spanish and Portuguese censorships on saints and Fathers of the Church to avoid misinterpretations of their texts written in other times. Despite his highest prestige (the word martyr appears six times, once "Santo Martyr Tomas Moro"), even More had to be corrected. But Porres warns his reader about the dangers of the chapter that deals with diversity and variety of religions. More, a true Catholic, was misinterpreted on this point by atheists and politics ("el Ateista, i Politico", More, 1637, viii vº). Skepticism and machiavelism are the enemies he points out. This note, similar to a theologian's opinion, clearly reflects the problematic reception of More's *Utopia*. From the Spanish side, legalism, that is obedience to the partial prohibition of a martyr's writing, and fear of the effects of his originality. This note confirms, *a posteriori*, the Portuguese fear of the *«abhorrenda»*, possibly not only of religious kind.

In a fascinating article, Sanford Kessler argues that "More wrote *Utopia* partly to promote religious freedom for Christians" (Kessler, 211). He insisted that on this point the English humanist preceded from long John Locke's letter on toleration (1689). The strong Portuguese opposition to *Utopia* and its cautious and late diffusion in Spain are facts that tend to consolidate his thesis.

5 CONCLUSION

Comparing the Indexes of 1612 and 1624 and taking into account the conclusions of Bujanda on the sixteenth century Indexes, Payan Martins disagrees with I.S. Révah claim that the Portuguese did not participate significantly in the book censorship (Martins, 2011, 71). Taking into consideration the Indexes, none is a mere compilation or plagiarism of the previous. If we calculate the number of new entries of authors and works, limited to the field of science and humanities, in the six expurgatories published until 1624 (two Spanish, two Portuguese, one Dutch and one Roman), we can observe that the Peninsular is responsible for 47% of the total (484 entries until 1640). In this percentage, both Portuguese Indexes added almost 16% of new entries, the Spanish approximately double this (31,19%). In relation to the dimensions and capacities of these two markets, the censorship of Thomas More by the Peninsular inquisitors provides a good example of a clear rehabilitation of Révah's conclusions, at least according to the sixteenth and seventeenth centuries. The Portuguese strongly and originally contributed to the catholic crusade against heterodoxy. They significantly in-creased the list of works to cleanse with names like Kepler, Sannazaro or Cervantes. Vipers hissed all around.

ACKNOWLEDGEMENT

I wish to thank Russell Williams (The American University of Paris) for his corrections to this paper.

BIBLIOGRAPHY

Bataillon, Marcel. *Érasme et l'Es-pagne*. Genève: Droz, 1989, vol. 1. 903 p.

Baudry, Hervé, "Tuto lege". A microcensura dos livros de medicina em Portugal (séc. 16–17). Balanço estatístico-metodológico e perspeti-vas (to be published)

Érasme de Rotterdam. *La Philosophie chrétienne*. (Edited by Mesnard, Pierre). Paris: Vrin, 1970. 399 p.

Gibson, Reginald W., St. Thomas More, a preliminary bibliography of his works and of Moreana to the year 1750, New Haven (Conn.), London: Yale University Press, 1961. 499 p.

Kessler, Sanford. «Religious Freedom in Thomas More's *Utopia*». In *The Review of Politics,* Vol. 64, No. 2 (Spring, 2002), pp. 207–229

Lopez Estrada, Francisco. *Tomás Moro y España*. Madrid: Editorial de la Universidade Complutense, 1965, 120 p.

Manning, Patricia. Voicing dissent in seventeenth-century Spain: Inquisi-tion, social criticism and theology in the case of El Criticón. Leiden; Boston: Brill, 2009. 323 p.

Martins, José V. de Pina, «L'*Utopie* de Thomas More au Portugal (XVIe et début du XVIIe siècle)». In Moser, Fernando de Mello; Martins, José V. de Pina. *Thomas More au Portugal*. [Braga: Barbosa & Xavier], 1983. pp. 37–91.

Martins, Maria T. Payan. « O Índice Inquisitorial de 1624 à luz de novos documentos ». In *Cultura. Revista de História e Teoria das Ideias*, vol. 28 (2011), pp. 67–87.

More, Thomas. Utopia de Thomas Moro, traducida de Latin en Castellano por Don Geronimo Antonio de Medinilla i Porres. En Cordova: Salvador de Cea, 1637, in-8º

Moser, Fernando de Mello, «More's Early Reputation in Portugal». In Moser, Fernando de Mello; Martins, José V. de Pina. *Thomas More au Portugal*. [Braga: Barbosa & Xavier], 1983. pp. 25–33.

Pierrot, Claire. *La fortune de l'*Utopie *de Thomas More en France à la Renaissance*. PhD Dissertation, University of Paris X Nanterre, 2002. 631 f.

Tomás, José Pardo. Ciencia y Censura. La Inquisición Española y los libros científicos en los siglos XVI y XVII. Madrid: CSIC, 1991. 390 p.

The utopia of a healthy land: Leprosy reports in Portuguese colonial America

Ana Carolina de Carvalho Viotti

Department of History, São Paulo State University, Franca, São Paulo, Brazil

ABSTRACT: Pero Vaz de Caminha has written, in the first letter about America that "the land itself has a good atmosphere". This statement would be repeated throughout the colonial period in chronic, letters, travel reports, and other books, until the paradisiac image of Brazil was replaced by an idea of an unhealthy place to be in. One of the elements that corroborated with such turn around on the descriptions about the tropics was the emergency of all sorts of diseases: fevers, smallpox, syphilis, measles, animal bites made the life in the South become tough. One of them, well known of the Europeans—and believed not to exist on such land—called the attention and the worries of the settlers' leprosy. Believed to be a sign of God's anger with human behaviour, leper meant both insalubrity to the place and punishment to men.

The purpose of this brief article is to follow the paths of the procedures taken against such contagious disease in colonial Brazil, highlighting the main actions to prevent the spread of decease and the sins all over the Portuguese America. It is argued that slaves were blamed as the main responsible for the dissemination of that evil. Leper challenged the utopia of a healthy land, and the utopia of a cure was nurtured by the isolation, a sort of "death in life".

Keywords: Health, leper, Colonial Brazil, Leper Hospitals

And the leper who has the disease on him is to go about with signs of grief, with his hair loose and his mouth covered, crying, Unclean, unclean. While the disease is on him, he will be unclean. He is unclean: let him keep by himself, living outside the tent-circle. (Leviticus 13: 45–46)

[...] here, everyone lived ceaseless horror; people have been seeing the propagation and communication of something as a voracious fire, it is a harm that turns even the most beautiful body in something misshapen and disgusting. It is the morphea, the leper, or St. Lazaro's disease, and as there is no proper hospital to heal them, the leper kept roaming all over the streets, frightening and horrifying all the inhabitants. (Carta... 25 Jul. 1788)

Both extracts belong to texts and authors from distinct periods: the first one is part of the Holy Bible's Old Testament, specifically from the Book of Leviticus—part of a whole chapter devoted to leprosy, indeed—and the second, a letter sent in 1788 by the *Câmara de Pernambuco* officials (or Senate House), in Brazil, to the Queen Maria I. Although the chronological and geographical distance between those writings, it is possible to find some approaches on their perceptions about the lepers, now identified as those who have the Hansen's disease[1]: horror, revulsion, aversion. The need for isolation of these individuals, not to offend or infect other people or the environment around them, seems very high. Even with the charitable example of Francis of Assisi—who preached that loving those who were taken as impure was a sign of holiness[2]—and the action of religious orders in gathering these individuals at the early modern period, the physical distance of these diseased had been reaffirmed over the centuries. In Brazil, at least until the eighteenth century, this pattern of action against the infected seems to have persisted, especially when the places where the lepers should stay or where they would be healed were chosen.

1 LACK OF MENTIONS

In the first writings about the tropics, there are no register of such morbid manifestation. The absence of descriptions of lepers in the main documents produced by the colonists is, certainly, the pattern among the authors, expressly in the letters of the Jesuits and on general chronic. In the writings from André Thevet (1502–1590) and Jean de Lery (1536–1613) this subject simply does not appear; Gabriel Soares de Souza (1540–1591) spoke of *boubas*, wounds and fevers, but not leprosy; William

Piso (1611–1678) scrutinized Brazil's flora, fauna and diseases without even mentioning the *morphea*, just to name a few of the most important authors about colonial Brazil.

The lack of mention about ill Indians has opened a gap for discussion about the origin of leprosy in the tropics, mainly among the nineteenth and twentieth historians. While many authors had considered such illness as a heritage from Portuguese settlers, many others had stated that the slaves had direct relation with that evil disease, a disease associated since ancient times with dirtiness, impurity, and lust. Luís dos Santos Vilhena (1744–1814), in one of his many letters about America, had written that:

[...] Another principle for the forfeiture of health in Bahia is the twenty vessels from Africa coast that land here, with infected blacks because of their scurvy, bladders, measles, *boubas*, syphilis, scabies, and finally the leper plague, as we all believe." (Vilhena, 1922, 155–156; 171–172)

Furthermore, the data that the most affected by the disease were the slaves was used as some kind of proof that it had an African origin.

2 BUILT AND IMAGINED HOSPITALS IN THE NORTHEAST

Either in the province of Bahia, we are told that, in 1640, in the city of Salvador, one Christian brotherhood has created the Field of Lazarus. That would be the oldest site to lepers in Brazil, built in a specially chosen place where those patients could dwell. The first official representation of the Câmara to the King D. José requesting permission to charge the St. Lazaro Royal tax, which was essential to maintain the isolation hospital, however, dates from 1755 (Representação..., 5 Jul. 1755). Five years later, the Conselho Ultramarino (Overseas Council) refers to the king on such representation, indicating that the place would "collect people suffering from contagious diseases" (Consulta..., 6 mai 1760). D. Rodrigo de Menezes inaugurated St. Christopher's Leper Hospital, finally, in 1787.

There are also some accounts (Souza-Araújo, 1946, 19–25) on the foundation of an asylum for lepers in 1714 in Recife, Northeastern Brazil, which was a strategic port. This first leper hospital in Recife would be the result of Father Antonio Manoel's efforts, and it was based in his own home. It seems, however, that it has become too small for the high number of patients in a short time. In 1754, there is the request to continue to:

... work on the property [...] because it would result in so many good things in this Bishopric [...] and because the great amount of money there was invested there [...] if that was not made, all the incurable poor people there would be injured [...]. (Consulta..., 6 mai 1760)

Apparently, these works have not been carried out, as shown in another letter from the Câmara to the Queen Maria I in 1788: they complain about the previous administration and say that:

Now the leper hospital is going to be built, because the far-sighted governor [José Tomás de Melo] is doing all the necessary to establish such a useful place". (Carta..., 25 Jul. 1788)

Two years later, Melo himself writes to the Secretary of State of the Navy and Overseas. He explains "the need to separate the other people from the many lepers who are on the streets" and that had ordained "to all the counties to seek for the sick people and send them to the proper place". Rather than inform the Crown and the Overseas Council about his achievements as governor, he wanted to remind that it was a hospital "whose income was not right", (Ofício..., 1 Jul. 1790) and that the doles have all been used for the site's construction. He also claims that the money was needed to guarantee the help to those miserable lepers, as the hospitals of Bahia and Rio de Janeiro had been doing.

3 EFFORTS IN RIO DE JANEIRO

Taken on that occasion as a model to be followed, Rio de Janeiro struggled to find a place proportional to the large number of infected patients in the streets. In 1697, Dom Pedro II, the pacific, King of Portugal, authorizes the foundation of a leper hospital in the town as long as the governor Artur de Sá e Menezes and the Câmara handled with the funding and maintenance of the establishment. Without such amount of money, they said, the hospital was not built. After nearly half a quarter of a century, in 1740, the Overseas Council issues a favourable letter to the granting of funds for that purpose: they had been noticed about an unprecedented epidemic of leprosy in the city. The new request brings alarming statistics about the contagion: they talk about three hundred injured—from an estimated twenty thousand people population[3]—, about the urgency to provide medications to the dying people and also the opinions of physicians and surgeons, who warned that patients should be "assisted briefly with proper medicine or Rio's inhabitants would all get infected" (Parecer..., 24 Nov. 1740).

About this hospital, it is only known that a Royal Order (Souza-Araújo, 1946, 33) authorized its construction and that Gomes Freire de Andrade, Count of Bobadela, Governor and General Captain of Rio de Janeiro, built it with by his owns expenses (Pinheiro, 1893). The place chosen was located at the outskirts of the city, in the neighbourhood of Saint Kitts, in order to take them out from the city centre. In the Overseas Archives, repository of the exchanged correspondence between Brazil and Portugal, there is a record of a new consultation in 1760, requesting the foundation of a (new?) leprosy hospital in Rio, including a proposal for architectural layout with two large houses—one for men and one for women—, separated houses for the chaplain, the pharmacy, the surgeon and the other servants, two altars for the celebration of Masses. The place should also have running water and it must be finished within four years (Consulta..., 29 May 1760). This one, indeed, seems to have been created, but soon found some difficulty to continue working: according to another letter, in 1793, the Viceroy reveals that, although the construction of the previous hospital had been very useful, its location not provided the necessary safeguards to the other inhabitants remain immune, since this hospital was very close to the city centre (Ofício..., 18 Feb 1793).

Other requirements concerning this matter can be found in the captaincies of Maranhão, Espírito Santo, and São Paulo (Ribeiro, 1971, 29), as well as reports of the many lepers in Minas Gerais, Mato Grosso, and probably in Pará (Souza-Araújo, 1946, 108; 133–134).

4   THE NEED OF ISOLATION

Among them all, the clearest message is the need to provide the stigmatized with some kind of treatment, bringing them together in one place not to spread their illness, and that place, to be suitable, should be far from the city. It is important to note that the organization of urban areas, previously not valued by the Portuguese Crown, starts to play a special role on the political aspects, of course, but also on disease control and maintenance of public health: in Rio de Janeiro, for example, Brazil's capital and seat of the Viceroyalty since 1763, the second half of the eighteenth century became the stage to "the first systematic efforts to control the city and the population related to the interests of the State". Here we see the very first beginning of notions such as public health, derived from the eighteenth-century Europe. This model will replace the individual attention model to an accurate observation of collective, to the prevention: both will be the basis of health policy

in nineteenth century Brazil, namely the "social medicine" (Machado).

From the many procedures taken by the Marquis of Lavradio (1729–1790), after there was featured in and reaffirmed by scholars and travellers that the city where he established residence was unhealthy— because its fetid air, wetlands, moisture, heat, waste and lepers on the streets (Cavalcanti, 2004)—one is particularly interesting.

5   BLAME ON THE SLAVES

It is also Lavradio who says that the slaves, "as soon as they give their input on the Customs by the sea, should be boarded for a site called Valongo, which is on the outskirts of the city, and they must be separated from all communication", so they would not infect Brazil with their plagues. It is interesting, essentially, because this kind of quarantine recommended to the slaves also appear, for example, in the aforementioned letter on the lepers, sent by the officers of Pernambuco in 1790, which asked for annual receipt of 200$000 to ensure that slaves did not would transmit the "plague" to others.

It is known that many slaves came from endemic regions of leprosy. However, is it a strong reason to affirm that they have brought leprosy to America? Here we combine the question asked once about the responsibility for the introduction of the stigma in Brazil, with the purposed quarantine only to slaves. Well, as leprosy is not an easy-to-hide disease, it seems obvious that a slave, who was carefully examined by the potential buyer, would be passed over in case of apparent pustule.

The slave with active leprosy was deformed, disgusted by his appearance and judged as extremely contagious by scholars and laymen, which suggests that even with lower prices, few would be interested on buying them, and those who perhaps purchase them, they were more aware of the "product" in hand (Souza-Araújo, 1946, 11–12). There are some accounts on runaway slaves that say that they sought refuge among the lepers, in the colonies, to avoid capture and return to captivity, due the fear of contamination and spread of leprosy among masters and other slaves (Moura, 2004, 106). In addition, the leper slaves were usually freed, because keeping them in a proper hospital was expensive and no slave owner would pay for that. In this sense, it is unlikely that the introduction of the disease has been made by slaves and more plausible to claim that they have acted as major agents of internal dissemination, due the poor living conditions and high contact with all sorts of people and things to which they were exposed.

6 CHANGES IN THE NINETEENTH CENTURY

Throughout the nineteenth century, leprosy became a research subject to doctors, as a disease that should be observed under the light of new scientific parameters, in Europe (Gould, 2005, 2) as much as in Brazil. From the pages of memories about Brazil, from a wound in travel reports or public calamity in the letters to Portugal, leprosy emerges as a theme in "discussions of scientific associations, medical journals, and theses submitted to the Faculty of Medicine of Rio de Janeiro" (Cabral, 2006, 35–44). It is a new therapeutic scenario with schools, universities, academies and specialized journals in the medical field, a fruitful environment to the discussion of Brazilian diseases, the healing possibilities not yet explored, the study of new drugs, climate, plants of the tropics. Altogether, it is a new relationship between health and disease, which included rethink leprosy issues. Even so, exclusion and isolation will only be questioned as an essential means of preventing or "treatment" of leprosy in the early decades of the twentieth century, at least in Brazil.

7 "DEATH IN LIFE" *VERSUS* THE UTOPIA OF HEALTHINESS

It is observed, therefore, with the isolation of the stigmatized, in general, and the leper slave, in particular, a kind of death in life, at least regarding to the social one, caused by visual disgust of their illness and the removal of these people, voluntary or compulsory, from the common living. The slave, nonetheless, was doubly victimized with this social death. The leprosy hospitals will eventually become a different and particular place to establish social relations, limited to stigmatized and their caregivers—usually members of Christian charity and slaves under punishment—where families were raised, as much as groups and labour relations, all internal to the asylum.

By isolating lepers from the view of passers-by, the idea that the disease and the sins were being contained and cured took shape. It would not be wrong to assume, in this sense, that the leper colonies could give the impression of a more wholesome town. Given that the disease, as was signalled, continues to be seen as a kind of punishment until the mid-nineteenth century and that healing was possible only by a miracle, a sort of prevention and cleaning utopia is supplied with insulation.

From the idea of a land "of good atmosphere" to an unhealthy one, many reports full of diseases have emerged, some culprits were elected as vectors of evil and some steps were taken to contain the spread of morbidities. The lepers had the signs of disease and blame enrolled on their bodies, and, to their peers, the city and the ideal society did not have room for them.

NOTES

[1] Hansen published his findings "Causes of leprosy", as part of his annual report for 1873 to the Norwegian Medical Society. He says that he identified the bacillus in 1871 (Hansen et al., 1895, p. 31).
[2] Lepers were unclean, but loving them was, as Francis of Assisi had shown, a sign of sanctity (Edmond, 2009, 1).
[3] The population of Rio de Janeiro increased from 12,000 in 1706 to 24,397 in 1749. For the latter number Southey, R. 1970, 813). Southey's first book was published in 1822.

BIBLIOGRAPHY

Cabral, Dilma. Lepra, morféia ou elefantíase-dos-gregos: a singularização de uma doença na primeira metade do século XIX. *História Unisinos* 10(1): Janeiro/Abril 2006, 35–44.

Carta dos oficiais da Câmara do Recife à rainha [D. Maria I], sobre as providências do [governador da capitania de Pernambuco], D. Tomás José de Melo, relativas ao dique do Recife, à Casa dos Expostos e ao hospital dos lazarentos, à administração da justiça e do abastecimento. 25 Jul 1788.

Carta dos oficiais da Câmara do Recife à rainha [D. Maria I], sobre as providências do [governador da capitania de Pernambuco], D. Tomás José de Melo, relativas ao dique do Recife, à Casa dos Expostos e ao hospital dos lazarentos, à administração da justiça e do abastecimento. 25 jul 1788.

Cavalcanti, Nireu. *O Rio de Janeiro Setecentista*. Rio de Janeiro: Zahar, 2004.

Colônia de leprosos, escravos fugidos na. In: Moura, Clovis. Dicionário da escravidão negra no Brasil. São Paulo: EDUSP, 2004.

Consulta do Conselho Ultramarino ao rei D. José sobre a representação dos Oficiais da Câmara da cidade da Bahia em que solicitam autorização para imporem a contribuição do real de São Lázaro, para com o seu produto fundarem um Lazareto para recolha de pessoas portadoras de doenças contagiosas. Anexos: Requerimentos, pareceres, consultas, ofícios, termos, lembretes. 6 mai 1760.

Consulta do Conselho Ultramarino ao rei D. José sobre a representação dos Oficiais da Câmara da cidade da Bahia em que solicitam autorização para imporem a contribuição do real de São Lázaro, para com o seu produto fundarem um Lazareto para recolha de pessoas portadoras de doenças contagiosas. Anexos: Requerimentos, pareceres, consultas, ofícios, termos, lembretes. 6 mai 1760.

Consulta do Conselho Ultramarino ao rei D. José, sobre o requerimento de Francisco José da Fonseca, solicitando autorização para fundar um lazareto, no qual se possam recolher as pessoas que contraíram a doença da lepra no Rio de Janeiro. Anexo: cartas (cópias), carta, provisões (cópias), termo, auto de vereação (cópia), ofício e proposta. 29 mai 1760.

Costa, Jurandir Freire. *Ordem médica e norma familiar.* Rio de Janeiro: Edições Graal, 2004.

Edmond, Rod. *Leprosy and Empire.* A medical and cultural history. Cambridge, 2009.

Gould, Tony. *Don't face me in. Leprosy in modern times.* London: Bloomsbury, 2005.

Hansen, G.; Carl, Armauer and Looft. *Leprosy: in its clinical and pathological aspects* (translated by Norman Walker). Bristol, John Wright, 1895.

Machado, Roberto et. al. *Danação da norma: a medicina social e constituição da psiquiatria no Brasil.* Rio de Janeiro: Edições Graal, 1978.

Ofício (cópia) do [vice-rei do Estado do Brasil], conde de Resende, [D. José Luís de Castro, ao [secretário de estado da Marinha e Ultramar], Martinho de Melo e Castro, sobre o tratamento dos pacientes no lazareto, em São Cristóvão, criado pelo vice-rei, conde da Cunha, [D. Antônio Álvares da Cunha]; indicando a mudança do referido hospital para outro local devido a proximidade com a cidade do Rio de Janeiro. 18 fev 1793.

Ofício do [governador da capitania de Pernambuco], D. Tomás José de Melo, ao [secretário de estado da Marinha e Ultramar], Martinho de Melo e Castro, sobre as providências para construção de um hospital para os lazarentos a fim de os recolher e manter. 1 jul 1790.

Parecer do Conselho Ultramarino sobre a carta dos oficiais da Câmara do Rio de Janeiro, acerca do contágio da lepra que se tem alastrado naquela cidade, solicitando a criação de um lazareto, recomendando o conselho que se aplique no tratamento das pessoas já contaminadas, não só o acréscimo do donativo, como também a sobra do produto dos soldos dos governadores, utilizados na reforma da casa da Câmara. 24 nov 1740.

Pinheiro, F.B. Marques. Hospital dos Lázaros. *Irmandade do Santíssimo Sacramento da Candelária,* Rio de Janeiro, 1893, vol. 2.

Representação dos oficiais da Câmara da cidade da Bahia rei [D. José] solicitando a faculdade para imporem a contribuição Real de São Lazaro para poder fundar um Lazareto. 5 Jul 1755.

Ribeiro, Lourival. *Medicina no Brasil Colonial.* Rio de Janeiro: Sul Americana, 1971.

Southey, Robert. *History of Brazil: part the third.* New York, Burt Franklin, 1970.

Souza-Araújo, Heraclides-Cesar de. *História da Lepra no Brasil: período colonial e monárquico (1500–1889).* Rio de Janeiro: Imprensa Nacional, Vol. I, 1946.

Robertson, Jo. Leprosy and the elusive M. leprae: colonial and imperial medical exchanges in the nineteenth century. *Hist. cienc. saude-Manguinhos,* Rio de Janeiro, v. 10, supl. 1, pp. 13–40, 2003.

Terra, Fernando. Lepra no Rio de Janeiro: seu aparecimento, freqüência e formas. *O Brazil-Medico,* 33(6), 1889, pp. 41–44.

Vilhena, Luiz dos Santos. Notícias soteropolitanas e brasílicas (1780–1802). 2 volumes. Imprensa Official do Estado da Bahia, 1922.

A project of an ideal agriculture for Brazil in the 18th century

Milena da Silveira Pereira
Post Graduation Program in History, School of Human and Social Science, University of the State of Sao Paulo, Sao Paulo, Brasil

ABSTRACT: The Portuguese 18th Century became notable, among other happenings of political and social order, for the attempt of its men to forge a scientific knowledge, linked to reason and experience. On this stage, agriculture was on the main pillars of discussion and yielded many of the pages produced at the Academy of Science of Lisbon. Based on the analysis of the first volume of *Economic Memories* published by that institution, this paper intends to produce a brief essay on the indications for improving agriculture in Brazil in the 18th Century. In other words, it intends to map out what was reported about the use of natural resources in that colony at a time when knowledge in Portugal sought to support itself on new parameters. Therefore, this proposal is developed in the sense of apprehending some of the ideals of agriculture written on the pages of the Academy of Science of Lisbon to promote the kingdom and its main conquered land.

Keywords: 18th Century, ideal agriculture, Brazil, Academy of Science of Lisbon

In 1789, the first volume of *Economic Memories for the Development of the Portuguese Agriculture, Arts and Industry*, the famous Italian naturalist Domingos Vandelli presented a study on the natural products of the Portuguese colonies, highlighting that, among those production activities, "the gold mines are those that are the dearest, and that are universally more cared about than agriculture" (Memorias, 1789, p. 187). To him, the gold mines should not be the main concern and objective of work in Brazil, but rather "its other natural production obtained through agriculture". (Memorias, 1789, p. 188). All of the most "wise politicians", carries on the erudite in his diagnosis, are well aware of the mistake that is in the exaggerated appreciation of the gold mines; in other words:

> Those who have the mines of the most precious metals and use them to make their patrimony, are less rich than those who care on Agriculture, Arts and Trade.

The tone of the memories of that professor, who was ahead of the major educational reform and of the introduction of studies on natural history in Portugal, denounced the state of the Portuguese economy and warned about the need to value and take advantage of agriculture in its main colony, considering the gradual exhaustion of the mines of precious metals at the end of the 18th century. Hence, it is known that the interest in the Brazilian nature appeared on the Portuguese writings since the previous century, having been intensified during the second half of the 18th century, among other reasons, due to the royal sponsorship of memories and handbooks on agricultural practices, with the goal of promoting the production of raw materials and the revival of agriculture in Portuguese-Brazilian lands (Serrão, 1988; Dias, 2009). In this context, roughly put, the writings on natural history have gained a growing place in the Portuguese publications about Brazil, and nature took a position of holding wide, and yet unexplored, economic possibilities. Making up this scenery, in 1772 an academy of science was founded in Rio de Janeiro. Named Scientific Academy of Rio de Janeiro, that society aimed to exam all natural productions belonging to plant, animal and mineral kingdom, making "all sorts of analysis and observations that could be possibly made, in order to give the public, every month, the full notice of the findings as they are achieved". (Lavradio, 1978). Despite its short existence, there is no denying the pioneering of this initiative of an illustrated viceroy who, somehow, was carried out by the Royal Academy of Science of Lisbon, founded in 1779.

Despite being notorious the changes in the Portuguese intellectual atmosphere and interests in the land of the Portuguese America, it was only with the Royal Academy of Science of Lisbon that the scientific knowledge began to echo with greater vibration and those writings on the use and the rational knowledge about the land gained strength and became published systematically. Such scientific conception announced by the erudite men of that society, which favours the desire to learn and to adopt new points of view on arts and sciences, became affiliated, among other names, to the stream of thinking of the French encyclopaedists, which played a decisive historical role in establishing pragmatic relationships between the literate and the rest of society. Diderot (2015), for instance, claimed "an increase in natural science, anatomy, chemistry, and experimental physics as a first step in the reform of society". Science, in this stage, won a social dimension and the so-called scientific studies came to be understood as capital for the progress of nations. In that sense, the Royal Academy of Science of Lisbon sought to develop a knowledge that would instruct the people and promote the State and its economy. That new scientific approach, as remembered by the Academy members, argued that such a course not only offered "abundant field to the operations of chemists and anatomists, but [enriched] the farmer, the merchant and the artist, and with them the State" (Memorias, 1797, prologo); in other words, it would be a sort of exaltation of the literate and the scientists as the practical man and the man of action, who would carry out the improvement of society through the inventions and discoveries that are useful to the welfare, health and benefit of the country. Thus, that approach would accomplish the "public utility", as defended by the Academy members.

In this quest of Portugal for a scientific knowledge linked to reason and experience, agriculture was one of the main pillars of discussion and yielded many of the pages produced at the Royal Academy of Science of Lisbon. Hence, let's see what kind of mention the Portuguese have made about agriculture in Brazil; more specifically, what was reported regarding improvements in the use of the natural resources of that colony at a time when the knowledge in Portugal sought affirmation based on new parameters. Although this space is rather small to carry out a mapping of the many volumes of the published memories, we attempt to draw a brief outline of the projects of an ideal agriculture for the Portuguese America that are presented in the first volume of *Economic Memories*.

From start, it is noteworthy that the first appearances of Brazil in the production of the Academy of Science were a little shy, but anyhow revealing of the new scientific interests of the Portuguese for their colony[1]. Getting back to the initial words of Domingos Vandelli, the concern in reversing the economic model resulting from the so considered mistaken Portuguese strategy that privileged the extraction of precious metals at the expense of agriculture was constant in those first writings of *Economic Memories*. Not only the tireless Italian

naturalist registered in those memories the losses caused by such economic choices, but also the famous minister D. Rodrigo de Sousa Coutinho, in Memory *about the true influence of mines of precious metals in the industry of nations which possess them, specially Portugal, and* the specialist in astronomy and populations José Joaquim Soares de Barros, in *Memory about the causes of different populations of Portugal in different times of monarchy.* The first one, in spite of presenting a detailed study of such a scenery does not directly address the Portuguese America, our focus of analysis. Soares de Barros, on his turn, when analysing the major causes of the most notable variations amongst the Portuguese population, highlights the discovery of precious metals and stones when referring to the population of Brazil, and to whom:

> … imagination fired up and a growing number of individuals started going from this kingdom to that country, and as well as those who were already there, they despised Agriculture for risky fortunes. (Memorias, 1789, p. 135)

For that Portuguese man of science, the discovery and the abundance of gold had brought a "misleading rest" to the Portuguese nation, since activities such as agriculture and ways of occupying the space which effectively would make Portugal and its colony develop had been left in the background by the interest and greed of the kingdom for the "most precious metals and the finest stones." (Memorias, 1789, p. 135). As it turns out, these findings about the neglect of agriculture denounced the interest of those scholars for presenting practical solutions for this key sector of the Portuguese-Brazilian economy.

And it was Vandelli, living up to the ideas of "prompt utility", who described in more detail some practical measures and notes of an ideal agriculture in Brazil. In *Memory about some natural productions of the conquered lands, which are either little known or poorly used,* the Italian naturalist, as the title itself suggests, established a sort of guideline on how to use the natural resources and to obtain more profits in the Portuguese conquered lands. Having said that, among the potentially advantageous animals for the economy in Brazilian lands, Vandelli highlights "the little use of the soft *Caviacobaya do Brasil* skin"—a sort of guinea pig—, besides the "little use of the *Porco Tajacú* meat"—wild pig. More detailed is the analysis of the use of cattle, which could "produce cheese and butter for the whole kingdom and for international trade, thereby avoiding the large extraction of money that each year leaves Portugal after these goods". (Memorias, 1789, pp. 190–191). From the cattle slaughtered in Brazil, continues the erudite, the "largest part is for taking out the leather", while "more could be gotten out of the penis, which when dried and shredded serves as thongs" and "out of the tendons or the ligaments of the neck" with which one could make "strings for the springs of carriages much better than those made of iron or wood". (Memorias, 1789, p. 191). From the birds, "dear for their colours", we have feathers of the Emu, for example, "for the ornament and fluff to the hat factories", and suggests expanding the fishing of whales and sperm whales, which, up to that time was made only on the island of Santa Catarina and in the Todos os Santos Bay. Presented as a "fishing genre of much use", Vandelli, making use of the strategy of describing successful foreign examples to serve as a model to the Portuguese, demonstrates the case of the Dutch who with whaling "in the year 1697 earned more than two million Florin". (Memorias, 1789, p. 191)

Regarding the ideal plantations, the scholar highlights the rice, a "plant also natural from Brazil," which appeared in abundance in these lands thanks to the incentives of Mr. Martin Mello e Castro and had generated good savings to Portugal, given that earlier the rice was purchased from "Carolina, which in the year 1740 earned 800 pounds with the crop, most of which had been paid by Portugal." (Memorias, 1789, p. 189). And "with the efficiency and zeal" of this Minister Secretary of State for the Marine Business and Overseas Dominions, the anil (indigo plant) culture of excellent quality prospered, "not only to our factories but also to trade with other nations." (Memorias, 1789, p. 189). Thereby, Vandelli exalted the initiatives of some men linked to the Kingdom for the knowledge and use of Brazilian lands in order to generate income for Portugal and to prevent the import of large amounts of goods from foreign nations.

Concerning the "vegetable kingdom," Domingos Vandelli presents, among other descriptions of the flora and even medicinal plants, a table of the main wood for construction in Brazil, highlighting trees such as the Sucupira that "serves for ships"; the Jequitiba "for masts, grupės, spars and topmasts"; the Jataipeba "for doors and windows of houses"; the Jacarandá "for all sorts of things for houses"; the Cedar "for the figures on the wreath, lions and more carvings and images on the stern" (Memorias, 1789, p. 195); as well as the woods and bushes for dyeing, as the Urucu and the Arariba tree of Pará. In addition to highlighting the pursuit of naturalists for the quinaquina in Brazil, a "very useful" tree, and a new kind of "Puchari", "called a precious fruit [...] may well replace the nutmeg ". (Memorias, 1789, p. 197).

Still on the "Vegetable Kingdom", the Portuguese philosopher and physician José Henriques Ferreira presents a study on the economic feasibility of a bush said to be "a pest" in Brazil. The author of *Memory about the Guaxima* tells that as soon as he learned that with the bark of this plant "the country men made strings to hold their beasts, cattle and other ordinary uses", he went after the Marquis of Lavradio in order to present him the Guaxima and its potential for making fibres for navigation and other uses. In addition, this illustrated viceroy, aware of the news, did not delay in sending cables made of this plant for analysis at the Royal Cordage. Henriques Ferreira, in that memory, describes and shows on a table the experiments of strength and resistance conducted at the Royal Cordage with the guaxima-made strings compared with those made with hemp plant. The results were favourable to hemp, however, the Portuguese physician did not hesitate to point out that he had information from "reliable officials" who have tested the guaxima strings and came to the conclusion that they "suffered a lot more work and lasted longer than hemp", meaning that the fibres of guaxima could offer advantages over hemp as they were less heavy and had greater resistance.

In describing that comparison, José Henriques Ferreira wanted to present to the members of the Academy an economically viable alternative to the incipient industry of sacks and cordage in the colony, given the difficulties of cultivating hemp plant and the fact that the Portuguese navy was almost dependent on the Dutch cordage industry. Assuming the defence of the use of this plant for the Portuguese-Brazilian economy, Henriques Ferreira suggested the use of strings or other materials produced with guaxima at least in some activities, since "we already have it, and in such abundance." (Memorias, 1789, p. 6). The proposal of Ferreira, therefore, was to present a study on the possibilities of economic uses of nature in Brazil. In his own words:

A good economy relies on what each one has at home, avoiding [having to] get outside; each genre has its usefulness, and because some have better quality one should not overlook those of lower quality [...] [what means] the richest country is the one that has the most genres. (Memorias, 1789, p. 7)

And it was precisely those images of the rich nature of the colony across the Atlantic that the Portuguese physician sought to make public to the Academy members, confirming to the saying, so dear to the 18th century and to the members of the Academy: to enjoy the advantages and potentialities of nature it was necessary to know the land and the richness it has to offer.

The Portuguese projects for the Brazilian agriculture in the 18th century, finally, craved for an ideal society in the tropics that would kwon how to take advantage of the land and generate income for both the Kingdom and for its colony. Much more could be said about the suggestions and prescriptions for an ideal agriculture in Portuguese America; however, what deserves to be emphasized is the search by Portugal for understanding and proposing solutions to the path slips and answering why some initiatives in Brazil have not worked out. Thus, would the learning of the natural sciences and the economic returns from natural resources be another Portuguese utopia for overseas lands?

NOTE

[1] Out of the twenty writings that make up this volume, eight make some mention to Brazil or are dedicate to it: *Memoria sobre a Guaxima*, by José Henriques Ferreira; *Memoria sobre a transplantação das arvores mais úteis de paizes remotos*, by João de Loureiro; *Memoria sobre as causas da differente população de Portugal em diversos tempos da monarchia*, by José Joaquim Soares de Barros; *Memoria sobre a agricultura d'este reino e das suas conquistas*, by Domingos Vandelli; *Memoria sobre algumas producções naturaes das conquistas, as quaes ou são pouco conhecidas ou não se aproveitam;* by Domingos Vandelli; *Memoria sobre as producções naturaes do reino e das conquistas, primeiras matérias de differentes fabricas ou manufacturas*, by Domingos Vandelli; *Memoria sobre a verdadeira influencia das minas dos metaes preciosos na indústria das nações que as possuem, e especialmente na portuguesa*, by D. Rodrigo de Sousa Coutinho; and *Ensaio de descripçao physica e econômica da comarca dos llheos na America*, by Manoel Ferreira da Câmara.

BIBLIOGRAPHY

Amaral, Ilídio do. *Nótulas históricas sobre os primeiros tempos da Academia das Ciências de Lisboa*. Lisboa: Edições Colibri, 2012. 65 p. ISBN 978-989-689-261-6.

Dias, Maria Odila Leite da S. *A Interiorização da Metrópole e outros estudos*. São Paulo: Alameda Casa Edtorial, 2005. 163 p. ISBN 85-98325-08-02.

Diderot, Denis; D'Alembert, Jean le Rond. *Enciclopédia, ou Dicionário razoado das ciências, das artes e dos ofícios*. Translated by Pedro Paulo Pimenta, Maria das Graças de Souza, Luís Fernandes Nascimento. São Paulo: EdUnesp, 2015. ISBN 978-85-393-0560-5.

Kury, Lorelai; Gesteira, Heloisa. (org.). *Ensaios de história das ciências no Brasil: das Luzes à nação independente*. Rio de Janeiro: EdUERJ, 2012. 328 p. ISBN 978-85-7511-239-7.

Lavradio, Marquês do. *Cartas do Rio de Janeiro, 1769–1776*. Rio de Janeiro: Instituto Nacional do Livro, 1978 (carta de 6 de março de 1772).

Memorias Economicas da Academia Real das Sciencias de Lisboa, para o adiantamento da agricultura, das artes e da industria portuguesa e suas conquistas. Lisboa: Officina da Acdemia Real das Sciencias, t. 1, 1789. 421 p.

Memorias da Academia Real das Ciências de Lisboa. Prólogo. Tomo I desde 1780 até 1788. Lisboa: Typografia da Academia, 1797.

Serrão, José Vicente. O pensamento agrário setecentista (pré-'fisiocrático'): diagnósticos e soluções propostas. In Cardoso, J.L. (org.). *Contribuições para a história do pensamento económico em Portugal*. Lisboa: D. Quixote, 1988, pp. 23–50.

Silva, Maria Beatriz Nizza da. (coord.). *O Império Luso-Brasileiro*. v. VIII. Lisboa: Estampa, 1986. 613 p. Nova História da Expansão Portuguesa.

António Feliciano de Castilho: Towards an agricultural utopia

Cecília Barreira

CHAM, FCSH, Universidade NOVA de Lisboa, Lisboa, Portugal Universidade dos Açores, Portugal

ABSTRACT: António Feliciano de Castilho (1800–1875) belongs to the generation of the early Portuguese Romanticism where Herculano and Garrett stand out as major figures. The work *Felicidade pela Agricultura* (1849) proposes a utopia of happiness and well-being in rural areas. As of Thomas More's *Utopia* or Campanella's *The City of the Sun*, Castilho has built an earthly paradise based on farming. Essentially, the book puts advocates the establishment of agricultural societies. Around this idea of Utopia, I would like to draw attention to several ways to think about reality in the 19th century.

Keywords: Utopia, agriculture, rural societies, romantism

1 CASTILHO AND THE AGRICULTURE UTOPIA

António Feliciano de Castilho (1800–1875) belongs to the generation of the early Portuguese Romanticism where Herculano and Garrett stand out as major figures.

Castilho is currently remembered for his vast poetical production, probably too influenced by Arcadia literary model, However, his notoriety was not exclusively due to poetry. The name of is inextricably linked to the famous cultural controversy known as "Bom Senso e Bom Gosto" when he was highly criticized by Antero de Quental, young student in Coimbra at the time, mostly inspired by German idealism and French authors.

It is very interesting that Antero, during his childhood, was a Castilho's pupil, when the latter was a primary school teacher on the island of São Miguel, Azores.

Some year later, faced with the impetus of the initial Antero's literature, Castilho showed a fundamental misunderstanding of his value. Perhaps it could explain the acrimony between them.

Not so much as a result of his age, and more because of his ideological position, Castilho was part of the so-called "Geração de 50", facing the inevitable consequences of social changes from his perspective, as a man embedded in the *status quo*.

His work *Felicidade pela Agricultura* came out in 1849, utopically preaching the wonders of farm work. It consists of several chronicles published by Castilho in *Agricultor Micaelense* magazine, collected in two volumes.

The main idea behind these pages is his rejection of politics, as is the fact of being a politician was to entail a stain or to participate in a web of dark interests.

Cities were aging and diminishing, becoming a place of perdition, sin and gloom, but countryside was a space of happiness and well-being. This ancient antinomy (rural world versus urban world) also constitutes an obsession for romantic literature: Almeida Garrett's *Viagens na minha terra*, Júlio Dinis's *Pupilas do senhor Reitor*, Cesário Verde's poems or Eça de Queirós's *A Cidade e as Serras*, are all examples of a generic nostalgia hardly facing the reign of engines, steam and gas lighting.

Castilho builds an earthly paradise based on farming. He primarily advocates the formation of farming companies within each administrative capital or Episcopal seat. The Prefect or the Bishop would dignify each community by their presence. Begging and unemployment would disappear. He also proposed the creation of public libraries in the areas of veterinary and agricultural sciences, as well as public sessions on agricultural issues of each region. Through the edition of a farming newspaper, sowings, and plantings, animals' importation and the use of agricultural machinery would be a priority subject. There would be a rural event, which would coincide with the main pilgrimage or religious feast.

None of those considerations could, in fact, be inscribed in a modern understanding of the nineteenth century society, but they were consistent with the Enlightenment values, where order, peace, and social equilibrium were part of a predominantly agricultural economy. For example, to finance those farming communities, Castilho conceived a district system of annual lotteries and charitable donations. According to him, the funds for rural societies might also come from theatrical events, philharmonics, clergy's incomes, or animal rental.

Men like Manuel Adelino de Figueiredo or João Andrade Corvo were studying agriculture in its meteorological, geological and economic dimensions, to promote, for instance, the use some chemical fertilizers, but Castilho was really far from this kind of concerns. At some time, he envisaged a political Parliament solely formed by farmers chosen by each district Agricultural societies would elect the members of the aforementioned parliament.

How ironic that Castilho, the author of *Cartas de Eco e Narciso*, as spent most part of his life in Lisbon, the corrupt and immoral capital.

Happiness. What would it be in this kind of romantic imaginary? Would it be an appeasement, a total absence of conflict, a sort of agony, an equilibrium...? If we follow Castilho's ideas, we have to say yes. Will not be romanticism, in his odd-

ness, an awkward journey through different places, by unbalanced platforms, therefore a conflicting dystopia?

In *Felicidade pela Agricultura* Castilho fiercely proposes a concept of a devoted, chaste, and married woman; a pure offspring and a virtuous nation.

Castilho sees money as an irremovable factor; he does ignore his reproduction and the way it operates in modern societies.

In this utopian world, "the land will make us rich"; we will be powerful by education, united by morality. Wealth, Power and Fraternity will make us happy; they are Civilization.

BIBLIOGRAPHY

Castilho, A.F. de. *Felicidade pela Agricultura*. 2ª Edição. Lisboa. Empresa da História de Portugal. 1903. Depósito legal 18895/8.

Utopia and dystopia in Jorge Barbosa

Hilarino da Luz
CHAM, FCSH, Universidade NOVA de Lisboa, Lisboa, Portugal Universidade dos Açores, Portugal

ABSTRACT: Jorge Barbosa's utopia is disclosed intermittently, alternating between what is essential or desired and the possibility of it being satisfied. The author seeks to ground the origin of the islands in a kind of imaginary expedition returning to those origins, thus attempting an existential void of the archipelago with a myth. In this way, he enhances a tradition recovered by the **Cape Verdean** intellectuals who studied at the S. Nicolau Seminary-Secondary School—mainly Pedro Cardoso and José Lopes—influenced by a teaching model that esteemed Classical culture and Romantic nationalist ideas. This was also due to the handling of the New *Luso-Brazilian Almanac of Recollections* and to the presence of exiles in the archipelago. **Dystopia**, in turn, is marked by the tragic situations Cape Verdeans are forced to face, namely droughts that result in hunger and high mortality rates.

Keywords: Cape Verde; Jorge Barbosa; utopia; dystopia

This article aims to make a brief analysis of utopia and dystopia in the literary production of Cape Verdean writer Jorge Barbosa, also known as *Nhô Jorge*.

Barbosa was born in the island of Santiago on 22 May 1902 and died in Portugal, in Cova da Piedade, on 6 January 1971, from a heart condition. His literary work includes poetry, several *Crónicas de S. Vicente*, letters, some articles, two short stories and an unpublished novel called *Bia Graça*, which was left unfinished. He published the poetry books *Arquipélago [Archipelago]* (S. Vicente, 1935), *Ambiente [Environment]* (Praia, 1941), *Caderno de um Ilhéu [Notebook from an Islet]* (Lisbon, 1956) and left a number of poems unpublished, scattered in journals, magazines and newspapers, such as *Jornal da Europa, Seara Nova, Presença, Diabo, Cabo Verde, África, Mundo Português, Momento, Descobrimento, Cadernos de Poesia, Atlântico, Aventuras, Notícias de Cabo Verde, Mensagem, Boletim de Cabo Verde, Fradique, Claridade* and *Diário de Notícias*.

As is known, Cape Verde is a small insular state composed of ten islands and some islets divided into two groups, according to their position in relation to the prevailing north-east winds: Barlavento (Santo Antão, S. Vicente, Santa Luzia (uninhabited), S. Nicolau, Sal, Boa Vista, and the uninhabited islets of Pássaro, close to S. Vicente, Branco and Raso, close to Santa Luzia) and Sotavento (Santiago, Maio, Fogo and Brava, as well as some equally uninhabited islets such as those of Santa Maria, close to Cidade da Praia, and the islets of Luís Carneiro, Sapado, Grande and Cima, close to Brava).

It is a country where biological miscegenation, the healthy getting together of people and the merging of values and cultural elements turned the archipelago into a Luso-tropical utopia. Therefore, it would become an exemplary prototype in which the revival of insularity prevailed. Thus the «claridosos», in their identity search, namely Jorge Barbosa, felt close to Atlantis. Not only did they reconsider the designation given to the legendary submerged continent, but Jorge Barbosa also opened the book of poems *Arquipélago* (1935), connecting it to Atlantis, with the poem «Panorama» (Panorama), as we will see further ahead.

Adalberto Carvalho believes that:

> When we speak of utopia, we are faced with an area full of ambiguities, of different opinions and views concerning its status, and even with very diverse and subjective value judgements. (Carvalho, 2001, 145).

In that line of thought, Carvalho feels there are:

> … at least three—the *literary*, the *political* and the *philosophical*—that, despite enjoying a certain number of common features and some instability or precariousness concerning their status, making it impossible to ever find a particular utopia perfectly delimited in terms of its identity, are still distinguishable from one another. This circumstance should avoid their exclusively ideological

mobilisation with the inherent waste, namely of their critical and aesthetic aspects. (2001, 145)

Such a discourse appears in the interval that emerges between what is necessary or desired and the possibility of it being satisfied, which is why we cannot avoid recalling Thomas More and his famous *Utopia* in which he criticises the institutions of his time and projects an ideal, imaginary society, where there would be no private property but the absolute community of goods and land, no antagonism between the city and the country, no paid employment, no superfluous spending and excessive luxuries, and the state would be the body administering the country's production.

In the case of Cape Verde, Jorge Barbosa, one of the leading exponents of Cape Verdean literature, embodying the history of the concept and the myths connected to it, sought to justify in a utopian way the origin of the islands in a kind of imaginary journey returning to those origins and in an attempt to fulfil an existential void of the archipelago with a myth. In the author's «imagination», « dripped dots/appeared/almost invisible/upon the globe's/bright, Atlantic blue ...» (Barbosa, 2002, 176).

This is why the author asks himself whether the islands exist on the map, in view of their smallness and that of the islets, even though they have a millennial history:

Would they all/have been/marked on the map?//Ten islands still/waiting for the end/of this destiny/of us all/that half a millennium ago/one day started! (Barbosa, 2002, 176)

Historical and geographical constraints make him write a poem about the expectations of Cape Verdeans, as confirmed by the following passage:

Then burst/in the poet's soul/the desperate poem/of our expectation! (Barbosa, 2002, 176)

Divided into ten cantos with different titles, Jorge Barbosa presented in this poem some of the facets of the islands' geology and history. In the first canto, «Existência» (Existence), he introduces the archipelago as being:

Tiny dots/dripped upon the map's/maritime blue/– it's us!/ten islands! (Barbosa, 2002, 177)

In the second canto, called «Contraste» (Contrast), Jorge Barbosa deals with the geographical location of the country and continues to do so in the third, «Destinos» (Destinies/Destinations), fourth, «As Ilhas e a Amplidão» (The Islands and Amplitude), fifth, «Meio Milénio» (Half a

Millennium), and sixth, «Problemas» (Problems), cantos of the same text: 2nd Canto/Tiny dots/left there/forgotten on the map/– it's us!/ten islands! [...]//3rd Canto/Tiny dots/– it's us!/ten islands!// Emerged boundaries/At the intersection of routes/ that once belonged to seamen/to the kings of the discoveries. [...]//4th Canto/Tiny dots/somewhere in the ocean/– it's us!/ten islands/melancholically/ and geographically possible/surrounded by the sea/as they come in textbooks...//[...] 5th Canto./ Tiny dots/there was no remedy/and they were left on the map.../– It's us/ten islands! [...]//6th Canto/ Tiny dots/dropped there/upon the map./It's us/ten islands/waiting» (barbosa, 2002, 178–182).

Concerning the poem «Panorama», dedicated to Manuel Velosa, the poet raises a series of questions, seeking to justify the geological existence of the islands. He concludes that they emerged from the eruption of a volcano. With that view, he proposed a mythical solution for their origin, influenced by the above-mentioned myth of Atlantis:

remains of what continent,/of what cataclysms,/ of what seism,/of what mysteries? (Barbosa, 2002, 35)

Whereas in the poem «Panorama», the author questions the geological origin of the archipelago, given its mysterious surroundings, he himself assumes the responsibility of answering this question in the poem «Descoberta» (Discovery), referring that those «dramatic islands» were changed: «virgin and red/by the thick lava/of a submerged volcano» (Barbosa, 2002, 309).

In the poem «Destinos» he also deals with the geological origin of the islands «lost, forgotten in a corner of the world», by raising, once again, a series of questions regarding their formation: «Remains of what continent,/of what cataclysms,/ of what seism». (Barbosa, 2002, 49). He answered those questions saying they emerged from the remains of a «shipwreck» that continues, in view of the problems Cape Verdeans are forced to face:

Islands lost,/forgotten/in a corner of the world.../ Remains of a shipwreck!...//...But the shipwreck continues... (Barbosa, 2002, 49)

Jorge Barbosa confirms his answer in the third and tenth cantos of the poem «Expectativa» (Expectation), entitled «Destinos» and «Regresso» (Return), by describing geological remains. Thus, in the third canto they are presented as «Volcanic shards/destinations of stone» (Barbosa, 2002, 179) and in the tenth canto as:

volcanic shards/countless basalts/remote seashells and whelks/hills with remains/from the bottom of the sea/buried craters/barren hills/with their peaks and crests/sharpened and honed/by the strong trade winds//[...]//they seem to be waiting/enigmatic and prophetic/one day perhaps/for Cyclopean bangs and earthquakes/in which everything will be/wrapped once again/in thick, slow waves/ of lava and vapour/marching through the plains/ dripping along the hillsides/they seem to be waiting/(who knows?)/to return one day/to the bottom of the sea (Barbosa, 2002, 190–191)

Pedro Cardoso also wrote about the origin of the islands by referring to remains of the mythical continent of Atlantis or of the Hesperides:

Ancient legends say/that in the depths of the sea/ The Hesperides stayed/With the famous orchard trees//Paradise of Fortune,/Filled with charms!/ It was the most graceful land/Covered by the rose of the sun.//Palaces with doors of gold/And ivory balconies/They could be spotted everywhere/They were countless [...] (Cardoso, 1930, 15–17)

According to that theory, the archipelago would have originated from the remains of the Atlantis continent. It was considered rich because of its flora and mineral treasures. It had a lot of iron, copper, gold and «oricalco» (a type of brass with gold and silver), a metal that glowed like fire. Its inhabitants were highly advanced, which had made them rule over great part of the peoples of the Mediterranean and of the world. However, they were defeated by the Athenians and started to decline, having finally submerged, in a single day, because of a strong cataclysm from which the Azores, Cape Verde, the Canary Islands, Madeira, and the Savage Islands emerged.

This was a mythical tradition recovered by the intellectuals who had studied at the S. Nicolau Seminary-Secondary School, namely Pedro Cardoso and José Lopes, influenced by a type of teaching that esteemed aspects of Classical culture and Romantic nationalist ideas, also due to the handling of the New *Luso-Brazilian Almanac of Recollections* and to the presence of exiles in the archipelago.

In this way, Jorge Barbosa's use of the myth, initiated by José Lopes and Pedro Cardoso, works as a way of compensating for a historical discontent by means of a reinvented reality, communicating his intention to approach the utopian reality of the Cape Verde archipelago, an approach he conveyed along with dystopia.

That Barbosian dystopia lies in the physical constraints and bitter experiences undergone by the Cape Verdeans over time, among which droughts, hunger, and death stand out. Death was an emblematic act of a penal archipelago within an archipelago and the closest to an imaginary concen-

tration camp, since everyday life on the islands was marked by suffering because it was a «land a little forgotten, of low budget», punished by «droughts, sometimes partial and having little impact, other times widespread and tragic» (Barbosa, 1953, 3–4).

Rain symbolises a «gold mine» for Cape Verdeans because it means contentment when it falls, but also anguish when it does not. Thus, Jorge Barbosa, when faced with rainfall, presages abundance, and new perspectives for the islands because the fields become covered with fruits and there is plenty of food, as noticed in the following passage of the poem «Panorâmica» (Panoramic View):

If heavy rainfall/comes our way/foliage and fruits/sprout and grow/the poor delude themselves/ in their optimism/that abundance came/to stay forever. (Barbosa, 2002, 394)

The years of «good rain» are followed by times of partial or total droughts and dry spells, and by the consequent exodus of people to Angola and S. Tomé, as the author witnessed in his unpublished chronicle «Depois das Chuvas» (After the Rains):

But, unfortunately, it cannot always be like this... The years of *good rain* are inevitably followed by the anguished years of partial or total droughts. Thus, the traditional tragedy of dry spells—responsible for so many calamities!—will once again fall upon the people, forcing it once more into exodus, towards S. Tomé and Angola. (Barbosa, 1953, 30)

Droughts and dry spells disrupt the archipelago's everyday life, as they destroy cultivated fields, taking joy and hope away from all Cape Verdeans, as can be seen in the poem «A Terra» (Land):

«If there is no rainfall,/—the low spirits/the tragedy of dry spells!—[...]—Oh the tragedy of rain,/oh the low spirits/of dry spells!//—Oh the whirlpool/ of hunger/taking lives! (...the sadness of the lost cultivated fields...)/—Oh the tragedy of rain! (Barbosa, 2002, 41–42)

Hunger, because of droughts, in turn resulting from the lack of rain and from the abandonment of the islands by the Portuguese authorities, highlighted a dramatic picture for the history of the people of the islands, particularly in 1927 and 1947. Those two years were marked by long periods of droughts during which not a single drop of rain permitting the harvesting of food fell upon the little «fertile land» that existed on the islands. Hence the restlessness caused by the «agricultural tragedies», caused by the rain «that does not come» (Barbosa, 2002, 44) because, if it does come, there is plenty of food and, if it does not come, there is malnutrition and many deaths. Those were painful years that laid the groundwork for claiming better conditions and testifying to the experience of life on the islands, marked by the fight for subsistence and many deaths.

Children were the worst victims of those tragedies by being immediately struck at birth:

> How much disgrace is not/spread throughout our islands/at the start of life!//For there are children who die/many children who die/at the start of life (Barbosa, 2002, 270).

After two years of droughts, during which once again not a single drop of rain fell, the crisis set in and with it the lack of food. Skinny women, men, and children left the countryside, formerly abundant, in exchange for the city, where they begged on the streets to survive. Compared to funereal figures, the desperate children sucked the breasts of their mothers, in a sad image that was part of the archipelago's everyday life, at several times in its history, as noticed in the poem «Seca» (Drought):

> Two years of droughts/lived/as/only God knows!/ Skinny children/wander through the city (Barbosa, 2002, 66–67)

Overall, despite being physically weak, Cape Verdeans have a stoic attitude and, even in the worst situations, continue to have hope and seek to solve their problems, mainly because, at such times, they received no support whatsoever, neither from the international community nor from the Portuguese authorities. According to the author, Cape Verdeans are always hoping for rainfall that will quench the population's thirst and make agriculture possible. These are the situations, along with the return to the historical origins of the islands, which underlie the choice to approach utopia and dystopia in Jorge Barbosa.

BIBLIOGRAPHY

Carvalho, Alberto Dias, *A contemporaneidade como utopia*. Porto. Edições Afrontamento, 2000. ISBN: 972-36-0550-3

Barbosa, Jorge, *Obra Poética*. 1.ª ed. Lisboa: Imprensa Nacional-Casa da Moeda, 2002. ISBN: 972-27-1185-7

Barbosa, Jorge. *Cabo Verde*. Praia: 1953.

Cardoso, Pedro, *Hespérides. Fragmentos de um poema perdido em triste e miserando* naufrágio. Cabo Verde: Edição do Autor, 1930.

Ferreira, Manuel, *Aventura crioula*. 2ª ed. Lisboa: Plátano Editora, 1973. 441 p.

Luz, Hilarino, *O imaginário e o quotidiano cabo-verdianos na produção literária de Jorge Barbosa*. Lisboa. Tese de Doutoramento apresentada à FCSH, UNL, 2013. 288 p.

Pereira, Daniel, *Estudos da história de Cabo Verde*. 2ª ed. Praia: Alfa-Comunicações, 2005, 366 p.

Thomas More and Fernando Pessoa: The aesthetics of utopia in *The book of disquiet*

Cláudia Souza

Department of Philosophy, University of São Paulo, São Paulo, Brazil
FAPESP/Center of Philosophy, University of Lisbon, Lisbon, Portugal

ABSTRACT: In this article we intend to approach *The Book of Disquiet* project to the utopian aesthetics. In 1516, with his book *Utopia,* Thomas More inaugurates a new type of discursive narrative: the utopian narrative. Many scholars have been working on this genre in an attempt to demarcate this literary space. Regarding Pessoa's literary work, some researchers approached the concept of utopia from poems of his book *Message* and his written papers about Sebastianism and the Fifth Empire. In our work we analyze how his theatrical play *The Sailor* and the project of *The Book of Disquiet* have aspects that can be related to an aesthetics of utopia.

Keywords: Utopia, The Sailor, *Book of Disquiet*

> *Literature is the space par excellence of utopia*[1]
> Jacinto Prado Coelho
> *To think is to create utopias. Unconditionally*[2]
> Adalberto Dias Carvalho

In the sixteenth century, Thomas More opens with his work *Utopia* a new kind of philosophical, political and literary discourse: the utopian discourse. In his text the real and the imaginary intermingle. The dividing lines between fictional and political/philosophical accounts are tenuous. The reader finds it difficult to categorize the book *Utopia*, because it cannot be considered a political/philosophical narrative for one finds many fanciful elements in this narrative; and it cannot be considered a novel, because the theme and many elements refer to real and specific data of English history. To counteract the real and the ideal world, More's narrative blends the literary and the non-literary, the fictional and the historical. The author's participation in the text expands the dialogue between fiction and reality.

In *Utopia*, two islands make up different scenarios: in the English island reigns injustice and oppression, it is a society governed by a system whose vitality overlaps with social welfare; the Utopia island society is based not on an oppressive regime, but in a collective welfare regime: besides work—reduced in this island to six hours per day as opposed to the endless hours that Renaissance man worked at that time in England—Utopia's resident has time for leisure and for other unproductive tasks according to the perspective of the economic system. The real and the imaginary are present in the text, not only in the opposition between England and Utopia, but also in the writing that reports the British context, switching at the same time, story with fictional elements.

Thomas More works extensively in the social, political and economic areas with respect especially to the critique of the economic system. Apart from this fact, More's work inaugurates a new literary genre: the utopian genre. This genre is characterized by a textual presentation of another reality, where there is a clash between the utopian world and the real world, and makes an alternative proposal. The imagination, the dream and the need for idealization structure the utopian genre, which aims to build another world, an alternate history, as Trousson argues:

> The utopia—as utopianism—supposes the will to build against the existing reality, another world and an alternative history. (Trousson, 2005: 128)

If the work *Utopia* serves as support to the literary genre, however this comes down to a proposal of an alternative reality. In this work there is a strong political, social and economic criticism and a proposal to build a new society structured in other paradigms. But beyond this aspect, the work in question has a philosophical and literary dimension, one built by imagination, able to create another reality, enigmatic, mysterious, a non-place. Scholars hold different perspectives regarding the criteria to define the utopian genre, as shown by

Geraldo Witeze Junior in his article, "Where is nowhere? A journey in search of utopia":

If we consider that the founding work of the utopian genre actually proposed something to the society of his time, do we need to accept that all other works that have arisen in its wake did it too? That is, can we consider this as a criterion for defining the utopian literary genre? For some, like Trousson, the answer is no, but for others it is positive. (Junior, 2012: 364)

In our work, we are in line with Trousson's thinking, that is: regarding the utopian genre, it does not have to propose something to society in political, social and economic terms. According to the researcher Maria Luisa Malato Borralho, in her article "Are there no Portuguese utopias?", utopian genre would be characterized by four elements: the lack of real, fictionalization, sociability and reflection (Borralho, 2004, p.1).

Fernando Pessoa published in 1915 in *Orpheu* magazine, a theatrical play entitled "The Sailor". This narrative starts in a castle tower, where three watchers talk to pass time. Then appears the image of a shipwrecked sailor in a desert island, creating for himself a reality more real than reality itself. Pessoa reproduces, in this text, the four characteristics listed by Maria Luisa Malato Borralho: the lack of real, fictionality, sociability and reflection. The reader is led to two instances of the unreal: the first in the tower with watchers and the second on the island with the sailor. It is also worth mentioning that the place or rather the non-place where the sailor finds himself is just an island, as the non-place of More's second book. In this play the question of language is the main point, that is, it would be possible to create through language another reality, more real than the one in which we live. This play, "The Sailor", is an utopian layered writing, expressing the wish to reach another reality. The question that arises is: must an utopia always obey an ideological regulator? That is, as the play "The Sailor" does not have political/social content, can it be considered an utopian work? The answer seems to be in the quotation at the beginning of this article (a sentence that ends the book chapter, "Utopia: concepts to renew senses", authored by researcher Adalberto Dias Carvalho) "To think is to create utopias. Unconditionally." If thinking generates utopias, or generates other places, which are non-places—since in them we are not in concrete reality, but we occupy them with our imagination—we can through this perspective assert that "The Sailor" is an utopian work.

According to Jacinto do Prado Coelho, in his book *Camões and Pessoa, poets of utopia*, literature is the space par excellence of utopia (Coelho, 1983: 19). Literature is inscribed in a field where the categories of truth and lie don't make sense. Literature is inscribed in the field of fiction. Thus, referring to an utopian literature is referring to a particular literary field where the imagination of the artist creates a space that points to another location, a non-place. Regarding Pessoa's literature, we see this movement in the play "The Sailor", in the book *Message* and in his writings on Sebastianism and on the Fifth Empire. In addition to these texts, we think it is also possible to consider Pessoa's project of *The Book of Disquiet* through an utopian aesthetic perspective.

The Book of Disquiet was first published in 1982, by the publisher Ática, an edition coordinated by Jacinto do Prado Coelho, Teresa Sobral Cunha and Maria Aliete Galhoz. During his lifetime, Fernando Pessoa published only some excerpts from *The Book of Disquiet*. The very structure of this project points to an utopian aesthetics, because it is subjected to several changes, occupying therefore a non-place in this regard. The first published excerpt, "In the Forest of Estrangement"—presented in the magazine *The Eagle* [*A Águia*] in August 1913—is signed by Fernando Pessoa. In this text the atmosphere is pervaded by dream, imagination, utopia. The narrator, in the first few lines, says that he is between dream and waking, between two worlds, between the sky and the sea:

Half awake and half asleep, I stagnate in a lucid, heavily immaterial torpor, in a dream that is a shadow of dreaming. My attention floats between two worlds, blindly seeing the depths of an ocean and the depths of a sky; and these depths blend, they interpenetrate, and I don't know where I am or what I'm dreaming. (Pessoa, 2003: 417)

We have in this passage a perfect description of utopia's non-place. The narrator, embarked on a journey between dream and waking, does not know where he is, liing between two worlds, in a place that cannot be named. Gradually the imaginary landscape will be outlined and the narrator recognizes to be in a forest, the forest of estrangement, imagined, dreamed with all the details, created from a productive imagination. In this other world, this non-place there was no time or purpose, because it is not a real place:

We belonged to no age and had no purpose. For us the ultimate purpose of all beings and things had remained at the door of that paradise of absence. The souls all around us, as to feel us feel them, had became perfectly still: from the nubile soul of flowers to the dangling soul of fruits... (Pessoa, 2003: 422)

In this passage we see the criticism of the real world, where there is time and purpose, where the man is, in a certain way, trapped by his context, where there is a purpose for everything, because society is led by the notions of loss and gain, utility, finality and purpose. In another world, of dream, the whole purpose of things and beings were at the door of paradise. The *Disquiet* project starts with the writing and publication of this first text, immersed in an aesthetics of utopia. This aesthetics is drawn through a journey, accomplished through the imagination that puts the narrator in another space—the forest of estrangement—where it is oblivious of the real world and its demands. This forest approaches the idea of paradise. And in this paradise, paradigms transform themselves, in this non-place, a place of dreams and imagination where the subject is freed from social constraints.

Pessoa continues to write stretches for the project of *The Book of Disquiet* until possibly 1934, a year before his death. Through the documents left in his archives we know that parts of this project were written from 1913 until 1919, suffering a break of ten years from 1919 to 1929. The project was resumed between 1929 and 1934. The structure of this project is involved in an atmosphere of dream and doubt. In the first phase of the *Disquiet* project the fragments are signed by Fernando Pessoa, as indicated by the signature of *The Book of Disquiet* published in *The Eagle* [*A Águia*] in 1913. In a second step, Fernando Pessoa passes the task of writing the book to his literary personality, Vicente Guedes. In the third phase— after 1929—Soares assumes the *Disquiet* project.

It's interesting that in the *Disquiet* project, the narrator seems to hover between a lived and a dreamed reality. He imprinted in his writings an atmosphere of mist, and he is always traveling in his imagination, changing places, plunging into abstractions. On the other hand, there is in several fragments of the project a criticism of the existing capitalist society. Central in this regard is the figure of boss Vasques because he embodies the role of the explorer, the businessman who is pure externality, and that somehow holds Soares in a daily reality, ordinary, productive, as is said in the following excerpt:

> Today, in one of the pointless and worthless daydreams that constitute a large part of my inner life, I imagined being forever free from the Rua dos Douradores, from Vasques my boss, from Moreira the head bookkeeper, from all the employees, from the delivery boy, the office boy, and the cat. In my dream I experienced freedom, as if the South Seas had offered me marvelous islands to be discovered. It would all be repose, artistic achievement, and the intellectual fulfillment of my being. (Pessoa, 2003: 17)

This excerpt seems to be in full agreement with Adalberto Dias Carvalho's phrase: "To think is to create utopias. Unconditionally". The narrator is transported, through thought, to a non-place where the daily and materialistic reality does not make sense, where he would be free of all references to the company: Rua dos Douradores, the boss Vasques, the co-workers and the cat. In his imagination, the narrator felt he could tame wonderful islands. The reference to the islands is interesting for it is a brand of utopian writing. In these islands the narrator finally has rest and intellectual fulfillment of his being. Following on the text, the narrator confesses that he would be sorry to be free from social constraints, oscillating between the desire for freedom and the desire to continue in a reality that would generate a certain security for himself.

In another fragment, he demonstrates the central role of boss Vasques in the office and in his bourgeois life:

> The capitalist partner of the firm here, always ill at large, I do not know by what whim of disease's interval, wanted to have a picture of the whole bureau staff. And so, yesterday, we were all aligned, by order of the joyous photographer, against the dirty white barrier that divides with fragile wood boss Vasques' from the general office. At the center the same Vasques; and in two wings, first in a definite and after in an indefinite distribution, other human souls who gather here every day for little doings whose ultimate aim only Gods' secret knows. (Pessoa, 2010: 235)

In the center of the imaginary picture is boss Vasques, who owns the hours of their employees (as confirmed by other the *Disquiet* excerpt). Every day the employees gather at that space in Douradores street for little doings whose ultimate aim only Gods' secret know. The passage begins with reference to the capitalist partner. There is a fragile line that separates the general office and boss Vasques' office. Fragile perhaps because the employees who make up the backdrop of the general office and boss Vasques are parts of a labor market occupying a productive task. It is interesting to note that the landscapes imagined delimit at least two realities, one banal, boring, dull, utilitarian, and another composed of dream, utopia, imagination.

In many places of the *Disquiet* project the narrator writes about the dreamed, imagined life:

> I see dreamed landscapes plainly as real ones. If I lean out over my dreams, I'm leaning out over something. If I see life going by, my dream is of something.
>
> Someone said about somebody that for him the figures of dreams had the same shape and substance

as the figures of life. Although I can see why somebody might say the same thing about me, I wouldn't agree. For me, the figures of dreams aren't identical to those of life. They're parallel. Each life—that of dreams and that of the world—has a reality all its own that's just as valid as the other, but different. Like things near versus things far away. The figures of dreams are nearer to me, but… (Pessoa, 2003: 93–94)

Let us live by dreams and for dreams, distractedly dismantling and recomposing the universe according to the whim of each dreaming moment. Let us do this while being consciously conscious of the uselessness and of doing it. Let us ignore life with every pore of our body, stray from reality with all of our senses, and abdicate from love with our heart. Let us fill the pitchers we take to the well with useless sand and empty them out, and refill and re-empty them, in utter futility. (Pessoa, 2003: 342).

Dream has a major relevance in all the phases of the *Disquiet* project. As we can notice in the quoted texts, there is a pleasant dream, in living and in dreaming, in weaving another reality as real as a lived reality. From the first sentence to the last one of the *Disquiet* project, the space is made up of dream, fog, the narrator is always ready to make a journey through imagination (*To travel, one must only exist*). Pessoa, Guedes and Soares build on their *Book of Disquiet* an utopist aesthetics that points to a second reality, a non-place, a fictional space and a dream space, where are present the four characteristic elements of the utopian genre highlighted by researcher Maria Luisa Malato Borralho: the lack of real, fictionality, sociability and reflection. *The Book of Disquiet* criticizes the tedious reality, productive and monotonous and points to an aesthetic output, diving into utopia, dream and imagination.

NOTES

[1] The original phrase: "A literatura é o espaço por excelência da utopia." (COELHO, 1983, p. 19)

[2] The original phrase: "Pensar é gerar utopias. Incondicionalmente" (CARVALHO, 2004, p. 17).

BIBLIOGRAPHY

Borralho, Maria Luísa Malato. "Não há Utopias Portuguesas?". E-topia: Revista Eletrônica de Estudos sobre a Utopia, n°1 (2004).

Carvalho, Adalberto Dias de. "Utopia: Esclarecer Conceitos para Renovar Sentidos". *In*: Fátima Vieira e Maria Teresa Castilho (orgs.). *Estilhaços de Sonhos*: Espaços de Utopia. Vila Nova de Famalicão, Edições Quasi, 2004. pp. 13–17.

Centeno, Yvette K. *Hermetismo e Utopia*. Lisboa: Edições Salamandra, 1995.

Centeno, Yvette K (coordenação). *Utopia, Mitos e Formas*. Lisboa: Fundação Calouste Gulbenkian, 1993.

Cioran, Émile Michel. *História e Utopia*. Tradução de José Thomaz Brum. Rio de Janeiro: Rocco, 1994.

Coelho, Jacinto do Prado. *Camões e Pessoa, poetas da utopia*. Lisboa: Europa-América, 1983.

Junior, Geraldo Witeze, "Onde está o não lugar? Um percurso em busca da utopia." Revista Via Litterae. Anápolis. v.4.n2. Jul/Dez 2012. pp. 353–374.

Pessoa, Fernando. *Livro do Desasocego*. 2 vols. Edição de Jerónimo Pizarro. Lisboa: Imprensa Nacional-Casa da Moeda, 2010.

Pessoa, Fernando. *O Marinheiro*. Edição de Cláudia Souza. Lisboa: Editora Ática, 2010.

Pessoa, Fernando. *The Book of Disquiet*. Edited and Translated by Richard Zenith. 3rd ed. USA: Penguin Books, 2003.

Reis, José Eduardo. Do Espírito da Utopia: Lugares Eutópicos e Utópicos, Tempos Proféticos nas Literaturas Portuguesa e Inglesa. Lisboa: Fundação Calouste Gulbenkian, 2007.

Trousson, Raymond. *Voyages aux pays de null part. Histoire littéraire de la pensée utopique*. 2ème ed. Bruxeller: Ed, Université de Bruxelles, 1979.

Trousson, Raymond. "Morus—Utopia e Renascimento." Utopia e Utopismo. N°2. Campinas, 2005. pp. 123–135.

Vieira, Fátima; Castilho, Maria Teresa (Orgs). *Estilhaços de sonhos espaços de utopia*. Portugal: Quasi Edições, 2004.

Architectures of madness: Lovecraft's R'lyeh as modernist dystopia

Jorge Palinhos
Centro de Estudos Arnaldo Araújo, Escola Superior Artística do Porto, Portugal

ABSTRACT: Howard Phillips Lovecraft (1890–1937) was one of the most influential horror writers of the 20th Century, with a strong stylistic and thematic impact on numerous other writers of the genre but also on literary fiction and on film and videogames. In addition, architecture was one of his preferred tools to create atmospheres of dread and infuse in the reader a sense of alienation and loneliness. Most of his short stories and novelettes are filled with architectural descriptions and details. Either of the traditional architecture of his beloved New England, or of what he called "Cyclopean architecture", a name usually identified with Mycenaean historical architecture, but which Lovecraft used to qualify the architecture built by his fictional alien races and gods. This Cyclopean architecture seems to have much in common with the Modernist architecture, which Lovecraft detested and that in his time was on the rise.

In this paper, I propose to analyse the most well known example of such architectonic fantasy locations: the sunken "corpse-city" of R'lyeh, also called Arlyeh or Urilia, in Southern Pacific, which figures prominently in his short story «The Call of Cthulhu». I will try to determine which architectural features Lovecraft imbued in the "Cyclopean Architecture" to create his unsettling atmospheres and especially investigate de barrenness of references—historical, geometrical, and geographical—that seem to permeate the buildings and cities dreamt by Lovecraft to make them existential dystopias.

Keywords: modernist architecture, Howard Phillips Lovecraft, R'lyeh, imaginary places

1 THE VIRTUAL CITY

If we access Google Maps and search for the location of R'lyeh, we will find a city near Antarctica, near the coordinates of 47° 9'S, 126°43'O, in the middle of the Pacific, somewhere between New Zealand and Chile, under the sea.

If the user tries to use Street View, he or she will not be allowed to examine the architecture of the city, but the 37 reviews left by users at the time this paper was written tell us that it is a family-friendly destination, speaking about the non-Euclidian architecture, the Cyclopean towers, and the tomb of Great Cthulhu.

This on the positive side, because the negative side is focused on the high cost of drinks, some disappointment in comparison with other contemporary architectural landmarks and repeated episodes of insanity.

This is obviously another of the so-called internet memes, using an online service of mapping the Earth to describe and locate fictional locations. Because R'lyeh, or Arlyeh, or Urilia is arguably the most famous of the fictional locations in 20th century literature, created by Howard P. Lovecraft in his short stories and novelettes, and developed by other writers. It is on the league, but more famous that other fictional locations suggested in the stories of Lovecraft, like Dunwich, Kingsport, Innsmouth, or even Arkham.

This is due to the setting of one of his most famous short stories (Clute & Grant, 1996:596): «The Call of Cthulhu», a short fiction written in 1926 and published in the magazine *Weird Tales* in 1928, whose popularity I will not try to address or explain in here.

Lovecraft was one of the most popular and influential writers of the conventionally called pulp fiction, which was an escapist type of fiction that was commercially very successful in the USA in the first thirty years of the 20th century. It was mostly composed of short genre fiction published in cheap magazines that catered to a mostly male low-class reader, usually with a sensationalistic and chauvinistic bend, following the conventional escapist narrative of the male hero that saves the female interest while defeating the sinister foreigner and defending the traditional values of family, maleness, capitalism, and USA independence.

Some of the most famous authors writing for these magazines were, besides Lovecraft, Edgar Rice Burroughs, Ray Bradbury, Raymond Chandler, August Derleth, Erle Stanley Gardner, Dashiell Hammett, Fritz Leiber, Joseph Conrad, Jack London, C.L. Moore, Sax Rohmer, Clark Ashton Smith, Robert E. Howard, Jack Vance, among many others, writing under the conventions of the fiction genres of Western, Police procedural, Sword and Sorcery, Science Fiction, Fantasy, Horror, etc.

Although the popularity of these magazines and authors declined from the 1940s onwards, like Conan, Tarzan, Doc Savage, Phantom, etc. became quite influential in a series of popular entertainments, like comic books, genre films, videogames, and tabletop games. Just to give an idea of the popularity of some of these authors, a simple Google search for "Cthulhu" at the time of the writing of this paper returns "around 14 900 000 results" (Google Search), a number which, I believe, clearly shows the internet popularity of the work of Lovecraft.

«The Call of Cthulhu» is heavily inspired on Poe's «The Narrative of Arthur Gordon Pym». The story is purportedly told by Francis Wayland Thurston and concerns three different stories that the narrator is aware of: one taking place in Providence, Rhode Island, another in New Orleans, and the last part in R'lyeh, all around the time of the awakening of the alien being called Cthulhu.

Each of these different events, although affecting different characters and taking place in different locations all are defined by its atmosphere of dread and portentous omen, which will lead to the climax of the last story, about the awakening of Cthulhu, which takes place in R'lyeh.

The city itself is described as:

The men sight a great stone pillar sticking out of the sea, and in S. Latitude 47° 9′, W. Longitude 126° 43′ come upon a coast-line of mingled mud, ooze, and weedy Cyclopean masonry which can be nothing less than the tangible substance of earth's supreme terror—the nightmare corpse-city of R'lyeh, that was built in measureless aeons behind history by the vast, loathsome shapes that seeped down from the dark stars. There lay great Cthulhu and his hordes, hidden in green slimy vaults and sending out at last, after cycles incalculable, the thoughts that spread fear. (Lovecraft, 1996: 90)

According to the text, the crew of the schooner Emma should have briefly found the city, which was lost in the ocean. Such discovery had been witnessed by the Norwegian sailor Gustaf Johansen, the sole survivor of the crew, which also included a Portuguese sailor, Rodriguez (*sic*).

The city that Johansen, Rodriguez and their companions found in R'lyeh was described as:

... the unbelievable size of the greenish stone blocks, at the dizzying height of the great carven monolith, and at the stupefying identity of the colossal statues and bas-reliefs. (2000: 91)

and:

Without knowing what futurism is like, Johansen achieved something very close to it when he spoke of the city; for instead of describing any definite structure or building, he dwells only on broad impressions of vast angles and stone surfaces—surfaces too great to belong to any thing right or proper for this earth, and impious with horrible images and hieroglyphs.

Obviously, writing under the genre of horror—or weird fiction as Lovecraft described it (2000) he was trying to create an emotional impression on the reader, through a textual description that could cause discomfort, disquiet or even fear, and it is significant that for that he is clearly inspired in a specific architectonic style, a monumental architecture of high drama, causing disorientation, that breaks with the principles of classical architecture and geometry, yet, at the same time, has a clear affinity to it. It is clearly an architecture very close to the futurism architecture that was then appearing in Italy and inspiring the fascist architecture, or the Nazi architecture of Albert Speer, the Constructivist architecture in Russia or, later, the daring designs of the Frank Lloyd Wright.

This massive architecture, of neoclassicism taken to a superhuman distortion, is, for Lovecraft, the writer that lived most of his live in New England, and hated to travel, the close opposite of the picturesque architecture of his homeland, and that he lovingly described in «The Case of Charles Dexter Ward» like this:

Here he was born, and from the lovely classic porch of the double-bayed brick facade his nurse had first wheeled him in his carriage; past the little white farmhouse of two hundred years before that the town had long ago overtaken, and on toward the stately colleges along the shady, sumptuous street, whose old square brick mansions and smaller wooden houses with narrow, heavy-columned Doric porches dreamed solid and exclusive amidst their generous yards and gardens. (Lovecraft, 1999: 71)

Therefore, Lovecraft is comparing the "evil" futuristic architecture, associated to monsters and nightmares, being confronted to the traditional Georgian architecture of the Northeast of the USA, the place of family life, comfort, and intimacy.

There is no doubt about what kind of architecture was the favourite of the Providence writer, even if he, himself, was known to be a sympathizer of the Nazi ideology and Hitler, but I rather describe him as a conservative, than an outright adept of national-socialism. It is known that at the time a sizable part of the conservative parties were seduced by the ideas and flair of Mussolini and Hitler, as paladins of "Western" societies against the "danger" of communism.

Yet, this sympathy clearly was not extended to the ironically referenced "futuristic" architecture and art. Not only Lovecraft, writer of considerable knowledge, contemporary of James Joyce, Virginia Woolf, T. S. Elliot, was strongly against the literary modernism of his time—which, in the short story, is ironically referenced as being composed of "broad impressions" of a time and space described as impossible to understand, that Lovecraft could only relate to a stuff of nightmares.

According to Sayer (2013) the mysterious city is a typical convention of adventurous literature, but the idea of the massive architecture and disorientating city was drawn from the experience of Lovecraft in New York, where he lived briefly, and reflects the rejection of the author for cosmopolitan societies, made of mixed populations.

I would argue that this refusal of modernist architecture and literature were based on the fact that they were trying to display the lack of meaning in the shape and movement in a world devoid of divinity, teleological direction and machine dominance. Moreover, trying to find a way of expressing such a lack of sense, the modernists were promoting the hybridity of shapes, combining the internal and external time of fictional characters, the classical, and the industrial, the deformation of space through fantastic or massive volumes.

And Lovecraft, even being himself an sceptical atheist deeply interested in the processes of machines, and, being, as other pulp writers, like Clark Ashton Smith or Robert E. Howard, coming from small homogeneous North-American town, of conservative population and urban design—looked with suspicion towards these new literature, this new architecture and these new populations arriving to the USA as immigrants—and against the hybridism that they represented, which he denounced as aberrant and incomprehensible, only possible in space or the most remote areas of Antarctica. This is confirmed by Ingwersen (2013: 6), who notes that the non-Euclidian geometry of Lovecraft is an attempt of subverting the categories, which we apply to our worldview, to create a feeling of grotesque and fear.

Yet, Houellebecq (2005: 13) notes that, Lovecraft had no common world routine and no denouement, that is "open works of fear" that is, no lost or regained paradise, no healing story, but an endless stare at the meaninglessness of the world, which makes Houellebecq say:

An absolute hatred of the world in general, aggravated by a particular disgust for the modern world. This summarises Lovecraft's attitude well. (Houellebecq, 2005: 16)

Yet, it also does not. Because, even if the narrative structure of Lovecraft is one of revelation—not installation—of fear, there are still different spaces in his works that are clearly connected to different states of fear: Modernist architecture symbolizing the end of hope, and Georgian architecture, probably representing a mistaken state of ignorance, but yet more blissful than the emptiness and incomprehensibility of the universe that existed in the colossal futuristic architecture of R'lyeh is clearly a strong theme in Lovecraft.

Even the different use of locations, which this story exemplifies, is revelatory of the way Lovecraft uses space—more than time—as a way of building fear in the reader. As if emotions were not activated by events, alone, but also by the spaces where these events happen and the way the spaces define and configure these events.

In addition, I would like to draw attention to the way Lovecraft handles modernist architecture. As if he seems to recognize it as a form of truth, of a true vision of reality. Therefore, we have paradoxically Lovecraft apparently rejecting Modernism, not as a falsehood, but as an unnecessary and unwanted vision into the true reality of the world. Of a world devoid of meaning and the consolation of familiarity and order, which he found in the Georgian architecture of his childhood.

Therefore, Lovecraft is also a writer of nostalgia, of the spaces of nostalgia, which we can confirm, for instance, in some of his other works, like The Case of Charles Dexter Ward, with his nostalgic description of Providence, to that create feelings of comfort, of intimacy, of familiarity, of routine and closeness between beings and closing of beings towards other beings—in sum, nostalgia for the lifestyle that Lovecraft could clearly feel was vanishing from his time and being replaced by another lifestyle marked by hybridism, traveling and acknowledgement of the other and of the otherness that Modernist architecture and art encapsulated.

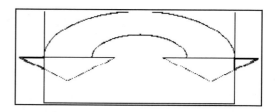

BIBLIOGRAPHY

Clute, John & Grant, John (ed). *The Encyclopedia of Fantasy*. London: Orbit, 1996.

Houellebecq, Michel, H.P. *Lovecraft—Against the world, against life*. New York: McSweeney's, 2005.

Google search for "Cthulhu" - https://encrypted.google.com/search?hl=pt&q=cthulhu—Acedido a 13 de fevereiro de 2016.

Ingwersen, M. (2013) *Monstrous Geometries in the Fiction of H.P. Lovecraft*. At http://www.inter-disciplinary.net/at-the-interface/wp-content/uploads/2013/04/Ingwersen.pdf—Access 15 February 2016.

Lovecraft, H.P. The Call of Cthulhu. In Lovecraft, H.P. *The H.P. Lovecraft Omnibus—The Haunter of the Dark*. London: Voyager, 2000, vol. 3, p. 61–98.

Lovecrat, H.P. The Case of Charles Dexter Ward. In *Lovecraft, H.P. The H.P. Lovecraft Omnibus—At the Mountains of Madness and other Tales of Horror*. London: Voyager, 1999, vol. 3, p. 61–98.

Sayer, F. (2013) *Horreur des villes maudites dans l'ouvre de H.P. Lovecraft*. - http://etc.dal.ca/belphegor/vol3_no2/articles/03_02_Sayer_Lovecr_fr_cont.html (Access on the 15 February 2016).

Hermann Hesse's *Das Glasperlenspiel*: At the frontier of a modern utopia

Fernando Ribeiro

CHAM, FCSH, Universidade NOVA de Lisboa, Lisboa, Portugal
Universidade dos Açores, Portugal

ABSTRACT: This paper is divided into 5 parts: 1: Introducing the paradigm of ancient utopia based on More, Campanella and Bacon's models; 2: Ancient utopia paradigmatic principles aiming its fulfilment in future time; its matrix: Plato's *Republic;* 3: Modern utopia paradigm according to fundamental aims such as peculiar regeneration, collective rebuilding, and detached consciousness among individuals looking forward to an open society; 4: Hesse's *Das Glasperlenspiel* modern utopian model: dynamics inside-outside Kastalia in order to surpass its frontiers with the help of fantasy and creativity while building a community implementing: dignity, humbleness, and freedom in every citizen based on the consolidation of principle-of-responsibility within a present time lapse; 5: Modern literary work of art: dance metaphor. Knecht's wisdom plus Tito's creativity: the basis of consciousness towards a modern utopia.

Keywords: Utopia, Frontier, Modernity, Open-Society, Principle-of-Responsibility

1

"Discourses of Raphael Hythlodaeus of the best state of a commonwealth" so declared Sir Thomas More in *Utopia* frontispiece (1516). Offering a model for production, fair administration, and distribution of goods among all members of a modern community in which they all would live under the principle of happiness respecting either usefulness or rationality within its social, political, and economic organizations. Existing though in a space and a time to be found—in order to raise its reader's belief in and amazement before the use of rational and spiritual capabilities during leisure time.

Circa a century later, in *La Città del Sole* (1602), Tommaso Campanella praises similar state of government More described in *Utopia* underling happiness and perfection as major principles to be implemented. According to these, all community's members should live by means of usefulness and rationality to consolidate autarky with the help of goods sharing and law, arts, and crafts reform, in order to implement culture and education among all citizens (Campanella, pp. 68; 91).

Two decades later Sir Francis Bacon's *The New Atlantis* (1627) defends how developing the study of applied and pure sciences college members, serving God and the community, would lead the community to charity and at the same time to longevity and happiness, always respecting usefulness and rationality, while pursuing a scientific method (Bacon, pp. 8; 12, 19; 22; 37; 43).

2

The utopian paradigm underlying such novels is driven by principles such as happiness, perfection, longevity presenting itself as the «fictional ability to re-describe life» (Ricoeur. p. 501). Those play the major role within such a community whose targets are socio-politic (*Utopia*), religious-metaphysic (*Città del Sole*), and social-scientific (*New Atlantis*). They are the means on which every inhabitant is dependent upon the acceptance of himself as a part within the whole and closed system. Herein every community member is not given any opportunity to accept changes, to develop his emotions since he was not meant to be educated to welcome any external (space) and present (time) influence too (Lapouge, 1990, 213; 223; Freyer, 1936, 36).

Space and time become vital factors, enough to the successful interpretation of this paradigm, taken above all as a criticism to western civilization. This aims to draw the attention back to the main targets already present in the *Acts of the Apostles* (17, 28–9) and in *Letter* of *St. Paul to the Thessalonians* (2nd 3, 10), to the *Romans* (12,4–16), and to the *Corinthians* (1st 13,4–13; 2nd 12, 19). Meanwhile fostering a model in which economic development should support scientific progress within an equivalent socio-political organization, modern enough to share the administration of the whole community with the *bourgeoisie*—a social *stratum* whose key role was becoming crucial to the future of the European culture for goodness' sake praxis ascribed to Christianity (Jaeger, 1961, 88–9).

The utopian model presented along these literary forms might be taken into account—the more the reader detaches himself from the work, the better he will understand its goals—as a narrative strategy to stress the value of humanist principles as essential prerequisites to a universal community balance with its already known asymmetrical developments. One could then analyse more accurately the results put forward inside this utopian lab. They are presented as delayable to a prospective, and real, space-and-time situation as long as happiness, perfection and eternity would be recovered too as goals of better community governed by a philosophic aristocracy, always concerned with an ethical and political development—either collective or individual respectively—just as declared by Plato in his *Republic* (1990, 592 b).

In any of these three utopias, the foreshadowed pattern is recognizable: economy starts by being exclusively agrarian; then based on a commercial and industrial basis, while incorporating the monopoly of war and finance administration handled by noble-warrior elite (Elias vol. 2, 100.). The exhaustive planning of all functions and its respective fulfilment; caring for mutual relationships between social, political, technical, and scientific spheres; already denotes to what degree the urge for modernity was imposing itself gradually within the limits of a community.

It should be offered a new political and social model to prevent every individual from been the victim of his instincts, taken as a natural cause of social disruption.

Power administration plays a central role in the formation of such an equalitarian-totalitarian model. In order to consolidate an intellectual culture, along with the help of education and training systems, regulation extended even to leisure time.

Central administration was also necessary for the developing of scientific investigation and experiment. Thus, it was possible to reveal how individuals of that epoch were aware of the need of a language, translatable in mathematical-rational code, to make understood the various principles of creation with equivalent *ad aeternum* time and space validity (Hall, 1983, 214–215, 252–3).

3

Notwithstanding, this utopian paradigm of the three narratives becomes object of accurate lab observation provoking, at the same time, a sudden questioning about such a collective existence. Present time is for instance seen as a timespan that is far away from getting its relevance, as space where all subject's experiences take place (Elias, 1990, 50–52,126).

The individual must live under a utopian welfare state aiming self-regulation wherein mutual interaction between man and his social and natural environment should be deprived of any inventiveness led by imagination, scientific creativity, and free from any fears (Elias, 1989, 180–1).

Shouldn't one therefore ask if constant interacting between the individual and the community wouldn't represent a surplus value either to the individual or to the community if the subject wouldn't be given the opportunity to cope with the definition not only of future but above all of present day goals?

Should not then so be better strengthened every capacity to resist against contingency and utter strangeness provided by current and strictly detailed regulations?

One wonders if then such a paradigm would not bear in itself the birth of an individual consciousness too, by means of which a crucial detachment would not make the subject feel the need for a regeneration and rebirth of society: a new and *open society* (Elias, 1989 184; Ricoeur, 487–8; Popper, 1972, vol. 2, pp. 201–3).

4

In *Das Glasperlenspiel*, the longest and best Herman Hesse's novel, the Swiss writer reacts against a dangerous situation Europe was passing through as a result of the totalitarian policy Hitler succeeded to impose in Germany (Hesse, 1973, vol. 1, 60).

This literary work of art deals with a utopian model within the requisites: *Dichtung zur Utopie* followed by Hesse after analysing the path of modernity.

He therefore points out to the urge felt by his fellowmen to create and play a game (glass bead game) through which **contemplation** and **meditation** (GW IX: 38–9) of cultural masterpieces (literature, music, painting among others), the result of every creator's mind and soul inventiveness, would allow every player to oppose alienation and superficiality promoted in such a «feuilleton» epoch.

Moreover, with the aid of mass culture masterpieces which were very typical of that highly mechanized and industrialized civilization shaped under western culture canon (Hesse, 1970, IX, 38–9; X, .579; IX. 29–31; IX. 12; 13; 14–15; 17; 19; 44).

According to the narrator's point of view an elite formed by proficient and responsible inventors was improving the standard of the glass bead game and striving towards an abstract and philosophical language to achieve a mutual-relational synthesis (*Kunst–Wissenschaft*) which was carried out by applied individual skills such as imagination and inventiveness.

The complete perfection and happiness already attained by his pedagogical province in a bygone age are put into question, in fact, since the reader is acquainted with the game consequences, the hero and elite member lived through during his lifetime experiments.

Das Glasperlenspiel deepens the meaning of Modernity while representing the interaction between *diegesis* and *addressee*; the way how the game and its elite evolves is also shown therein while one is told about the hero's, *Joseph Knech*t, evolution through the words of the novel narrator: his biographer and historian.

The reader is therefore given the chance to: judge the fact the sublime game player, and *Magister Ludi* lived his life in leisure too along; and to appreciate the value of such game that was created and played within and for Europe *grosso modo* before and after the 20th century.

Its aim was to foster happiness, perfection and equanimity supported by functionality and rationality implemented by an elite focused on the preservation of the community's social welfare.

In it, every individual ought to care about his own transformation as well as the transformation of the mass culture community he serves without despising his intellectual and mental skills (Hesse, 1970, IX, 7; 32–3, 23, 31).

Hesse makes the readers face up their historical reality while leading them from a defined spatial area, the pedagogical province of *Kastalia* (similar to every utopian space), to a psychical and singular space, from which everyone is invited to depart to reach society.

Time will be defined, in turn, as an inter-relational unity driven by an individual rhythm, which is then presented to be understood as a way to apprehend aesthetically and emotionally reality during present time individual experience. Making use of imagination in form of concrete objects, aiming to enforce a process resisting every attempt of ideology and to strengthen individual consciousness on present day and personal capability to change the course of history. Moreover, *The Glass Bead Game* shows literally a new kind of utopian model whose importance lies on the social and historical dynamics literature is able to propose aiming a vital change within nowadays society where modern and mass culture prevails (Hesse, H., 1977, vol. 2, 905; 885–6, 868–9; Hesse, 1970 GW. X, 585, 583,580).

Joseph Knecht, the modern hero in this fictional universe, will be a full member of the pedagogical province of *Kastalia* whose self-segregated geniuses are forever faithful to the cult of intellect and culture as represented by Fritz Tegularius.

As modern hero he will succeed in occupying in proper (and before due) time the highest level of the Province Hierarchy and become utmost defendant of Glass Bead Game under the form of true combination of disciplines like poetry, music, and algebra.

These are the essential requirements, in his point of view, for a language applicable to a «space» where social and historical reality play a role, though chaotic and contingent, good and evil: **the** effective one. This is his strategy shown above all as needed to reform a community life within *Kastalia* ancient utopian paradigm understood as cause of a *non-open* society.

Knecht embodies then the attempt to reform the ancient utopian model according to which community lives peacefully and happily in an enclosed space to escape present day singularities under the paradigm of such a community that is always longing for a better and in future time postponed life.

That is why Hesse turns his hero into a traveller. Constantly on the frontier of reality between the models personified by the *älterer Bruder* and his secular friend and counterpart *Plinio Designori* (Hesse, 1970, GW IX, 294–5, 303–5, 239–40, 259, 179–180–1; 142, 352–3, 111–2.). Thus, he makes readers able to understand the rich dynamics brought by a comprehensive understanding of history in the figure of Pater Jakobus; and underlines Knecht's urge to commit himself to the secular society as he decides to be a schoolmaster.

Reality gains then its proper sense as a space where «something magical» coming from the individual's soul takes place: his deepest and emerging sensibility, which ought to result from the cult of fantasy and creativity. This is the very counterpoint to the cult of rationality by means of which the world was explained intellectually and systematically in *Kastalia*.

The hero becomes space of freedom and serenity, of translation, i.e. space for an individual dynamical and real commitment within an *open society*. In it, the subject achieves his consciousness through a contemplative acting with the help of a culture always willing to interpret reality and to seek constantly truth and beautifulness.

Knecht, in so doing, remains at the frontier zone where he can overcome every contradiction reflected upon the object every time he invents a game summoning elements out of community life and at the same time creating/making art-meaning/sense perceptively felt.

An interpretation of reality, not a sense derived from high culture, e.g. high spirituality but from every subject's life experiment (Hesse, 1970, GW. IX, 178, 259, 205, 311, 84, 238–40, 85, 347–8, 239, 440–41, 418–9). This kind of modern hero becomes an acting subject: a space-platform of a brand new modern Utopia, which paradigm lies on the subject's development since the individual unfolds himself while serving secular society by educating a young and rebel of its members.

Thus representing *in actu* the magnitude of emotionality born of a handed out culture—not of a self-segregating and philistine one but out of a promoting humbleness one—whose influence over every member of society will make him give up his *status quo* for the sake of present day generation.

Tito—ready to be part of a brand new utopian model—acts/dances presenting simultaneously during a moment-in-time—symbolically summoned—a new paradigm. He does not speak; he dances/acts hearing a silent melody showing how able he is to refuse a speech built on a principle of causality preferring instead being driven by the principle of synchronicity.

He then personifies a dialectical sensibility constantly moving between rationality and emotionality aware of a *continuum* of facts and experiments whose relationships are to be practiced by every subject who understands himself as the proper space of an autonomous interiority, free from space and time stereotypes boundaries: personifying a brand new utopia paradigm for an *open society*.

Hesse drives his literary strategy towards Modernity creating a masterpiece—example of a «superior piece of art» wherein a utopian paradigm emerges out of a unique experiment by expressing a correspondence between reason (*Ratio*) and imagination (*Magie*).

At the same time, by refuting a «dogmatic program» by which every reader could be persuaded to accept how achievable a «platonic idea» within his diegesis could be, no matter what contradictions it might be made of.

On the other hand, Herman Hesse cannot help «showing» a new utopian model by refusing it as a space for evasion. Moreover, he does not deny the strong influence imagination plays proving how every fiction turns every diegesis the more resilient, the more magical. Thus caring about the way an «Idea» might be put into practice in a near and prospective future, reinforced by his putative reader.

This is Hesse's peculiarity while giving birth to a modern literary masterpiece: the more influent the less abstract (Hesse 1973, vol. 1, 301, 98, 57, 301, 189–190, 241, 93; Hesse, 1982, vol. 3, 223–4, 232–5, 165; Hesse, GW. X, 579); the more supported

by historical facts the hero's existence will be, the more persuasive a modern literary work of art will also be whenever its hero's peculiar reality is thoroughly and pertinently depicted.

The more every glass bead game creator/educator will be able to summon ludically facts of a social and individual living with the help of his creative imagination, the more, under the influence of interrogation, hope, under the dynamic stimulus of culture in progress; will manage to put an end to the absurdity of existence by overcoming it.

Such a narrative will express then the intensification characteristic of the way individual experiments might build the very matrix of utopia. A true reality aware of its contingency but without any kind of space for «simplifiers» - felt as a surplus value to reinforce a main ideology/culture as long as it will incorporate the sage-saint (*Heiliger*) profile as paradigm of an elite member who cannot help learning wisdom out of universal culture and history prospectively for the sake of the preservation of human culture (Hesse, 1982, vol. 3, 218, 51, 107–8, 252, 334, 476, 45; Hesse, 1970, GW. X, 90, 436, 438–9; GW. XI, 90; Hesse, 1973, vol. 1, 232, 293).

Hesse depicts not only what in fact happened but also what might happen if such a model would be put into practice again. He makes then the narrator stimulate the addressee's imagination to be more practically engaged in his present time. In so doing he manages to provoke a dialogue between the narrator and the addressee about changes they might begin to feel vital to their community.

This novel presents then an approach to the modern novel paradigms aiming to deal with a utopian model by starting making the reader follow an «individual-subject» who tells shows and reflects upon a civilization model. Moreover, under every narrator's speech, directs one's attention to the main actions-documents perpetuated by a hero-Knecht who dared to propose changes in his present time epoch: applicable to the one during which the addressee lives (Ziolkowski, 2003, 53, 71, 41).

Inside the narrative hero J. Knecht lies the foundations upon which a modern and *open society* may grow up as long as at the basis of a modern utopia stands—not like a work of philistine art and culture promoted by an elite such as *Kastalia*'s paradigm of an ancient utopia.

A culture promoted by such elite's paradigm—and expected to be later applied by the best society's political administration—, as happened with *Designori*, didn't succeed: *Designori* and his son *Tito* didn't get along well with one another. The addressee understands immediately the new function culture is given when he sees Knecht assessing and judging—keeping in mind at the same time every effect it would have inside society—while watching deeply impressed the beautiful dance ritual performed by Tito amidst real nature impressive scenery.

Inside *Kastalia,* culture played a functional role for the sake of the community's welfare, in the form of glass bead games. Instead, *Tito's* art expresses for the sake of its own being simultaneously a projection of Knecht's feelings who gives in to his pupil's *Weltanschauung.*

The hero hands over his legacy to Tito; he refuses his self-segregationist cultural model, thus crossing the frontier and accepting the new paradigms according to which culture includes interaction with everyday real time and space in order to get freedom out of its achievements endowed with social skills to preserve community's welfare. The more reality will be reflected back into art the more «taste» will stimulate social, political, and historical dynamics; as *Magister Ludi* J. Knecht ends by rejecting every authority awarded by culture since he understands himself as a humanity's servant responsible for balanced co-existence.

Every work of modern art exhibits its alternative against absolute and infallible values in either science and philosophy or art, offering instead a platform for subtle and constructive criticism.

Das Glasperlenspiel is a true example of this, since its «publication» expresses the humbleness, through which work and narrator's dialog constantly with the Other (reader-addressee) upgrading the natural and necessary common sense every «town square» desperately longs for (Arendt, 1972, 286–8, 283).

Every work of modern art recovers even amidst a mass culture and society its rebel role because like every cultural good it still keeps its resilience against every social *modus vivendi.* Every «market» sometime unable to appreciate prospective wealth of emotionality, sensibility and creative imagination a work of modern art may pass on—even under every present day mass dynamics because only used to detect functional, commercial, exchangeable surplus value attested by its cultural players—won't remain the same.

Therefore, *Tito's* choreography «puts an end» to the high consideration institutionalized and philistine *kastalian* culture enjoyed—as it is symbolized by *Magister Ludi's* death. The narrator instead «updates» humankind's goodness through «emotionality-by-the-moment» dance scene he makes *Tito* arouse by the end of the «Lebensbeschreibung»-narrative (Arendt, 1972, 260–1, 264, 256–7, 266).

The Glass Bead Game gives literary shape to the existing frontier between Ancient and Modern Utopia assigning every community's individual a mission according to which he has to develop himself respecting his authenticity.

Just like a modern human being's paradigm also one should keep up with: 1—dignity, as long as he/she doesn't care only about rationality; 2—humbleness, as long as he/she does not get satisfaction only out of progress and technological achievements; 3—freedom, as long as he/she does not deprive himself of living in present time, postponing instead opportunities to a hopeful future already ontologically formatted.

One should develop one incomplete nature because every human acting lies on an absolute present time dimension (Jonas, 1979, 410,409). Then people start to recognize their own predisposition for a perfect and new tenure, dependent of their historical consciousness while feeling every moment as unique with the help of any work of art expressiveness and its exclusive ability to stand for a paradigm worthy of re-shaping reality. Dance stands for an individual and collective singular *replica* (it cannot be postponed) while giving expression to immediate and interactive bio-psychical density in present time, symbolizing the urge of one's natural sensibility, and above all historical consciousness every individual expresses whenever assimilates the principle-of- responsibility (Jonas, 1979, 412, 418–9, 420–1).

It is the case of Tito after receiving the baton from the hands of ancient *kastalian* utopia reformer: J. Knecht. Hesse stresses the importance of studying every masterpiece culture has brought to us though by means of explaining/understanding every method behind its form-structure: its strategies. Only then could culture be piece of an engaged act-of-transforming/«serving» humanity (Unseld, 2003, 187–8).

5

What better than a ritual dance amidst natural scenery performance—against the clock—so that addressee and reader might understand time vigorous dynamics? The same Tito is dealing with, in order to continue shaping Knecht's wisdom, according to which only out of organic growth— contemplating an actualized and spoken dialog between logical-deductive intelligence and active and applied imagination—can the social utopian building towards a plural (Human) Being emerge.

To help reinforcing literary fragment shape— amidst a work of modern art—to foster a social-political configuration led by modern time-moment dynamics; finally to reinforce—during the «last» scene of Knecht's life—contingency's strength «shown» by a modern work of art aiming to keep the literary addressee in silence, once dazzled by the dance metaphor as a secular *replica* to the polysemy meanings of music enhanced in *Kastalia*.

All of this simply by suggesting choreographically what words would not be able to achieve even through the boldest metaphor (Steiner, 2012, 30, 24–5, 77, 16, 13–4).

The Glass Bead Game makes us shift value and culture centre from learning and intellectual competences to the space of nature in which dynamical centre stands individual paradigm whose competences become greater: the more one plays with the four elemental *space-nature*, the more wealthier the community gets.

Commitment is therefore the adequate concept not only to characterize the hero, Knecht, but also *mutatis mutandis* the whole narrative as a modern utopia paradigm (Ziolkowski, 2003, 72; Swales, 2003. 150). Both prove how important is to start learning from every *infinitesimal* piece of reality, obeying the principle-of-responsibility.

This narrative upgrades thereby our attention to how capital modern art can be as long as it will not let itself be deprived from tradition, and will not reject any fantastic dimension of the real while upraising its addressee's attention to every normal contradiction a utopian model contains, by making him/her keeping their eyes wide open on how to see and to overcome every western *weltanschauung*'s contradiction (Rancière, 2011, 34).

BIBLIOGRAPHY

———. *Bíblia*, Lisboa: Difusora Bíblica, s/d, 1692.

Bacon, F. New Atlantis, London: University Tutorial Press, s/d.

Campanella. *La Città del Sole*, (tradutor Alvaro Ribeiro) *A Cidade do Sol*. Lisboa: Guimarães, 1990.

Hesse, H. *Das Glasperlenspiel*, G W IX, F/M: Suhrkamp, 1970.

———. *Gesammelte Briefe* (1936–48), vol. 3, F/M: Suhrkamp, 1982.

———. *Gesammelte Werke*, 12 vols. F/M: Suhrkamp, 1970.

———. *Materialien zu H.Hesses Das Glasperlenspiel* 2 vols. F./M: Suhrkamp, 1973.

———. "Politik des Gewissens"-*Die Politische Schriften*, 2 vols. F/M: Suhrkamp, 1977.

More, Th. *Utopia*, Cambridge: CUP. 1989.

Platão. *A República*, (translator M-H. Rocha Pereira) Lisboa; FCG, 1990.

Secondary Sources:

AA VV, *Utopien; Die Möglichkeit des Unmöglichen*, Zürich: Verlag der Fachvereine, 1989.

Arendt, Hanna. *Between Past and Future*, (translator Patrick Lévy) Paris: Gallimard, 1972.

Bloom, H. (ed.) *Hermann Hesse*. Philadelphia: Chelsea House Publishers, 2003.

Elias, N. *Der Zivilizationsprozess*, (translator: Lídia Rodrigues) Lisboa, D. Quixote, 1990.

———. *Über die Zeit*, F/M, Suhrkamp, 1990, pp. 196.

———. «Thomas Morus und die Utopie» in AA VV, *Utopien- Die Möglichkeit des Unmöglichen*, Zürich: Verlag der Fachvereine, 1989.

Freyer, H. *Die politische Insel: eine Geschichte des Utopien von Platon bis zur Gegenwart*, Leipzig: Bibliographisches Institut A.G., 1936.

Hall, A. Rupert, *A Revolução na Ciência 1500–1750* (Translator: Teresa Perez). Lisboa: Edições 70, 1988.

Jaeger, W. *Paideia* (translator: Artur Mourão.). Lisboa: Ed. 70, 1961.

Jonas, H. *Le Principe Responsabilité* (translator: J. Greisch), Paris: Flammarion, 1979.

Lapouge, G. *Utopie et Civilizations*, Paris: A. Michel, 1990.

Popper, K. *La Société Ouverte et ses Enemies* (translator: J. Bernard; Ph. Monod). Paris: Seuil, 1972, 2 vols.

Rancière, J. *The Politics of Aesthetics—The Distribution of the Sensible* (translator: G. Rockhill). London/N.Y: Continuum International Publishing, 2011. available from: https://selforganizedseminar.files.wordpress.com/2012/10/rancic3a8re-jacques-politics-aesthetics-distribution-sensible-new-scan.pdf.

Ricoeur, P. *Lectures on Ideology and Utopia*, (translator: Teresa Perez), Lisboa: Ed. 70, 1991.

Steiner, G. *A Poesia do Pensamento: Do Helenismo a Celan*, (translator: Serras Pereira, M.)1ª edição, Lisboa, Relógio d'Água, 2012.

Swales, M. «Hesse: The Glass Bead Game (1943)» in Bloom (ed.), *Hermann Hesse*. (Philadelphia, Chelsea House Publishers, 2003.

Unseld, S. «Hermann Hesse's Influence: Ethics or Esthetics?» in Bloom (ed.), *Hermann Hesse*. (Philadelphia, Chelsea House Publishers, 2003. pp. 177–193.

Ziolkowski, Th. «The Glass Bead Game: Beyond Castalia» in Bloom (ed.), *Hermann Hesse*. (Philadelphia, Chelsea House Publishers, 2003, pp. 39–77.

Tlön: Journey to a utopian civilisation

Aristidis Romanos
Architect Planner, Athens, Greece

ABSTRACT: A 14th century Lithuanian explorer named Ladislas travelled in remote regions of the East and discovered the ruins of an unknown utopian civilization.

Ladislas recorded his impressions in an illustrated Manuscript that remained in obscurity until the middle of the 19th century when Leonid Krk, a notable Baltic literature scholar, translated it into English. A copy later came into the possession of a Scandinavian urban planner, a friend of the author. Access to this interesting piece of discovery reveals the language, philosophy, social values, and history of the civilization discovered including its architecture, town design, and structure.

Ladislas discovered a civilisation whose main philosophy was idealism and whose highest social value was tolerance and the pursuit of harmony among the different sections, creeds, and races and attempted to explain how these principles were expressed in town building.

The civilization is named 'Tlön' after a short story by Jorge Luis Borges, (*"Tlön, Uqbar, Orbis Tertius"* Labyrinths, New Directions Book, N.Y., 1964), which describes the ambitious undertaking of a secret society to write an encyclopaedia about an imaginary planet. As the reader travels through the story, it becomes increasingly clearer that the four Ages in the history of Tlön resemble, in some ways, humanity's stages of evolution. In particular, the third, alluding to the environmental crisis, forewarns a Huxley-like scenario of overcoming it.

Keywords: Dystopia, Utopia, tolerance, town building

1 AIM OF THE STORY

Some of the 'nightmarish' probabilities that haunt our world are: proliferation of unsustainable exploitation of natural resources, ecosystem approaching collapse, intensification of inequalities between social strata and nations, wars of domination over energy sources, wars of desperation over means of subsistence and water, large scale migration of economic and climatic refugees.

A balance between the whole and the parts is a prerequisite for establishing a peaceful world society in future. A way has to be innovated, whereby the Parts can pursue their own aims within an agreed structure of the Whole, designed to set the limits between 'individual' goals and life-expression and reciprocal tolerance and co-existence, i.e. accepting racial, religious, political, or other differences. It sounds simple, but we all know that it requires a struggle of Utopian scale.

A train of thought following the two hypotheses set above prompted me to write this story.

It is about a utopian civilisation (named Tlön, after Borges) revealed in a manuscript written by Ladislas, fictitious 14th century Lithuanian explorer, who travelled in remote regions of the East and discovered ruins of its, hitherto unknown, existence. He described a society whose philosophy was idealism and whose ethics were peaceful co-existence, balance, and tolerance. To reach its Golden Age Tlön had to survive and overcome a terrifying dystopia marked by environmental crisis, total anomy, and disastrous wars.

2 TLÖN'S HISTORY

Four eras are distinguished: Infantile Age. Man, subordinate to the primordial forces of nature, worships and fears the divine, the king, the magician, and the father. Man tries to protect against nature, savage beasts, and hostile tribes. Emancipation Age. Man attains spiritual liberty and knowledge; he creates a civilisation distinguished from the natural order. With ingenuity and crafts, he subordinates and finally exploits nature. Settlements evolve in industrial cities and grow out of proportion. Age of Redemption. Forfeiture from the natural order; man faces dangers unthinkable before. World Governance follows a dark period of generalised wars. Golden Age. "The period when the spirit of Tlön triumphed".

3 THE ENVIRONMENTAL CRISIS

At the end of the second Age Tlön lived a 4-stage Dystopia. Very few listened to wise men's warnings about the dangers of insatiable production and to the consumption of ever growing quantities of material goods, although the signs were already visible: depletion of natural resources, unsustainable food and energy production, unclean air, high temperatures, drought, water shortage and poverty. A series of panicky but ineffective environmental technology devices aiming to protect human environment followed. Some city governments went as far as to cover large urban areas with a new layer of 'nature' presumably to protect the community from foul air and the scorching sun. (Figure 1).

Strong countries unwilling to curtail their own consumption of resources and waste decided to reduce world energy consumption by cynically destroying the productive and technological infrastructures of up-coming populous empires. Poor countries used terrorism tactics to resist. Hostilities between states ended up in catastrophic armed conflicts known as Environmental Wars (*"land"* *"virus"* and *"water wars"*).

The victorious powers pressed for world governance to control excessive consumption (of energy, services, and material goods). They transformed the powerful intra-state agency against terrorism of the preceding period, the notorious Science and Security Service (SSS), into a planetary law and order instrument enforcing strict environmental controls. The SSS invented the Summa Project imposing maxima levels of emissions and energy consumption—calculated for the planet as a whole, for each state, each region, each province, each community, and each individual within a community! Its extreme tool was the 24/7 surveillance and monitoring of thought processes using miniscule drones and invisible implants. 24/7 was dropped when the SSS discovered that the subversive ideas to world order were more common in the minds of the political and military elite appointed to safeguard order. The population reacted to Big Brother methods in many ways; a mild, passive reaction was that of the No-walls Communities (Figure 2).This era was rightly named Environmental Authoritarianism.

In the Recovery decade the world government managed to steer the social vessel back to normality. A new mature generation with a sense of collective responsibility emerged. It was a bitter disappointment of liberal and left-wing intellectuals that the responsibility shown by the population in observing austerity measures was not a result of rationality and free will, but of coercion—sometimes cruel oppression.

The world experienced the final clash between two tendencies manifest in the previous eras: The 'materialismus alternative' which, fearing doom, conceived survival through the ambitious plan of 'self-propulsion communities' leaving mother earth to reach galaxy destinations; and the 'idealismus alternative' which began the foundation of small high-tech autonomous urban farms in a clinically healthy and environmentally controlled earth. (Figures 3, 4)

4 TLÖN'S PHILOSOPHY AND SOCIAL VALUES

Common to all was the desire to relieve life from the burden of time. This ethical aim was considered fundamental to man's happiness and to his attitude to death. It became main preoccupation

Figure 1.   Cities covered by artificial nature.

Figure 2.   The reaction of No-Walls communities.

Figure 3.   Self-propulsion communities.

Figure 4.   Sustainable high-tech urban farms.

of Tlön's education to disassociate death from time and to abolish fear of death by insisting on its biological necessity.

All states of mind enjoyed by men were treated with a mystic reverence. Anyone enjoying these virtuous states should refrain from boasting or impressing his joy onto others. One philosophical school taught that happiness is primarily a property of remembering the past, or of contemplating the future; that when we are enjoying a happy event, we are not really aware of the fact, as enjoyment itself precludes the consciousness of happiness. This comes as a memory at some future time ('how happy we were then'); or it is achieved as precedent contemplation ('how good it shall be when...').

The "materialistic nausea" of previous eras led to voluntary restriction of material needs, a return to more communal ways of life and the decentralisation of energy sources—a true "people's

technology", based on renewable energy. A few large automated industrial plants were responsible for the stable production of a limited number of basic "raw-material" goods. Goods required for the sustenance of daily life were made at home, with no primitive technology. Sophistication of technique was not the prerogative of centralised industrial process; it was in everybody's hands and could be achieved in everyone's back garden.

Due to material affluence and to idealism, "power struggle" was limited to the sphere of ideas. Intellectual conflict was sought after and elevated to virtue favouring social progress.

The highest philosophical values and practical objectives were (a) the ability to keep one's self-awareness alive, be constantly in a state of mental and intellectual alertness, (b) sharp memory and (c) balance. Balance everywhere: between the groups constituting society, in "reconciling intellectual and ethical individualism with ordered social life", between the values of the Greek City state and the need to abide by a new world order, between chance and necessity. Balance was the cardinal value leading to the primordial community goal, Harmony.

5   THE ARCHITECTURE OF TLÖN

Some principles that apparently pervaded the theories and practice of Architecture in Tlön were discovered, engraved on a megalithic Urartian 'vishapkar' near Tesebaini:

a. "In spite of the world being successive and temporal, architects should be taught to design for static space and for eternal buildings";
b. "The design of Public Buildings is a collective work; Architects act as knowledgeable advisors to the interested community groups";
c. "In a society of frugal material needs Architects discard all determinisms and teleologies (functional, technological, productive). The Architectural object, having satisfied the material needs, enters the realm of Art. As in literature, where poetic objects were formed by combination of words and monosyllabic prefixes, architectural design also became pure play."

Contrary to the liberal forms of private dwellings public buildings followed strict rules. Labyrinths, for example, were integral parts of public buildings due to their symbolisms and connotations, known to all citizens. They were classified according to the number of entry points, the number of access paths, and the existence or not of a goal (centre), and the existence of blind alleys. One labyrinth type implies that you may follow different paths, leading to impasses, before you discover the one and only door revealing the true access to the centre. If, from first try, you fall upon the true path, know that you have missed the essence of the moral lesson. The allegory of another one is that no doctrine is recognised as true over others; they are all valid as didactic methods of inquiry. All methods are copious and require toil to overcome.

Figure 5.   The Plateau of Ustyurt from the pass.

6   TOWN BUILDING

Ladislas writes (excerpts are marked "....."):

"I covered a vast area of what we term "the East"; partly I followed a route similar to that of Alexander. The most exhilarating cases of town building I discovered in two plateaus: Ὑψίπεδον των Ελίκων" (Plateau of the Helices or Spirals, in Urartu) and "Υψίπεδον των Χθονίων Ερειπίων" (Plateau of Underground Ruins, in Ustyurt)".

Figure 6.   Weird ruined structures at regular intervals.

The towns at the plateau of the subterranean ruins

"My first view of the Plateau of Ustyurt from the pass gateway was breath-taking. Amazement overtook me. This weird landscape was an artificial ground, constructed over the original layers of ground and human settlements lying beneath. A vast urban space, totally covered by this 'artificial nature' (indistinguishable from 'natural' nature) extends underground."

It surpasses our contemporary imagination that an ancient civilization could command such overwhelming power—material and organizational. Did the people of Tlön wish to safeguard their historical past from the ravages of time? What were the causes for such an almost superhuman enterprise? Did the sun or other natural element turn from blessing to calamity, one that threatened their civilization?

The Subterranean Towns were presumably constructed during the Environmental Crisis; an enormous investment by the society of Tlön to protect its populations from environmental cataclysmic disasters. These were Dystopian, not Utopian creations. (Figures 5, 6, 7, 8)

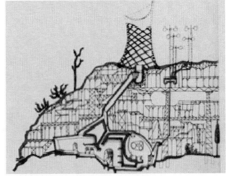

The highest value of towns

"In Tlön they believed that the design of human settlements incorporates the primordial god-sent value of balance; a notion difficult to objectify, but from certain references it was concluded that it has a connotation of justice: absolute justice is never achieved, but social balance is the aim of justice. The whole history of cities showed a continuous process towards equilibrium and harmony."

Figures 7 and 8.   Gigantic structures marked the entrance to vast underground spaces.

Town design at the plateau of helices

"A stone plate in Urartu reveals a Canon that prescribes the rules of proper design of cities. Two are primordial: balance between the founding groups and subjugation to an overall structural plan.

The city is the image of an ideal situation, where groups live in harmony within a community, this harmony being the end of all cities; the population constantly pursues this end, but no town will ever achieve perfection.

The structure's geometry is based on the helix, a shape typical of earthly and celestial objects, identified with evolution (ex-helix-is in Greek means evolution). It increases by accumulation; older parts remain intact while the total changes. Its form incorporates its history." (Figure 9)

Radial and Helicoidal roads

"There are two types of road in the helix town: spiral and radials. Radial roads, or 'utilitarian' take you from point A to point B in a straightforward way. The spiral road, or 'road of historical memory' or 'recreation road' takes you from one sector of the town to the other reminding you that the town owes its existence to the effective communication between the different cultures that created it. You travel indirectly, not pressed by time, having the opportunity to perceive the materialized spiritual history of the town. As you visit successive sectors, you marvel at their architectural and urban design styles and you appreciate the effect of diversity of social ideologies in the formal diversity and beauty of your town." (Figures 10, 11)

"Radiating lines subdivide the area in sectors (equal to the number of founding groups). Each group develops, according to its code of ethics, its own sector without upsetting the overall community structural balance." (Figure 12)

"The helix cannot grow indefinitely. Towns would not expand without limit. Some stopped at a predetermined natural element as final barrier to expansion. Others were abandoned, by consent or by arbitration, when one sector grew to a size endangering the principle of balance; no town had prospered, if one social constituent showed overwhelming dominance over all others. Slight growth

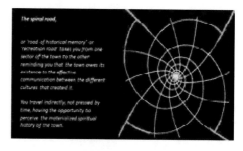

Figure 10. Composite helicoidal town structure.

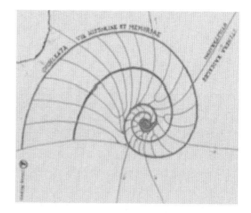

Figure 11. Radial and helicoidal roads.

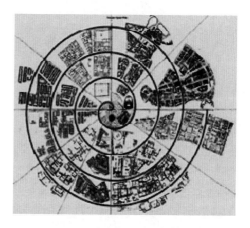

Figure 12. Town sectors for symbiotic development.

unevenness was corrected with concerted spilling over onto adjoining sectors.

In July 1398 I visited the ruins of Irene near the valley of Araxes; a town where the development of the spiral started from the outer area and proceeded inward towards the pole! In their mentality, centrality was just a geometric characteristic with no economic connotation. Urbanisation of the pole was considered grand social project and extremely idealistic conception."

Figure 9. Equiangular helix.

The Aegean townships

Some thinkers accused "the design of helicoidal towns as mechanistic and teleologic; they praised the Aegean towns as:

"...seemingly individualistic and anarchic, but pleasant and surprisingly interesting" "permeated by innate unwritten laws, implicit rather than explicit, and lenient rather than dogmatic".

Moreover, they concluded that

"...such an approach seems to unleash the creativity of the builder and to give the visitor the joy of surprise". (Figure 13)

Tesebaini and the Arame dam

"September 1398, visit of the town of Tesebaini, built on a mound, located in a mountainous landscape with successive ridges and peaks vanishing in the horizon. The steep mountain slopes are meticulously terraced, cultivated, and irrigated. Water is collected at a dam, built at high altitude; channelled through an underground canal it reaches the top of the hill-town and rolls down an artificial spiral river bed (in lieu of the helicoidal road of other helix-towns). Irrigation canals carry it to the terraced orchards and vineyards on the mountain slopes." (Figure 14)

Conclusions from the Agora

"In May 1399 I visited a site, which I tend to identify with Zernaki Tepe, and explored the ruins of its Agora. It is a calm landscape, a small valley between two moderate mounds, joined by a wide alley, but the design of the Agora incorporates an enigmatic concept. Except for two square buildings on the mounds, there are no other important edifices. Public life, for some mysterious reason, seems to have taken place at extensive spaces, provided at various underground levels. Only relatively small structures are on the surface: signs and entry points to the reality of public activity taking place in large under-

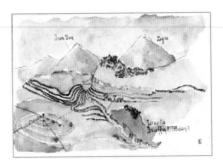

Figure 14.   Tesebaini and the Arame Dam.

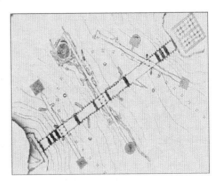

Figure 15.   The Agora at Zernaki Tepe.

ground spaces. Piles of rubble forbid my entry to all these underground spaces. I only gained access to the labyrinthine basement of the south edifice; I remain puzzled about its scope and function. It looks to me like a kind of prison, but whether it is for imprisoning people, treasures, or perhaps memories, that, I do not know." (Figure 15)

The text of Ladislas ends here. In the following paragraph, Caspar Amorson tried to answer some of the puzzles of Ladislas:

"I fully understand the meaning of the omnipresent duality "surface-underground"; I maintain that it is a symbolic expression of the theory of Idealism in civic space. The visible world at the surface is the idea or the 'mind-object' comprehensible through Reason, which prevails over the 'matter-object' that lies in a subterranean level, the real field of our activities, accessible to our senses. As for the labyrinthine basement and windowless rooms that Ladislas came face to face with, I venture to propose that they were specially designed storage spaces for codified data; the precious stones lining the walls of the rooms are nothing less than electronic storage chips.
  I conjecture that the civilization of Tlön, facing an Armageddon and the end of time, took measures to record the totality of its history and its civilisation and culture in order to safeguard it for survivors."

Figure 13.   Seemingly anarchic Aegean settlements.

Thinking against utopia: A comparison between Thomas More and Emil Cioran

Paolo Vanini

Department of Literature and Philosophy, University of Trento, Trento, Italy

ABSTRACT: The purpose of this article is to underline the relation between Utopia and carnival imagery in More's dialogue. Additionally, we aim at clarifying Cioran's definition of Utopia as "the grotesque *en rose*". Subsequently, we try to define the concepts of "seriousness" and "futility" in Cioran's interpretation of utopian literature. When Cioran says that "futility is the most difficult thing in the world", he is actually pointing out a problem that will enable us to outline the paradoxical condition of the inhabitants of Utopia.

We will see that in Cioran's writings the utopian idea of building a perfect society is *necessary and ridiculous at the same time*. It consists in a human attempt to renew the world, which reveals the double nature of this renovation: *serious and grotesque* at the same time. In fact, Utopia is a *revolutionary city* that looks like a *carnivalesque overturning* of this unrighteous reality.

Thanks to the enlightening analysis of Carlo Ginzburg, we will examine the complex game that More entertains with Lucian of Samosata and the "ritual of inversion" in his *Saturnalia*. Finally, we will see how the concept of "inversion" operates in Cioran's reflections about history and politics, where he defines utopia as the desire "to remake Eden with the instruments of the Fall".

Keywords: Cioran, More, Utopia, carnival

1 A DOUBT ABOUT UTOPIA

In *History and Utopia* Emil Cioran asks himself why Thomas More had decided to follow Plato and to become "the *founder* of modern illusions". Such a question not only does imply an accusation toward More, but also represents an effort to understand the ambiguous status of any utopian project. In Cioran's writings, the utopian idea of building a perfect society is *necessary and ridiculous at the same time*: this human attempt to renew and to improve the world reveals the double nature of this renovation: serious and grotesque at the same time. In fact, Utopia is a *revolutionary city* that looks like a *carnivalesque overturning* of this unrighteous reality.

In order to underline a two-faced profile presupposed by any plan of ideal city, Cioran states "utopia is the grotesque *en rose*". As he affirms:

The very notion of an ideal city is a torment to reason, an enterprise that does honor to the heart and disqualifies the intellect. (How could a Plato condescend to such a thing? He is the ancestor, I was forgetting, of all these aberrations, revived and aggravated by Thomas More, the *founder* of modern illusions.) To construct a society where, according to a terrifying ceremony, our acts are cataloged and regulated, where, by a charity carried to the point of indecency, our innermost thoughts are inspected, is to transfer the pangs of hell to the Age of Gold, or to create, with the devil's help, a philanthropic institution. Solarians, Utopians, Harmonians—their hideous names resemble their fate, a nightmare promised to us as well, since we ourselves have erected it into an ideal. (Cioran 2015)[1].

In this quotation, the dream of a "new world" is converted to the nightmare of "the worst of all possible worlds". This happens because the ideal of a perfect equal society requires a forced homologation of its citizens. One the one hand they cannot do anything by themselves. On the other hand, they have to act exactly as the rest of their fellow citizens. We have the feeling that these cities of nowhere could be illuminated by a light that does not allow any space for shadows, ruled by a logic that does not admit any reasons but its own. Bronisław Baczko has noticed that Utopia is the city of *transparency*, a fully transparency, which reminds us of a brightly prison of glass: an uncomfortable, well-lighted place (Baczko 1979).

Cioran has always tried to think against himself. Therefore, he does not want to give up of thinking against the city and thus he denounces the risk of this kind of inequality. In his criticism, the structure of an ideal city reminds of the hierarchy of Hell, as if it belonged to a Golden Age dominated by the Devil—and not by Saturn. Such a metaphor recalls an upturning of roles who characterized the playfulness ritual of carnival: according to Cioran, in the streets of Utopia the Devil plays the role of Saturn because we are attending a carnival of unhappiness. Hence, a question needs to be answered: What is the true "mask" of Utopia: the island of joy or the island of sadness?

However, before answering the question we need to analyse the relation between carnival rituals of inversion and More's utopian imagery.

2 UTOPIA AND CARNIVAL

Erasmus edited the first edition of Utopia in 1516. More and he had previously translated some tales by Lucian of Samosata, which were published in Paris in 1506 and were known as Lucian's *Opuscula*. The influence of Lucian on their works, above all on *Praise of Folly*, has already been studied and we know that Utopians were "delighted with the witty persiflage of Lucian" (More 2011: 69)[2]. However, as Carlo Ginzburg has observed, the philosophical and literary values that More and Erasmus found in the Latin ironist writer are very important to comprehend the sophisticated approach to reality operated in the text of *Utopia*[3].

In some letters and dedications that follow his translations, Erasmus defines Lucian as a *useful, pleasant* and *elegant* author. In fact, he was able "to mix seriousness and frivolousness" in his humoristic depiction of the superstitions and the iniquities of human society (Ginzburg 2002)[4].

These three adjectives (*useful, pleasant, and elegant*) describe a verbal constellation that we can recover in the complete title of the first edition of More's masterpiece: "A truly golden handbook no less beneficial than entertaining [*nec minus salutaris quam festivus*] concerning the best state of a commonwealth and the new island of Utopia". In the second edition, the expression "nec minus *salutaris* quam *festivus*" (no less *beneficial* than *entertaining*) was substituted for "non minus *utile* quam *elegans*" (no less *useful* than *elegant*). The third edition, instead, came back to the original title. In any case, we find again the values that Erasmus used to characterise Lucian's writings: the incomparable example of transforming the most ridiculous joke in the most serious criticism, through the intelligence of the most elegant style.

We have just translated *festivus* with "entertaining" even if its main meaning is linked to the religious word of *festum* (holyday). From this point of view, it is important to note that in 1514 Erasmus published a new edition of Lucian's *Opuscula* with the addition of three other tales that he had just translated: *Saturnalia; Cronoloson, id est Saturnalium legume lator; Epistolae Saturnales*. All these stories are focused on social inequality. The narrative setting is the festival of Saturnalia: the week of celebrations dedicated to Cronus—the father of Zeus who was identified with the Roman Saturn—which took place in the second half of December. During this carnivalesque week, social hierarchies were subverted and class differences were (theoretically) abolished: slaves were treated as masters, and masters even waited on their servants. Lucian turns upside down this idyllic portrait of the holyday and images Saturn as a god without authority, unable to guarantee justice even for a single day. When a priest of Saturn asks to his god why he entrusts the administration of justice to the arrogant Zeus and why he allows the strong to abuse the weak, letting his poor worshippers to live in misery for all their life, Saturn simply replies:

> … my own charge is confined to draughts and merry-making, song and good cheer, and that for one week only. As for the weightier matters you speak of, removal of inequalities and reducing of all men to one level of poverty or richness, Zeus must do your business for you. (Lucian 1905: 121)

In his *Utopia* (printed two years later of Erasmus translation of *Saturnalia*), More takes care of those "weightier matters" that Saturn did not want to address, dealing with the "removal of inequalities" and of private property. Therefore, we can say that More's *libellus* is *festivus* in a double meaning: it implicitly evokes the festival of Saturnalia in order to tell an entertaining story about the very serious problem of justice. More himself reveals this aspect in the second letter to Peter Giles, which follows Book II of *Utopia*. In this passage, More replies to a "very sharp fellow" (who had criticized his *Utopia* defining it as a story either too absurd to be a fact or too inconsistent to be a good fiction) with this argument:

> I do not deny that if I'd decided to write about a commonwealth, and a tale of this sort had occurred to me, I might have spread a little fiction, like so much honey, over the truth, to make it more acceptable. But I would certainly have tempered the fiction a little, so that, while it deceived the common folk, I gave hints to the more learned which would enable them to see what I was about. (More 2011: 143)

By using the unreal conditional, More is ironically confessing what he *actually* did in writing his tale: he gave special names to the parts of the city in order to offer an explanatory key for his educated readers, thus enable them to realize that Utopia was a non-existing place. He states:

> Unless *I had a historian's devotion to fact*, I am not so stupid as to have used those barbarous and senseless names of Utopia, Anyder, Amarout and Ademus. (More 2011: 143)

The last quotation reminds Lucian's affirmation at the beginning of *A true history*. Before the narration of his voyage to the Fortunate Isles, the Latin satirist states that his lies are more credible than philosophers' theories:

> … as I have no truth to put on record, having lived a very humdrum life, I fall back on falsehood—but falsehood of a more consistent variety; for I now make the only true statement you are to expect— that I am a liar. This confession is, I consider, a full defence against all imputations.

More, who is as historian as Lucian was, describes a travel to an island, which is probably similar to the ones visited by the Latin writer. In another commendatory letter published in the third edition of *Utopia*, the preeminent French humanist Guillaume Budé notes that, if the rest of civilized nations imitated Utopian constitution:

> … greed [...] would been gone from hence once for all, and *the Golden Age of Saturn would return*. [...] But in truth I have ascertained by full inquiry that Utopia lies outside the bounds of the known world. *It is in fact one of the Fortunate Isles*, perhaps very close to the Elysian Fields. (More 2011: 138–139)[5]

In his remarks on *Utopia*, Budé regards More's project of a perfect society in the context of a carnivalesque representation of equality. Utopians' legislation is as ephemeral as the carnival overturning of social hierarchy: an illusory fleeting week swiftly bound to end. In this sense, symbolic-subversive connotation of carnival imagery offers to More the instruments to reflect on a revolution which is compelling but impossible to achieve. When Hytloday says "there is no place for philosophy in the councils of kings", More answers back that it actually exists:

> … but not for this school philosophy which supposes that every topic is suitable for every occasion. There is another philosophy that is better suited for political action, that takes its cue, adapts to the drama in hand, and acts its part neatly and appropriately. (More 2011: 33)

More was looking for "another philosophy", a philosophy that understood the *ridiculousness of this world* because it has been able to imagine *the serious possibility of another reality*. In his thought, comic tone and serious tone work together, allowing him "to look at reality from an unusual point of view, to pose oblique questions to reality" (Ginzburg 2002: 44). Saturnalia rituals of inversion offered him a model to imagine a *paradoxical* city where gold and silver were used to make chamber pots and where jewellery was considered a sign of slavery. However, this kind of paradoxes (typical of Lucian's tales) is the *conditio sine qua non* of utopian thinking, because paradox is a philosophical way to find a dimension of possibilities still unknown. In order to see the carnivalesque side of real history, firstly we have to imagine a history ruled by the law of Carnival. If we want to discover what is actually wrong, *here and now* we have to overturn our orthodox image of reality. Hence, the description of Utopia allowed Thomas More to highlight the absurd condition of his contemporary England: a real island where sheep "have become so greedy and fierce that they devour human beings themselves" (More 2011: 19).

3 CIORAN AND THE "URGENCY OF THE WORST"

In a famous essay dedicated to Rabelais, Michail Bachtin demonstrates that Carnival rituals of inversion celebrated the recurring renewal of the world. Conversely to other religious festivals, which sanctify the past to crystallize the present, Carnival desecrates the past and jokes about the present in order to change the future (Bachtin 1979). During this festival, any traditional structure of society is overturned for a few days, revealing the irrational logic that rules the world. This is the very reason that allows the slave to be transformed into king, the servant to become master, the fool to be turned into wise man. The law of Carnival claims that we have to laugh in front of the world before overturning it in a joyful spectacle. At the same time, this joyfulness is the effect of a cheerful representation of the scariest and most grotesque aspects of reality, Bachtin underlines that a carnivalesque vision of humankind is not a naïve interpretation of the present, but it is rooted on the historical consciousness, which defines the very essence of Renaissance age. The dreadful aspects of human life are not neglected; they are seen from a different angle: the truths of power and religion (and their punishments, dogmas and anathemas) are fought by the revelation of a new world, from which emerges the comic ground of darkness. Carnival looks like a grotesque celebration of life because life itself is a Janus-faced reality, made up of contradictions and astonishing absurdities (Bachtin

1979: 91–128). If we want to live truly, we have to experience the grotesque and dynamic essence of existence. Therefore, the chaotic aspect of Carnival can be seen as a reaction against a mournful established order of things which has to be improved (Stoichita and Coderch 2002).

When Cioran reflects on the utopian thought, he asserts that human history is a Carnival of blood and there is no possibility to disavow such an appetite for destruction. In the first paragraph of his first French book, he says:

> Idolaters by instinct, we convert the objects of our dreams and our interests into the Unconditional. History is nothing but a procession of false Absolutes, a series of temples raised to pretexts, a degradation of the mind before the Improbable. Even when he turns from religion, man remains subject to it; depleting himself to create fake gods, he then feverishly adopts them: his need for fiction, for mythology triumphs over evidence and absurdity alike. (Cioran 2012a)[6]

In the rest of the paragraph, history is painted as a masquerade of wars and crimes, as an "indecent alloy of banality and apocalypse".

At this point, if Cioran defines the project of a perfect city as the most improbable thing that a society could ever realize, he obviously concludes that the utopian hope of human perfection is co-essential to historical development. In fact, in the utopian literature he discerns:

> The [fruitful or calamitous] role taken, in the genesis of events, not by happiness but by the *idea* of happiness, an idea that explains—the Age of Iron being coextensive with history—why each epoch so eagerly invokes the Age of Gold. Suppose we put an end to such speculations: total stagnation would ensue. For we act only *under the fascination of the impossible*: which is to say that a society incapable of generating—and of dedicating itself to—a utopia is threatened with sclerosis and collapse. Wisdom—fascinated by nothing—recommends an existing, a *given* happiness, which man rejects, and by this very rejection becomes a historical animal, that is, a devotee of *imagined* happiness. (Cioran 2015: 157)

In these quotations Cioran does not reflect on More's *Utopia*; he deals with the relation between utopia, ideology, and fanaticism in our contemporary history. Actually, he ponders over the destiny of Western society, which had believed in the illusion of Reason and Progress and has realized nothing but the less righteous society ever imagined: the very negation of the philosophical ideal of Humanism. In the chapter *Urgence of the worst* he states:

Since the Renaissance, humanity has merely evaded the ultimate meaning of its progress, the deadly principle manifest within it. The Enlightenment, in particular, was to furnish a fair contribution to this enterprise of obnubilation. Then in the next century came the idolatry of the Future, confirming the illusions of the one that preceded. To an age as disabused as ours, the Future persists in displaying its promises, though those who believe in them are rare indeed. Not that such idolatry is over and done with; but we are obliged to minimize, to disdain it—out of caution, out of fear. This is because we now know that *the Future is compatible with the atrocious*, that it even leads there or, at least, that it gives rise to prosperity and horror with equal facility. (Cioran 2012b: 73)

By writing *Utopia*, More had ironically contested the 16th century Western society. On the contrary, Cioran, by criticizing the utopian presumption of human perfection, wants to deny the philosophical portrait of Man as a rational being capable of building his own salvation. Both of them react to same necessity of understanding human history, but from two different perspectives. On the one hand, More wants to address the platonic question on what a philosopher can do to create a good city. On the other hand, Cioran defends the antiplatonic thesis that a philosopher is incapable of doing anything to improve human destiny. In his opinion we should do nothing, seeing that "Man moves only to do evil" (Cioran 2015: 101).

This statement is not a mere provocation: it represents an ethical principle of behaviour. In Cioran's view, men have always done wrong because they have never known their real dimensions: they created an image of themselves that does not suit the reality. Philosophy is perhaps the main responsible of this falsified image of humanity. This is the very reason that leads Cioran *to subvert* any metaphysical definitions of Man, which take too seriously his rational essence. In *The temptation to exist,* he writes:

> My purpose was to put you on guard against the Serious, against that sin which nothing redeems. In exchange, I wanted to offer you... futility. Now—why conceal it?—*futility is the most difficult thing in the world*, I mean a futility that is conscious, acquired, deliberate. In my presumption, I hoped to achieve it by the practice of skepticism. [...] Each time I catch myself assigning some importance to things, I incriminate my mind, I challenge it and suspect it of some weakness, of some depravity. I try to wrest myself from everything, to raise myself by uprooting myself; in order to become futile, we must sever our roots, must become metaphysically *alien.* (Cioran 2012c: 211).

Cioran operates a philosophical overturning of humankind, which reminds of the Carnival rituals of inversion. Far away from being "the king" of the universe, the Man has become the most miserable animal ever existed: "he has definitely failed, but masterfully failed". This failure defines his uniqueness and his foolishness, his "privileged" condition among the rest of beings. Therefore, Utopia is "the grotesque *en rose*" because there is no happy conclusion for human history, this strange "Odyssey of Rancor".

When Cioran repudiates Utopia, he is actually doing a utopian exercise of "defatalization" of history. In his very description of the "mechanism of utopia", he compares the dream of a faultless society with the fear of *the end of the world*, the conclusion that logically follows "the Fall into time". He suggests that:

> The great majority of mortals [...] never renounce the quest for *another* time; they devote themselves to it, on the contrary, with desperation, but locate it here on earth, according to the prescriptions of utopia, which seek to reconcile the eternal present with history, the delights of the golden age with Promethean ambitions, or, to resort to biblical terminology, to remake Eden with the instruments of the Fall, thereby permitting the new Adam to know the advantages of the old one. (Cioran 2015: 198)

4 CONCLUSION

More had used his *Utopia* to propose a truthful interpretation of the Christian principles of life as the foundation of an admirable commonwealth; Cioran takes advantage of a Christian metaphor to show the diabolic absurdities of any utopian project. In *Urgence of the worst* (a chapter of *Drawn and quartered*), he imagines what would happen after a hypothetical apocalyptical catastrophe which would put an end to civilization. A few men will survive and they will decide to forget the past and to destroy the last signs of their old cities:

> So radical will be the rejection of history that it will be condemned *en bloc*, without pity or nuance. And so it shall be with time, identified with a blunder or with a profligacy.

These men will be free and equal, redeemed by the weight of our evil tradition. However, Cioran finally confess:

> Let us leave off these divagations, for it serves no purpose to invent a *comforting interlude*, wearisome feature of all eschatologies. [...] Boredom in the midst of paradise generated our first ancestor's appetite for the abyss which has won us this procession of centuries whose end we now have in view.

That appetite, a veritable nostalgia for hell, would not fail to ravage the race following us and to make it the worthy heir of our misfortunes. (CIORAN 2012b: 98)

It is difficult to establish if Utopia was created by the nostalgia for hell or by the hope of a paradise; in any case Cioran would say that it is not a very good place to live. However, this vagueness, this ambiguity is a utopian aspect of cioranian thought. As he says in his *Notebooks*, "I looked for my salvation in the utopia and I only found some relief in the Apocalypse".

NOTES

[1] Original title: *Histoire et utopie* (1960). For a critical edition of the complete works of Cioran see: Cioran, Emil. *Œuvres* (Edited by Cavaillés, Nicolas; Demars, Aurélien). 1° ed. Paris: Gallimard, 2011, "Bibliothèque de la Pléiade".
[2] See also David Marsh (2001).
[3] For this analysis about More and Lucian we follow: Carlo Ginzburg (2002).
[4] See also P.S. Allen (1910: 430–431)
[5] For an analysis of Budé's letter, see: Dominique Baker-Smith (2000: 229–230).
[6] Original title: *Précis de décomposition* (1949). For an accurate analysis of this text see, Nicolas Cavailles (2011).

BIBLIOGRAPHY

Allen, P.S. "Letter n. 199." In *Opus Epistolarum Des Erasmi Roterodami*. Vol. I. Oxonii: 1910. 430–431.
Bachtin, Michail. *L'opera di Rabelais e la cultura popolare*. 2°ed. Translated by Milli Romano. Torino: Einaudi, 1979.
Baczko, Bronislaw. *L'Utopia. Immaginazione sociale e rappresentazioni utopiche nell'età dell'Illuminismo*. 1°ed. Translated by Margherita Botto and Dario Gibelli. Torino: Einaudi, 1979.
Baker-Smith, Dominique. *More's "Utopia"*. 2° ed. Toronto: Toronto University Press, 2000.
Cavailles, Nicolas. *Cioran malgré lui. Écrire à l'encontre de soi*. 1°ed. Paris: CNRS Éditions, 2011.
Cioran, Emil. *History and utopia*. 2°ed. Translated by Richard Howard. New York: Arcade Publishing, 2015.
——. *A short history of decay*. 2°ed. Translated by Richard Howard. New York: Arcade Publishing, 2012a.
——. *Drawn and quartered*. 2°ed. Translated by Richard Howard. New York: Arcade Publishing, 2012b.
——. *The temptation to exist*. 2°ed. Translated by Richard Howard. New York: Arcade Publishing, 2012c.
Ginzburg, Carlo. "Il vecchio e il nuovo mondo visti da Utopia." In *Nessuna Isola è un'isola. Quattro sguardi sulla letteratura inglese*. Ed. Ginzburg, Carlo. 1ª ed. Milano: Feltrinelli, 2002. 17–44.

Lucian. *The Works of Lucian of Samosata*. Trans. Francis George Fowler and Henry W. Fowler. Vol. 4. Oxford: Clarendon Press, 1905.

Marsh, David. *Lucian and the Latins. Humor and humanism in the early Renaissance*. 4°ed. Michigan: The University of Michigan Press, 2001.

More, Thomas. *Utopia*. 3°ed. Translated by George M Logan. New York: Norton & Company, 2011.

Stoichita, Victor, and Coderch, Anna. *L'ultimo carnevale. Goya, Sade e il Mondo alla rovescia*. Milano: Il Saggiatore, 2002.

Island societies and architectural imagination: From Deleuze's desert island to Plato's Atlantis and More's Utopia

Aikaterina Myserli

Department of Architecture, Faculty of Engineering, Aristotle University of Thessaloniki, Thessaloniki, Greece

ABSTRACT: The contemporary crisis of spatiality issues lies largely in the sense of the uncanny, the existential anguish. This thesis studies the way in which fictional island societies became the defensive refuge of man, a separate reality. Deleuze in his text "Desert Islands" studies this separate reality, which he highlights as a fundamental driving force for the re-creation of the world. Plato's Atlantis and More's Utopia are typical examples of such separate island societies that function as virtual places of re-birth and re-creation.

A renewed reading of these classic works has the potential to bring contemporary architecture close to what Deleuze saw in the 'Desert Islands', that is to be the concept of collective in the individual. This approach is related to the recognition that island societies such as More's Utopia are conceived as imaginary worlds where we discover the collective soul of the place that we lost in contemporary cities and elucidate a spatial model of common good where architecture and urban planning emphasize on collectivity—and thus, public space—and convey and draw values from the society.

Keywords: island society, re-creation, collectivity

1 DELEUZE'S DESERT(?) ISLAND

As always with Deleuze, the desert island is an imaginary place, selected and intended to be isolated. The process of its selection and isolation is crucial as it revitalizes our imagination and prepares us to redefine the world by making space in constant transformation, within the forces of flow, repetition, and re-creation. In his effort to reflect on the causes of the island, the French philosopher begins with the description of the two kinds of islands—continental and oceanic—according to geographers.

Continental islands are considered as unsubmerged parts of the continental shelf that are surrounded by water. They are created due to erosion, fraction, or disarticulation[1]. If we observe carefully, we see that they are usually found close to the original land from which they were detached. Continental islands are always reminiscent of the continent and their cultural roots. On the other hand, oceanic islands are definitely free from any cultural or social reference. Oceanic islands are those that rise to the surface from the floors of the ocean basins[2] and are considered as the most isolated lands on earth, surrounded on all sides by vast stretches of water.

These two kinds of islands, continental and oceanic, reveal, according to Deleuze, the deep opposition between earth and water. We can reasonably

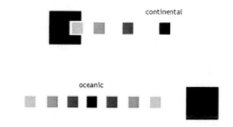

Figure 1. Concept diagram of continental and oceanic islands illustrated by author. Continental islands are detached from the continent whereas oceanic emerge from the depths of the ocean.

assume that these elements are in constant struggle and that the very existence of islands represents this struggle between them. As result, people will inhabit islands only after they have forgotten what they really represent. It is in this point that Deleuze refers to another way of comprehending the island, entirely based on imagination.

What makes the story of humanity so uncertain and full of charm is that people live in two worlds, the inner and the outer, with the first being subject to transformations including assumptions, illusions, images and beliefs, based on which people shape their behaviour. People persist in this clear separation of the land from the ocean, of the island from the continent because of their

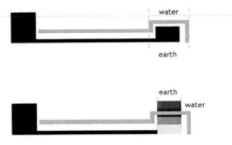

Figure 2. Island formation and opposition between the two elements, earth, and water. Either the sea is on top of the earth or the earth gathers its strength to pop up out of the water. Concept diagram by author.

primitive instincts and their inability to understand the never-ending struggle between water and land. When they dream of islands, they actually dream of being separate and beginning anew[3]. In the realm of the imaginary, it is not the island that has been created from the bowels of the earth but people who choose the island as the ideal place where they could re-create the world.

When the island is inhabited however, it is still deserted, as the French philosopher claims in his text[4]. In fact, people could occupy the island, if they coordinate with the élan vital of the island, its vital impetus. It is this vital force that would transform life to an indivisible continuity and the inhabitant of the desert island to a picture of the whole, a reflection of the island. This is the moment when imagination and geography are pieced together[5]. However, what kind of imagination? It is doubtful whether the individual imagination, without help or guidance, would ever consider such a remarkable identity for humanity. The idea of the man as absolute creator, totally separated, demanded the collective imagination, the concept of a "common soul", similar to what we find in rites and mythology. This "common soul" creates a new kind of space, which passes from the finite to the infinite, through a constant process of metamorphosis, a procedure of successive formats, which each time leaves a trail to penetrate it. Here, separation and re-creation are two warring—yet mutually complementary—forces and the reconstruction of the world is no longer a work of God, but it is entrusted to man.

2 ATLANTIS-CONTINENT OR ISLAND?

Plato's Atlantis is a typical example in mythology, where society is created anew on an island. In Plato's text, Atlantis is presented as an island larger than Libya and Asia together, that is to say bigger than an entire continent. Both its extent as well as its socio-political formation refers directly to the continent. Nonetheless, we could hardly consider Atlantis as a continental island, mainly because we miss the original piece of land from which it should have been detached. On the other hand, it cannot be considered as an oceanic one, because we encounter traces of human act, populated towns, and organized social institutions. The myth here defines a new kind of land, something between continent and island.

Nobody can say with certainty that Atlantis existed or not, let alone whether it was island or continent. This is of little importance for architecture after all. What matters is that in Atlantis human intention and knowledge crush the barren passivity of nature. Reversing the dominant perception that every cultural reference comes by default from the continent and then is disseminated to the island during the process of detachment is what makes this island society so important. The inhabitants of Atlantis have developed organized social structures equivalent to those of the continent. Atlantis is an isolated piece of land where there is an organized social and political framework and a comprehensive system of beliefs that allowed inhabitants to build cities and even to clash with the continent and claim the extension of their sovereignty. In *Timaeus*, Athens seems to symbolize the "perfect society" and Atlantis its opponent, representing the very antithesis of the "perfect" features described in the *Republic*.

Figure 3. Illustration of Atlantis by author. The island of Atlantis is presented as a piece of land where a socio-political model is developed and architecture as par excellence symbol of cultural action plays a significant role.

3 UTOPIA-A THIRD KIND OF ISLAND

One would expect that More would place his Utopia on an originary, oceanic island, as those described in the text of Deleuze, in order to establish his fictional island society and its religious, social, and political customs on a place without any cultural reference. However, the social utopia presented in this work has not lost its reference to the functional regulation of its archetype that is to be characterised by features of the real world. We find here the city and its region as a unit of political life (which refers more to the urban organization of the continent). On the island, natives live in fifty-four cities, each of which are thirty kilometres away from its neighbour city and has jurisdiction in the surrounding countryside within these thirty kilometres. In other words, More's Utopia appears

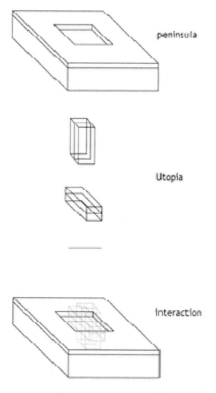

Figure 4. Concept diagram of the formation of Utopia by author. The island was originally a peninsula but the community's founder King Utopus to separate it from the mainland dug a 15-mile wide channel. In the illustration, Utopia is depicted as a box detached from a large surface (the peninsula) which can rotate in different directions while the surface remains motionless. In this way, it is pointed out that Utopia occupies an island that is as isolated as it takes so that its inhabitants interact with the world on their own terms.

as a continental island. Could it be inferred as a result that Utopia is indeed such an island?

The island of Utopia did not detach randomly from the continent due to erosion of soil structures from water neither emerged from the depths of the ocean as a product of an underwater explosion. Instead, the natives constructed it. From this critical ascertainment a third type of island appears to emerge, which is entirely manufactured and has characteristics of the continent (as had the Atlantis of Plato), but its creation is not due to geological forces or natural phenomena. Utopia owes its existence to human intentionality, strength, and inventiveness and of course, to the stock of human and natural capital. The man here takes responsibility of the reconstruction of the world. In this context, the very structure of the island aimed at understanding that the contemplation of the ideal society cannot be achieved in the real world but will always serve as a reference and benchmark for our society.

4 FROM SPACES OF SINGULARITY TO THE NEED FOR PUBLIC SPACE

What characterizes strongly the postmodern era is nostalgia that drives us towards an illusory flight to the imaginary, to the dream of community and the lost common soul. The related loss of any collectivity, the sense of the uncanny, the rise of individualism, consumerism, and mass culture of virtual communication systems are some of the characteristics inherited from the 20th century to the 21st. In the introduction of "The poetics of space", Bachelard examines a mental capacity almost identical to the one that runs through the works of Heidegger: the poetic imagination. According to the French philosopher, the space conceived by imagination offers a lived experience of architecture.

In the same way, Tom Conley argues that Deleuze's desert island is a place where man comes to realize primary myths and the forces that generate them and this phenomenon makes the island a space of singularity[6] [6]. When it comes to Utopia, the fictional society of More developed its self-consciousness after the "construction" of the island whereas the island of Atlantis served from the beginning as an alternative socio-political model against the perfect society envisioned by the Athenians for their city. The singularity of the path that leads to the myth of the isolated island society hides in fact a desire to redefine the lost collectivity to a new stable place. Here is invoked the aspect of the imaginary that relates to the ideal of a collective, "common soul", where man will join the élan of the island, establish an identity and become member of a community. The mythical desert island and its society, separated from the continent

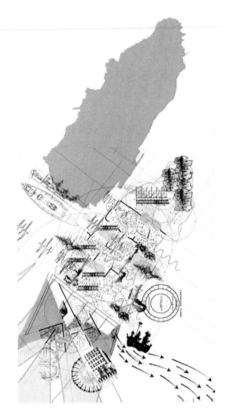

Figure 5. Conceptual map of the Alcatraz Island by author. The small island was developed with facilities for a lighthouse, a military fortification, a military prison (1868), and a federal prison from 1933 until 1963. In this illustration, a new program is proposed, where architecture focuses on public space, whereas the former military buildings and prisons are transformed into places of memory and of potential cultural use.

and its influences could revitalize the driving force to discover the lost collective soul of the place and build a model based on common good.

In such a model, architecture has a dual role: (i) to convey the meanings and functions of the community, and (ii) to derive needs, values and expectations from this community. This role evolves through the process of architectural composition and produces space in harmony with both the environment and the place. The value of the collective spirit expresses the urgent need for public space as a common good of the whole community. The poetic geography of the mythical island serves for a while as a fictional field of shared principles, goals, ideas, worldviews, and moral values and then could provide a proportion of programs and functions for the city space. The concept of the myth will no longer be a subjective picture, but to the extent that it is based on shared thoughts, concepts and

planning conditions of the place, it could enhance the design of the collective public space of the city with meanings, ideas, and values.

NOTES

[1] "Continental islands are accidental, derived islands. They are separated from a continent, born of disarticulation, erosion, fracture; they survive the absorption of what once contained them." (Deleuze, 2004, p. 9).
[2] "Oceanic islands are originary, essential islands. Some are formed from coral reefs and display a genuine organism. Others emerge from underwater eruptions, bringing to the light of day a movement from the lowest depths. Some rise slowly; some disappear and then return, leaving us no time to annex them." (Deleuze, 2004, p. 9).
[3] "Dreaming of islands—whether with joy or in fear, it doesn't matter—is dreaming of pulling away, of being already separate, far from any continent, of being lost and alone—or it is dreaming of starting from scratch, recreating, beginning anew." (Deleuze, 2004, p. 10).
[4] "An island doesn't stop being deserted simply because it is inhabited." (Deleuze, 2004, p. 10).
[5] "Then geography and the imagination would be one." (Deleuze, 2004, p. 11.
[6] "Deleuze's desert island becomes the emblematic place where man perceives originary myths and the concurrent forces that generate them. It may be, in the fashion of a projective conclusion, that the island could be imagined as a space of singularity." (Conley, 2005, p. 217).

BIBLIOGRAPHY

Bachelard, Gaston. *The poetics of space.* (Translated in English by Jolas, Maria). 2nd ed. Massachusetts: Beacon Press, 1969. Translation of *La poétique de l' Espace.* ISBN 0-8070-6473-4.

Bergson, Henri. *Creative Evolution.* 1st ed. New York: Cosimo Classics, 2005. Translation of L'évolution créatrice. ISBN 0-7607-6548-0.

Conley, Tom. The Desert Island. In Buchanan, Ian & Lambert, Gregg. *Deleuze and Space.* 1st ed. Edinburgh: Edinburgh University Press, 2005, p. 207–219. ISBN 0-7486-1874-0.

Deleuze, Gilles. *L'Île déserte et autres textes. Textes et entretiens, 1953–1974.* 1st ed. Paris: Minuit, 2002. ISBN 978-2707-3304-13.

Deleuze, Gilles. Desert Islands. In Deleuze, Gilles. *Desert Islands and Other Texts, 1953–1974.* 1st ed. New York: Semiotext(e), 2004, pp. 9–14.

More, Thomas. *Utopia.* (Translated in Modern Greek by Karagiannis, Giorgos). 1st ed. Athens: Kalvos.

Mumford, Lewis. *The Story of Utopias.* (Translated in Modern Greek by Tomanas, Vassilis). 1st ed. Skopelos: Nisides, 1998. ISBN 960-8480-26-4.

Plato. *Timaeus.* (Translated in Modern Greek by Kalfas, Vassilis). 2nd ed. Athens: Polis, 1998. ISBN: 960-7478-11-8.

Vidler, Antony. *The Architectural Uncanny, Essays in the Modern Unhomely.* 1st ed. Cambridge: The MIT Press, 1992. ISBN 0-262-22044-6.

Eros and the struggle for *reason*: Our mind as the vehicle of the Universal Mind's enfolding meaning

Maria Saridaki

Architecture, Edinburgh College of Art, The University of Edinburgh, Edinburgh, UK

ABSTRACT: In this paper, I will attempt to unfold the journey taken by the human mind in its evolving maturity and progressive complexity. This journey has taken humankind and the cosmos from a primordial undifferentiated consciousness through to a dualistic alienation and back towards a redemptive synthesis and reunification of the individuated self within the universe. Philosophy and science, these valuable pursuits of the human mind, emerged with the primary objective to attempt to understand the universe as a whole. Through centuries of evolution, the compiled mass of complementary as well as contradictory theories moved beyond the object of the physical world and its scientific explanation encompassing the our moral, aesthetic and spiritual consciousness. Humankind, through its development of philosophical and scientific thought, initiated the struggle to apply reason to existence. Our philosophical and scientific thinking developed hand in hand, with each area provoking adaptation within the other. Our advanced consciousness of self is thus bound in a continuous cycle of reference, examination, and attempts of independence with the otherness. Finally, this creative force is revealed as the archetypal *Eros*. The universal mind and humankind are thus seen as constituting a living totality, a reality unfolded in our consciousness. We are unconsciously aware of it as a creative boundary that invites us to engage with it in a perpetual dialectic.

Keywords: Eros, progressive complexity, perpetual dialectic

1 SCIENCE, PHILOSOPHY AND THE STRUGGLE FOR REASON

Starting from the existential beginnings of science in ancient astronomy and cosmology, I was drawn to the pre-Socratic philosophers/scientists who first attempted to analyse the world in a scientific way by attaching individual reason to it. Probably because of the dynamic essence in their religious and mythical traditions, which placed them in stark contrast with the fully developed and constant religious worldview of Mesopotamia and Egypt, the Greeks felt compelled to approach the heavens with an enquiring mind. They allowed themselves to question their beliefs since their religion did not offer to them ready answers to the workings of the cosmos and so they directed the evolution of their thinking to merge the scientific and the philosophical approach.

The Pythagorean worldview unified the spiritual and physical world, conceiving nature as a structured system ordered by mathematical forms and humankind's understanding of this cosmic order showed its essence as intrinsically linked to that of the universe. Heraclitus saw the cosmos as an intelligent system regulated by Logos, a single intelligent order. He presented a unity in opposites and Flux, paradoxically, as a necessary condition of constancy. Anaxagoras proposed a universe constituted of an infinite number of seeds and postulated that matter was brought to motion by a primordial Mind (Nous); a notion that influenced greatly the philosophies of Socrates, Plato and Aristotle.

Science and philosophy are seen from now on, throughout the ages, to go hand in hand, provoking adaptations within each other. The advanced consciousness of the self and its growing humanism became bound in a continuous cycle of reference, examination, and attempts of independence with the otherness and our role within it or separated from it.

2 PLATO AND ARISTOTLE'S DUAL LEGACY AS SYNTHESIS

In this evolution of the mind, humankind was seemingly aiming towards a higher level of consciousness.

Socrates was a man consumed by a passion for truth and placed the soul central in achieving excellence. Through his unique and influential dialectic form of argument, he attempted to dismiss conventions and challenge opinions in his search for the absolute essence of things. Socrates' synthesis of *eros* and *logos* in his desire for truth was complemented by a powerful belief in the immortality of the soul and its longing for a reunification with an immortal cosmos. The divine and universal archetypes of the pre-Socratics found in Socrates and Plato a renewed essence; an independent reality of their own.

Through the supreme discipline of philosophy, these universals could be again contacted by the human mind in its attempt to regain the divine wisdom that was once its possession and to unite with the universal mind. Only in this way could he glimpse ultimate reality where true being resides. Accessible only by the mind this reality was put in opposition with the world of the senses, initiating dualism, the object/subject debate and the questioning of the validity of truth and experience.

The realm of Ideas for Plato revealed the true essence of things. However, the journey towards the revelation of this real world was seen as evoking an intense emotional and mystical response. The sublime form of Eros overcomes the philosopher (the lover of wisdom) and drives him to restore the unity with the divine. In the Republic, Plato refers to this as freeing man from the chains of his sensory body towards an archetypal light, a divine existence where true reality can be experienced (the well-known cave story). Through philosophy, this divine wisdom can be regained and the soul's eternity revealed.

The Platonic worldview of universal forms, according to Tarnas, revealed this transcendent intelligence, this divine wisdom of the universal mind as coming from within us as an innate structure of the mind. The mind's ability to penetrate and access this reality implied that they were both governed by the same intelligence.

In the person of Aristotle, the scientific and philosophical scene of the Platonic world is endowed with a powerful opposition, a creative dialectic that will continue to the present day. The interaction of science and philosophy gathers speed as it brings to light an instinctive representation of the duality, the dialectic, and its creative energy. Aristotle was the initiator of empirical knowledge, uniting *logos* with *aitia* (cause) and basing his scientific method on observation and deduction. He employed reason in order to discover an inherent order within the empirical world. He transmuted the principles of formal and material causality into those of actuality and potentiality within a unified universe in which each substance is involved in constant and unending motion. His scientific methodology is to this day, extremely influential.

The dual legacy of Plato and Aristotle is everpresent in human reason. Their opposition has provided a unique synthesis, a guiding light, generating the independence and actualisation of our mind within the world. On the one hand, we have the notion of a transcendental world order accessible to the mind through intellectual development and on the other hand, the belief of a tangible knowable universe accessible through empirical observation. Their search for meaning implied a dialectical reference to the cosmos.

One thing, however, has remained the same throughout history. "All men desire to know", this is how Aristotle starts his *Metaphysica*. As advocated by Socrates, philosophy questions views that have been established without the consent of our reason, our independent thought, the ones that are mere products of the conventions and beliefs of the society and time we happened to have been born in. Philosophy liberates through the process of raising doubts about what has been taken for granted, it restores and revives the mystery of the world, by challenging what the world is, in its pursuit to unveil what it may be. In this endeavour, science and philosophy are united.

Philosophy is unique in the sense that there is no stage of its evolution that can or should be regarded as falsified. Indeed, one could say that Plato's theory of Ideas is utterly absurd and unnecessary but his insights have been instrumental in the way we

Figure 1. The boundary in human reason portrayed by the two central figures of 'The School of Athens': Socrates, pointing to the heavens and Aristotle motioning us to remain on earth. (by Raphael, ca 1510, Vatican) Source: www.italian-renaissance-art.com.

think about the world. It is for this reason that I disagree with this kind of compartmentation of thought. A student of chemistry may not need to study the evolution of his science to move forward, a student of philosophy, however, should. The journey of the mind is so great because its course is mysterious. Every stage is valuable not only in itself but because it will propel us to the unexpected.

The true legacy of the philosophy of Aristotle and Plato is that it has been instrumental and influential throughout the entire course of philosophy. Much of what has been written or thought has been as a reaction to their unique minds and prolific writing that ranged from physics and metaphysics to ethics and politics.

3   THE CHALLENGE OF THE CHRISTIAN WORLDVIEW: AUGUSTINE AND THE SELF AS A PRIVATE INNER SPACE

A defining factor in the evolution of thought in the West was the growing authority and pervasiveness of the Christian faith. At a time of great upheaval and faced with the imminent distraction of the Roman Empire, Augustine of Hippo (354–430 AD), with his Confessions, directs his view towards the self and the soul. In his Confessions—his autobiography-, Augustine's personal experiences and sinful past, recognised in the self, becomes the medium towards finding God. According to Cary, Augustine first affirms the existence of the human ego in the soul. Calling upon the Socratic and Platonic doctrines, Augustine, a Neo-Platonist, asserts that self-awareness and introspection were necessary for the salvation of the soul but places God and his divine grace as the instrument of this

Figure 2.   Boundary of life and death. The painting divided into two sections, the earthly, and the heavenly, with the soul of the deceased ascending as though born into heaven. The Burial of the Count of Orgaz, El Greco (1541–1614 AD) Source: wikipedia.

salvation. He attempted to locate God within the soul, without affirming the divinity of the soul; God was other and external to the self.

According to Cary, Augustine first conceived of the self as a private inner space, an inner world of human memory and thought, starting the Western tradition of inwardness. This ability to look within, which we now conceive of as self-evident, influenced the way we interpret ourselves in relation to our now external world: "[f]or once you have conceived of an inner self, you think differently about the external and the other. If there is an inner world, then the world we all live in can be 'merely external', and you can worry whether it is lifeless and meaningless like a Newtonian mechanism. You can even worry whether the external world exists." (Cary 2000, 141)

Throughout the centuries, the mind has gone through a series of staggering stages in evolution in both the scientific as well as the philosophical realms. The areas presented, though by no means unique, are still powerful examples of the underlying search for a dialectical unity driving humankind in its scientific and philosophical quest. The realms of theology and science eventually reached a critical threshold, a point of no return, a point of emergence, which brought about probably the greatest paradigm shift that humanity has encountered.

4   THE POST-COPERNICAN PARADIGM SHIFT: THE UNIVERSE AS A MEASURABLE MODEL

An important stage in both the scientific and philosophical realms—with enormous religious implications-, according to McClellan & Dorn, was the paradigm shift that occurred in science as a direct result of Copernicus' planetary model. Living in the height of the Renaissance, Nicolaus Copernicus (1473–1543) attempted to resolve the problem of planetary movements realising that classical astronomy's imperfections were based on a fundamental error. Having reviewed all accessible scientific literature from the past, he found that the Pythagoreans and Platonists had proposed a moving Earth; Aristotle's geocentric universe was, therefore, not the only option to consider. Copernicus thus set forth to form a heliocentric model for the universe with the Earth rotating around the Sun.

Johannes Kepler (1571–1630) further vindicated the Pythagorean belief of a mathematically ordered universe, by conceiving elliptical orbits for the planets and Galileo Galilei's (1564–1642) telescope provided tangible and qualitative new evidence, his observations pointing towards a much larger than previously conceived universe. Unfortunately, the Catholic Church chose to retain the cosmology suggested in the scriptures, favouring a stationary Earth, thus alienating the great minds of the day. It lost its influence and claim to represent human aspiration towards full knowledge of the universe, thus widening the divide between the spirit and the mind.

René Descartes' (1596–1650) mechanistic philosophy, drawing parallels with the basic principles of ancient atomism, visualised an image of nature as an intricate impersonal machine strictly ordered by mathematical laws. The notion of the existence of an attracting force that governed both falling bodies and planetary motion completed the synthesis of a mechanistic universal model. Finally, with his three laws of motion (inertia, force and equal reaction), Isaac Newton (1642–1727) formulated what appeared to be the governing laws of the entire cosmos. Newton thus fulfilled Descartes' mechanistic vision of nature establishing the foundation of a new worldview.

5   THE ADVENT OF DUALISM: MANKIND NO LONGER AT THE CENTRE OF CREATION

The Scientific Revolution had a multitude of consequences. On the one hand, creation was finally viewed with empirical proof justifying measurable knowledge. On the other hand, however, the mystery of the world was removed. Humankind's intelligence prevailed. We had penetrated the universe's essential order. God, the perfect being, existed as the creator of this perfectly explainable world, but was separated from it; the universe was self-sufficient. God's existence was affirmed through reason and the universe was robbed of its Aristotelian teleological purpose. In this environment, Descartes' dualism took hold and Kant's critical philosophy attacked humankind's remaining security; its ability to achieve knowledge. Could we really know anything as it really is? This formed a question that was to haunt and challenge the human mind to this day. The dichotomy between mind and matter, the thinking substance and the physical body was established. Upon this dual foundation philosophy proceeded and science conquered.

The scientific implications of this era essentially removed the Earth and thus humankind from the centre of creation. Up until that point humankind was always assured of its role and importance as the centre of reference. According to Tarnas, a peculiar contradiction ensued. On the one hand, our intelligence was vindicated in revealing the laws of the universe; on the other, our cosmic position was no longer fixed or absolute. The new freedom enjoyed in this new worldview had also stripped us of our cosmic meaning. One could say that Darwin's theories delivered the last blow.

Humanity was no longer at the centre of creation and was little more than an animal whose consciousness rose accidentally through evolution.

6   THE REVOLT AGAINST DUALISM: TOWARDS A UNIFICATION OF SCIENCE AND PHILOSOPHY

The mechanistic worldview escalated to a dichotomy between human culture and nature, robbing our individuality of its context; the cosmos. This alienation is often described as the curse of modernism. By endorsing this mechanistic worldview, modernism ultimately failed because it did not reveal a greater, ordered coherence. It was, however, a necessary part in the evolution of our mind in its attempt to understand the cosmos and our purpose within it. Becoming aware of our own hubris in our self-created seclusion, our mind was again yearning to reunite with the cosmos. Opposing Kant, Hegel saw the world as a single, unified coherent structure of thought. Seeking understanding of anything leads us to relate it to something else. This reference, this relation between things not only reveals but also partly constitutes their nature. For Hegel, the knowing mind and the object known cannot be considered as two separate things; they are related. His ontological dialectics support the coexistence of thesis—antithesis, with synthesis as the interaction of two dialectically opposite definitions whose common denominator is their unbreakable unity. The world was thus revealed essentially as a construct and human knowledge as interpretive. Human endeavours could no longer establish their ground on an independent reality. "Meaning is rendered by the mind and cannot be assumed to inhere in the object, in the world beyond the mind, for that world can never be contacted without having already been saturated by the mind's own nature." (Tarnas 1996, 419)

In this way, the phenomenology of human experience started to enter formal philosophy. Instead of analysing an objectified world the philosophers of this age focused their attention on 'being' itself; they focused on the lived world of human experience and its complexity.

In our age something indeed is ending; something new, however, has begun; a new paradigm shift in the scientific developments of the 20th century. Crushing the Newtonian model, Modern Physics and especially Quantum Theory showed that there is no objective reality that the physicist can know. The rise of quantum physics brought about the realisation that man is an intrinsic ingredient of observations, an entity we are unable to separate from the phenomena. In fact, human consciousness, observation, and interpretation seemed to have an inherent influence in the phenomena.

This, in a sense, brought humankind back to the centre of reality. The theory of everything in physics as an attempt towards a unified theory of the universe, the anthropic principle and David Bohm's wholeness and implicate order in an unfolding/enfolding universe portrayed as an unbroken whole, point towards the same direction. Echoing Aristotle's formative and final causes, Bohm (1980) argues that life itself is enfolded in the totality and that even when it is not manifest it is somehow implicit to the totality. Human beings should therefore be viewed in the same way; not as an independent actuality that interacts with other human beings and with nature, but as a creative projection of a single totality.

Karl Popper saw humankind approaching the world as a stranger with a thirst for explanation and an ability to create myths, theories, conjectures that at times turn out to be successful. His unsatisfying explanation was that these successful theories created by man were lucky guesses. Tarnas, however, offers another possible rationalisation, one that detects the quest for knowledge at a far deeper source. He believes that these bold conjectures and myths come "... from the wellspring of nature itself, from the universal unconscious that is bringing forth through the human mind and human imagination its own gradually unfolding reality". (Tarnas 1996, 345)

Insights into the quantum self further uphold this boundary between matter and consciousness leading us to a coherent worldview of undivided wholeness. We call our conceptualisations of these realities, theories, but they might best be described as insights. According to Bohm, "...our theories are not 'descriptions of reality as it is' but, rather, ever-changing forms of insight, which can point to or indicate a reality that is implicit and not describable or specifiable in its totality." (Bohm 1980, 17)

7   THE UNIVERSAL MIND REVEALING ITSELF THROUGH OUR MIND

The relation of the human mind to the world can ultimately be seen not as dualistic but participatory. According to Tarnas, this conception acknowledges the validity of Kant's critical insight that all human knowledge of the world is in some sense determined by subjective principles. These principles, however, can be seen as sharing their essence with the world, in a way not unlike the one supported by the ancient Greek heritage. They can be seen as "... an expression of the world's own being, and that the human mind is ultimately the organ of the world's own process of self-revelation. Nature's unfolding

truth emerges only in the active participation of the human mind. Nature's reality is not merely phenomenal, nor is it independent and objective; rather, it is something that comes into being through the very act of human cognition. Nature becomes intelligible to itself through the human mind." (Tarnas 1996, 433–434) Our mind is thus invited to engage in a perpetual creative opposition with the cosmos, as part of a dynamic unity.

Tarnas believes that there is a reciprocal existence between mind and world. The human mind does not reveal an objective truth in the world but rather that the world's truth achieves its existence when it comes to birth in the human mind gradually in the form of new stages of human knowledge much as a plant that grows and blossoms. "[A]s Hegel emphasised, the evolution of human knowledge is the evolution of the world's self-revelation." (Tarnas 1996, 435)

This thought process sees the mind as sharing a radical kinship with the cosmos, which would also reflect the pivotal role of our mind as the vehicle of the universe's unfolding meaning. The universe and humanity are thus seen as constituting a living totality. This reality is unfolded in our consciousness, it exists within us.

8   EROS AS THE ARCHETYPAL DIALECTIC

It becomes clear that through this paradigm shift we are witnessing a powerful archetypal dialectic in the modern mind, our rebirth, a profound boundary between self and other, experience and reality and ultimately us and the cosmos. There is a sense of a renewed consciousness of self. Humankind is searching for its legitimate place in this unity, this cosmic scheme of things. We are now reaching towards a new synthesis with our world through a dialectical opposition that has taken us from the primordial undifferentiated consciousness through to the dualistic alienation and back towards a redemptive synthesis and reunification of the individuated self within the universal matrix.

According to Anshen, we are beginning to accept our convergence with the Cosmos by allowing the mystery of our surrounding world to be gradually revealed to us. "Mind and matter, mind and brain, have converged; space, time, and motion are reconciled; man, consciousness, and the universe are reunited since the atom in a star is the same as the atom in man. ... We have reconciled observer and participant. For at last we know that time and space are modes by which we think, but not conditions in which we live and have our being. Religion and science meld; reason and feeling merge in mutual respect for each other, nourishing each other,

Figure 3. Unifying Creation Mythology and Science: Shiva's Cosmic Dance a symbol of creation and destruction, birth and rebirth, quite appropriately placed at the European Centre for Research in Particle Physics (CERN), in Geneva. Source: www.fritjofcapra.net/shiva.html.

deepening, quickening, and enriching our experiences of the life process. We have heeded the haunting voice in the Whirlwind." (Anshen 1983, xxii)

The scientific and philosophical paradigm shifts can thus be seen as a necessary part of a larger evolutionary process. No stage should be dismissed as a mistake; they are all integral parts of our unfolding being in its dynamic interaction with the cosmos. According to Tarnas, the Cartesian hero that has attempted the separation is a masculine, suffering hero; seeking redemption and unity, he embraces the 'female' cosmos. Could we not see this as the creative force that it is? Could we call this the archetypal Eros? Is longing for reunification not an act of love?

In Plato's *Symposium*, Socrates and his guests attempt to decipher the nature and meaning of Eros. Socrates calls on Diotima, a priestess, who claims that the highest fulfilment of Eros is the philosopher's conjugal union with the Idea of Beauty, which brings forth the birth of wisdom. Setting Eros against Beauty, Diotima identifies Eros as the in-between, something that is neither beautiful nor ugly, good or evil, ignorance nor wisdom; instead, she insists that Eros is capable of being either one. Answering her own question, Diotima identifies the function, the ergon of Eros as 'giving birth in beauty, both in body and soul', binding together, according to Hyland, the divine, and the mortal into a whole. Ultimately, Eros is the philosopher whose quest, whose *telos*, though unattainable, is wisdom. Eros is thus manifested as the creative in-between.

This emerging synthesis, this archetypal boundary is indeed the purpose, the direction, the Aristotelian 'aitia' of our being. Moreover, it is finally leading us to reunite with our alienated but not lost other. In the words of Wolfgang Pauli, from his 'Writings on Physics and Philosophy': "Contrary to the strict division of the activity of the human spirit into separate departments—a division prevailing since the nineteenth century—I consider the ambition of overcoming opposites, including also a synthesis embracing both rational understanding and the mystical experience of unity, to be the mythos, spoken and unspoken, of our present day and age." (Pauli 1999, 3) Have we not always needed a myth to live by?

BIBLIOGRAPHY

Adler, M. (1991) *Aristotle for Everybody, Difficult Thought made easy*. First Collier Books Edition.

Anshen, R.N. ed. *Introduction to Convergence*. In: Salk, J. (1983) *Anatomy of Reality: Merging of Intuition and Reason*. Columbia University Press. pp. xi–xxiv. ISBN: 9780231053280.

Bohm, D. (1980) *Wholeness and the implicate order*. 9th ed. London and New York, Routledge. ISBN: 9780415289795.

Cary, P. (2000) *Augustine's Invention of the Inner Self: The Legacy of a Christian Platonist*. New York, Oxford University Press. ISBN: 9780195158618.

Hawking, Prof. St. (1998) *The Theory of Everything: No boundary Proposal*. My notes from Prof. Hawking's lecture (Professor Emeritus of Mathematics at Cambridge University) at the University of Crete, August.

Heisenberg, W. (1989) *Physics and Philosophy: The revolution in Modern Science*. Penguin Books. ISBN: 9780141 182155.

Honderich, T. ed. (1995) *The Oxford Companion to Philosophy*. Oxford, Oxford University Press. ISBN: 97801992 64797.

Hyalnd, D.A. (2008) *Plato and the question of beauty*. Bloomington, Indiana University Press. ISBN: 9780253219770.

Kuhn, Th. S. (1970) *The Structure of Scientific Revolutions*. 2nd ed. Chicago, University of Chicago Press. ISBN: 97802264 58083.

Mcclelland J.L. & Dorn, H. (2006) *Science and Technology in World History; An Introduction*. Baltimore, The Johns Hopkins University Press. ISBN: 9780801883606.

Pauli, W. from his *Writings on Physics and Philosophy*. Quoted in: Brockelman, P. (1999). *Cosmology and Creation: The Spiritual Significance of Contemporary Cosmology*. New York: Oxford University Press ISBN: 9780195119909.

Pojman, L.R. (2002) *Philosophy: The Quest for Truth*, New York. Oxford, Oxford University Press. ISBN: 9780195156249.

Saridaki, M. (2012) PhD thesis in the College of Humanities and Social Science: '*BOUNDARY*' an expression of the dynamic unity between man and environment. Building a paradigm to unravel the mind's fundamental kinship with the cosmos and its role as the vehicle of the universe's unfolding meaning. Edinburgh, The University of Edinburgh.

Tarnas, R. (1996) *The passion of the western mind*. London, Pimlico edition Random House. ISBN: 9780345368096.

Vlastos, G. (1996) *Studies in Greek Philosophy, Volume 1: The Presocratics*. Graham, D.W. ed. Princeton, Princeton University Press. ISBN-13: 978–0691019376.

Utopia seen through the symbolic and through anthropotechnics

José Carlos Vasconcelos E Sá

Departamento de Ciências da Comunicação, ECATI, Universidade Lusófona de Humanidades e Tecnologias (ULHT), Lisboa, Portugal

ABSTRACT: This text aims at examining the place of utopia in present times. Initially, we base ourselves on the analysis of two moments of the utopian discourse, namely the time of the sources with which one can identify the following texts: Plato's *Republic*, More's *Utopia* and Campanella's *The City of the Sun*. Following that, we analyse what may correspond to Bloch's theories, which constitute a kind of "re-foundation of utopia" by proposing its connection to Marxism.

From an analytic point of view, utopian discourse is conceived as a communication device that, contrary to what is generally accepted, contains as one of its main effects the mobilization in History. The theoretical apparatus shall convene the virtual, real, current, and possible concepts understood as the production components of the real.

Ultimately, we shall question the place of utopia and the symbolic, at present, before a world increasingly dominated by anthropotechnics, in which the pressing question is no longer representation but creation.

Keywords: Utopia, communication, anthropotechnics, virtual, language

1 UTOPIA: THE NON-PLACE THAT WANTS TO BECOME THE PLACE OF GOODNESS...

The structure of the utopian thought relies mainly on the devaluation of the present, ephemeral, transitory time, valuing future time as being substantial, true, complete, and eternal. This structure has been announced since Plato in the *Republic*, whose influence was important in Western thought, and later in Christianity[1]. This transformed the present into a time for the expiation of guilt given the disobedience that led to the Fall. For humanity, there was only remorse and penance so that salvation and eternal happiness could be attained. With Marxism, one of the strongest utopias of our Western world, the utopian structure, as we shall see, shall be strengthened. Conceived by Ernst Bloch as "practiced utopia", "should be" turns into "can be".

Utopias are traditionally intellectual creations motivated by dissatisfaction, nonconformity or even the feeling of helplessness in the face of rampant injustices in our societies. Thus, utopias are first a kind of individual, intellectual, and symbolic catharsis, assumed as a revolt. From a narrative point of view, utopias are normative since they radically refuse the present, conceived as violence (motivated by privileges, corruption, scarcity, by the arbitrariness of the power over the weak) projecting it into an idealized future.

The utopian seminal texts refer to from Plato's *Republic*, to More's *Utopia* and to Campanella's *The City of the Sun*, proposing a world under the guise of "should be", radically refuse the world "as is". In this sense, the utopian discourse is a critical discourse on the reality that considers Campanella's *The Sun*, proposes a world under the guise of "should be" radically refusing the world "such as it is". Therefore, the utopian discourse is a critical discourse on reality since it assumes that the present has betrayed the fundamental human values, striving to show its restoration in a perfect society.

The figures of the city and of the island are the spatial representations inside this utopian world where human societies live in a perfect communitarianism, based on an irreproachable order illustrated by a rigid social hierarchy. The traditional utopian narrative, from Plato's *Republic* to More's *Utopia* or Campanella's *The City of the Sun*, strives to describe minutely and obsessively such an ideal society: the *modus operandi* of social and political institutions, justice, interdictions and permissions, the adoption of ethical, civic and aesthetic values, the form and function of religion, family structure, the form of interpersonal relations, up to the regulation of sexuality itself. Utopian discourse in its narrative transforms the figure of a leader into an action maker, a leading force, usually an exceptional legislator responsible for the production of a perfect legislation. Law acquires, beyond its prescriptive function, a pedagogic character in order to educate the citizens through the correctness of its rules, which should be strictly enforced. The goal is to achieve a perfect identification among the individual, the law, and the state. On the other hand, the "ideal city" exerts on its citizens a permanent surveillance using a panoptic system, through which one may see without being seen. Peace is a function of order, and happiness can only reach a space where war and physical violence have been permanently banned.

Utopian narrative relies on the creation of a complete and perfect world, closed in upon itself, a world without otherness, where conflicts between mankind and nature, the individual and society, society and the state have been resolved; a society where the transparency of the social becomes total, and where all social mechanisms and political decisions are not concealable.

This first utopian discourse is thus in perfect harmony with the Greek root of the word utopia[2], as a non-place and, simultaneously, as the "place of goodness", as the utopian idealization is built upon the idea of a perfect harmony between freedom and order, the synthesis that will allow the experience of true happiness. Because that harmony cannot take place in existing societies, utopian thinking conceives it as occurring at an undetermined time and in a deterritorialised space, i.e., a non-place.

...but...

On this issue, Zigmund Bauman stated that human societies seek—without having achieved it so far—a compatible balance between security and freedom, yet, whenever they achieve satisfactorily one of these, the other is in deficit[3]. This unresolved equation generates, according to Freud, the malaise in civilization. Utopia realizes the scale of the issue and tries to show that it can be resolved. This constitutes a characteristic feature of many utopias in particularly during the Renaissance. However, the place occupied by Order, that is, by law, is, in these utopias, completely dominant, which means that freedom tends to be "smothered" as the narrative of the "ideal city" becomes obsessively prey of the description of its structure.

The dominance that law and order manifest in the utopian writings also results—that is no doubt our conviction—in the cognitive effects operated through writing, reading, and the book, upon the Renaissance individual. Following Pierre Lévy analysis, in the text *The Technologies of Intelligence* (1993), writing is defined as a highly regulated, logical, linear, reflective, and operationalized structure

through a variety of codes, with respect to an "extended" and durable time, unlike oral speech, which is inscribed in a "real-time" caption. Writing is more removed from people because it has to be enough by itself, in order to be "well read", that is, unambiguous. This new writing paradigm combines itself with the new ways of reading that a book requires. On the one hand, a silent and isolated reading; on the other, the new internal organization of the text that the book presents, with its much defined spatial structure (titles, captions, an index, a bibliography, footnotes, citations). All these aspects train the writer/reader towards cognitive routines in which order acquires a key role. On the same note, McLuhan also underlines the effects that writing, associated with the press, had upon the modern individual, emphasizing the split they cause in human perception, developing its rational mechanical and orderly aspects at the expense of sensitivity and the imagination. In sum, what we intend to show is that these aspects contribute so that the value of order dominates freedom, which, in practice, is being subsumed by order. In this moment, what we want to underline is the fact that the symbolic order contributes to the dominance of the order and law rather than the value of freedom. This argument will be developed later when we treat the way that the culture is sign up in symbolic order. A utopian narrative assumes, from our point of view, a major illustration of how the symbolic strength acts potentiating the logic "should be" and weakening the logic of "could be".

Nonetheless, as we shall later analyse, it is the foundation of the symbolic itself that will form the basis of the rule of law and order, relegating freedom to a secondary place. This is a problem affecting not only the first but also all utopias. Freedom, the value that largely motivates them is systematically cancelled, transforming utopias into their opposites, i.e., into dystopias. In this sense, utopias, since they degenerate into dystopias, can no longer stand in place of the goodness to which the root of word, as we have stated, also refers.

2 BLOCH AND THE FOUNDATIONS OF UTOPIA

This is what also occurs in Marxism, a different kind of utopia than the ones we have been addressing, since it presents itself as finished political programme. The theorizing establishing the connection between the utopian universe and Marxism is conceived by Ernst Bloch, a German philosopher of Marxist affiliation (Ludwigshafen, 1885, Tuebingen 1977) in his monumental *The Principle of Hope* (2005–2006), a three-volume work written in exile in the US, between 1938 and 1947, considered one of the finest analyses of utopia. One of the usual criticism on the utopian discourse is that its refuses to face reality by positioning itself outside time and space. "The realm of utopia is not of this world" is the phrase that best illustrates this argument. We have seen that even in utopian discourses, in which there is no action programmes but rather an imaginary exercise, such as in the case of Renaissance utopias, this criticism is unfounded for the reasons already stated. Nevertheless, in the case of Marxism, a programme of action is proposed, in order to attain communism and, therefore, the relationship between Marxism and History becomes clear.

However, let us see, for the sake of argument, how Bloch establishes the relationship between utopia and Marxism. The philosopher argues that the central element of utopia is hope and that this one is what constitutes his true reason of being, its élan. The place of utopia lies between realism and dream, between knowledge and imagination. Bloch proposes a kind of a "dialectic of goodness" governed by the "principle of hope". Far from being "outside of History" utopia constitutes one of the most mobilizing dimensions of History. In this sense, it "produces a world", i.e., it organizes the sense of History, framing the mobilization of the masses within the series dissatisfaction-critical-hope-revolt-realization.

From Bloch's point of view, utopian discourse and its imaginative component is conceived as one of the fundamental components of Marxist thought, without which Marxism runs the risk of being understood—as indeed it does according to an orthodox interpretation—exclusively, as a scientific programme reduced to the observation of the "laws of history", having a merely economist character. *The Principle of Hope*, while rehabilitating utopia as a fundamental category of politics and history, seeks to recover Marx's Marxism in its entirety, giving imagination the privileged place that it has always occupied in the philosopher's thought.

However, Bloch's utopia is radically different from the Renaissance ones. Bloch's has a positive character since it seeks an achievement of the future, of what has not-yet-happened. Hope, inspired above all by the imagination, motivates the struggle in the present and the building of the future. The revolution, the action required for change, is incremented primarily by hope, by a subjective privilege, which, together with reason, whose function is to guide hope towards the communist project, builds a positive, rationalized hope. The refound character of Bloch's thought, on the one hand, reaches utopia because it does not forget the present, anticipating the future in it. Utopia becomes what is possible, that which may

have a place, and such a place remains the place of goodness. On the other hand, it also impinges on Marxism itself inscribing it in the universe of humanism through the importance that the categories of imagination and hope attain for its project of change. By inscribing utopia in History as the possibility of an achieved goodness, Marxists were challenged by unexpected results. In fact, the existing socialist societies practically do not correspond at all to the hopes deposited in them. Again, utopias become dystopias, as Orwell as masterfully illustrated in his *1984*. Once again, the place of goodness and has not lived there.

3   UTOPIA AND VIRTUALIZATION

Nonetheless, how does utopian imagination work? How can it be thought, as we have stated, as a producer of worlds? What is the mechanism by which it becomes incorporated in reality, creating History? The clarification of these issues leads us to the contributions of P. Lévy, and to Deleuze and Guattari presented respectively in their books, *What is Virtual?* and *What is Philosophy?* (Deleuze and Guattari 1993; Lévy 1996), which we shall analyse. The conception of the constitution of the real, expressed by Deleuze and Guattari, is greatly influenced by Bergson's philosophy of perception. The authors conceive it as the result of an actualization of virtual images that surround it and enhance it. Such figurations "emerge" under the pressure of specific social circumstances and policies incorporating themselves in the universe of experience. According to the authors of *Mille Plateaux*, the virtual is a dimension of the real, one of its constituents, a power that actualizes itself in our experience according to the circumstances in which it is questioned. Ideas, objects, finally, all existing things, are the result of a contamination of images that are virtually associated with them and whose achievements "appear" motivated by given social, cultural, and historical circumstances. In this conception, the real is involved in a universe of virtualities, "ready" to be actualized. The virtual should also be distinguished from the possible since this one would be already constituted in a latent, static, complete state, ready to be incorporated into the real. Contrarily, a kind of tensional environment, comprising a complexity of connections, forces, tendencies, which involve an idea, an event, and an object, constitutes the virtual. Its actualization is thus more problematic and complex than the universe of the possible.

In the same manner as utopia, the virtual has as its matrix the idea of a "non-present" that opens itself to an actualization, that is, to the real. Utopian narratives of the Renaissance are incorporated into the real through fundamental figurations that will be later incorporated into Modernity. One could point out as examples the figure of mankind and his central role in the world, the place of religion in society, the centrality of the figure of Reason, the subject as a principal category of History, and the preponderance values of order, freedom, solidarity and equality. These utopias are stories whose main function it to highlight the importance of these figures in reality through a narrative structure. The value of these utopias reveals itself in the importance of these figures for human development. They do not inscribe themselves into a logic of the possible, that is, of the realizable, of the "could be", as is the case of the Marxist utopias, as we have seen, but in the purely virtual, in a "not present", narrated as an imagined environment which is not without tensions with the experienced. Thus, they are discourses that do not fail to question History, showing a latent dissatisfaction upon the present, through the representation of the main figures of novelty of change. They are not destined to make a revolution; in fact, there is no conceived project of change. Here, in the context of Renaissance humanism, we are rather within the imaginary experimentation of virtual worlds. Nonetheless, they were privileged vehicles of controversy, as stated by Antonio Monclús[4]. These utopias are imagined and written for a literary society, for elite, a club, for a club of the humanist movement, as Sloterdjik designates it[5]. They are texts that work History believing in time, believing, eventually, that time shall prove them right.

4   CHANGE, SPEED AND TECHNOLOGY OR *ALL THAT IS SOLID MELTS INTO AIR*

Technology, a new force that Modernity saw arising within the industrial world, has raised vast hopes by promising formidable achievements for which all individuals yearn. The association between science and technology (Techno-science) has increased humankind's transformative capacity in such a manner, that it was impossible, given the emerging possibilities, that human societies were not able to bring forth transformations able to benefit all. Moreover, technological expansion made possible one of the most productive periods of utopias and dystopias in the Western world. Lewis Mumford (2007) confirms this when he points out that more than two thirds of utopias were written in the nineteenth-century. Marx also underlined, in 1848, the importance of technology for the industrial world, in the *Communist Manifesto*, when describing the alterations in the pace of change in liberal societies. The phrase that became emblematic—"all

that is solid melts into air"—synthesizes one of the effects of technology: speed[6]. With it, Marx was drawing our attention to one of the most important achievements of Modernity: the acceleration of History. In the nineteenth-century context, the phrase clearly referred to the speed of change in societies, stressing the importance of such reality in order for the bourgeoisie to maintain power. However, it has a great relevance nowadays, to the point of assuming astonishingly prescient characteristics. With the technologies of communication and information, developed in the post-industrial era, this acceleration has been greatly intensified. In this context, Marx's phrase can be interpreted not metaphorically but literally, in fact, ICT (information and communication technologies) transform "atoms into bits" (to use an expression by Negroponte (1999)[11]), and make the individual—should we continue citing Negroponte—into a "digital being". It is in this context that, more clearly, the social effects of digital technologies cut deeper than it may have appeared. They affect the axial axes of human experience—time and space. On the one hand, we are faced with the creation of a new space—the virtual space. This is the space of computer networks generating a huge attraction since it is also the space of desire, where everything can be carried out, contrary to "natural" space, governed by the principle of reality, where law, obligation, responsibility, order, and censorship become masters. On the other hand, they have compacted time, constricting it, i.e., accelerating it in such a manner that they almost nullified its duration. Our social time is not a human biological time, but an instantaneous time, a "real time", the time of "intelligent machines". With the spread of digital language and the expansion of the media universe, authentic prolongations or extensions of ourselves, as McLuhan anticipated, human societies have radically transformed themselves.

5   FROM ANTHROPO-LOGICAL TO ANTHROPOTECHNICS

The issue of acceleration is directly related to the theme of utopia. As we have analysed, the utopian structure always projects the good society into the future. Be it under the form of "ought to be" or under the mode "may be". Nonetheless, the present is always denied by dissatisfaction. It should be pointed out that acceleration produces the tension to get out of the present, bringing us closer to the future. If we combine this effect with another—simulation—a digital manipulation strategy that allows us to work on a model anticipating its effects, we realize that the utopian structure is now permeated by questions that challenge its operational capacity. Yet that is not all. The application of technology to human beings—referring to the universe of action of biotechnologies—has sparked a debate on complex decisions that contemporary societies are called to undertake and that are central to the future of humankind. We intend to show that this problem, involving key issues of an ethical, political, social, and cultural nature, also has a substantial relevance concerning the place that utopian thinking occupies in contemporary times. In this essay, we follow the thought of Gilbert Hottois (2005), which raises the question of the ban on the use of biotechnology in humans, in particular, in the trials related to their cloning. Establishing the difference in attitude between Europe, with a more closed position, and the US, taking a more open one with which Hottois agrees. The author frames the issue noting that Western culture has always considered technology as a reality inscribed in the universe of the *homo faber*. They should therefore only be applicable to the environment in which humans live, arguing that they have nothing to do with the individuals themselves, seen as rational and free beings. That is, technology has always been conceived as being definitely outside culture, when considered in its most noble sense, that is, when identified with the symbolic order. Against this postulate, one should open an exception concerning medical technologies, to which only certain interventions are in fact allowed, namely those whose purpose is to restore the physical condition of the human being, its balance, and its natural order. Therefore, any medical procedure aimed at improving or exploiting with a view to this improvement is, reprehensible in principle, although many areas of medicine take as problematic this axiom, as is the case of sports medicine, genetics, breeding, and neurosciences, among others. This way of thinking is itself what Hottois designates by the *anthropo-logical*, meaning with this term that it is only through the *logos*, the word, that the human being is and evolves. This means that human beings are allowed to be creative from a symbolic point of view, i.e., to create images, languages, representations of all kinds, even though techno-physical creation is completely banned, as well as calling into question the order of nature and, above all, modifying one's own nature. In terms of a cosmic order, God or nature would be legitimately creators, but individuals should remain creatures. Their transcendence would be restricted to the symbolic order.

It is against this conception of man that anthropotechnics reacts. It may be argued that allowing, without limitations, for an intervention of technology in our human becoming is to run very serious risks, and this is completely true. No one with a minimum of common sense would agree with such a solution. The problem is that limits should not be absolute, as the defenders of the "untouchable" understand, but contingent and modifiable since they depend on the circumstances. Hottois wants to leave this quite clear: we do not pretend to dispute the existence of limits, but only their absolutism, their essentialism. That is, their position regarding experimentation of biotechnology in humans should be possible, but properly regulated or controlled.

What we want to show with this apparent drift on biotechnology is that the refusal of the present and the appreciation of the future by utopia are related, as we shall see, with its subordination to the symbolic order (language) and to its foundations, while the anthropotechnical project is based on completely different assumptions, which offset the utopian logic. Let us look then at the principles of the symbolic order. First, it is based on the conception of man as a being defined as a linguistic entity. This definition goes back to the Greeks, more specifically to Aristotle, who defined us as animals who speak. Contemporary thought departs from this principle, adducing other arguments, among others, of a psychoanalytical and anthropological order, developed by Freud and Lacan. It is language that constitutes and institutes humankind as a subject, that organizes it by imposing norms (the function of law) that limits it and, at the same time, that stabilizes it. The incest interdiction is the archetypical institution founded on the Father-Law-Order synthesis. It is this decree that allows for our access to the symbolic order, i.e., the order of culture, which enables social organization and the procreation of the species. This prohibition is responsible for overcoming the chaotic state in which humans lived, using, however, their power, the fullness of their strength, although not controlled. There, everything was possible, everything could be. The order of culture, the order of the symbolic, came to place boundaries on this state, establishing the virtual, and a mediation structure that operates as a kind of "censorship", adjusting what is in power in the real, to the various codes and conventions, i.e., to culture. Without law and order, anything could happen. Anything would be possible. It should be pointed out that anthropotechnics recovers, at present, all that logic that preceded the incest interdiction, where everything could be possible, awakening the human forces still not controlled, proposing a world organized no longer by symbolic representation, but by creation. Now, we return once more to our initial subject: utopia can only be possible as it uses a linguistic strategy able to produce a "gap" between the present and the future. Designing or representing the future in the present is what you can do with language, with the symbolic. However, humankind can nowadays adopt the logic of anthropogenics, liberate itself, finally, from such limitations and create, create in fact, possible futures.

NOTES

[1] Nietzsche states that "Christianity is a type of Platonism for the people" (Nietzsche 2001).
[2] The origin of lhe word utopia has its roots in ancient Greece. Derives from topos which means place and the prefix *u* is used with a negative meaning. Thus, utopia means "no place" or "nowhere". However, as the utopia is the realization of a happy life, the word has correspondence with another Greek prefix *eu* meaning "what is good" or "doing what is good".
[3] Consulted interview on 24 February, 2016 on https://youtu.be/POZcBNo-D4 A.
[4] "Renaissance utopias appear as response, together with all the critical and liberal thought that is becoming important in Europe, until they end up being a privileged protagonist. Sometimes utopianism is the medium through which this response will be exercised that will end up undermining the foundations of the established structures" (Monclús 1988: 41).
[5] "[a literary society] in which participants discover through canonical readings their common love for inspiring senders. We may find in the humanistic core thus understood, the fantasy of a cult or club—the dream of the predestined solidarity of those who were elected to know how to read" (Sloterdjik 2000: 10).

[6] "All fixed relations, strengthen with their flavour of antique and venerable prejudices and opinions, are banned; all new relationships become antiquated before they get to ossify. All that is solid melts into air, all that is holy is profaned (…)" (apud. Berman 1986: 15).

BIBLIOGRAPHY

Berman, M. *Tudo o que é Sólido se Desmancha no Ar*. S. Paulo: Editora Schwarcz, 1986.

Bloch, Ernst. *O Princípio Esperança*. Vol. vol. 3. Rio de Janeiro: Contraponto/Ed. UERJ, 2005–2006.

Deleuze, Gilles, and Guattari, F. *O que é a Filosofia?* Rio de Janeiro: Ed. 34, 1993.

Hottois, Gilbert. "De L'Anthropologie á L'Anthropotechnique?" *Tumultes* 2.25 (2005): 49–64.

Lévy, Pierre. *O que é o Virtual?* S. Paulo: Ed. 34, 1996.

—. *As Tecnologias da Inteligência; O Futuro do Pensamento na Era Informática*. Trad. Fernanda Barão. Lisboa: Piaget, 1993. 263 p. ISBN: 972-9295-99-9.

Monclús, A. *El Pensamiento Utópico Contemporâneo*. Lisboa: Circulo de Leitores, 1988. IBSN: 84-226-2645-4.

Mumford, Lewis. *História das Utopias*. Tradução de Isabel Donas Botto. Lisboa: Antígona, 2007. 267 p. ISBN: 978-972-608-190-6.

Negroponte, N. *Ser Digital*. Lisboa: Caminho, 1999.

Nietzsche, Friedrich Wilhelm. *Além do Bem e do Mal*. Trad. Márcio Pugliesi. Curitiba: Hemus, 2001.

Sloterdjik, P. *Regras para o Parque Humano*. S. Paulo: Estação da Liberdade, 2000. ISBN: 85-7448-021-5.

The utopian pact or there is no alternative

Szymon Wróbel

Institute of Philosophy and Sociology of the Polish Academy of Sciences, Warsaw, Poland

ABSTRACT: In this paper, the author advances the thesis that utopia has always been an indispensable horizon to philosophical thinking and political praxis and the lack of utopian imagination would render both philosophy and politics dead. This is what the author refers to as "utopian pact". The author traces genealogy of the utopian pact from Thomas More's *Utopia* and paying tribute to Theodor W. Adorno. For Adorno the "utopia of knowledge" embodies the desire to reach sensuality without resorting to a conceptual framework. For Fredric Jameson "dialectics of identity and difference" is the source of dynamics which eventually turns any given utopia into a "program" or "impulse" in three distinct areas— that of the body, time and community. And in Ernst Bloch's approach, utopia is cast as "the principle of hope" embracing all life domains to such an extent that "being" either becomes utopia or blends with it. Corroborating a claim that "there is no alternative to utopia", the author elucidates its true meaning. The promise of happiness and the expectation of a kingdom come though both unthinkable yet formally admissible are in fact compulsory; we must categorically request the utopia for it is not a Chimera but a kind of "protean investment" and "political practice".

Keywords: Anti-utopias, heterotopia, philosophical imagination, real utopia, utopian pact

1 WHAT IS THE UTOPIAN PACT?

Dreams are the essence of sovereign humanity and the foremost dream of humanity represents a desire for self-establishing a community that is in fact a definitional part of humanity itself. Should such a dream be lacking we would speak of "population" instead of "humanity", or, in other words, we would represent "naked humanity". However, utopian dreams are not as innocent as they may seem. Dreams of a different society seek not so much "imaginary islands" but project our "hopes" by relating them to the future. Therefore, the "ideas-images" fuse with "myths" of State-Nation, Progress, and Revolution. Consequently, utopian imagination becomes a particular form of organizing a wider set of ideas. Only in schematic terms utopia is "always subversive" and collective memory is "always conservative"; historical reality proves to be much richer and much more complex because there are "invented traditions" along with "archaeologies of the future". Similarly, there are "times past" which have never happened before and there are "times to come" which have already taken place. Utopia is therefore conceivable only as a distortion; distortion of time and the order of history.

Utopias have been bandied about in the past and still are bandied about nowadays, although not in all life domains. We have thus acknowledged "dynamic utopia", i.e., the ethical and political utopia aiming at discovery of Rousseau's "authentic man", and along with that we acknowledge "elite utopia", i.e., the social-aesthetic utopia of a Madame de Staël. The prerequisite of the latter is social inequality; the existence of idle, leisure class attempting to synthesize intellectual debate and political intrigue. Elite does not, of course, eliminate social status but it is only able to suspend it. Elite may only attempt to create a life form by indulging in worldly leisure activities. In turn, Rousseau's dynamic utopia is based on belief that abolishing social and economic inequality is possible. While elite utopia is private space dedicated to public service, Rousseau's dynamic utopia is public space catering to private needs. New Heloise warrants life fulfilment from work and provincial life. Socialized individualism, gentle manners, play and idleness—these are the qualities in elite utopia. Work, communism, and utility, simplicity of manners and morals, commitment to work towards community—these are the features of New Heloise. To Madame de Staël the appropriate life is consumed in talking about politics and literature. To Rousseau appropriate is action. To Madame de Staël all evil stems from lack of social ranks and distinctions. To Rousseau inequality and the lines between "mine" and "yours" are evil. Considering the above inconsistencies allow me to question what indeed constitutes a utopia.

363

The answer the questions: "what is utopia?" and "what is its philosophical and political meaning?" let us first review the genealogy of utopia. To begin with, a paradigmatic text to all utopian imagination is Thomas More's *Utopia* published in 1516. The full title of this important book is *Libellus aureus salutaris minus quam nec festivus de optimo Reipublicae stat de que nova insula Utopia* that translates to "A truly golden little book, no less beneficial than entertaining, of a republic's best state and of the new island Utopia". The protagonist of this golden little book, a traveller Raphael Hythlodaeus is a clever storyteller, philosopher and a sailor, so to say an amalgam of Plato and Ulysses, who testifies to the existence of a country with the capital city in Amaurotum, which is a city-mirage in itself, a city-in-the dark, and a city in *felix locus*. At first glance, Utopia is short of a sheer lunacy with the potential to lead all history of utopia into the territory of "political madness". Yet, this golden book lays foundation for a paradigmatic topic in literature; a fictionalized account of a travel. Along with this topic we get another one; the account of a fictitious society living "nowhere" in this world. This contexture establishes what I refer to as the utopian pact.

Our key word literally translates to "not-place" (Greek *ou-τόπος*), but it may well stand for "good-place" (Greek *ευ–τόπος*) if not in fact "best-place" or "best-country/state/regime" imaginable, which in Moore's rendering eclipses that of Plato's. One may easily conclude that the best political community, even if conceivable, is set nowhere else but in human imagination and is a derivative product of knowledge. In this interpretation utopia does not represent reality but nonetheless it comes into existence upon it being conceived and founded in text. It is this liaison of foundational text and both geographic and ontological hyper-reality of a given space that constitute "the utopian pact".

It is the intrinsic characteristics of the genre to refer to More's text. History and succession of utopian generations are exemplified in later utopias and incorporated into all utopian imagination. Utopian discourse, just as philosophical, carries with it the obligation to respond to the whole lot of utopian tradition, which has preceded it and created it. Consequently, the history of utopia brings with it its entire history. The social entirety is neither Law, nor Love, nor Nature; it is neither Money, nor Delight, nor Relativity, nor even Indifference. The social entirety is the utopia of unification of all and everything and reconciliation between all and everything. This is why utopia always claims the Social Whole (without the Law, or without Marriage, Money, Necessity, Labour, Procreation, institutions, etc.). This dream, as we can imagine it, can only be fulfilled outside the Whole, on a particular island, on an isolated territory—be it Corsica, Utopia, Haiti, Poland, or Nowhere. Utopia is a paradoxical dream of entirety outside the Whole.

What separates the narrative of More and Hythlodaeus from Plato's theorizing myth narrative or from St. Paul's revealed truth narrative is "engineering" of the utopian narrative along narrating the story. They are neither prophets who through the illumination wish to restore paradise on Earth, nor they are philosophers who wish to convey truths beyond the grasp of reason. Their intellectual effort is to construct representations and render them as artefacts. The prerequisite for creation of a utopian paradigm is the creation of a particular place to be occupied by the intellectual who may then claim the right to imagine an alternative social and political reality. Let us try to understand this invention concurrently with the invention of a space of thought. Set aside the critique of existing reality or design of a perfect reality, their endeavour is to discover territories inhabited by people living a good life.

In this respect, utopians have something of geographers, explorers, and anthropologists. Utopia and satire are not necessarily their own inverses. Satire is not in fact anti-political genre as it challenges mindless, stupid life, life controlled by habits, existing laws, and rules of a fallen world. In the light of utopia whoever lives that way is a fool.

Of course, nothing stands in the way to invent new and perfect community, but the mere fact of inventing "real communities" living in harmony with their own desire has the potential to establish a paradigm for the utopian imagination and inspires one to "play utopia". When Benedict Spinoza in the preamble to his *Theological-Political Treatise* ridicules the entire philosophical tradition by bringing it either to the formula of utopian design of an ideal political order or to the formula of a satirical commentary on human nature he only bears witness to his not understanding a new concept of "concrete utopia". Intellectual exercises within the paradigm of "concrete utopia" contribute in their own way to answering the grand question of modernity about the possibility of a self-establishing society, a society whose self-determination is based on no external order, yet it binds individuals into community, a community exercising full control over itself. So conceived it is a society not so much disenchanted as it is prepared to maintain a new type of touch with reality.

Utopia is not a Chimera as it is often inaccurately claimed. It is only in everyday speech where we may identify, not without reason, some utopias with "chimeras". This said, concrete utopias are certainly not the case. Utopian imagination, and in particular its tendency to include in *one* project the *entirety* of social otherness is parallel to presenting people as they *should be* instead of what they really *are*. In other words, utopias are "chimeric" as far as they stem from desire for logical, consistent, transparent society and a belief that social life can be freely transformed and rationalized. Well, it is simply beyond possible. What we call a "real utopia" is not a representation of our longing for a perfect society or for complete rationality and it neither is a satire on existing society. "Real utopia" is an incentive for inventing "other spaces" and putting them into action by incorporating "imaginary realities" into real and material political order.

At one time, in its history, Utopia was called "Abraxa" and it was inhabited by uncouth and fractious horde. Abraxas is a term of the Supreme Deity in Persian and Gnostic mythologies. Hellenistic texts on magic the word exemplifies magical logic and is synonymous with fullness. It probably derives from the Hebrew *ha-berak-hah* meaning a blessing, or it is a superposition of two Hebrew words—*abh* and *bara* that could mean "the Unbegotten Father." In More's narrative the land of Abraxa was conquered by King Utopus who, as the community's founder, ordered to dig a 15-mile wide channel in order to separate it from the mainland. Later Utopus, we are told, the tyrant and colonizer, raises the quality of man's life there to unprecedented level. Interestingly, Utopia is constituted only when Utopus separates it from the Whole. Utopia is therefore the political form of a blessed region, of unbegotten territory, a full part, and not Whole. In other words, utopia is representation a project of social life other than that remaining under the rule of the Father or in the name of the Father. In this sense, "real utopia" is always an anti-Oedipal project.

The birth of utopia is closely related to the birth of modern, enlightened forms of tyranny. In 1514, at the time More is writing *Utopia*, Machiavelli publishes *The Prince*. At first glance, *The Prince*, whose motive is political effectiveness devoid of any scruples, is in every way opposed to Utopia, whose inspiration was primarily moralistic and evangelical humanism. However, one may wonder whether the Utopians, trying to win the war by cunning, bribery, intrigues, etc., are not in fact "machiavellists". Likewise, there are elements of utopia in *The Prince*. After all, we get there a glimpse of a fair and orderly community founded on rational practice of policy. The modern forms of stupidity diagnosed by Desiderius Erasmus, utopia discovered by More, and modern politics designed by Machiavelli are all "inventions" of the same time.

We come across Machiavelli's great theme in Rousseau's Great Legislator along with the problematics of antinomies that go along with it.

A mythical founder of the community must be free from all human weaknesses because only such a figure may transcend corrupt and anomic society to pursue its reform. Where such a figure relies on political status or prestige, it would make of it a fatherly figure (such as Stalin) or a charismatic fascist-type superman. *Übermensch* of Nietzsche, Freud's Moses, Modern Prince of Antonio Gramsci, the big Other of Lacan are figures that successively struggle with the same dilemma.

Images of a new society irresistibly permeate the sphere of human life where public dreams are collected, compiled, and devised. Utopian paradigm, once it is set and begins to take effect on social imagination is both inert and it gains momentum. By virtue of imitation utopian stories multiply in series and the term "utopia", becoming generic and broadens its scope of meaning without compromising on accuracy. Over the years, the genre spawned para-utopias and anti-utopias. Further, we shall trace the history of utopian pact by the milestones in the history of human dreams (of ideas).

2 GENEALOGY OF UTOPIAN PACT AND ITS DECAY

It would be worthwhile to trace the genealogy of utopian projects from Charles Fourier's *Théorie des quatre mouvements et des destinées générales* published as early as 1808, then move on to Henri de Saint-Simon's late work *Catechism des industriels* published in years 1823 to 1824, Victor Considérant's *La Destinée Sociale* of 1836, Barthélemy Prosper Enfantin's *La Vie eternelle passée, presente, future* from 1861, and finally Frederick Engels' famous text *Socialism: Utopian and Scientific* from 1880, Georges Sorel's *Reflections on violence* from 1908, Ernst Bloch's *Geist der Utopie* from 1918 and Karl Mannheim's *Ideology and Utopia* from 1929. Such was the blazing trail of the utopian pact project. I put dates of these works on purpose so that we can fully appreciate the timeline.

To comprehend fully the utopian pact we should at all cost refrain from associating it with messianism or millenarism. After all, we refrain from tracing crypto-theology in liberal and conservative projects just as we do without it inspecting communism and socialism. Truly, no known utopia pronounces the coming of the Kingdom of God. When Thomas Müntzer made public his *Prager Manifesto* in 1521, entwining the vision the coming Kingdom with the advent of the New Man and initiated German Peasants' War (peasant uprising) spanning the years from 1524 to 1526. What inspired the peasants was the revealed Word of God rather than a work of fiction or a fantasy novel. For More, on the other hand, the founder of utopian discourse, the idea of social diversity was a product of the written word and imagination that allowed setting on a non-existent island a narrative space wherein ideas-images can mingle with other forms of imagery. Millenarian ideas only support the idea of "Earth-without-Evil"—there is no mention to a perfectly designed, isolated place on Earth where people live a good life. Messianism was it ever compared to utopia, is at best much-generalized utopianism as it pertains to all humanity and to all world. This, as we have already stated, stands in sharp contrast to the regionalism and material setting in the utopian pact.

In the course of the 18th century, the utopian pact gets gradually abandoned. It is the time when utopias give way to para-utopias. The milestone works in this transformation are *Micromégas*, a 1752 pioneering work of science fiction by Voltaire, Rousseau's 1761 epistolary novel *Julie, or the New Heloise* depicting life of a community in Clarens, Donatien-Alphonse-François de Sade's *Aline et Valcour ou Le Roman philosophique* published in 1786 and, of course, Jonathan Swift's novel *Gulliver's Travels* 1726. Micromégas (from Greek *mikrós*—"small" and *megas*—"great") is a resident of a planet in the solar system of Sirius, who in his travel across the universe happens to visit eighteenth century Earth where he encounters a group of scholars returning from a scientific expedition. In his naivety, Micromégas mistakes Earth for a place of perfect happiness. Philosophers manage to prove it otherwise by presenting men as "fools, rogues, and unhappy people". Micromégas confuses hell with paradise, large with small, senses with reason, madness with presence of mind, and fiction with reality. In *Micromégas* utopia is therefore distorted and Voltaire's book embodies what we may refer to as *para*-utopia.

It is for the first time in history that the seals on utopian pact are broken. The reader encounters two opposite civilizations based on offsetting principles. As a result, the idea of universal standard suffers damage. It is for the first time when fictitious societies can be looked upon as *para*-societies. Rousseau and Voltaire, Sade and Swift, each author in its own way conjures up fantasies that undermine the power of travel experience along with the possibility of a site separate from the rest of the world. It cannot be ruled out that Michel Houellebecq's *The Possibility of an Island* brings the paradigm of para-utopia to the end. *Para*-, a loanword from Greek meaning "at or to one side of, beside, side by side" does not make of para-utopia a sheer contradiction of utopia. Likewise, adjective "paranormal" does not contradict "normal" and the epithet "paramedic" does not imply inept imitation of medicine. Para-utopia is a utopia "at the side of island" and "as if on an island".

Two works herald the coming of anti-utopia. It is *The Brave New World* by Aldous Huxley's published in 1932 and a 1949 George Orwell's *Nineteen Eighty-Four*. With these two major works the search for a paradigmatic territory, for normative legislation and for exemplary constitution are all abandoned. Anti-utopias are not in search of secluded islands, they abandon geography and space altogether. In return, however, the heroes of these works need to be shifted in time. Anti-utopias are set in "times different" and "beyond present". Of course, these a settings do not share the characteristics of "proto-time". Utopian ideas are set in history; for it is in the course of time, that utopia may come to fruition.

There are reasons to talk about conversion of space into time, ou-*topia* into ou-*chronos*. The imaginary society which before was set somewhere-else is now situated in imaginary time. It is tempting to speak of the contemporary crisis of enclosed spaces—such as that of prison, hospital, school, or factory in terms of this more fundamental shift. It seems that the very idea of a happiness stands in opposition to a "closed space" such as that of an island. Today, closed places are seen as unfortunate. Similarly to the aforementioned text by Michel Houellebecq, Peter Sloterdijk's *In the World Interior of Capital: Towards a Philosophical Theory of Globalization* is likely the last chapter of in the history of "architectural" utopia which eventually transforms closed space into a human conservatory subjected to political inbreeding.

For both Orwell and for Huxley the principle of a community is that "everyone belongs to everyone else". Huxley's Directorate of Hatcheries and Conditioning ensures that through conditioning people accept their inescapable social destiny. Socializing therefore, instead of being a boon to civilization turns out to be its greatest curse. Politics is a constellation of actions undertaken by directorates, which pursue transformation of basic concepts in language. Politics is narrowed down to the politics of collective semantic madness. It is the policy of madness even more so, when we consider that instead of being detached from reality it questions the reality of intersubjective meanings. Earlier, Swift confronted human nature with the ideal and concluded that it is beyond human reach. In contemporary anti-utopia, however, it is utopia that turns out to be below the standard of human abilities. The focal point of both these novels is the conflict between imperfectly realized utopia that turns into a nightmare and irreducible values found in humans. Here, human nature is no longer the norm; it is the target for subsequent techniques aimed at thinking humanity anew.

Could we go wrong in concluding that humankind today fears dreaming? Does humanity fear the radical dream? Is Huxley's and Orwell's warning strong enough to cause fear and constrain human imagination? As it was once aptly noted, the onset of anti-utopia made of utopia nothing but a story, in other words, utopia became known as fiction. Is it not the reason why 20th century presented us with a plethora of works of fiction in this respect? Let us take, for example, anti-utopian writings of Philip Dick and Ursula Le Guin, Isaac Asimov or Stanislaw Lem along with the abundance of science fiction films such as *Surrogates, Men in Black, The Matrix* trilogy, *The Alien* series and more recently *Prometheus* to mention but a few? Utopia today is not a "concrete utopia" and it is no longer appears a necessary component of the human project. The slogan "utopia or death" has been replaced with "utopia is death".

3 HETEROTOPIA OR PHILOSOPHY IN THE SERVICE OF UTOPIA

Perhaps the most astounding is the relationship between utopia and myth. There is a clear trade between myths and utopias: utopias expand and radicalize the potentials and promises of the future (although they can as well put them into inertia in a closed system of ideas), and at the same time utopias draw the dynamics and symbolic material from myths. On the other hand, the mythical resource easily carries utopia away and brings dangerously close to totalitarianism. In this world of global voyeurism, in the times of universal surveillance, utopia loses its liberating and subversive potential. Utopia is often the enemy of freedom, and all the more dangerous that it is masked by the allure of myth. Utopia therefore serves as the anticipation of totalitarian world, not to say: the concentration world.

This new utopian formula led to comparisons of More's *libellus aures* and Fourier's *phalanstère* to Gulag. The images of utopia were hastily subjected to prosecution. Utopia was declared dead. Certainly, the haste all too quickly gave rise to false accusations. Utopia faltered but never gave in for one important reason. Utopia was never intended to conserve time but it neither was meant to pursue radical acceleration of time. Utopia is a shake-up of all times available, all available stories, histories, and lands. It is myth that remains at the service of conserving time and making history inert. Therefore, it is chiefly myth that should be sabotaged and protested against, not the utopian imagination.

In order to account for the relations between utopias and totalitarianisms we need to ask questions that are historically accurate. We need to restrict our questions to this or that particular utopia in the context of any given totalitarian system, and especially in its ideology. Having adopted the thesis that there exist concrete utopias we simply cannot proceed otherwise. For example, communist utopia paves the way to revisionism because it remains autonomous from the use that authorities make of it. Consequently, the authorities can be subjected to criticism on the grounds of utopia. Let us recall Nikolai Berdyaev's warning addressing the utopian imagination of the totalitarian and post-totalitarian policies: "Utopias have today become more real than ever." The reality of utopia does not mean that the fiction of anti-utopia can be confused with reality. More than ever, utopias are now susceptible to manipulation by means of modern communications technologies, powerful means of propaganda, and new media. Today, the historical appeal of utopia can easily be fabricated by totalitarian power or by the forces wishing to monopolize or confiscate social imagination. Utopia has evolved into nothing but a story at the very moment when we saw it turning real. It may seem a contradiction, but only if we forget about the new stage of virtuality in modern times.

As utopia transforms into anti-utopia it becomes a projection of fear of anomaly, disturbance, general strike, or of unknown future, and thus it begins to fear its own self. The image of other society is placed in the future at hand's reach. True, it is future imagined but by no means it is chimerical nor it belongs to fiction. The advent of this future is supposedly predestined by the inevitable course of history known as Progress, culminating with a utopia becoming a reality. Bright, utopian future is the manifestation of the Enlightenment. Utopia belongs to the era of light. In modern times, however, our light is artificial and mediated. People live behind the soft bluish glow of their smartphone and tablet screens, which provide for a new synthesis of time and space, media and body but also mediate false utopias. The main feature of utopia of the Enlightenment was its transparency. In modern and postmodern world the one and only "concrete utopia" is to be found in the shadow. It is shadow where one retreats to in order to avoid overexposure in the real world that is constantly illuminated by the electric glow. Let us remember that before the conquest by Utopus the land of Abraxa was left in the dark.

I return to the question of our fear of utopia. What is it that makes us seek the revolutionary and progressive base in the "real" and deny the power of abstractions? Is it the expression of our thirst for concrete reality, fear of being disconnected from the world, or yet a trauma of thus far realized utopias? Michel Foucault wrote: "First there are the Utopias. Utopias are sites (*emplacements*) with no real place. They are sites that have a general relation of direct or inverted analogy with the real space of Society." After utopias there come heterotopias. Heterotopias are real places (*lieux*) determined by their social and political organization which are like counter-sites (*contre-emplacements*); effectively enacted Utopias. Heterotopias are not "abstractions" subject to "imaginary order", they do have a "real agenda" and are a sort of "effectively played utopia". Heterotopias are abstractions reversed and realised. Rousseau depicts heterotopia in *The New Heloise* where Clarens community live "invented tradition" and "naturalized culture". The protagonists, Wolmar and Julie celebrate a bountiful grape harvest and live in eternal feast day. In a choir, singing in unison there are no leaders and everyone is equal, everyone sings on their own but in one voice, which allows them to forget about their loneliness. Utopia is played out "univocally" and "unanimously".

Theodor W. Adorno presents us with the "utopia of knowledge" embodying desire to reach sensuality without resorting to a conceptual framework; without the help of concepts. Fredric Jameson addresses utopia in terms of desire and the "dialectics of identity and difference" being the source of dynamics which eventually turns any given utopia into a "program" or "impulse" in three distinct areas—that of the body, time and community. We also have Ernst Bloch's approach wherein utopia is cast as "the principle of hope" embracing all life domains to such an extent that "being" either becomes utopia or blends with it. Myself, I would not hesitate to admit, "there is no alternative to utopia". Paraphrasing Bloch, the promise of happiness, the expectation of a kingdom come, and our hope for dreams come true—though all unthinkable are formally admissible and are in fact compulsory; we must persist and categorically request a utopia. Utopia is real—it manifests human will and as such it a kind of "protean investments" and "political practice".

BIBLIOGRAPHY

Carta del Carnaro—between utopia and politics

Daniela Spinelli

Department of History, Universidade Federal de São Paulo (UNIFESP), São Paulo, Brazil
Department of History, Archaeology, Geography, Arts, Entertainment of the Università degli Studi di Firenze (UNIFI), Florence, Italy
Fundação de Amparo à Pesquisa do Estado de São Paulo (FAPESP), São Paulo, Brazil
Research Group "Renascimento e Utopia (CNPq/UNICAMP), U-TOPOS—Centro de Pesquisa sobre Utopia (IEL/UNICAMP)
Research Group "História, Memória e Patrimônio do Trabalho" (CNPq/UNIFESP)

ABSTRACT: This presentation comprises the investigation of *Carta del Carnaro*, written by Gabriele D'Annunzio and Alceste De Ambris, with the objective of analysing the utopian, literary, and modernist aspects of the fiuman constitution. Promulgated in 1920, during the military occupation of the city of Fiume, *Carta del Carnaro* is the fundamental law of the new Italian state. Written in a poetic form, this document is considered a very curious literary exercise to build an imaginary ideal state. The transfiguration of the State form in a prose *d'arte* evokes a magnanimous dream that through politics, man can repair their miseries, offering dignity and social justice to the society. The State is conceived as an artificial being, formed from the will of a people; it would represent the structure inside of which man could develop its faculties and fully achieve its nature. *Carta del Carnaro* is considered an extremely important document for its aesthetic and literary value, but also its political significance. The fusion of mystical and religious radical elements offered an immeasurable contribution to modern political theory.

The aim of this text is to evaluate how the historical events experienced by the generation that lived during the Great War served as a background so that we can understand the direction of the Italian vanguard and how the aesthetic formalization of their ideals transformed the early 20th Century Italy into a political laboratory.

Keywords: Carta del Carnaro, Gabriele D'Annunzio, Alceste De Ambris, utopia, fascism

1 BETWEEN UTOPIA AND POLITICS

At the end of the Great War and with the dissolution of the Austro-Hungarian and the Ottoman Empires, the winning countries got the opportunity to redesign the political and geographical world map. Italy, which fought along with the Triple Entente, presented its delegation in Paris on January 18th, 1919. The London Pact, firmed on April 26th, 1915, determined that in the case of victory, Italy would have the right to the territories of the Trentino, High Adige until Brennero, Trieste, Gorizia, occidental part of Istria, Dalmatia and its islands and the part of the colonial estate of Germany and the Ottoman Empire. These territorial conquests matched the irredentist's claims for the conclusion of the process of Italian unification and the imperialist yearnings of nationalistic groups.

The American president Woodrow Wilson was not part of the Pact and openly opposed the fulfilment of the private agreements signed at the beginning of the war. As a diplomatic alternative, he proposed 14 points that should be followed by the winning nations with the objective to guarantee world peace. As far as Italy was concerned, the American president declared that its borders would have to be readjusted following clearly recognizable nationality dividing lines. The city of Fiume was not enclosed in the area to be annexed by Italy, according to the London Pact, and since the end of world war, was occupied with battalions of American French, British, and Italian armies. An atmosphere of tension between the diverse ethnic groups that composed the local population took the city of Fiume. Without ever belonging to the claims of the Italian irredentist movement, the "question of Fiume" only became a central diplomatic stalemate when the just-instituted *Consiglio Nazionale Italiano di Fiume*, presided by Antonio Grossich, declared that, on the basis of the principle of self-determination of the peoples, the majority of the population of the city (Bertotto, 2009, p. 50) desired to join the Kingdom of Italy.

In the Peace Conference, the Italian delegation, directed by the ministry Vittorio Emanuele Orlando and Sidney Sonnino, demanded the fulfilment of a maximum program, that is, of the London Pact and the city of Fiume. This proposal was slandered in such a way both by the allies and by President Woodrow Wilson, for being considered an imperialist proposal and transgressive to the order. The problem was that the Italian delegation tried to reconcile two incompatible diplomatic conceptions. After all, it requested that the right of self-determination of the people should have to prevail as a criterion in the case of Fiume, but not in the other territories established to Italy by the London Pact.

Pressured by the public opinion, Vittorio Emanuele Orlando and Sidney Sonnino abandoned the Peace Conference temporarily. The gesture of hostility with regard to the allies would not go unnoticed. At the end, Italy got the territories written in the London Pact but, due to the behaviour of its delegation, it did not participate of the allotment of the former-colonies of the dissolved Ottoman and Austro-Hungarian Empires. The diplomatic deadlock on the territories of Fiume and Dalmatia kept on. Since it became so difficult to establish a border between Italy and the just-formed Yugoslavia, on the 8th of August of 1919, it was determined that the administration of Fiume would be run by an Inter-allied Military Commission, composed by French, English, American and Italian military troops, while the question remained undefined.

The poet Gabriele D' Annunzio, taking advantage of the atmosphere of resentment to exercise a nationalistic and imperialist campaign, created the slogan "vittoria mutilata". The expression synthesized a diffuse sensation of disillusionment between the masses, that the war was not really won since its conquests had been embezzled. During the negotiations promoted in the Peace Conference, the idea that the allies had not recognized the effort of Italy was disseminated. It was believed that Italy had been deceived by its allies and that, therefore, the Italian victory had been mutilated. Gabriele D' Annunzio, 'poet vate' of the nation, publicly demanded the right of the State for the city of Fiume, based on the right to self-determination of the people and in the trend of the italianity. Thus, Fiume became a martyr city and symbol of an incomplete triumph.

In the morning of September 12th, 1919, the celebrated poet and war hero Gabriele D' Annunzio and an army of legionaries (composed by former-combatants and defectors) had taken the city of Fiume by assault. The entrance became a great theatrical act. Acclaimed by local crowd, the poet took ownership of the government of the city and publicly declared the annexation of Fiume to the Kingdom of Italy. The act of Gabriele D' Annunzio was not only a challenge to the authority of the chairperson of the board of Ministers, Francesco Saverio Nitti—that ironically was called "Cagoia" by the poet as a nickname—and a spectacular example of international violence, but a great gesture of insubordination of the army and the navy to the Italian crown.

Therefore, not solely for this reason the occupation of Fiume must be considered a fact of great importance in the history of modern Italy. Federico Chabod calls attention to the fact that the dannunzian expedition was considered by leaders of the just-formed fascist movement a form of trial for the *Marcia su Rome*. The fiumanism, as a mass movement originated from the Great War and as radical reaction to the socialism and the old Italian liberal class, offered a new political form to the fascist movement. Dino Grandi and Italo Balbo, together with the other archetypes of the fascist movement, envisaged Gabriele D' Annunzio as the true *Duce*. They believed that the poet was the *spiritual priest of fascism*, since he idealized a cultural and political revolution through the construction of a totalitarian State. After the end of the political experience in Fiume, Gabriele D' Annunzio was contacted a few times by directors of the fascist movement, with the objective of being convinced to assume the position of leader of this movement. Due to his status as a national hero, bestowed with five silver and a golden medals during the Great War, the poet was considered the most qualified person to lead the movement. Only when he resigned to this position Benito Mussolini was able to assume as the only fascist *Duce*.

Through the *coup d'état*, Gabriele D' Annunzio took the city, which he administered for sixteen months under the heading of "Commander". Daily, the poet dressed with the uniform used by the *arditi*, delivered speeches from the balcony of the Palazzo Del Governatorato, which were followed by a multitude that chanted back to him with the hymn "*A noi... Eja, eja, alalà*". Hymn that was later incorporated by the fascist regime ideology. Efficient in combining the interests of heterogeneous groups, Gabriele D' Annunzio used to his advantage the power of beautiful words to mould the crowd and to invoke the city as a centre of a new world-wide revolution. The troops of legionaries, dressed with black shirts of *arditi*, promoted military parades and awarded medals to those that according to dannunzian mysticism, should be considered the new heroes of the Italian homeland. Spectacle politics was thus inaugurated.

The occupation of Fiume was characterized by a climate of adventure and revolt against the old order. The atmosphere was a festive one and deferment of rules. Moreover, while the poet isolated himself from reality in a universe imagined by him, the political and economical situation in Italy deteriorated. Scarcity of charcoal, the lira devaluation, and inflation created general discontent that culminated in a series of strikes, which halted the country. The fascist's squadrons, paramilitary organization led by Benito Mussolini that opposed violently the worker´s movements generated a civil war atmosphere, which prevented the political and economical recovery of the country. These grave political episodes slowly started to diffuse the attention and interest for the Fiume question.

The post-war economic situation was also disastrous in Fiume. Ousted internationally, the city was surviving on Italian government resources, donations and pirate acts. Gradually the local population became dejected with the constant chaos and the inconvenient deficit caused by the dannunzian government. Political divergences between the radical factions that supported the Fiumian occupancy also harmed the cohesion of the movement. Under economic pressure, internal and external politics, the Commander, already weakened by the crisis, nominated the revolutionary syndicalist Alceste De Ambris for the position of cabinet head.

The presence of this experienced politician gave a new hope to the poet´s government, when he proposed the establishment of an organic project. Revolutionary action, as a non-legal phenomenon, was the prerequisite for the birth of a new constitutional order. Against the loss of effectiveness of the previous Constitution, and with the need to provide stability to the process of social change, the *Carta del Carnaro* represents the last attempt to give form to the Fiume political experience.

Publicized on September 8th of 1920, this constitution, written by the poet and the revolutionary syndicalist Alceste De Ambris and destined to regulate the *Reggenza Italiana del Stato Libero di Fiume* (Name given by Gabriele D'Annunzio to the new State), launched the idea of a corporative State, established in the consecration of work as fundamental to life in community. The basic law was a manifesto that would have to serve, simultaneously, as a solution to the national problem (creating a model for the Italian modernization and the inclusion of the masses in the political life of the country) and as catalyst of a Latin revolution.

The *Carta del Carnaro* remains as an obscure object in the Italian constitutional conscience. After almost one hundred years of its publication, the fundamental Law of the Free State of Fiume resists as an enigmatic document, which eludes a precise historical definition. Because of a political experience of a military nationalistic movement, which under the leadership of the poet promoted the armed occupation, and the establishment of a temporary government in the city of Fiume, the *Carta del Carnaro* is a *constitution-manifest* that endorsed a radical revolution of the society.

Written by Gabriele D'Annunzio and Alceste De Ambris, the decree condensed the normative trend of the modernist vanguard and Italian nationalists in trying to solve the social problems and to modify radically reality. Under the whim to create a new world, the two authors attempted to surpass the existing forms of organization of human beings and to suppress all evil. Inspired by the dannunzian idealism and by Alceste De Ambris theoretical ideas (that discussed with the bases of revolutionary syndicalism -political movement that developed in the turn of the 20th century as a radical expression against the bourgeoisie and the reformist left, to promote a new form of labour action), the statute of the Free State of Fiume proposes to give the world an example of constitution which abides to the most audacious ideas of modern philosophy, the varied forms of freedom and retrieves the glorious traditions of Latin culture. Through a set of basic rules, where political doctrines mingle with historical experiences and literary constructions, the image of an 'ideal' State that would allow the existence of a harmonious collective life materializes.

Under the form of *imaginative art*, the drawing of the *Italian Reggenza del Carnaro* is configured as product of the technique. Thus, the individual capacity to construct human society from freedom is celebrated, endowing it with an aesthetic dimension. Devoted to the idea of beauty, the commander-poet Gabriele D' Annunzio presented to the people of Fiume—in the terms used by the historian Jacob Burckhardt—the "State, as conscious and calculated creation, as a work of arte" (Burckhardt, 2013, p. 37). The transfiguration of the State form in a prose *d'arte* evokes a magnanimous dream that through politics, man can repair their miseries, offering dignity and social justice to the society.

The State was conceived as an artificial being, formed from the will of a people; it would represent the structure inside of which man could develop his faculties and fully achieve his nature. It was believed that only the State could guarantee freedom and peace, offering excellent conditions of existence for man. Aiming at the common good and Justice, the State registers the definite exit of man from his natural state. Thus, Gabriele D' Annunzio and Alceste De Ambris conscientiously searched to perfect the world, through the new formularization of the social institutions and politics. Reclaiming alternatively political forms from Athens, from the Roman Republic, the early Middle Ages, the Serene Republic of Venice, and the Renaissance, they proposed a State derived from the combination of idealized historical models, to present the synthesis of an *excellent* form of government.

According to the *Carta del Carnaro*, the city-state of Fiume should be administered as a direct democracy based on the productive work and as criterion, the widest functional and local autonomies. The statute recognized the collective sovereignty of all the citizens without sex, race, language, class or religion distinction, but it granted greater rights to the "producers". This generic term contemplated the formation of a sole class, including workers and employers who contributed for the prosperity of the State.

The constitution would legitimize the perfect civil and political equality of the two sexes and would guarantee the freedom of thought, of word, of meeting, of association and of press. It was to be a secular regime that safeguarded religious tolerance and tutored the right to the construction of temples. However, religion could never be invoked or serve as justification for those citizens who did not fulfil the duties prescribed by law. Thus, the statute anticipated the disruption with religious dogmatism and with the interference of the Church in the State. All citizens, without discrimination, would have the right to primary education in laic public schools, to physical education and to paid work.

The statute also assured a minimum wage to live well, medical assistance in case of illness case, retirement and the use of the goods legitimately acquired. Issues that were certified: the inviolability of the home, the *habeas corpus,* and the liability compensation in case of judiciary error. Property was not considered as an absolute dominion of a person over a thing, but as its social function. The citizens, the corporations, and the communities would be something as souls that formed the bases of the *Italian Reggenza del Carnaro*. All the citizens that joined forces for the material prosperity and the civil development of the State would have to be enrolled in one of the ten corporations to be instituted.

According to its creators, the *Carta del Carnaro* was essentially a Latin constitution that would perpetuate the following mission: "to be a luminous lighthouse in the darkness, for those who advance onto the unknown, probing a way with their hands" (De Ambris apud Fressura and Karlsen, 2009). Launching itself ahead of those who looked for an alternative life different from that imposed, the statute of the *Italian Reggenza del Carnaro*, as an authentic form of philosophy, took the risk to jump over abysses to illuminate the track.

It was indeed, a limitless exaltation of the possibility of man to recreate himself continuously and to reorganize his social relations through politics. However, the *Carta del Carnaro* did not happened as a fantasy but because of the recognition of the abysses through which humanity would be condemned to march. The drawing of a new order for the establishment of the *Stato Libero di Fiume* was born of an acute conscience of the social rupture generated by the Great War and of the impossibility of accomplishing the political projects dreamed by the *nationalistic modernism*. The *nationalistic modernism* started in the beginning of the 20th century, from the meeting of the crisis culture and the need to nationalize the people. As an expression of ideological and cultural efforts for the modernization of the State, modernism was the result of a conservative state that kept aside from the great European scientific and technological progress. After all, it was expected that after the war a great Italy would finally emerge, which would become a world political leader and stimulate cultural and economic renaissance movement. Once these aspirations became frustrated and victory was of a mutilated type, the nostalgic leader of new Italy proposed an action prompted by force that found substance in the historical matter.

Therefore, it was a limitless exaltation of man's ability to continually recreate and reorganize its social relations through politics. The design of a new system for the establishment of the Free State of Rijeka was born from the perception of the social fracture created by the Great War and the impossibility of implementing projects dreamed up by political modernism nationalist. The poet and head of the Cabinet attributed to the new Constitution the function manifest bearer of a new historical and cultural consciousness, announcing the formation of a "new" man. For this reason, Article XVI of Foundations, decrees:

La vita è bella e degna che severamente e magnificamente viva. Uomo deve essere rifatto intero dalla libertà. L'uomo intiero è colui che sa ogni giorno inventare la propria virtù, per ogni giorno offrire ai suoi fratelli un nuovo dono. Il lavoro, anche il più umile, anche il più oscuro, se sia bene eseguito, tende alla bellezza e orna il mondo.

In other words: the State should provide the necessary tools so that the nature of its citizens could be fully reform. Becoming a whole man, capable to invent their own virtues every day, this new man would have to shape constantly his actions for the benefit of collective life.

Therefore, the *Carta del Carnaro* can be seen as a curious literary exercise of imaginary state building. The social design of the statute formalize the contradictions of the historical moment through the spectacular contrast of a state structure that radically opposed all existing forms with the ultimate aim of building a state as a work of art. However, this does not imply reducing the analysis to a playful game with no relation to reality, but on the contrary, recognize on it the opportunity for critical re-evaluation of a concrete historical reality. Finally, since it has a constitutional, political, and social meaning the *Carta del Carnaro* is a document of great importance for the formation of modern Italy.

BIBLIOGRAPHY

Bertotto, Alberto. *L'Uscocco Fiumano—Guido Keller fra D'Annunzio e Marinetti.* 1ed. Firenze: Ed Sassoscrito, 2009.

Burckhardt, Jacob. *A cultura do Renascimento na Itália.* 1ed. São Paulo: Cia das Letras, 2013.

Fressura, M. And Karlsen, P. *Gabriele D' Annunzio—La Carta del Carnaro e altri scritti su Fiume.* 1ed. Roma: Ed Castelvecchi, 2009.

Sensate utopia—experiencing unreachable in the state spectacle of socialist Yugoslavia

Jelena Mitrović & Vladan Perić

Department of Architecture and Urban Planning, Faculty of Technical Science, University of Novi Sad, Novi Sad, Serbia

ABSTRACT: This paper aims at exploring the notion of reachable utopia understood as one of the tools used in the state spectacle by the government of socialist Yugoslavia. Utopia and utopian ideas are commonly understood as unreachable moral and aesthetic ideals, outside of the grasp of society, placed both in its golden past and glorious future. While acknowledging the fact that as a physical reality utopia remains unreachable, it is the goal of this paper to examine the notion of utopia that can be lived in during the state spectacle.

In order to do so, this paper relies on concepts of liminal space, porosity of space and architectural chronotopes. These theoretical constructs and their relation to utopian ideas are explored in the geopolitical context of Socialist Federal Republic of Yugoslavia, on the case of memorial complex Slobodište. Spatial characteristics of Slobodište as well as dramaturgy of and ideology behind Days of Freedom festivities that took place in it make it the ideal candidate to explore relations of utopian ideas and state spectacle on. Through exploring Slobodište, we hope to point to the possibility of the utopia that can be sensed inside a specific space/time frame.

Keywords: utopia, state spectacle, Yugoslavia, Slobodište

1 INTRODUCTION

For as long as the human kind exists on this earth there also exists the concept of the perfect, ideal world, of the best moral and aesthetic values, that does not succumb to deterioration or disease. "What makes human history such an uncertain and fascinating story is that man lived in two worlds—the world within and the world without—and the world within men's heads has undergone transformations which have disintegrated material things with the power and rapidity of radium (Mumford 2003)." This world within is the world of ideas that are solid facts "as long as people continue to regulate their actions in terms of the idea" (Mumford 2003), as people did in the light of the concepts of the utopia throughout the ages.

The origin of the word utopia refers to a place that is not, and the similar word, in Greek *eu*-topos, means a good place. So in order to be possible, utopia cannot have reachable geographical coordinates. Also, utopia cannot succumb to the passing of time (Mumford 2003). The form of this no place/good place is the city (Mumford 2003), all of its aspects—from the way it functions to the way it is built, to the form it has. The city is a projection of the way a society is regulated. Utopian city corresponds to

the picture of the world within, to which humanity aspires. It is therefore no wonder that these qualities of utopia are employed over history as a means of promoting the current ruling apparatus of a state. This paper is precisely concerned with the implications of using these utopian qualities in the forming of the state spectacle. Mechanisms of state spectacle have developed from the Hellenic and roman times and had reach the peak of their manipulative power on the observer in the time of the monarchs of the baroque epoch (Todorović 2010).

Both temporal and spatial qualities of utopias are employed as means of state spectacle. "The time of the celebration was supposed to renew the primordial, ideal time, the time of the utopia" (Todorović 2010). This was achieved through a majority of means, from constructing the specific spatial framework to carefully selecting the character and dramaturgy of the events that are to take place within it. This research will look into the spatial framework, multiple relations this framework forms with the events, and the effect it has on the observer of the spectacle. The main hypothesis it sets to prove is that Yugoslavia is actualized as a utopia in the liminal field of the observer, during the festivities of the Days of Freedom, at the location Slobodište, near Kruševac, Serbia.

Figure 1.   Sun gate of Slobodište.

In order to achieve this, this research relies on the interdisciplinary theoretical framework, mainly on the work of Jelena Todorović concerning the baroque state spectacle, and the ways through which it is connected to the utopia, and the definition of liminal space as introduced into the field of art and architecture by Miljana Zeković. Dealing with space-event, this research will also rely on the definition of the architectural chronotope as an inseparable union of the event and place, and the definition of the porosity of space, as a spatial quality enabling spatial memory. Recognized theories will serve as a ground for analysing the paradigmatic example of a realized utopia, Days of Freedom held at the monumental complex Slobodište, built by the architect Bogdan Bogdanović in 1965.

2   SETTING THE FRAME

"State spectacle as a key medium of the new age political propaganda had two basic functions—to glorify the bearer of the power and his order, and to, through the interdisciplinary union of all arts create an uncanny perfection, from the otherwise average reign" (Todorović 2010). Since they share the similar purpose, of creating an image of an ideal state, the terms of state spectacle, and utopian thought are tightly connected, in that there had always existed "a need to transcend the political event from the domain of the real, in order to celebrate the power, into the domain of the eternal, mythical and imaginary" (Todorović 2010). Creating this illusion presupposes a specific relationship to time and space, space without time, a space where time stays frozen, thus removing the event from the everyday. There are three types of time in the spectacle, Todorović claims: the now, the past time, and timelessness. The past time of the spectacle evokes a utopia, a projection of a Golden era that is recreated in the spectacle, as the achievement of the ruler. It is there to show the image of the ideal future. The represented

golden era is often closely tied to classical times in its symbols and rituals, which function as a "guarantee of the ideal moral and aesthetic values" (Todorović 2010). The time of the past is also present through the idea of the eternal circle of time, which is another characteristically utopian thought—"it supports the belief in the new utopia" (Todorović 2010). This new utopia, spectacle claims, is possible through the new government regime.

Another important aspect of the spectacle, tied to the utopian thought, is, as mentioned, the notion of space. The spatial framework supports the perception of time as belonging to utopia. There are also three key categories of ceremonial spaces (Todorović 2010). There are constant places, which in every reign become the loci of state spectacle, such as the church/temple and the residence. In the second category are the ever-changing places tied to the territory of the city, such as procession paths. Third category of ceremonial spaces are the meta ceremonial spaces that are "first and foremost a concept rather than reality" (Todorović 2010). This category is the most interesting for this research since it is the one most connected to the utopian thought. "They contain a projection of the ideal, an ellipse that moves the narrative of the spectacle in the past and in the future" (Todorović 2010). Since they exist on conceptual levels, these spaces are mainly seen as symbols, such as labyrinths and symbolic landscapes.

Using elaborated types of time and space leads to the realization of the utopia in the mind of the observer of the state spectacle. Here the connection with the notion of liminal space is established. "Liminal space is a specific spatial category emerging in the mind of the spectator facing the work of art or an artistic/performative event" (Zeković 2014). The unity of space, time, and event is therefore a precondition in the forming of the liminal space. Furthermore, "liminal space could be any type of space occurring in the real space-time continuum that provides a spectator with an illusion of the Great Idea, spanning beyond its limits" (Zeković 2014). This illusion of a Great Idea, looking from a perspective of spectacle, is actually a utopian ideal, that spectacle claims to give an illusion. Liminal space is therefore is a path to incepting the ideas of the spectacle to its observer/participant.

In the triad of space/time/event of the spectacle, and forming of liminal space is real spatial framework engaged by the event is extremely important. The attention given to building places that will serve as the representation of power is emphasized through the history. These places must possess a number of values that can easily be recognized by the observers through affecting their primordial impulses. The qualities of places that displace the spectators from the present into the past time affect precisely the utopian thought of the world within. The main quality of these places is connected to the notion of spatial memory. Spatial memory is enabled by the existence of the porosity of space as a super quality enabling the accumulation and reflection of meaning in space (Mitrović 2014). Porosity is a result of the complex relations of various functions and ambient qualities of space.

Accumulation and reflection of meaning happens when the event is introduced into a place—this event, while bringing its own pallet of values, is also influenced by the atmosphere (the reflection of meaning) of that place (Mitrović 2014). This leads to the notion of architectural chronotope. In previous research architectural chronotope was defined as "spatially and temporally limited connectedness of meaning of architectural space and an unexpected event that is introduced into that space" (Perić 2014). As such, architectural chronotope is a concept whose characteristics are dependent on both the space and the event, and is recognized as a suitable frame for exploring the connection of utopia and state spectacle.

3 EXPLORING SLOBODIŠTE

Throughout the history of Socialist Federal Republic of Yugoslavia, monuments and monument complexes have represented an important and unique element of establishing Yugoslavian identity through the built environment. As a highly narrative synthesis of architecture, sculpture, urbanism and landscape these vocal points scattered across the six republics that made up Yugoslavia, helped to establish historic, mental, emotional and cultural space, layered on top of physical spaces of the union. Form, aesthetics, materialization, and language of these works changed as the ideology behind them altered, reflecting the position of Yugoslavia on the global political scene, balancing between east and west, Soviet communism and American capitalism. Roughly, all of Yugoslavia's monuments can be divided into three periods, with determined with distinct style and ideology: the first phase that lasted from 1945 until 1948, was characterized by a strong influence of SSSR, resulting in monuments that celebrated the fallen soldiers of the Red army; second phase was marked by estrangement from Soviet ideals and focus on the battle and sacrifices of the Partisans during fight against Nazism; final stage in development of Yugoslavian monuments shows a radical change in both the narrative of the pieces and their spatial characteristics. This shift resulted in works

of art that celebrated civil victims, members of the diverse masses of Yugoslavia that laid their lives for freedom, and thus became the common ground for development of joined Yugoslavian identity. On other hand, these memorials took place outside of cities, conquering landscape, and concretized space of Yugoslavia as a whole.

One of the most impressive and significant examples of a monument from the third phase of this development is memorial complex Slobodište, designed by famous Yugoslavian architect Bogdan Bogdanović. As a memorial complex, Slobodište was built between 1960 and 1965 at a location near Kruševac in Serbia, where, during the occupation estimated 1650 people were executed. Slobodište was built in order to commemorate both soldiers and civils that lost their lives. It was conceived as a complex that will, at the same time have a commemorative, cultural, educational, and didactic role in the Yugoslavian society. Located outside of the city, Slobodište acts as a goal of a pilgrimage. Unlike the monuments built in the urban fabric of cities in Yugoslavia that allowed for everyday honouring of heroes of the resistance, Slobodište demanded an active and willing visitor, which dedicated time in order to be fully immersed in the narrative of this place. That same desire for immersion guided Bogdan Bogdanović to a form of a monument that blurs the lines between art, architecture, urbanism, and landscape. Spaces of the complex emerged as a dialogue between the historic and imaginary narrative of a heroic past and golden future that needed to be conveyed and the topology and atmosphere of the nature where monument was being built. Slobodište is divided into three parts: Vestibule with Mounds, Valley of the Living and Valley of Homage. Those three parts of the complex house several sculptural interventions, an amphitheatre, places for gathering, gardens, gates and places for meditation. These segments are all part of a horizontal plane over which the monument starches, departing from the usual architectural languages of emphasized verticals that mark places of mourning, remembering, and celebration. Architecture of Slobodište was never thought without a specific event that will take place in it, Dani Slobode (Days of Freedom). The manifestation lasted every year from 28th June until 7th July. During those days thousands artist and members of the audience gathered in this democratic theatre in order to celebrate both the fallen brothers and fathers of the resistance, but also the living sons and brothers that lived the glorious future for which the former gave their lives. The Days of Freedom consisted of national dances of all the nations of Yugoslavia, theatre plays of national and classic authors, and rituals that replaced religious ceremonies both in form

Figure 2.   Stone birds of Slobodište.

and in providing a much-needed sense of belonging to an idea greater than oneself.

As a work of art, that transcends visual and spatial arts, and theatre, Slobodište can be considered a paradigmatic example of an architectural chronotope. The full potential of the spatial frame of the monument to convince its visitor in the possibility of a socialist utopia was achieved only through the event that took place in that spatial frame every year. The periodical repetition of the event that inhabited Slobodište in annual cycles exploited the porosity of the spatial configuration in order to create a more vibrant liminal space in which utopia was incepted.

4   REACHING THE UTOPIA

If we observe Todorović's meditations on nature of relationship between state spectacle and creation of utopia (Todorović 2010), we can deduce that Slobodište, first and foremost satisfies two ground conditions, not just formally but in totality of the essence of its being. Through the language of the spatial frame and sculptural interventions that inhabit it, Slobodište stands as a fertile ground for introduction of different kinds of didactic and artistic events that take place in it during the Days of Freedom. Ideology behind these events, its dramaturgy, and aesthetics are in complete harmony with the ideology and language of the spatial frame. It is through them that the space of Slobodište comes alive, becomes a place, and serves its purpose in creating the liminal space of the observer necessary for realization of the utopian ideas.

As Zeković states:

> … liminal space is an active, dynamic, uneven and unstable space created in synergy of real architectural space, an actual artistic/performative event that takes place in this space and the ability of the spectator to feel, recognize and employ all relations that occur in the given space-time framework, leading to full experience of the artistic vision. (Zeković 2014)

With this in mind, it is no wonder that the creators of Slobodište and Days of Freedom used a pallet of means, which have symbolism, language and aesthetic values that were used regularly in the everyday life of Yugoslavia of that age. By relying on the familiar iconography, the creators of the spectacle made sure that the observers would feel connected and convinced in the ideas that the spectacle conveyed.

On the other hand, state spectacles performed during the Days of Freedom crossed temporal boundaries and mixed near past, distant Hellenic past and a glorious future of Yugoslavia that was promised to the observer. At the same time, in the same place, observers of the spectacle were exposed to national dances of all the Yugoslavian nationalities, plays that glorified Partisan battles in the Second World War, but also the staging of classical Greek plays, such as Oedipus Rex, Antigone, Prometheus, Medea, and others. Here we recognize the three types of time present in the state spectacle, as noticed by Todorović (Todorović 2010).

While Yugoslavia was a secular, atheistic society, it used iconography and dramaturgy of religious rituals to perform secular rituals that had the same strength of conviction as religious ones did in the past. The ceremonial lighting of the fire by Light bearers (Lučonoše), in addition to strong religious references offers a strong relationship with the golden heroic times of ancient Greece. With the words of the poem Letter for the Living, written for this ceremony, "light the fires of Partisans and Prometheus" the poet makes a clear connection between the ideals behind the Yugoslavian regime and the Hellenic times. Todorović also notes that the basic characteristic of the Utopia is its timelessness, its ability to exist beyond our time and space that is in the state spectacle achieved though invocation of Eden and Jerusalem (Todorović 2010). By displacing Slobodište in the landscape beyond the city, Bogdanović crated not only the Arcadia, a utopia of man and nature, but also a goal for a Biblical pilgrimage that lead the procession from the mundane reality of the city to the timeless, utopian world of Slobodište.

Together with the iconography and dramaturgy of the spectacle that took place in it, the spatial frame of Slobodište itself possesses the characteristics that Todorović emphasizes as essential for ceremonial places. In its position, morphology and decorative language Slobodište carries a great symbolic weight. It is displaced in the nature, but the demarcation of the monument and the surrounding landscape is clear through the use of stone boundaries and ceremonial gates. By crossing this boundary, the observer of the spectacle finished its transition from the reality of everyday into the time/place of the spectacle, where he was presented with the idea of Socialist utopia. In planning of the complex Bogdanović relied on strong axes and primal shapes, most of all circles, which directed movement and pointed to the places of gathering. The use of stone and it rustic finishing implied that the complex preceded its builder, that it was found as such in nature, a forgotten relic of an ideal civilization long gone. All of these spatial characteristics have strongly and clearly translated from the existential plane to the real physical space, coming very close to the idea of meta ceremonial place that Todorović argues is only possible as a concept, and is most suitable for utopia, as a place that is not (Todorović 2010).

These same spatial characteristic enable Slobodište to be a porous place. The values and meanings of each annual event bleed into the stone, and through it remain permanent in the collective consciousness of Yugoslavia. In every next Days of Freedom, the observers encounter the meanings established in all the previous spectacles that are now reflected by the same spatial frame that absorbed them one cycle ago. This circular repetition of the spectacle served as a tool for establishing collective memory and identity that were the main goals of state spectacle. The porosity of the spatial frame and circular nature of the events help to create a perfect cumulative chronotope, whose values got more interconnected and became larger with each passing cycle (Perić 2014). Such a chronotope served as an ideal spatial and temporal frame for fulfilment of the role of liminal space. It offered the observer the idea of a greater story that exceeds its boundaries (Zeković 2014). By blending different spaces and times in one real temporal and spatial frame, the chronotope of Slobodište and Days of Freedom managed to establish the liminal space of people of Yugoslavia that attended these spectacles, and placed in it the ideas of socialist Utopia. In the liminal space of the observer of Days of Freedom of Slobodište, through morphology, aesthetics and language of the space, combined with the dramaturgy and ideology of the state spectacle

Figure 3. Cenotaph in Slobodište.

Utopia became a reachable reality for dedicated brothers and sisters of Yugoslavia.

5 CONCLUSION

From its conception in ancient Greece until today, in the core of the idea of utopia stands its unreachability. As a moral and aesthetic ideal utopian idea always remain outside personal and collective human reach, giving faith in and hope for a better future, by evoking a glorious past, seductive and didactic at the same time. However unreachable as an actual and permanent state of human society glimpses of utopias can be anticipated and felt in the intersection of carefully selected states of space and time. As it was shown in this research, this fragile possibility of experiencing the state of a citizen of utopia was commonly used as a tool of representation of state power by rulers, from the Hellenic times, across baroque period, until the great socialist societies of the twentieth century. One of the societies that relied heavily on the evocation of idea of utopia in order to demonstrate and celebrate the power of state was Socialist Federal Republic Yugoslavia. This claim was explored on the case study of Slobodište memorial complex and Days of Freedom ceremonies.

Displaced from the city, located on the place of a great national tragedy, embedded in nature, but still clearly demarked from it, Slobodište served as a perfect spatial frame for placement of utopian ideas. Its morphology points to a clearly ritual purpose, while sculptural elements used in it, language of the architecture and the material suggest that it is older than it claims to be. It is built as a highly narrative place that simply waits for an event to take place in it and complete its message. Days of Freedom that took place in it combined highly ritual ceremonies, national dances and theatre plays to pay respect to the fallen, to celebrate the living, and to foreshadow a glorious future ahead. By invoking at the same time the near past of the Partisan battle, the Myth of battle of Kosovo and the golden Hellenic period these events broke down time barriers and inhabited Slobodište with past, present and future, thus making it timeless.

In the synthesis of this spatial frame and spectacle that took place in it, in the liminal space of the observers an idea of utopia was placed. During the Days of Freedom in Slobodište, the observers of this event, mostly citizens of Yugoslavia, were convinced in the existence of utopia. This perfect society was shown to them in this dream-like place, during this dream-like event. Utopia was in the past of their society, and with the current leadership, it was in their future as well. Until that future arrived all that people of Yugoslavia had was Slobodište and Days of Freedom (along with other memorial complexes and state spectacles) to give them fragments of the perfect world, that was for them real and possible and reachable in the here and the now.

BIBLIOGRAPHY

Mitrović, Jelena. "Porosity of Space: Accumulation and Reflection of Meaning introduced by Event." In *Dramatic Architectures: Places of Drama—Drama for Places Conference proceedings.* Eds. Palinhos, Jorge and Maria-Helena Maia. Porto: Centro de Estudos Arnaldo Araújo CESAP/ESAP, 2014. 310–324.

Mumford, Lewis. *The Story of Utopias.* Whitefish: Kessinger Publishing, 2003. ISBN: 0766127907.

Perić, Vladan. "Dramaturgy of Space: establishing Ephemeral Chronotope in Architecture." In *Dramatic Architectures: Places of Drama—Drama for Places Conference proceedings.* Eds. Palinhos, Jorge and Maria-Helena Maia. Porto: Centro de Estudos Arnaldo Araújo CESAP/ESAP, 2014. 337–348.

Todorović, Jelena. *Entitet u senci: mapiranje moći i državni spektakl u Karlovačkoj mitropoliji.* 1st ed. Novi Sad: Platoneum, 2010. ISBN: 978-86-85869-43-3.

Zeković, Miljana. "Liminal Space in Art: the (in)security of our own vision." In *Dramatic Architectures: Places of Drama—Drama for Places Conference proceedings.* Eds. Palinhos, Jorge and Maria-Helena Maia. Porto: Centro de Estudos Arnaldo Araújo CESAP/ESAP, 2014. 70–82.

Retaining and re-creating identity among Polish migrants in the United Kingdom

Renata Seredyńska-Abou Eid
University of Nottingham, Nottingham, UK

ABSTRACT: After Poland's accession to the European Union in 2004, the United Kingdom experienced an unprecedented immigration wave. Thousands of Polish people decided to settle in Britain with hopes of an improved quality of life. With an aim of highlighting cultural aspects of migration as opposed to statistics and economic data, this paper presents selected findings of a qualitative research study *Translating Cultures—Adapting Lives* with regard to cultural, national and social identity issues Polish migrants experience in the host culture. The concluding remarks emphasize the fact that the Polish community in the UK retains and re-creates their identity in everyday life and through well-established online communities[1].

Keywords: cultural identity, national identity, social identity, migration, Polish migrants in the UK

1 INTRODUCTION

When Poland joined the European Union in May 2004, thousands and then millions decided to migrate to other EU countries in search for a better life. The United Kingdom (UK) experiences the largest migration wave, though exact estimates of the number of Polish migrants are not available due to less than perfect data collection tools available within united Europe. Nonetheless, the 2011 Census revealed the vast scale of the Polish language spoken in England and Wales.

The unprecedented in its scale migration to Britain affected not only the British and Polish economies, but had an enormous influence on the socio-cultural aspects of the lives of both migrant and local communities. The issues of cultural values, adaptation, and identity are very often disregarded in official data and political discourse. Nonetheless, they constitute a basis for every individual who has experienced a migrated life. Identity therefore plays a significant role in migration, both for the migrant and the host community. Although individuals often have multiple identities, when crossing the borders, three types of identity become prominent, i.e. cultural, national and social identity, while the context of migration provides convenient setting for either idealising (utopia) or demonising (dystopia) the values within home and host cultures.

This paper considers the issues of cultural, national, and social identity with reference to Polish migrants who live in the East Midlands, UK. The

data was collected and observations performed within a doctoral research project *Translating Cultures—Adapting Lives* between 1st May 2013 and 31st March 2014 in the East Midlands. The mixed-methods study was designed to collect qualitative data through semi-structured interviews, a focus group, and an online and face-to-face questionnaire. All collected data was also supplemented by everyday observation of the Polish community in the East Midlands between October 2008 and March 2014. The target group were Polish post-2004 adult migrants residing in the East Midlands. Apart from individual interviews, a set of institutional interviews (e.g. with the Polish Consul) were conducted in order to establish the response of the authorities and adaptation support options that migrants could possibly consider in constructing their lives in the East Midlands.

2 IDENTITY IN MIGRATION

In the simplest terms, *identity* refers to the representation of an individual; hence, it signifies who one is. Aspects such as culture, language, nationality, religion, ideology, beliefs, a socio-political setting, etc. will further influence the *identity* of an individual and their sense of belonging. According to Castells (2004:6), *identity* as "the process of construction of meaning on the basis of a cultural attribute, or a related set of cultural attributes, that is given priority over other sources of meaning." Hence it can be seen as a "people's source of meaning and experience". Hall (1996/2002, p. 17) accentuates further that identity 'is never singular but multiply constructed across different, often intersecting and antagonistic, discourses, practices and positions.' Hence, people of the same nationality may observe different religious festivals or follow different sets of traditions. Therefore, the concept of multiple identities can reflect the complexity of an individual and collective life. This interpretation can be further extended upon migrants and their migrated lives as their identities are inevitably confronted with the host culture(s).

Furthermore, in diasporic terms, *identity* is mentioned with reference to *cultural, ethnic, social*, or *national identity*, though those categories usually represent an identity of a group. Therefore, the meaning of the term is applicable to its collective aspects. In addition to the challenge of confronting the distinguishable Other, an individual recognizes their identity within their group, with which they share values. Therefore, *cultural* and *national identity* may occupy a prominent position in migrants' representation, everyday life activities, and thought processes. Its transcendence, on the other hand, renders the migration experience unique. Hence, the extent to which maintaining and/or re-creating home identity is utopian depends on migrants' individual circumstances if utopia is understood as 'the expression of desire for a better way of living' (Levitas 2003, p. 4).

3 CULTURAL AND NATIONAL IDENTITY

With regard to the collective aspects of identity, Boski (2003, cited in Lewandowska 2009, p. 211) emphasizes the concept of cultural identity that "refers to the content of values as guiding principles, meaning of symbols and lifestyle that individuals share with others, though not necessarily within recognizable groups." Hence, although the group becomes a loose concept and a fluid requirement, it is seen as essential to enable individuals to recognize their characteristics.

Hall (2003), however, marks two types of *cultural identity*. The first kind is *stable*, associated with race and ethnicity and is understood as a shared history of individuals while the other type of identity seems to be *changeable*, that is full of contradictions and affiliated with similarities and differences. The stable type of *cultural identity* is therefore partly related to *national identity* as experience or memories constitute a common ground for the *cultural* and *national* types of identity. At the same time, Hall's changeable identity fits the situation of migrants who are amidst a complex process of adaption in the host culture and their acculturation occurs due to comparing,

contrasting, accepting, and/or rejecting either their old values or the newly encountered ones. Also White (2011, p. 10) points out that the process of migrant integration is complex and comparable to social groups with the host society; therefore, "not every aspect of migrant integration should be seen through an ethnic lens".

Moreover, identity formation is a process that is triggered by confronting the Other and enabled by the initial recognition of differences. Since "the idea[l] of a cultural norm that is ascribed to or prescribed by those occupying the boundaries of the nation-state" (Kalra, Kaur & Hutnyk, 2005, p. 30), diasporic consciousness may emerge to confront and/or idealise the notion of the nation. Hence, *cultural identity* intermingles with *national identity*. The latter may be a relatively complex term although in political understanding, it is "a subjective or internalized sense of belonging to a nation" (Huddy & Khatib, 2007, p. 63). However simplistic such a definition of *national identity* may appear, it is widely recognized by people(s) and governments.

On the other hand, the cultural and philosophical perspective includes the typology of national identity that involves more than the single aspect of belonging. Smith (1991) and Kołakowski (1995, cited in Wodak et al., 2009) hold a belief that, apart from citizenship, there are other main elements that contribute to the construction of *national identity*. First, a national spirit—a set of collective cultural forms and behaviours that are shared by a group of people and exist in their minds. Secondly, the historical memory of a group and common myths are a necessary element of *national identity* regardless of the fact whether the memory is true or only legendary. Thirdly, a national body (also called homeland) in the form of national territories, landscapes and a nameable beginning identified as a founding event contribute to the notion of *national identity*. Finally, Kołakowski (1995, cited in Wodak et al. 2009) emphasizes the importance of anticipation and future orientation as elements indispensable to the survival strategy of the group while Smith (1991, p. 14) highlights "common legal rights and duties and a common economy with territorial mobility for members". The latter seems to be much more complex in the times of the common market and free movement of people in the European Union and can, therefore, trigger the pursuit of national ideals and an elusive perfection of the home culture.

Furthermore, collective identity, a term synonymous to national identity, holds certain features of national culture. Hall (2006) in his description of a narration of national culture, in addition to Kołakowski's classification, also distinguishes the element of the national character, which includes origins, continuity, tradition, and the aspect of timelessness. That additional element is expected to be unchanging and rather uniform. Migrants can frequently refer to their country of origin in their conversations in either positive or negative way, which can prove their attachment to the sending country, which can imply potential idealising (utopia) or demonising (dystopia) of the home culture in comparison to the host one.

In addition, particular codes of behaviour, certain interpretation of the surrounding and human relations and memories of 'the old days' are shared by groups, whether social or migrant ones. Politeness can be an example that shows a certain degree of sensitivity to the issue of interpretation of actions. With reference to Polish migrants in the UK, some participants of the TCAL project emphasised that it can be disturbing to hear 'no problem' or 'it's fine' when a problem occurred. While for some Polish respondents this could be read as two-facedness, for an average British person this would be a standard polite behaviour and a symptom of a positive problem-solving attitude. Similarly, being direct in expressing criticism could be read as boldness while in Britain it could be seen as rather impolite.

Although the list of examples of misinterpreted words and behaviour as a symptom of *collective* or *national identity* may seem endless, Wodak et al. (2009, p. 27) state that "[t]he process of *national identification* is promoted by the emphasis on national uniqueness." Consequently, *collective identity* through its features of special character and distinctiveness resembles *individual identity*, though they are not alike. Smith (1991, p. 17) stresses the element of individual-collective aspects of national identity through emphasising that a sense of it "provides a powerful means of defining and locating individual selves in the world, through the prism of the collective personality and its distinctive culture."

In terms of the perception of Polish cultural values, the language, family, and religion seem to be of highest importance (Dyczewski 2002). While the language is a distinctive feature of the nation as few foreigners speak Polish, the important role of the family and religion in everyday life could be compared to other nations. The role and importance of the Polish language for migrants was also confirmed by the results of the 2011 Census, which indicated that Polish is the second spoken language in England and third in Wales. Therefore, regardless of migrants' proficiency in English, it can be assumed that a vast majority of them speak Polish at home. Furthermore, the use of the Polish language enables migrants to retain the bond with their culture of origin and allows them to develop a strong sense of belonging to the Polish community during their stay abroad regardless of the length of time they have decided to spend out of Poland. Similar observations were made in Australian studies (see Leuner 2007 and Besemeres 2007). Nonetheless, although speaking Polish can be a potentially limiting strategy for migrants who want to adapt in Britain as it may prevent them from acquiring or improving their English, the use of the native language with built-in English names, abbreviations or short phrases creates a new language that can be fully understood only by the community of Polish migrants in Britain as it is subject and place specific type of communication. Therefore, a new identity is created while the identity of origin is retained. Such a new identity might be classified as a hybrid; however, I would argue that it is simply a new, evolved identity because being a hybrid is only a temporary state while identity formation is a long-term process of constant validation, reformulation, and alteration of the system of values that results in a gradual change and formation of new structures.

Similarly, since national uniqueness and distinctiveness are important features for national and individual identity, the history of a nation, next to the language and traditions, must play an important role. Hence, Polish communities in the UK eagerly operate Polish Supplementary (Saturday) Schools for all children of Polish origin. It is a well-known case in the UK and other countries of the world where Polish migrants reside. Since the 1950s, the Polish community in the UK emphasized the need and significance of educating the younger generations about Polish history, literature and the Polish language in Polish. This type of education has been offered up to the baccalaureate level; however, in the local system of education, currently there is an opportunity to take Polish at A-Levels.

Although Polish education for migrant children in the East Midlands is available in most cities and some towns (Boston, Derby, Leicester, Lincoln, Loughborough, Mansfield, Northampton, Nottingham, Skegness and Worksop), according to the Polish Consul in Manchester, there is a noticeable deficiency of Polish Supplementary (Saturday) Schools in comparison to the number of Polish children who live in the UK with their families and would be potential beneficiaries of such schools. The Polish government supports Polish families living abroad through their cooperation with the Polish Educational Society (Polska Macierz Szkolna). Nonetheless, the Polish community initiates educational enterprises on their own, as the language is a strong unifying factor for a community. The Skegness Polish Educational Association (SPEA) is a prominent example in the East Midlands. A needs analysis exercise within the Polish community in Skegness indicated a demand for a Polish school as parents were determined for their children 'to keep the language, to keep the culture;' however, the successful completion of the school project (opened in September 2013) was due to the determination and strong will of the then president of the Association, a Pole born in the UK—a second generation migrant who received such education himself. It is therefore apparent that the level of determination within the Polish community and authorities reflects the prominent role the Polish language, culture, and history play in preserving cultural and national identity among Polish migrants in the UK.

Another way of retaining those two types of identity are traditions, celebrations, festivals, and certain typical Polish customs. Although Christmas is celebrated across European cultures, national and regional traditions differ from country to country. Hence, Polish migrants cherish Polish traditions. Polish style Christmas meetings, which are very different from British Christmas parties, are organized in local communities in December. A majority of the participants clearly stated that celebrating Christmas the Polish way, Christmas Eve in particular, is very important for them. Although some respondents mentioned the English tradition of the Boxing Day as an interesting one, they still emphasised that a family gathering and traditional Polish dishes on Christmas Eve are necessary. Notwithstanding, none of the respondents indicated a possibility of participating in festivals typical of faith other than Christian.

Furthermore, an example of maintaining Polish traditions while adapting to the social conditions and organization of life in the UK is the celebration of All Saints Day (1st November) to commemorate the late family members through lighting candles and laying flowers on family graves and the Independence Day (11th November) to celebrate regaining national independence in 1918. Since most migrants do not have family graves in the UK (only the war-time generation of migrants and their descendants), they do not cultivate the custom as such; however, they can take part in an official ceremony at the Polish Cemetery in Newark to commemorate Polish pilots who fought in the Battle of England during the Second World War (see Fig. 1). This ceremony usually takes place on the last Sunday of October or the first Sunday of November and is organized by the Polish communities, local Authorities, and the Polish Embassy.

Polish usually attend such events along with British officials, Polish combatants, and Polish migrant families. From the cultural point of view, participation is such events can be seen as sustaining cultural and national identity as attending the ceremony in Newark allows migrants to preserve and propagate their tradition while holding

Figure 1. All Saints Day at newark cemetery—Polish war graves. Source: own photograph [27 Oct 2013].

memories of the tragic events from 20th-century history of Poland.

Another aspect of maintaining national identity and distinctiveness within British society is the use of the national adjective, i.e. Polish delicatessen, Polish restaurant, Polish hair salon, Polish travel agency, etc. The first two examples refer to food and cuisine, which are always culturally unique (see Fig. 2—Kantyna: Polish Traditional Dishes). Certain products might not be available as there is no such tradition in the host culture, e.g. pickled cucumbers, sauer kraut, or pretzels. Even if equivalents can be purchased, very often the value of the products is diminished due to a difference in the place of origins. Various natural factors affect the taste of food; however, mental approaches play a role as well. Often an individual can negatively evaluate an equivalent product by saying that "It's not as tasty as ours." Rabikowska and Burrell (2009) also emphasise the importance of food, Polish cuisine and home-cooked food for Polish migrants. In their study, a majority of Polish migrants stressed that replacing Polish cuisine with local food is a physiological and emotional challenge.

The respondents of the TCAL project were asked about what dishes or kind of food they miss from Poland most. While some participants named a few dishes, e.g. cabbage and peas (PL: kapusta z grochem), stuffed cabbage leaves (PL: gołąbki) or carp in jelly (PL: karp w galarecie), many respondents felt that all Polish food or components are available in the UK. A few responses were similar to participant Q167 who stated 'Nothing I can get all Polish food in the UK' while others simply said like participant Q181: 'Nothing. If I want some [dish], I can make it' (orig. PL: 'żadnej [potrawy], bo jak chcę, to sobie zrobię') or Q185: 'Now you can make or buy everything' (orig. PL: 'teraz wszystko jest do zrobienia lub kupienia'

Figure 2. Polish fast food Kantyna in Nottingham. Source: own photograph [22 Feb 2014].

Figure 3. Tymbark, Polish juice, in Tesco. Source: own photograph [19 Feb 2014].

Although some products are available in large supermarkets, such as Tesco or Asda (see Fig. 3), local shops with a sense of Polishness can make migrants feel less uprooted and seem to be of greater cultural and emotional value to many. It needs to be pointed out here that the overuse of the adjective Polish is noticeable. Such an observation may be consistent with the theory of banal nationalism; however, in terms of the Polish cuisine, the omnipresence of the national adjective can be justified by the food and diet being significantly different from the local cuisine.

4   SOCIAL IDENTITY

Apart from the concepts of *cultural* and *national identity*, scholars have also developed the notion of social identity. According to Tajfel's theory quoted by Berry et al. (2002, p. 359), "social identity is 'that part of an individual's self-concept which derives from his knowledge of his membership in a social group (or groups), together with the value and emotional significance attached to that membership'. " Therefore, elements of cognition and evaluation that relate to social knowledge and sense of belonging play a role in understanding the concept of identity in intercultural relations (Berry et al., 2002). Moreover, Jenkins (2003) argues that there are two types of identity, namely external—involving social relationships and internal—a self-image that is signalled to the group. The process of identity formation itself involves categorization, which is the influence of the external (others) on the internal (self). This can only strengthen the identity of an existing group through the processes of resistance and reaction. Therefore, classification and categorization may effectively contribute to the formation of group identity, which seems to be a key element of communities.

The individual—collective continuum and the constant confrontation of the self and the other are unavoidable elements of the formation of identity. Jenkins (2003) explains that to define us, one

needs to split off them first. Only by contrasting them and us can the process of identity formation takes place. It seems that categorisation is inevitable although people rarely neatly fit into prescribed categories, thus, multiple identities may occur. It is worth noticing that also other aspects of life may play a role in defining who one is. Therefore, migration experience provides an ideal setting for confronting native cultural values with those of the host society. Moreover, in a new environment migrants' social identity is reshaped, reconfigured, and recreated due to their exposure to a new setting and different values. The level of retention of migrants' original values depends on their life experience in the host country and their open-mindedness. It seems that the question to what extent Polish migrants remain Polish and live the Polish way is relatively complex as the British lifestyle undoubtedly influences the attitudes of migrants to various degrees.

In order to alleviate the challenges of integration and maintain the national and cultural bond with the homeland, a number of online communities (websites and fora) were established. Numerous elements of *cultural, national*, and *social identity* can be observed among virtually active Polish migrants in the UK. Online communities seem to play a complementary role to migrants' actual everyday issues in terms of maintaining and recreating Polish identity and providing advice and support to their members. The Polish language is a unifying aspect of online communities while interaction in the native language significantly contributes to sustaining the original identity of Polish migrants and maintains the community spirit among them. What is more, the language makes online forums unique, as they are exclusive to the speakers of the Polish language; therefore, they enhance the sense of belonging to the Polish community. The latter is also maintained through historical memories of migrants that are observable in forum threads and through the classification of online members in accordance with their place of origin in Poland.

The other classification resembling the place of living in the UK, however, represents the creation of migrants' new identity in the host culture. Furthermore, *cultural* and *national identity* is maintained and recreated within a new society through cultivating Polish traditions that are incorporated to the migrants' UK lifestyle. In addition, advice on issues related to UK life of migrants serves a double purpose of helping migrants to adapt to the host culture and allowing them to reflect upon their own background.

5 CONCLUSION

With regard to the three types of identity, i.e. cultural, national, and social, it seems that Polish post-2004 migrants in the East Midlands, UK have developed mechanisms for preserving their Polishness while conforming to the local lifestyle. They achieve that through insisting on speaking Polish and teaching it to the younger generation, through cultivating Polish traditions and consistently choosing the Polish cuisine over the local one, though with exceptions. The processes of retaining and re-creating identity can be observed in everyday life of migrants and in online Polish communities. The latter play a complementary role to real-life communities; however, seem to equally efficient in their ultimate mission.

NOTE

[1] Results of the desk-top research on Polish Online Communities are available in my article 'The Role of Online Communities for Polish Migrants in the UK' in Isański, J. and Luczys, P. (eds.) *Selling One's Favourite Piano to Emigrate*. Newcastle: Cambridge Scholars, 2011.

BIBLIOGRAPHY

Berry, J.W., Poortinga, Y.H., Segall, M.H., Dasen, P.R. *Cross-Cultural Psychology*. 2nd ed. Cambridge: Cambridge University Press, 2002.

Besemeres, M. Between żal and emotional blackmail: Ways of being in Polish and English. In Mary Besemeres & Anna Wierzbicka (eds.), *Translating Lives: Living with Two Languages and Cultures*, University of Queensland Press, St Lucia Australia, 2007, pp. 128–138. Accessed at: http://site.ebrary.com/lib/uon/Doc?id=10415716&ppg=153 [25 November 2011]

Boski, P. Jarymowicz, M., Malewska-Peyre, H. *Tożsamość a odmienność kulturowa [Identity and cultural differences]*. Warsaw, Poland: Wydawnictwo Instytutu Psychologii PAN, 1992.

Burrell, K. *Moving Lives: Narratives of Nation and Migration among Europeans in Post-War Britain*. Aldershot: Ashgate, 2006.

Durovic, J. Intercultural and Ethnic Identity. *Journal of Intercultural Communication*, 2008, vol. 16. Available at: <http://www.immi.se/intercultural> [25 January 2010].

Dyczewski, L. "Values and Polish Cultural Identity." In Leon Dyczewski (ed.) *Values in the Polish Cultural Tradition. Polish Philosophical Studies, III.* Washington: Council for Research in Values and Philosophy, 2002.

Hall, S. "Cultural Identity and Diaspora." In Jana Evans Braziel and Anita Mannur (eds.) *Theorizing Diaspora*. Malden, MA: Blackwell, 2003.

Hall, S. "Who needs 'identity'?" In Paul du Gay, Jessica Evans and Peter Redman (eds.) *Identity: A Reader*. London: Sage, 1996/2002.

Huddy, L. and Khatib, N. American Patriotism, National Identity, and Political Involvement. *American Journal of Political Science*, 2007, vol. 51 (1), January, pp. 62–77. Available at: <http://www.jstor.org/stable/4122906> [12 January 2010].

Jenkins, R. Rethinking Ethnicity: Identity, Categorization, and Power. In John Stone and Routledge Dennis *Race and Ethnicity: Comparative and Theoretical Approaches*. Oxford: Blackwell, 2003.

Kalra, V., Kaur, R. and Hutnyk, J. *Diaspora and Hybridity*. London: Sage, 2005.

Leuner, B. *Migration, Multiculturalism and Language Maintenance in Australia: Polish Migration to Melbourne in the 1980s*, Peter Lang AG, International Academic Publishers, Bern, 2007.

Levitas, R. Introduction: The Elusive Idea of Utopia. *History of the Human Sciences*, vol. 16 (1), pp. 1–10, 2003.

Lewandowska, E. More Polish or More British? Identity of the Second Generation of Poles Born in Great Britain. n.d. Available at: http://ebooks.iaccp.org/xian/PDFs/4_5 Lewandowska.pdf [2nd Sept 2011].

Rabikowska, M and Burrell, K. 'The Material Worlds of Recent Polish Migrants: Transnationalism, Food, Shops, and Home.' In Kathy Burrell (ed) *Polish Migration to the UK in the New European Union*. Farnham: Ashgate, 2009.

Smith, A.D. *National Identity*. London: Penguin Books, 1991, reprinted by University of Nevada Press, 2002.

White, A. *Polish Families and Migration since EU Accession*. Bristol: Policy Press, 2011.

Wodak, R., de Cillia, R., Reisigl, M. and Liebhart, K. *The Discursive Construction of National Identity*. 2nd ed. Edinburgh: Edinburgh University Press, 2009.

Exploring the realms of utopia: Science fiction and adventure in *A red sun also rises* and *The giver*

Iolanda Ramos

Department of Modern Languages, Cultures and Literatures, FCSH-NOVA University of Lisbon, CETAPS, Lisbon, Portugal

ABSTRACT: Drawing on the parameters established by Thomas More's *Utopia* for the making of a literary tradition, this essay aims to demonstrate that a blending of genres and subgenres has contributed to the revisitation and revision of utopia for a wider audience. I will examine two novels—one published in 2012 and the other in 1993, the latter adapted to cinema in 2014—which share features of science fiction and of the adventure genre, so as to exemplify how these have borrowed from convention while presenting original developments, thus growing in popularity amongst contemporary readers.

Keywords: Canon, science fiction, adventure, utopia, dystopia

From the establishment of the utopian tradition to the present day, literary utopias have known numerous adaptations which have enriched and problematized canonical references. In addition to the work of Bloch, Mannheim, Trousson, Mumford, Morton, Jameson and Kumar, to name but a few, scholars and researchers who are engaged in the field of Utopian Studies today—Sargent, Claeys, Levitas, Moylan, Vieira, for instance—continue to add relevant contributions to the discussion of the concept, method and tradition of utopia and utopian thought.

Thomas More's *Utopia* (1516) is widely recognized as having set a new literary genre, although the text mirrors and adapts conventions and motifs from other literary forms, namely travel literature. As is known, the narrative takes the form of a long conversation between the narrator (the character Thomas More, to be distinguished from the real-life author) and the traveller (Raphael Hythlodaeus, a figure inspired by the Renaissance discoverers) who has returned from a distant land, in this case an island. In spite of being a fictional text, the traveller departs from a real place, visits a non-place where he is welcomed and assisted by the locals, and goes back home to recount what he has witnessed about the society, economy, customs and so on of this other people. In spite of a terminology that points to a discourse of nonsense (Noplace, Nopeople, Nowater), the plausibility of the story and the improbability of its realization is made clear in the famous final words: "(…) I freely admit that there are many features of the Utopian Republic which I should like—though I hardly expect—to see adopted in Europe"(More 2002: 132; Claeys 2011a; Vieira 2010).

To this extent, like two sides of a coin, the utopian perspective encloses pessimistic or negative elements within the framework of an alternative model of society. Even More's *Utopia* is ambiguous on the utopia/dystopia line if we consider, for instance, the practice of slavery and euthanasia by the Utopians. In turn, it must not be ignored that many texts that fit within the dystopian lines sought to find solutions to injustice and inequality in the existing social order—only to present a new order in which, most of the time, the principle of freedom was sacrificed for the sake of the happiness of the majority of people. In general, Orwell's *Nineteen Eighty-Four* (1949) is widely evoked as a standard dystopian novel, although it must be noted that an anti-utopian attitude goes back to the significantly titled *Mundus alter et idem* (c. 1605), a work written in Latin and published by Mercurius Britannicus, who was eventually identified as being the Anglican bishop Joseph Hall. Over the centuries, several features have permeated the dystopian literary inspiration. Well's works, in particular *The Time Machine* (1895) and *The Island of Doctor Moreau* (1896), or Verne's *De la Terre à la Lune* (1865) illustrate primary trends to be followed by science fiction and dystopian texts, while Huxley's *Brave New World* (1932), set in London but in AD 2540, highlighted the importance of uchronia in utopian/dystopian fiction.

In point of fact, one gets the impression that, while the study of utopia and utopianism is expanding, it is becoming increasingly rare, nowadays, to find an example of a utopia written in accordance with the premises established by More. From the 1970s onwards, in particular, even the texts which are closer to the original matrix have assumed a critical dimension, as assumed in the subtitle of Ursula LeGuin's famous work *The Dispossessed: An Ambiguous Utopia*, published in 1974. The acknowledgement that even the much better societies would have problems was expressed in the coinage of several designations, namely in Tom Moylan's term 'critical utopias', Lucy Sargisson's 'transgressive utopias' and Lyman Sargent's 'flawed utopias' (Sargent 2010: 30). Not only do the texts alert one to the possibility that what appears to be a utopia may in fact be a dystopia, but they require an active participation of the reader in his/her awareness of what needed to be changed so as to be made better in his/her own society. As it has been emphasized, "Utopia is no longer the construction of an ideal society but rather a tool for criticism in the present" (Lancaster 2000: 112).

Hence, on the one hand, utopia seems to be evolving increasingly closer to dystopia, especially if this is taken as what Fitting coins "the model of 'if this goes on'" (Fitting 2010: 151), which points at aspects of present-day society that get worse in the future. On the other hand, utopian texts seem to have been developing within the science fiction paradigm and the(sub)genre of adventure fiction. For this reason, I have chosen to explore two novels that can be read as something else rather than utopian novels with the realization that they in fact both revisit and revise some of the utopian premises, the main one being that utopia presents a "non-existent society described in considerable detail and normally located in time and space (...) which takes a critical view of the utopian genre", as established by Sargent (1994: 9). For the purposes of this essay, Suvin's understanding of utopia as a community where *"socio-political institutions, norms and individual relationships are organized on a more perfect principle than the author's community"* (1979: 49. Original emphasis) must be also taken into account.

Published in 2012, *A Red Sun Also Rises* was written by the English steampunk author Mark Hodder, a former BBC writer, journalist and web producer who won the 2010 Philip K. Dick Award—an acclaimed science fiction award—with the novel *The Strange Affair of Spring-Heeled Jack*. This was the first volume of his six-part Burton & Swinburne series, the final novel of which was *The Rise of the Automated Aristocrats* (2015), all set in an alternative Victorian framework. The heroes in *A Red Sun Also Rises* are neither the real-life explorer Sir Richard Francis Burton nor the poet Algernon Charles Swinburne, but Aiden Fleischer, a British missionary who has lost his faith, and

Clarissa Stark, a crippled hunchback. It must be noted, though, that Hodder has stated (2014:n.p.) that his great uncle, James Leigh, was a missionary in the 1920s. Hodder himself came into possession of a journal written by an actual Anglican priest called Aiden Mortimer Fleischer, who took vows in 1882 and left Britain to go to Papua New Guinea in 1888, the year of the Ripper killings, when he was twenty-five years old. This is the last information Hodder was able to collect about Fleischer, but the journal also evoked the Stark family. In sum, we can say that, like Thomas More, Hodder was inspired by real people and plausible events in order to build his imaginary plot.

A summary of the novel is in order so as to make a connection to utopian paradigms. The whole story is narrated by Fleischer, who in 1885 meets the deformed but also "remarkably practical, resourceful, and inventive" (Hodder 2013: 12) Clarissa Stark, with whom he can discuss the merits of several writers in Latin, French, Spanish, Portuguese and Dutch. Wrongfully accused of having abused a young woman, Fleischer and his friend Clarissa escape to London, a city divided between incomparable opulence and terrifying poverty, an "overcrowded, filthy, noisy, stinking, and vicious" (2013: 23) place, where hope has been eclipsed by despair. Fleischer's and Clarissa's wish to be both somewhere else and someone else is answered when the priest is sent to Papua New Guinea. As Clarissa puts it, "The world is a wonderful place. It will rebuild you, and you'll be a better person for it" (2013: 27).

A quest for utopia, in my opinion, is therefore envisaged in the line of aspiration for an ideal or improved state of existence. The plot makes use of the typical literary device of a journey to a distant land—in fact, after arriving in Sydney, and due to bureaucratic problems, they are sent to the island of Koluwai and the coastal town of Kutumakau. Eventually, as the outcome of a tribal ritual, they regain consciousness in a different world, inhabited not only by the Koluwaians but also by creatures that speak Koluwaian, but resemble an amalgam of mollusc and crustacean (2013: 44–45).

In accordance with most utopias written before the nineteenth century, Hodder's text is not actually set in the future. Following More's narrative strategy, *A Red Sun Also Rises* depicts an alternative society already existing in the 1880s but in this case somewhere else rather than on Britain and on Earth. The plot transports both the characters and the reader to a version of Victorian London on the imaginary planet Ptallaya, under two bright yellow suns, known as the Eyes of the Saviour. The two protagonists are outcasts from their own society who travel to a place where society is shaped by impressions taken from Clarissa's mind by the Yatsill, and they are not only welcomed by the locals but integrated into their society—for Clarissa, who has been mended and looks beautiful—becomes a member of the Council of Magicians and Fleischer is enrolled in the City Guard. The reader becomes acquainted with a peaceful, utopian-like society on this planet where the two suns never set, but eventually finds out that there is also a red sun—the Heart of Blood—which, when it rises, brings with it evil and destruction by bringing to life terrible creatures called the Blood Gods. In the subsequent confrontation, Fleischer has to face his own demons so as to deal with psychological and moral questions.

If equality is "the crucial social dogma often regarded as definitive of the utopian agenda", as asserted by Claeys, (2011b: 8) the society in Ptallaya is anti-utopian, for it is far from being equal. There is, however, an implicit critique of the world the protagonists knew, built upon the Victorian capitalist society, which confirms the social critique intrinsic to any utopian text. Although the author does not state it, I find distinctive similarities to Well's scary vision of the future based upon a class-divided society, apparently ruled by the aristocrats but where the underground Morlocks are indeed the masters.

Mark Hodder thus combines elements drawn from science fiction, adventure, fantasy, steampunk, utopia and dystopia in this novel. Although he does not mention it, the title of his novel obviously borrows from Hemingway's work of 1926, based on real people and events, and about themes such as love, death and travels. Nevertheless, Hodder's blog provides a useful range of information about the making of his novels. In 2011 he posted that *A Red Sun Also Rises* was inspired by the work of Edgar Rice Burroughs but with a steampunk flavour. In fact, he asserted that this was much more a science fiction/fantasy novel than his previous books, though the themes he explored were "very much grounded in the real world" (Hodder) On its website, Pyr, the original publisher of the work, presents *A Red Sun Also Rises* as an adventure and a tale of good and evil. The entry is placed within the field of science fiction and fantasy, seemingly directed to a young adult audience[1]. For its part, the Del Rey edition I used for this survey transcribes an interview where Hodder considers his novel to be a Victorian adventure and confirms his fascination with Victorian history, as well as his homage to the planetary romances he read as a child (Hodder 2013: 278–279). If in his blog he asserts that steampunk is, for him, "the perfect arena in which to explore socio-economic policies that seem to have spiralled farther and farther out of control" (Hodder 2012) and a way to expose both the capitalist system and the distinctions between the ever richer 'upper class' and the 'lower' ever poorer, it is not, in my

opinion, rash to argue that *A Red Sun Also Rises* possesses an intrinsic utopian impulse.

By clearly intersecting science fiction, utopia and dystopia, *The Giver* (1993) is a novel that links our hopes and fears about the future with science and technology. Set in the future, it assumes the strategy of uchronia with what can be regarded as backward time travel to the lost memories of the past. *The Giver* is indeed part of a quartet in which the central character in *Gathering Blue* (2000) is a young orphaned girl who has a deformed leg. This is noteworthy because, as mentioned above, *A Red Sun Also Rises* features the disabled Clarissa, and there are not many other novels that have as their protagonists disabled young adults who overcome physical and psychological trauma.

The American writer Lois Lowry is the author of more than forty books. She received several awards, among them the International Reading Association Children's Book Award in 1978 and the Margaret A. Edwards Award for her contribution to young adult literature in 2007, and won the 1994 Newberry Medal for *The Giver*. She declared she "sat down in 1993 to write an adventure story", (Lowry 2014a)[2] though her work was not designed specifically for young people. She actually intended to explore the importance of memory in our lives after her elderly father began to lose pieces of his memory, especially the memories which were a source of pain for him.

Although the title of Lowry's novel points to the Giver of Memory as the protagonist, in fact the hero of the story is Jonas, a twelve-year-old boy who is appointed by the Committee of Elders at the Ceremony of Twelve as the Receiver of Memory. He is the narrator of the story and, in my opinion, acts as a traveller who undertakes a journey of both self-awareness and acknowledgement of the constructed 'reality' that surrounds him. In other words, he progresses from utopia to dystopia and eventually to a different utopia. It goes without saying that the book preceded the Harry Potter series written by J. K. Rowling, starring the most famous fictional child of our times—and who, by the way, is an orphan in the line of Dicken's character Oliver Twist. In the history of the novel, there are countless examples of children who play central roles in the course of some exciting adventure. It is suffice to think about 19th century literature, in particular within the British imperialist context, to single out Robert Louis Stevenson's *Treasure Island* (1883) and Rudyard Kipling's *The Jungle Book* (1894), or even Lewis Carroll's *Alice's Adventures in Wonderland* (1865) and Frances Burnett's *The Secret Garden* (1911), not the mention, in more recent fiction, fantasy novels such as C.S. Lewis's *The Chronicles of Narnia* (1950–56) and Philip Pullman's *His Dark Materials* (1995–2000).

At first sight, the conformist world depicted in *The Giver* is not prone to excitement. Everything seems to be smooth, orderly, perfect—especially because there is no suffering, no illness, no hunger, no violence, no war. The elderly and newest members of the community are lovingly cared for, there is much laughter and joy in everyday life, everyone is given a suitable job, everyone is peaceful and happy, and the climate and topography are scientifically controlled. On the other hand, humans are genetically engineered to stop seeing colour, so the utopian/eutopian world they live in is void not only of colour, but of freedom and individuality, for there is no change of seasons, no music, no books, no art, no questioning.

It is a world ruled by the Sameness plan, where people are neither preoccupied with the burden of making decisions nor overwhelmed by the responsibility of making changes. Above all, there are no memories, either pleasant or painful. Therefore, the Giver, an old man who is a former Receiver, is assigned to keep all the memories as lessons of history and use the wisdom they give him to make decisions for the community. As Jonas is gradually given all the memories of the world, through all of the senses, he becomes aware that their way of life is based upon lies and secrets, among which is the 'release' of all of those who question the established norms. He then devises a plan to break free, to go to the place called Elsewhere beyond the community's borders and liberate the whole community. The book ends with the words: "Behind him, across vast distances of space and time, from the place he had left, he thought he heard music too. But perhaps it was only an echo" (Lowry 2014b: 224). It is left for the reader to decide if Elsewhere is an allegory or an actual place. The open ending of the book, by preserving hope within the dystopian world through a strategy of "genre blurring" (Moylan and Baccolini 2007: 14), reinforces a conception akin to a critical utopia. This is also apparent in the film adaptation (2014), especially when, at the beginning, while the viewer is confronted with the image of a suspended futuristic island, where the community lives, Jonas says: "I'm asked if I should apologise for what I did. I'll let you decide" (Noyce 2014). In other words, the viewer is invited to become involved in the outcome of the story.

Both *A Red Sun Also Rises* and *The Giver* explore features and motifs from a wide range of literary genres and subgenres, the most obvious being science fiction and adventure. In as much as the criterion of plausibility helps to specify utopia and conceive its realizability, (Claeys 2011b: 15) I believe the society portrayed in *The Giver* is indeed frightening because it goes beyond the imaginary and impossible and actually expresses a plausible future world. There is no consensual answer to the debate on whether science fiction is a part of utopia or whether utopia is a branch of science fiction. Suvin, for instance, considers that utopia is not a genre but the "sociopolitical subgenre of science fiction" (Suvin 1979: 61) and Williams addresses science fiction, utopia and dystopia as belonging to a "contemporary structure of feeling" (Milner 2010)[3]. The SF designation itself has been questioned and it has been suggested that 'speculative fiction' is both a broader definition and a closer one to the also wider domain of 'utopian fiction' (Baker 2014: 18; Claeys 2011b: 163)[24]

It can be concluded that a postmodern canonical ambiguity, hybridity or permeability enhances the potentiality of utopia, which, as mentioned above, was conceived by Thomas More as a literary form inspired by various (sub)genres. From the very beginning, even eutopian texts were devised to contain dystopian features and dystopias—regarded as "the ideal site for generic blends" (Donawerth 2003: 29)—to point to optimistic possibilities. Dystopias may be read as anti-utopias as well as satirical utopias. Not only does utopian literature borrow from various literary forms—Platonic dialogues, didactic prose, satires, travel narratives and speculative fiction, for instance—but it presents various facets, often mixing opposing impulses, from a range of eutopian expectations and potentialities to dystopian projections and fears.

Utopia is thus the result of, as much as a source of, a blending of genres and a revisitation of the canon. In spite of their obvious differences, *A Red Sun Also Rises* and *The Giver* share the aspiration to overcome difficulties that arise even in a constructed, imaginary alternative way of life. In sum, both novels fulfil the ultimate strategy that lies at the core of utopia, which is to question reality and to follow the creative impulse characteristic of human nature.

NOTES

[1] "A Red Sun Also Rises". Available: http://www.pyrsf.com/RedSunAlsoRises.html. Accessed January 2016.

[2] See also Lowry's interview "2014 ALA Annual Conference—Lois Lowry on The Giver". Available: https://www.youtube.com/watch?v = A_OWwVwUJfs. Accessed January 2016.

[3] See in particular Williams's essay "Utopia and Science Fiction". Ibid, pp. 95–112.

BIBLIOGRAPHY

"A Red Sun Also Rises." Prometheus Books. http://www.pyrsf.com/RedSunAlsoRises.html. Accessed January 2016.

Baker, Brian. *Science Fiction*. London: Palgrave, 2014. 192 p. ISBN: 978-0230228146.

Claeys, Gregory. "A Genre Defined: Thomas More's Utopia." In *In Searching for Utopia: The History of An Idea*. London: Thames & Hudson, 2011a.

——. *Searching for Utopia: The History of An Idea*. London: Thames & Hudson, 2011b. 224 p. ISBN: 978-0-500-25174-4.

Donawerth, Jane. "Genre Blending and the Critical Dystopia." In *Dark Horizons: Science Fiction and the Dystopian Imagination*. Eds. Moylan, Tom and Raffaella Baccolini. London: Routledge, 2003. 29–46 p. ISBN: 9780415966146.

Fitting, Peter. "Utopia, dystopia and science fiction." In *The Cambridge Companion to Utopiam Literature*. Ed. Claeys, Gregory. Cambridge: Cambridge University Press, 2010. 135–153. ISBN: 978-0-521-71414-3.

Hodder, Mark. *A Word from the Author*. A Red Sun Also Rises. London: Del Rey, 2014.

——. *A Red Sun Also Rises*. London: Del Rey, 2013. 288 p. ISBN: 978-0091949815.

——. "Looking Back at 2011." http://markhodder.blogspot.pt. Accessed January 2016.

Hodder, Mark. "More Thoughts on Steampunk." http://markhodder.blogspot.pt. Accessed January 2016.

Lancaster, Ashlie. "Instantiating Critical Utopia." *Utopian Studies* 11.1 (2000): 109–119.

Lowry, Lois. "A Note from the Author." In *The Giver*. London: HarperCollins, 2014a.

——. *The Giver*. London: HarperCollins, 2014b. 240 p. ISBN: 978-0007578498.

Milner, Andrew, ed. *Tenses of Imagination: Raymond Williams on Science Fiction, Utopia and Dystopia*. Vol. Peter Lang. Oxford, 2010.

More, Thomas. *Utopia*. London: Penguin Books, 2002.

Moylan, Tom, and Baccolini, Raffaella. "Introduction: Utopia as Method." In *Utopia Method Vision: The Use Value of Social Dreaming*. Eds. Moylan, Tom and Raffaella Baccolini. Ralahine Utopian Studies. Berna; Nova Iorque: Peter Lang, 2007. 13–24.

The Giver. 2014. DVD.

Sargent, Lyman Tower. *Utopianism: a very short introduction*. Oxford: Oxford University Press, 2010. iv, 145 p. ISBN: 978-0199573400.

——. "The Three Faces of Utopianism Revisited." *Utopian Studies* 5.1 (1994): 1–37.

Suvin, Darko. *Metamorphoses of science fiction: on the poetics and history of a literary genre*. New Haven: Yale University Press, 1979.

Vieira, Fátima. "The concept of Utopia." In *The Cambridge Companion to Utopian Literature*. Ed. Claeys, Gregory. Cambridge: Cambridge University Press, 2010. 3–27. ISBN: 9781139798839.

In the Cyborg city: Utopias of connectivity, fictions of erasure

Teresa Botelho

Department of Modern Languages, Cultures and Literatures, FCSH-NOVA University of Lisbon, CETAPS, Lisbon, Portugal

ABSTRACT: Understanding that the relation between urban spaces and human corporality is governed by processes of interdependence invites a meditation on the protocols of utopian projections of the "good city" when both space and body are being reconfigured by the emergence of the trope of the cyborg. This metaphor, applied to both the deterritorialized speciality of the city and to the enhanced and hyper-connected human body, points towards a replacement of the organicist concept of the city that emerged in early utopias by the idea of the urban space as a network of diffused interfaces so that imagining the future urban space posits it increasingly as a prosthetic extension of the human body. This new relation-ship has been met both with enthusiasm and anxiety, generating a range of literary and filmic techno-utopias and dystopias that interrogate visions of the enhanced body in the city. This paper will discuss how tropes of human connectivity with urban spaces have emerged in recent post-singularity filmic and narrative texts examining in particular their constructions of possible future socialized spaces.

Keywords: Enhanced bodies, cyborg cities, techno-utopias

> *The city itself was transmogrified into an ideal form—a glimpse of eternal order, a visible heaven on earth, a seat of the life abundant—in other words, utopia.*
>
> Lewis Mumford

1   CITY TECHNOTOPIAS

When Google launched in 2015 *Sidewalk Labs* (Labs), a new initiative to improve city life, the new independent start-up described its intentions to develop technology at the intersection of the physical and digital worlds, invoking a quasi-utopian rhetoric of connectivity and sharing which, unlike "other technology solutions applied to cities" which "have failed to solve real-world problems", would "collectively transform city life", making it "responsive, equitable, innovative, and human". Its principles were enunciated in the kind of benevolent, hopeful terms that shaped the early communitarian and libertarian cyber-utopianism of the 1990s, when Howard Rheingold argued that cyberspace could be "one of the informal places where people can rebuild aspects of the community that were lost when the malt shop became a mall" (Rheingold 1994: 36) and William Mitchell described the democratic potential of the future on-line *agoras*, serving a *civitas* that would no longer reside on "a suitable patch or earth" but in a virtual "soft city" (Mitchell 1994: 160). If it is

now almost impossible to read the contemporary promises of the "dot.com barons", to use the term coined by Margaret Wertheim (Wertheim 2002: 222), with the same *naïveté* that was still sustainable in the inception of the cyberage, it is no less true that the myths of progress in which they are grounded are immensely appealing to the contemporary imagination. Vincent Mosco, who has deconstructed the implications of what he has termed "the digital sublime", identifies a rhetoric of change that promises, as did many other technological developments before it, the radical transformation of society and of the urban space, and a beneficial and happiness-inducing transcendence that "animate individuals and societies" by "lifting people out of the banalities of everyday life." In fact, as he points out, it is when technologies "cease to be sublime icons of mythology and enter the prosaic world of banality—when they lose their role as sources of utopian visions—that they become important forces for social and economic change" (Mosco 2004: 6).

Technological utopias of the 19th and early 20th century were equally concerned with the harmonious reinvention of the relationship between humans and the urban space, imagining societies not only improved by new tools and technologies but modeled by them. New means of communication and transportation that were often new and untried were anticipated to reorganize the very concept of the city. One of most creative utopias of connectivity, aiming to find a solution to the plague of the crowded metropolis, may have been *Roadtown,* proposed in 1910 by Edgar Chambless. Using the tropes of the human body, "where housing and transportation are fully coordinated by Nature" since "legs are her vehicle of passenger transportation, talons and arms are her freight system and the animal body is the house", Chambless imagined a linear city built on railway tracks, housing a thousand people per mile", "surrounded by farmland" so that "one need only to go perpendicular to the town to find (or grow) food". Described as "a single unified plan for the arrangement of these three functions of civilization—production, transportation, and consumption", and a workable way of coupling housing and transportation into one mechanism "eliminating all physical, mental and moral waste", the project was expected to create a happy utopian "environment where selfishness and inequality of opportunity will gradually disappear and where man will finally enjoy all the fruits of his labor" (Chambless 1910: 20, 22, 30).

In what terms utopian underpinnings such as these, centered on the premise of human mastery of space and technology, are able to survive a predicted future where, to invoke William Gibson's coinage, the separate realms of "meat space" and cyberspace are dissolving is an interrogation that requires an examination of the cyborgization metaphors used to make sense of present and future human/urban space dynamics.

2   THE BODY IN THE CITY

Elizabeth Grosz has argued how the two most pervasive models of the interaction of bodies and cities are fraught with equivocations. One posits that "the city is a reflection, projection, or product of bodies", made by sovereign and autonomous agents; the other suggests a "parallelism or isomorphism between the human body and the city". But rather than seeing this relation as just causal or representational, Grosz suggests that "a two-way linkage that could be defined as an interface" might be a more comprehensive way to understand how these "assemblages or collections of parts" cross thresholds, define and establish each other (Grosz 1999: 382–383, 385).

The recognition of how mutually constitutive and interdependent the relationship between the city and the city dweller has been invites a number of questions: What utopian urban dreaming will be available to the enhanced human body in a deterritorialized map akin to the non-space that the 1990s utopian *Declaration of Independence of Cyberspace* envisioned as "a world that is both everywhere and nowhere", "where a Civilization of the Mind",

"more humane and fair" than anything imagined before, might be created (Barlow 1996)? How will the human body, increasingly rewritten as "post-organic assemblages of viral, genetic and bacteriological data" in the process of being continuously enhanced and "retrofitted by technology" (Shaw 2013: 779), hope to relate harmoniously with the new cartographies of the city reconfigured as a network of diffused unmapped hubs of power and sociability? If the urban space, as Matthew Gandy suggests when he asks "what happens when the human subject is increasingly merged with the city itself?" (Gandy 2005: 34), becomes a prosthetic extension of the human body, so that it makes as much sense to think of the city in the body as it does to discuss the body in the city, how can this cyborg metaphor be imagined through the tropes of utopian hope?

The type of techno-optimism that leads urban scholars like Gandy to enunciate visions of "material interface between the body and the city", materialized in the "infrastructures that link the human body to vast technological networks" which will turn urban infrastructures into "series of inter-connected life support systems" (Gandy 2005: 28), might suggest that fictionalized imaginings of such new urban spaces would be inclined to explore the utopian potentials of human connectivity of future improved urban spaces, attuned to the needs of humans with which they are so intimately merged, but in fact in many contemporary narratives the urban space becomes a phantom, overlaid with networks of knowledge that isolate rather than connect human subjects.

A singularly powerful metaphor of this absence, of not-being in the city, is invoked, for example, in David Cronenberg's film *Cosmopolis* (2012), based on Don DeLillo's eponymous novel, that traces the navigations through a territorialized and recognizable city—New York—of a young financial cyber robber baron cocooned in a limousine that is also a global communications center, hermetically sealing himself off from contact with the physical urban reality, but permanently connected by touch screen displays to the insubstantiality of global capital flow. As New York bustles with material activity—the presidential visit and the inevitable traffic jams caused by its motorcade, the funeral of a popular Sufi rap star and an anti-capitalist demonstration, all demanding attention and emphasizing the tangibility of the embodied and territorialized occupation of space—Eric Packer, though visited by a few people in his bulletproof fortress, remains emotionally isolated and detached from the city and human others but hyper-connected with the cyber information that is instrumental to the currency speculations which he half-distractedly pursues (Cronenberg 2012).

Anxieties about the present projected into possible futures have always been the drive of science fiction and one might expect to find significant contradictory reflections on the promises of the cyborgization of the human/city interface, in particular in texts informed by utopian or dystopian readings of the post-singularity motif, a term which is used here to denote what lies beyond the horizon, commonly designated as the Technological Singularity. This topos of transcendent progress, first articulated by mathematician Vernor Vinge in a foundational paper delivered to NASA in 1993 (Vinge 1993), predicts an increased acceleration of technological progress that will cause a paradigm shift shaped by the exponential growth of genetic engineering capacity, nanotechnology and robotics (the trilogy of agents of radical change commonly known by the acronym NGR). For singularity theorists this future will necessarily change our understanding of what it means to be human, a prediction that may be viewed either with disquiet, if one agrees with Vinge's warning against the possibility of predictable extrapolation, or with utopian optimism, if one's views are aligned with futurist Ray Kurzweil, currently director of engineering at Google (Kurzweil 2006). This optimism is grounded on a cluster of "interlinked and overlapping topoi" pertaining to three main utopian mythologies of the singularity discourse: the posthuman future of the body, the reconfiguration of the social and material structures of society, brought about by post-scarcity economics and adhocracy organizational models, and the belief that what is to come is so radically different from what we know that the transformation is akin to the refoundation of human history (Raulerson 2013: 37). Of these prefigurations, the future of the human body has developed a particular hold on the contemporary imagination, captivated by the consequences of the transcendence of the organic–machine divide ending in the posthuman condition where, as Katherine Hayles describes, our "coupling with intelligent machines" will be "so intense and multifaceted" that it will no longer be possible to distinguish "between the biological organism and the informational circuits in which the organism is enmeshed" (Hayles 1999: 35).

3   WHAT BODIES? WHAT CITIES?

In *Accelerando*, Charles Stross proposes a vision of that posthuman body in its relation to territorialized space. The novel maps out a through-the-singularity vertiginous journey of three generations of Manfred Macx's family which begins in Amsterdam in 2010 and ends sometime in a post-Earth multiverse in the twenty-third century (Stross 2005). This narrative acceleration leads to the creation of a post-scarcity society where goods are available to all, assembled by combinations of artificial intelligence and nanotechnology, in an extrapolation from Eric Drexler's thesis in *Engines of Creation* (Drexler 2013), where mind-uploading and body reassembling have become the norm effectively abolishing involuntary aging and dying, where Reversibility, the process by which one can back oneself up, pick different life courses and choose which works best, is widely practiced, and where the creation of group-minds and distributed intelligence and the possibility of multiple simulated concurrent existences are no longer new, all in a vortex of deconstructions of the now that mirror directly the most outlandish premises of Singularity Theory as imagined by Kurzweil.

What is significant to the present discussion is how the concept of city as a social and territorialized space is emptied of actuality as the narrative moves beyond the post-singularity horizon. In part I, *Slow Takeoff*, which spans the first three decades of our century and is centered on the story Manfred Macx, an agalmic venture altruist committed to the disruption of the property regulatory system of copyright, cities are named and recognizable— Amsterdam, Rome, Edinburgh—and their airports, squares and markets are what you would expect them to be. But the technologically enhanced Manfred is not. In a section tellingly entitled *Tourist*, which takes place in the 2030s, Manfred finds himself lost in an Edinburgh that he is no longer able to recognize autonomously. After being mugged by a thief who steals his glasses where all his cyberware is stored (including his cyber memory), the city becomes immediately unreadable, just "a brightly colored blur of fast-moving shapes augmented by defining noises" as his "ear-mounted cameras reboot repeatedly, panicking every eight hundred milliseconds, whenever they realize that they are alone on his personal area network without the comforting support of a hub to tell them where to send his incoming sensory feed" (Stross 2005: 77). "Alone in his head" and without the storage prosthetics through which all information is normally processed, his questions "who am I?", "what has happened to my memory?" and "where am I?" cannot be answered by his organic "soft machine" self any longer (2005: 84–85).

Parts 2 and 3 follow Amber, Manfred's highly-enhanced cyborg daughter who, unlike her father, cannot be robbed of her cyber prosthetics because she has "grown up with neuro-implants that feel as natural as her lungs and fingers" (2005: 122), and her son Sirhan through the next two centuries when "place" loses all connections with the stable material and socialized markers associated with the city. As Earth deteriorates and is left to the few stubborn unmodified humans who reject the abolition of aging and of death, "home" for Amber may be either the small, privately-owned asteroid where she establishes the *Inner Ring Imperium* over which she rules, or the interstellar probe spaceship where her divided downloaded consciousness and other "formally physical humans" can compose virtual scenarios to fit temporary needs. In one particular instance, wanting to impress visitors who asked for an audience, the spacecraft recreates the Parisian throne room of Charles IX, "lifted wholesale from the film La Reine Margot", along with virtual extras but is immediately dissolved into the nothingness it came from as soon as the visitors leave (2005: 176).

In the eighth decade of the third millennium, Shiran is living in the artificially built, terraformed, networked habitat of a Saturnian floating lilypad city, which recovers the amphibious self-sufficient ecopolis for climate refugees designed by Vincent Callebaut (Callebaut 2008) and the Cloud Nine airborne geodesic habitat imagined by Buckminster Fuller (Sieden 2000). The flying city is fully served by assembler portals which, using an advanced version of nanotechnology extrapolated from the Automatically Precise Manufacturing predicted by Eric Drexler in *Radical Abundance* (2013: x–xii), cater to all the material needs of its inhabitants, freeing them from work and, subsequently, from the need for localized centers of socialization. The cyborg city itself, sentient and anthropomorphized, embodied by its self-generating avatars, serves as a circuit of information, interacting with its post-human residents, responding to questions and wishes that do not even have to be voiced.

Significantly, although readers are told that "post-humans are gregarious" and prefer not to live on the edge of the flying city, "huddling instead close to its hub" (Stross 2005: 271), free from material necessities and of the cooperation that work entails, they are very rarely seen using any public spaces to engage with each other beyond the family networks that persist, although highly modified since aging and dying have been superseded. This suggests that, emptied of their social multi-valences and significations, these ambiguous utopian cyborg spaces have been imagined as erasures, mere plastic projections of individual desires, emptied of all the markers of the democratic agora predicted by cyberutopians.

A similar theme of isolation generated by connectivity gives shape to the plot of *The Surrogates* (Mostow 2009), based on a graphic novel by Robert Venditti and Brett Weldele (Venditti and Weldele 2006). The plot imagines a generalized physical retreat from the city as people are replaced in all their dealings with the outside world by robotic human surrogates, which, as VSI, the company that produces them, advertises, "combine the dura-

bility of a machine with the grace and beauty of the human body". The origin of these androids is told in the opening credits of the film—engineered from the technology first applied in the development of intelligently powered bionic exoskeletons that enhanced the mobility of paraplegics, then used by the military to avoid human casualties in combat, they became mass produced and affordable. Their use, the viewer is told, has made the city streets safer, and eliminated violence, discrimination and other social evils.

Unlike traditional robots, these units have no autonomy as they are connected to and dependent on the conscious processes of their users who stay at home all day, sitting in their stem chairs, living vicariously through the younger, better-looking and fitter androids that their minds control. By means of the androids they go to work, walk the streets, amuse themselves and make love without ever exposing their real bodies to outside others, living, as the extensively quoted company publicity claims, "the life you always dreamed of without any risk or danger to yourself". Only at home, when the androids are recharging (though there are public recharging stations accessible everywhere), can their owners be their old *meatbag* selves. There are of course a few dissenters who reject this manufactured "utopia" living in reservations, surrogate-free zones, but in the city no one seen or met is really there.

The dystopian ethos of the narrative assumes an anti-technological telos here, symbolized by the appeal to a return to the natural voiced by the leader of the anti-surrogates (who turns out to be a replicant of their creator bent on destroying the monstrous technology he devised): "Look at yourselves. Unplug from your chairs, get up and look in the mirror. What you see is how God made you. We're not meant to experience the world through a machine."

It seems evident that these two narratives, coming from opposite ends of the utopian technofile/ dystopian techophobe spectrum, imagining either future cyborg cities that virtually merge with the post-human body, or the retreat from the city and the public space of cyber-connected real bodies, highlight the social isolation that may be fostered by the hyperconnected future. The contrast between the optimism of the promises of a world as William J. Mitchell describes in *Me++*, "of less rigid, more fluid and flexible relations—of knowledge to action, of shape to materials, and of people to places", marked by the dissolving of the virtual/ physical boundary (Mitchell 2003: 5), and its fictional projections suggests a puzzling imaginative gap. It may be that images of cities turning into deterritorialized and dematerialized entities, in a process that Gandy has called "the end of location" (2005: 36), may block the imaginative paths to the reconfiguration of future cyborgian public spaces that may survive the diminished role of the physical city. Fredric Jameson predicted a long time ago that accelerated technological progress might lead to the "atrophy of the utopian imagination" and to an incapacity to consider the future except in terms of "representations which prove, on closer inspection, to be structurally and constitutively impoverished" (Jameson 1982: 155). While this indictment may apply to both technological futurists and artists or writers, anxious fictions about the city, which seem incapable of conceiving the results of the transcendence of what we know now except through discourses of disengagement, erasure or apocalyptic estrangement, invite a question: if, as Lewis Mumford once memorably claimed, "the first utopia was the city itself" (1965: 271), will utopian dreaming overcome the nostalgia for the organic and post-organic physical city and find a way to project visions of the rearranged symbiotic body/city dynamics as vivid and compelling for their contemporary readers as Campanella's *City of the Sun* or Bellamy's year 2000 Boston once were?

BIBLIOGRAPHY

Barlow, John Perry. "A Declaration of the Independence of Cyberspace." https://projects.eff.org/~barlow/Declaration-Final.html Accessed April 2016.

Callebaut, Vincent. "Lilypad, a Floating Ecopolis For Climate Refugees." *Oceans World* (2008).

Chambless, Edgar. *Roadtown*. Melbourne: Leopold Classic Library, 1910.

Cronenberg, David *Cosmopolis*. 2012.

Drexler, Eric K. *Radical Abundance: How a Revolution in Nanotechnology will Change Civilization*. New York: Publicaffairs, 2013.

Gandy, Matthew. "Cyborg Urbanization: Complexity and Monstrosity in the Contemporary City." *International Journal of Urban and Regional Research* 29.1 (2005).

Grosz, Elizabeth. "Bodies-Cities." In *Feminist Theory and the Body: A Reader*. Eds. Price, Janet and Margrit Shildrick. New York: Routledge, 1999.

Hayles, Katherine N. *How we became Posthuman: Virtual Bodies in Cybernetics, Literature and Informatics*. Chicago: University of Chicago Press, 1999.

Jameson, Fredric. "Progress versus Utopia; or, Can We Imagine the Future." *Science Fiction Studies* vol. 9.27 (1982): 1–10.

Kurzweil, Ray. *The Singularity is Here: When Humans Transcend Biology*. London: Duckworth, 2006.

Labs, Sidewalk. "Why Now? http://sidewalkinc.com/ (viewed 5 April 2016)." http://sidewalkinc.com/ Acessed April 2016.

Mitchell, William. *City of Bits: Space, Place and the Infobhan*. Cambridge: MIT Press, 1994.

Mitchell, William J. *The Cyborg Self and the Networked City*. Cambridge: MIT Press, 2003.

Mosco, Vincent. *The Digital Sublime: Myth, Power and Cyberspace*. Cambridge: MIT Press, 2004.

Mostow, Jonathan *The Surrogates*. 2009.

Mumford, Lewis. "Utopia, the City and the Machine." *Daedalus* 94.2 (1965).

Raulerson, Joshua. *Singularities: Technoculture, Transhumanism and Science Fiction in the Twentieth Century*. Liverpool: Liverpool University Press, 2013.

Rheingold, Howard. *The Virtual Community: Homesteading on the Electronic Frontier*. New York: Harper Perennial, 1994.

Shaw, Debra Benita. "Strange zones: Science fiction, fantasy and the posthuman city." *City* 17.6 (2013).

Sieden, Lloyd Steven. *Buckminster Fuller's Universe*. New York: Perseus Publishing, 2000.

Stross, Charles. *Accelerando*. London: Orbit, 2005.

Venditti, Robert, and Weldele, Brett. *The Surrogates*. Marietta, GA: Top Shelf Productions, 2006.

Vinge, Vernor. "The Coming Technological Singularity. How to Survive in the Post-Human Era." *Address to VISION 21 Symposium* (1993).

Wertheim, Margaret. "Internet Dreaming: A Utopia for all Seasons." In *Prefiguring Cyberculture: An Intellectual History*. Eds. Jonson, Annemarie and Allessio Cavallero. Cambridge: MIT Press, 2002.

Committees

ANA CRISTINA
GUERREIRO
Senior Researcher,
(CIAUD) Centro
de Investigação
em Arquitetura,
Urbanismo
e Design—
Faculdade
de Arquitetura—
Universidade
de Lisboa, Portugal

Professor at the Faculty of Architecture, Technical University of Lisbon. She holds a Ph.D. in Visual Communication by the same Faculty and a Post-degree in History of Art by the FCSH, Nova University of Lisbon. She is graduated in Fine Arts, by the Faculty of Fine Arts in Lisbon. She works and researches in Fine Arts and Theory of Art, including the relation between Arts and Architecture.

ANA ISABEL
BUESCU
History
Department/
CHAM, FCSH,
Universidade
NOVA de Lisboa,
Universidade
dos Açores, Portugal

Position Associate Professor History Department Faculdade de Ciências Sociais e Humanas Universidade Nova de Lisboa.

anabuescu@netcabo.pt
www.fcsh.unl.pt/deps/historia
http://www.cham.fcsh.unl.pt/files/file_000992.pdf

Scientific domains & interests (15th-16th centuries) Portuguese History. Education of princes. Court culture. Royal ceremonies. Royal and aristocratic libraries. Biographical History.
Books (since 2005) Author:

2011—Na Corte dos Reis de Portugal. Saberes, ritos e memórias. Estudos sobre o Século XVI, Lisboa, Colibri, 266 pp., 1st ed., 2010
2008—D. João III (1502–1557), 2nd ed. Lisboa, Temas e Debates, 416 p.
2007—Catarina de Áustria (1507–1578) Infanta de Tordesilhas, Rainha de Portugal, Lisboa, A Esfera dos Livros, 491 p.
2005—D. João III (1502–1557), Lisboa, Círculo de Leitores, 344 p.Indexed Articles (since 2013)

- "Os Santos na Corte de D. João III e de D. Catarina", Lusitania Sacra, 2a série, 28 (Jul.-Dez.), 2013, pp. 49–72. ISSN: 0076-1508 | e ISSN: 2182-8822 ISBN: 978-972-8361-55-6, http://www.ft.lisboa.ucp.pt/resources/Documentos/CEHR/Pub/LS/2/Lusitania_Tomo28_indice.pdf
- "Utopia e profetismo no século das Luzes. D. José, Príncipe do Brasil, Imperador do Mundo", Revista de História da Sociedade e da Cultura (Coimbra), 13, 2013, pp. 283–310. ISSN: 1645-2259
- "Dimensão política e de poder da comida régia e do corpo do rei", Librosdelacorte.es (publicação electrónica www.librosdelacorte.es), 7, año 5, Otoño-invierno 2013 ISSN 1989-6425. http://www.librosdelacorte.es
- "Livros em castelhano na livraria de D. Teodósio I (c.1510–1563), 5o duque de Bragança", Revista de
Estudios Humanísticos. Historia (Univ. Léon), 12, 2013, pp. 105–126. ISSN: 1696-0300
- "A livraria de D. Teodósio (1510?-1563), duque de Bragança. A sua dimensão numa perspectiva comparada", Ler História, vol. 65, 2013, pp. 59–73. ISSN: 0870-6182

Current Research (Projects)

- De Todas as Partes do Mundo, O património do 5.o Duque de Bragança, D. Teodósio I. (PTDC/EAT-HAH/098461/2008), coord. Jessica Hallett/CHAM, FCSH/NOVA-Uac http://www.cham.fcsh.unl.pt/teodosio/ (researcher)

ANA MARIA
MARTINHO
Senior Researcher,
CHAM, FCSH,
Universidade
NOVA de Lisboa,
Universidade
dos Açores, Portugal

Ana Maria Martinho holds her faculty position at FCSH—Universidade Nova de Lisboa since 1989. From 2007 through 2010, she was at UC Berkeley, where she undertook a range of interdepartmental activities, bridging links between Africa and the Portuguese-speaking world. She spent a sabbatical at SOAS, London University, in 2011, strengthening her institutional links with Lusophone Africa. She is also a visiting faculty member at other universities, notably in Angola, Cabo Verde, and Mozambique.

She earned her Agregação in 2006, and her doctoral degree, in 1998, both at Universidade Nova de Lisboa, Portugal.

Ana Maria Martinho Gale has won awards and fellowships from the Hellman Family Faculty Fund, the Instituto Camões, Fundação Luso-Americana para o Desenvolvimento, the European Union, and the Science Ministry in Portugal.

She has supervised postgraduate students across the world (including in the United States, Angola, Portugal, Cabo Verde and Mozambique). A number of her postgraduate students have attained positions in academia and politics.

Having been involved in over fifty international missions, she visited multiple times all five Lusophone African countries. She also has established academic links with Brazil, South Africa, Argentina, Venezuela, Botswana, Canada, and the United States, having lectured extensively in many of these locations.

She serves on the editorial board of 4 international academic publications. Since 1990, she has authored 71 publications; books, editions, book chapters, international papers.

Her current research focuses on one monograph project: Material Culture and Literature in Sub-Saharan Africa, as well as on 2 papers for peer-reviewed publications.

ANA MARTA
FELICIANO
Senior Researcher,
(CIAUD) Centro
de Investigação
em Arquitetura,
Urbanismo
e Design—
Faculdade
de Arquitetura—
Universidade
de Lisboa, Portugal

Born in Torres Vedras, Degree in Architecture from the Faculty of Architecture of the Technical University of Lisbon in 1995, worked in the City Hall of Torres Vedras between 1996 and 1997, also exercising professional activity in Architecture and Urbanism.

In 2001 obtained the Master Degree in Housing Architecture from the Faculty of Architecture of the Technical University of Lisbon, with the dissertation "Habitação e Utopia nos Anos Sessenta; As Propostas do Grupo Archigram no Contexto de uma Década de Rupturas".

Between 2003 and 2007 hold a scholarship from the Foundation for Science and Technology and attended the PhD course Teoría y Practica del Proyecto in the Polytechnic University de Madrid. Obtained the European Doctorate in Architecture in 2008, with the the the thesis "La Metáfora del Organismo en las Arquitecturas Visionarias de los Años Sessenta; La Obra del Grupo Archigram como Reinvención de un Nuevo Habitar".

Since 1997, is professor and researcher at the Faculty of Architecture of the University of Lisbon, teaching theoretical and practical disciplines of Architecture Project.

Writes on Art and Architecture. Books:

– *A Metáfora do organism nas Arquitecturas dos Anos Sessenta; a Obra dos Archigram como manifesto de um novo habitar* (ed. 2014; ISBN 978-989-658-264-7).
– *A Casa Senhorial como Matriz da Territorialidade; a Região de Torres Vedras entre o Tempo Medieval e o final do Antigo Regime* (ed. 2015; ISBN 978-989-658-335-4).

ANA PAULA
AVELAR
Senior Researcher,
CHAM, FCSH,
Universidade
NOVA de Lisboa,
Universidade
dos Açores, Portugal

Associate Professor at Universidade Aberta (Portuguese Open University), researcher at CHAM. She has integrated several national and international projects, subsidized by the European Union. She is the author of several books and essays on History Studies, Asian Studies and Portuguese Culture, some of them published in indexed magazines. Among her books stand out *Fernão Lopes de Castanheda, cronista do governador Nuno da Cunha?* (Cosmos, 1999), *Visões do Oriente—formas de sentir do Portugal do século XVI* (Colibri, 2002), *Figurações da Alteridade na cronística da Expansão* (UAb, 2003), *D. João III- O Piedoso* (APH, 2009), *D. Luísa de Gusmão- A rainha mãe* (APH, 2011). She teaches undergraduate and graduate courses, and supervises numerous thesis and dissertations. She has been Visiting Professor at several universities, and she is a member of national scientific academies. She is working now at the critical edition of Fernão Lopes de Castanheda's History.

ANNA
KALEWSKA
Associate Professor
at the Institute
of Iberian and
Ibero-American
Studies, Warsaw
University, Portugal
Correspondent
researcher at CHAM,
FCSH, Universidade
NOVA de Lisboa,
Universidade
dos Açores, Portugal

Professor Anna Kalewska, Ph.D. (1996) and Habilitation degree holder (2006), is Full Professor in the Institute of Iberian and Ibero-American Studies of Warsaw University, Poland. Her major research interests are: Portuguese Literature, Lusophone Theatre, Comparative Literature, Theory and Methodology of Literary and Cultural Studies in Portuguese. Major works (books of her own): *Camões, czyli tryumf epiki* (Camões, or the triumph of the epic), 1999 and *Baltasar Dias e as metamorfoses do discurso dramatúrgico em Portugal e nas Ilhas de São Tomé e Príncipe. Ensaio histórico-literário e antropológico* (Baltasar Dias and the metamorphoses of the dramaturgical discourse in Portugal and in the Islands of Saint Thomas and Principe. A historical-literary and anthropological approach), 2005, both published by Warsaw University Press. In 2002, she published her translation of *The Return of Caravels* (As Naus) by António Lobo Antunes. She also published about two hundred articles, major and small translations, essays, reviews, some literary works, printed in Polish, Portuguese, and English (in Poland, Portugal, Brazil and some other countries). Member of ALP (Associação de Lusitanistas Polacos), CompaRes (International Society for Iberian-Slavonic Studies in Lisbon), NETCCON (UFRJ in Brazil), and A.I.L (Associação Internacional de Lusitanistas—International Society of Lusitanists). Lives in Warsaw. Sworn translator of Portuguese. a.kalewska@uw.edu.pl http://iberystyka-uw.home.pl

ANTÓNIO LEITE
Senior Researcher,
(CIAUD) Centro
de Investigação
em Arquitetura,
Urbanismo
e Design—
Faculdade de
Arquitetura—
Universidade
de Lisboa, Portugal

Degree in Architecture from the Faculty of Architecture of the Technical University of Lisbon in 1995, professional exercise since 1995 being responsible for several projects of Architecture and Urbanism.

In 2001 received the degree of Master of Housing Architecture in the Architecture Faculty of the Technical University of Lisbon. Between 2003 and 2007, was research fellow of Foundation for Science and Technology, obtaining in 2008 the degree of 'European Doctor' conferred by Polytechnic University de Madrid, with the thesis "La Casa Romántica; de la Matriz Romántica a un Concepto Acrónico y Operativo en la Contemporaneidad".

Is professor and researcher at the Faculty of Architecture of the University of Lisbon since 1996/1997, teaching theoretical and practical disciplines of Architecture Project, taught project workshops of Urban Planning and Architecture in seminars in Portugal and abroad.

Writes for Art and Architecture publications, of which stand out the books *A Casa Romântica; uma Matriz para a Contemporaneidade* (ed. 2014; ISBN 978-989-658-263-0)) and *A Casa Senhorial como Matriz da Territorialidade; a Região de Torres Vedras entre o Tempo Medieval e o final do Antigo Regime* (ed. 2015; ISBN 978-989-658-335-4).

CARLOS COELHO
Senior Researcher, Centro de Investigação em Arquitetura, Urbanismo e Design—Faculdade de Arquitetura—Universidade de Lisboa, Portugal

Born in 1962, Full Professor of Urbanism at the Faculty of Architecture of the University of Lisbon graduated in 1984 in Architecture at the Technical University of Lisbon and PhD at the same institution in 2002 with the thesis *The Complexity of the Irregular Urban Greeds*. Teaches Design Studio classes at the FAUL where he is part of the Forma Urbis Lab research group. Has been invited to teach in other institutions in Europe and Japan and Leader research on Urban Morphology. Author of many books, his academic activity and works in which he participated were recognised by different awards.

CARLOS EDUARDO BERRIEL
Universidade de Campinas S. Paulo, Brasil

Carlos Berriel is professor at the University of Campinas, in Brazil, editor of the academic journal *Morus—Utopia e Renascimento*, currently in the 10th issue, and coordinator of *U-TOPOS-* Centro de Estudos sobre Utopias. He held postdoctoral researches at the Università La Sapienza (Rome, 1996/7), Università degli Studi di Firenze (2006–7) and was Visiting Professor at the Dipartimento della Scienza dello Stato of the Università di Firenze (2015). He is the author of several papers on Brazilian literature and utopias. He is presently preparing an edition of the *Città del sole*, by Tommaso Campanella.

FÁTIMA VIEIRA
Associate Professor (with "Agregação") at Faculdade de Letras da Universidade do Porto/CETAPS, Portugal

She has been teaching since 1986. She is the Chairperson of the Utopian Studies Society /Europe since 2006. At ILCML—Instituto de Literatura Comparada Margarida Losa, she was the Coordinator for the three editions of the research project "Literary Utopias and Utopian Thought: Portuguese Culture and the Western Literary Tradition" (2001–2010) funded by FCT (Fundação para a Ciência e a Tecnologia). Now she is the Coordinator of the multidisciplinary research project (also funded by the FCT) "Utopia, Food and the Future: Utopian Thinking and the Construction of Inclusive Societies—A Contribution from the Humanities". Fátima Vieira is also the director of the collection "Nova Biblioteca das Utopias", of the Portuguese publishing house "Afrontamento", and the director of *E-topia*, an electronic journal on Portuguese utopianism since 2004.

(http://ler.letras.up.pt/site/default.aspx?qry=id05id164&sum=sim), as well as of *Spaces of Utopia*, a transdisciplinary electronic journal on Utopia written in English (since 2006)—http://ler.letras.up.pt/site/default.aspx?qry=id05id174&sum=sim). She is also Book Review Editor for the North-American Journal *Utopian Studies* (edited by Penn State) since 2006.

In 2013, the American and Canadian association Society for Utopian Studies selected Fátima Vieira as a recipient of the Larry E. Hough Distinguished Service Award.

FERNANDO MOREIRA DA SILVA
Senior Researcher, Director of (CIAUD) Centro de Investigação em Arquitetura, Urbanismo e Design—Faculdade de Arquitetura—Universidade de Lisboa, Portugal

Full Professor in Design at the Faculty of Architecture of the University of Lisbon (FA/ULisboa). MSc and PhD—University of Salford, UK; PhD—Technical University of Lisbon; Post-Doctorate from the University of Salford; President of CIAUD—Research Centre for Architecture, Urbanism and Design; Coordinator of the Design Scientific Area, Coordinator of the Masters and PhD degree in Design at FA/ULisboa; Member of the General Council of the University of Lisboa; Panel coordinator for the PhD and post-doc Scholarships in Design, Architecture and Urbanism at FCT; International consultant for CNPq, Brazil; Honorary Researcher at SURFACE, UK; Co-coordinator of the International PhD degree in Design and Innovation (general coordination of the University of Naples); Visiting Professor at several foreign universities in Brazil, Italy and Spain; Supervisor of 78 Master Dissertations and 48 PhD theses; Member of Scientific Committees of international scientific journals; Coordination and participation in funded scientific research projects; Publications in indexed scientific journals with peer review in data basis such as Scopus and ISI; several book chapters and three books, published by editors such as Caleidoscopio or Taylor & Francis.

IOLANDA RAMOS Senior Researcher, CETAPS, Universidade Nova de Lisboa—Faculdade de Ciências Sociais e Humanas, Portugal	Professor at Universidade Nova de Lisboa, Portugal. She has published extensively within the field of Victorian Studies, mainly on political, economic and gender issues under the framework of Cultural Studies and Utopian Studies. Her doctoral thesis on Ruskin's social and political thought was published by the Gulbenkian Foundation in 2002. She is the author of *Matrizes Culturais: Notas para Um Estudo da Era Vitoriana* (Colibri, 2014) and co-editor of *Performing Identities and Utopias of Belonging* (Cambridge Scholars Publishing, 2013). Her essay "'Tell me what you like, and I'll tell you what you are'—An Overview of the Victorian Political Economy of Art" (2003) was published in *Cahiers Victoriens et Édouardiens*, a journal indexed in Scopus. She is a member of the Editorial Board of the online and double blind-refereed journals *The Eighth Lamp*: *Ruskin Studies Today* and Spaces of Utopia: An Electronic Journal. She has been carrying out research as part of the project "Mapping Dreams: British and North-American Utopianism" within the CETAPS (Centre for English, Translation and Anglo-Portuguese Studies).
JORGE BOUERI Faculty Member, Faculdade de Arquitetura e Urbanismo, Universidade de São Paulo, Brazil	Architect, Urbanist and Designer. Full Professor at the University of São Paulo, Brazil. Visiting Professor at Osaka City University , 2005, and the University of Lisbon, 2014–2015. He wrote four books on architecture dimension and housing. He held several positions in the Public Administration of cities and produced several works of Urban Planning and Design.
JOÃO CABRAL Senior Researcher, (CIAUD) Centro de Investigação em Arquitetura, Urbanismo e Design—Faculdade de Arquitetura— Universidade de Lisboa, Portugal	He is architect (Escola Superior de Belas Artes, Lisbon), Honours Diploma Urban and Regional Development Planning (Architectural Association, London), PhD (Urban and Regional Studies, University of Sussex, U.K.) is associate professor at the Faculty of Architecture (FA), University of Lisbon and researcher at the Research Centre for Architecture, Urbanism and Design. At FA he is currently president of the Department of Arts, Humanities and Social Sciences. He is also the coordinator of the "Colégio dos Urbanistas" at "Ordem dos Arquitectos". His research interests and teaching activities range from urban planning, policies and methodologies to regional development, territorial governance, and spatial planning systems. He has been involved in research projects on governance and policies in urban regions. Recent relevant publications include, Crespo, J. L. & Cabral J. (2012) "The institutional dimension to urban governance and territorial management in the Lisbon metropolitan area" in Seixas J. & Albet A. (ed.) *Urban Governance in Southern Europe*, Ashgate; and Cabral, J. (2015) Portugal, *disP—The Planning Review*, 51, Taylor & Francis.
JOÃO LUÍS LISBOA Senior Researcher, Vice—Director of CHAM, FCSH, Universidade NOVA de Lisboa, Universidade dos Açores, Portugal	PhD in History and Civilization, European University Institute, Florence, (*Mots (dits) écrits. Formes et valeurs de la diffusion des idées au 18ème siècle au Portugal*). Researcher at the Portuguese Centre for Global History (CHAM) 2001–2002, Director of the Portuguese Institute for the Book and Libraries, Ministry of Culture 2004–2014, director of the Centre for the History of Culture. Fields of research: intellectual History, History of the book and reading, Modern European culture.

Some recent publications:

2016: "From publishing to the publisher—Portugal and the changes in the world of print in the 19th century" in Abreu and Silva (eds.), *The Cultural Revolution of the 19th century: Theatre, the Book-trade and Reading in the Transatlantic World*, London, I.B.Tauris Publ. (69–86 and 260–261).

2015: "Read, watch and laugh (with 18th century humorous books)" in Ferrão and Bernardo (eds.) *Views on Eighteenth Century Culture: Design, Books and Ideas*, Cambridge, CSP (346-357).

2011: "Ideas in(to) facts" in Irimia and Ivana (eds.), *Author(ity) and the Canon between institutionalization and questioning: Literature from high to late modernity*, Bucharest, ICR (144-152).

2009: "Facts being...", *Storia della storiografia*, 55, pp. 3–28.

JORGE FIRMINO
NUNES
Researcher,
(CIAUD) Centro
de Investigação
em Arquitetura,
Urbanismo e
Design—Faculdade
de Arquitetura—
Universidade de
Lisboa, Portugal
He graduated from the Faculty of Architecture of Lisbon Technical University in 1993, where he obtained his PhD in Architecture in 2012, with the thesis *A Razão da Arquitectura: A Ideia de Metrópole nos Pensamentos de Manfredo Tafuri e Rem Koolhaas*. He teaches at the Department of Arts, Humanities, and Social Sciences of the Faculty of Architecture of Lisbon University. Projects Editor of Journal *Arquitectos* (2002–2005 e 2009–2012). He was part of the editorial team of Ordem dos Arquitectos between 2005 and 2007. Coordinator, with Ana Vaz Milheiro and João Afonso, for the exhibition Habitar Portugal 2003–2005 (2006 Architecture Venice Biennale). Researcher of the project Arquitecturas do Mar at the Faculty of Architecture of Lisbon University, supported by the Foundation for Science and Technology.

LUÍS
BERNARDO
Senior Researcher,
CHAM, FCSH,
Universidade
NOVA de Lisboa,
Universidade
dos Açores, Portugal
He is a Professor in the Department of Philosophy of NOVA University of Lisbon's Faculty of Social and Human Sciences and a Researcher at the Portuguese Centre for Global History (CHAM). He has dedicated his research to interpreting themes and authors that bridge the philosophy of culture and the philosophy of knowledge. An expert in Eric Weil's philosophy, he is particularly interested in understanding how texts in the eighteenth and twentieth centuries defined the direction of modernity. He has published several books, articles and translations on different aspects of these textualities.

Some publications:

- Leonor Ferrão and Luís Manuel A. V. Bernardo (eds.), *Views on Eighteenth Century Culture—Design, Books and Ideas*, Newcastle upon Tyne, Cambridge Scholars Publishing, 2015, 408 pp. (ISBN 978-1-4438-8100-5).
- "Reasons of Violence, Violence of Reason: an Interpretation based on Eric Weil's Core Paradox", Diogo Pires Aurélio/João Tiago Proença (ed.), *Terrorism: Politics, Religion, Literature*, Newcastle upon Tyne, Cambridge Scholars Publishing, 2011, pp. 35–67 (ISBN [10]: 1-4438-2708-8; [13]: 978-1-4438-2708-9).
- "Martinho de Mendonça: un représentant des Lumières portugaises", Armelle St. Martin/Sante Viselli (dir.), *Les Lumières au-delà des Alpes et des Pyrénées—communications, transferts et échanges*, Paris, Hermann, 2013, pp. 123–150 (ISBN 978 2 7056 8689 5).
- "La nouvelle du tremblement de terre est arrivée à Königsberg: les écrits de Kant sur l'évènement", *Revue de métaphysique et de morale*, 2013/2 (N° 78), Paris, PUF, pp. 185–213 (ISSN: 0035-1571).
- "Une première dernière œuvre: quelques réflexions à propos des Pensées sur l'interprétation de la nature de Diderot", Bruno Petey-Girard/Marie-Emmanuelle Plagnol-Diéval (dir.), *Première Œuvre, dernière œuvre: écarts d'une écriture—Actes du colloque international*, Paris, Classiques Garnier, 2014, pp. 143–171 (ISBN 978-2-8124-2130-3).

LUÍS CRESPO ANDRADE Senior Researcher, CHAM, FCSH, Universidade NOVA de Lisboa, Universidade dos Açores, Portugal	Professor in the Department of Philosophy of the Faculdade de Ciências Sociais e Humanas of the Universidade Nova de Lisboa. Coordinator of the Early-Modern and Modern Thought Group of the Portuguese Centre for Global History (CHAM). Coordinator of the Seminário Livre de História das Ideias. Author of numerous studies on utopian thought and intellectual history, namely *Intellectuals, Utopia e Comunismo. A inscrição do marxismo na cultura portuguesa* (Lisbon, Fundação Calouste Gulbenkian—Fundação para a Ciência e a Tecnologia, 2010).

MARIA JOÃO DURÃO
Researcher, (CIAUD) Centro de Investigação em Arquitetura, Urbanismo e Design—Faculdade de Arquitetura— Universidade de Lisboa/Laboratório da Cor/ColourLab FAUL Coordinator

Prof. Maria João Durão, has a Diploma in Fine Arts, Faculty of Fine Arts, Lisbon University; obtained Ph.D. degree from The University of Salford, Manchester, UK (2000) funded by PRODEP; was granted equivalence to PhD in Scientific Area of Desenho e Comunicação Visual, Faculdade de Arquitectura de Lisboa, Universidade Técnica de Lisboa. Her post-Doctoral project funded by FCT- Foundation for Science and Technology: "Art, Science and Technology -Drawings of Santiago Calatrava, hosted at University of Salford Research Centre BUHU, Manchester, UK and Department of Architecture "Technion-Israel Institute of Technology Haifa", Israel (2009). She lectures courses of Drawing, and Architectural Drawing at Faculty of Architecture Lisbon since 2001, where she holds a wide range of scientific responsibilities within postgraduate courses:

Member of Scientific Board of the Doctoral Course of Design, since 2006; Master and Doctoral Programme Modules: "Anthropology of Vision" (2003-present), and the nuclear module "Seminar of Thesis Project". She set up 'Master Degree in Colour for Architecture', Faculty of Architecture-Lisbon Technical University (2003) holding its scientific coordination. In 2004, she founded Colour Lab-Faculty of Architecture where she supervises doctoral and post-doctoral research projects. In 2016, she became coordinator of Colour and Light Research Group –CIAUD—Centre of Research of Architecture, Urbanism and Design. Co-founder of the Portuguese Colour Association (2003): elected President (re-elected 2009) and presently its Honorary President. Internationally elected twice for Colour Association (re-elected 2015). Vice-President/Portugal- International Association of Colour Consultants/Designers, presided by Frank Mahnke, Geneve (2003).

Maria João Durão has a wide range of international conference participations and scientific committee/peer review invitations.

Her research interests in Space Architecture were established in 2002: Member of Team 11 and co-author of *The Millennium Charter—Fundamental Principles for Space Architecture* Houston, Texas, USA (2002). She is a senior member of AIAA- American Institute of Aeronautics and Astronautics and its Aerospace Architecture Subcommittee (Design Engineering); and invited member "Human Factors Subcommittee", Space Architecture Committee, AIAA-Aerospace Sciences Meeting, Nevada (2007). Presently her research is linked to the Education Subcommittee of the Space Architecture Technical Committee -American Institute of Aeronautics and Astronautics.

She publishes mainly internationally and serves as editor or scientific coordinator in *The Portuguese Colour Association Newsletter* (2003–2010); *FABRI-KART*, Universidad del País Vasco (2003-present); Scientific coordinator *Cor, Espaço Urbano, Arquitectura e Design*, Ed. Apcor & Robbialac (2013); ARCH-INEWS -02, *Revista de Arquitectura, Urbanismo e Interiores*. Coordinator of *Light and Colour-Interdisciplinary and Transversal Phenomenon* (2012). Lately, she was nominated Editor-in-Chief of the *AEROSPACE AMERICA Journal* for the yearly article of the Space Architecture Technical Committee. ISSN: 0740-722X (2014; 2015; 2016).

MARIA LEONOR GARCIA DA CRUZ
Professor and Researcher at Faculty of Arts and Humanities of the University of Lisbon, Portugal

Professor and Researcher at Faculty of Arts and Humanities of the University of Lisbon-FLUL. PhD. in Early Modern History (UL 1999).

Researcher in Early Modern History and European overseas Expansion, Maghreb, Asia and America, supervising master thesis, PhD. and postdoctoral dissertations in societies, mind-sets, institutions, political and financial management, representations.

Member of the Research Units Building and Connecting Empires and Cultural Encounters and Intersecting Societies at the History Centre UL, she supervises research projects in IMAGETICS: interdisciplinary studies on representations and identitarian constructions, and TREASURY: history of economic thought and management, taxation, social networks, politics and ethics.

Publications in RCAAP and in international journals indexed as *Vozes dos Vales* (Qualis-Capes, Latindex) and *O Ideário Patrimonial* (DRJI, Latindex).

https://lisboa.academia.edu/MariaLeonorGarciaCruz/CurriculumVitae

MARIA JOÃO PEREIRA NETO
Researcher, (CIAUD) Centro de Investigação em Arquitetura, Urbanismo e Design—Faculdade de Arquitetura—Universidade de Lisboa/CHAM, FCSH, Universidade NOVA de Lisboa, Universidade dos Açores, Portugal

PhD History, Master Sociology.

Professor at Faculty of Architecture University of Lisbon. Member of the Department of Arts, Humanities, and Social Sciences. Main domains of teaching and research: applied Social Sciences, Humanities, Art and Architecture History, Design, Scenography, Heritage. Since 1978 scientific research experience and professional consulting in the domains of Social Sciences and Humanities. Member of the Scientific Board Faculty of Architecture (2008–2011; 2014...). Senator of The University of Lisbon, elected for the Scientific Board (2013- 2017). Effective member CIAUD—Research Centre for Architecture, Urbanism and Design faculty of Architecture, University of Lisbon, associated member of CHAM—The Portuguese Centre for Global History—FCSH Universidade NOVA de Lisboa. Member of the Geographical Society—Lisbon, president of the Arts and Literature Section, and in 2016 also the Section of Heritage Studies, vowel of the sections: Tourism, Communication Studies, Education and Ethnology. Author of several indexed texts in Isis, Scopus, Web of Science

MARIA DO ROSÁRIO MONTEIRO
CHAM, FCSH, Universidade NOVA de Lisboa, Universidade dos Açores, Portugal

Professor of Comparative Literature at the FCSH/UNL (New University of Lisbon). Graduated in Modern Languages and Literatures in the University of Lisbon (1983), Master in Comparative Literary Studies by the Universidade Nova de Lisboa (1987) and PhD in Literary Sciences, specialty of Comparative Literature, by Universidade Nova de Lisboa (1997). Lectures Comparative Culture and Literature in graduate and postgraduate levels. Senior Researcher at CHAM. Editor of several books, author of the first Portuguese academic book on Tolkien and author of several published essays on Utopia. Further information: Maria do Rosário Monteiro https://fcsh-unl. academia.edu/MariadoRos%C3%A0rioMonteiro

MARIA DO ROSÁRIO PIMENTEL
Researcher, CHAM, FCSH, Universidade NOVA de Lisboa, Universidade dos Açores, Portugal

Maria do Rosário PIMENTEL graduated in History at Universidade de Coimbra and was awarded her PhD in Portuguese Studies at Universidade Nova de Lisboa. She is currently Associate Professor at Universidade Nova de Lisboa. She has carried out activities predominantly in the area of Cultural History and Society with a special focus on Cultural topics, Slavery and the Slave Trade. She has published, among other works and papers, *Viagem ao Fundo das Consciências. A Escravatura na Época Moderna*. Lisboa, Colibri, 1995 (338 pages), which is a version of her PhD thesis, and *Chão de Sombras. Estudos sobre Escravatura*, Lisboa, Colibri, 2010 (200 pages).

She is currently researcher in the fields of Portuguese Culture and History of Slavery and is senior researcher at CHAM. Member of, Sociedade Portuguesa de Estudos do Século XVIII, and its international branch, Sociedade Internacional de Estudos do séc. XVIII, as well as with the Portuguese Committee for the UNESCO Project Slave route. Author of two monographies on slavery, editor of 5 proceedings on Portuguese Culture and Literature, and more than 20 articles published in national and international journals.

MARIA TERESA VASCONCELOS E SÁ
Researcher, (CIAUD) Centro de Investigação em Arquitetura, Urbanismo e Design—Faculdade de Arquitetura—Universidade de Lisboa, Portugal

She has a Degree in Sociology at ISCTE-IUL, Lisbon University Institute, MA in the field Urban and Regional Planning at University of Lisbon, PhD in Sociology of work at ISCTE-IUL, Lisbon University Institute. She is Presently Phd professor at FAUTL, teaching disciplines in the field of sociology and anthropology to architecture courses, Urban Planning and Design. PREVIOUSLY TAUGHT at Autonomous University in the Departments of Sociology.

His main research areas are The relationship between the social sciences and architecture –urbanism; the participation of the population in developing the architectural project; The authors' study in the field of CS that bring contributions to the analysis of the territory, as Henri Lefebvre.

MÁRIO SAY MING KONG
Researcher, (CIAUD) Centro de Investigação em Arquitetura, Urbanismo e Design—Faculdade de Arquitetura—Universidade de Lisboa/CHAM, FCSH, Universidade NOVA de Lisboa, Universidade dos Açores, Portugal

Degree in Architecture AT FAUTL, PhD in Architecture in the field of drawing and visual communication at Escuela Superior Technical Architecture Barcelona—Universidad Politécnica de Cataluña (UPC-ETSAB). He is presently Phd Professor at FAUTL, Lecturer in ESELx and Visiting Professor at the Master Course in arts at ESBAL.

Previously taught at Lusophona University (ULHT) and Independent University, respectively in the Departments of Urban Planning and. Architecture.

In 2000 was Coordinator of the first year of the course in Urban Planning ULHT In 1998 Regent of the discipline Design/CAD/Geometry in ULHT.

He participated in scientific research studies and consulting work for outside entities. He has also participated in several publications, communications Indexed to TAYLER & FRANCIS and ISI, and training courses in order to disseminate the results of his research activities at national and international universities.

His main research areas are: Harmony and proportion in representation between West and East and its application to Sustainable Architecture, in particular by applying concepts of origami and Kirigami to materials such as paper and bamboo.

PAULO ALCOBIA SIMÕES
Researcher of Universidade Federal do Ceará, Brazil

He has graduation at Communication Design de by Escola Superior de Belas Artes de Lisboa (1989), master in Business Administration, specialization in strategic management by INDEG-IUL ISCTE (1995) and Ph.D. at communication Design by Faculdade de Belas Artes da Universidade de Lisboa (2012). Currently is teacher and investigator of Universidade Federal do Ceará. Has experience in the area of usability of diagrams, with emphasis on information design.

PAULO PEREIRA
Researcher,
(CIAUD) Centro
de Investigação
em Arquitetura,
Urbanismo e
Design—Faculdade
de Arquitetura—
Universidade de
Lisboa, Portugal

Art and architecture historian, Master in Art History (Univerdidade Nova de Lisboa), PhD in History of Architecture (Faculdade de Arquitectura da Universidade de Lisboa) with the thesis *A Fabrica Medieval. Concpeção e construção na arquitectura portuguesa, 1150–1550* (2012)

Invited lecturer in several academic meetings and seminars in Portugal, Spain, France (CNRS), Germany, Italy, U.S.A (George Washington University) and Brazil (USP). Curator of several exhibitions in Portugal and abroad Vice-President of IPPAR (National Heritage) between 1995 e 2002. Editor and author of *História de Arte Portuguesa*, (3 vols, Círculo de Leitores). Author of *Lugares Mágicos de Portugal* (8 vols., Círculo de Leitores, 2004–2005. Co-curator of the international exhibtion *Neue Welten* (Novos Mundos), held in the Deutsche Historisches Museum, Berlin (2007/2008). Author of *Arte Portuguesa. História Essencial* (Temas & Debates, 2012) and *Decifrar a Arte Portuguesa* (6 vols Círculo de Leitores, 2014) Teacher in Faculdade de Arquitectura da Universidade de Lisboa.

RUI ZINK
Researcher, Writer,
Professor DEP,
FCSH, Universidade
NOVA de Lisboa,
Portugal

Rui Zink (Lisbon, 1961) is a Portuguese fiction writer, as well as responsible for the Editing and Publishing disciplines in the M.A. in Text Editing in Universidade Nova de Lisboa. His Ph.D (1998) was on "*Comics as potential literature*", and he has been interested on alternative forms of writing in several media, as well as mainstream fiction. In 2001, he wrote the e-novel *Os Surfistas*, with the help and not-so-much-help from hundreds of readers who accepted to be co-writers. In 2005, his novel *Dádiva Divina* received Portugal's Pen Club award. His work is translated into several languages.

TERESA BOTELHO
Researcher,
CETAPS, DLLM,
FCSH, Universidade
Nova de Lisboa,
Portugal

She is Associate Professor at The Faculty of Social and Human Sciences, Nova University of Lisbon, where she teaches American Studies. She has published extensively on African American, and Asian American culture and literature, theatre and drama. Her current interests include technological utopias/dystopias and the post-human, post-black literature, identity theory in its intersection with utopia, visual culture and cinema, the collaboration between sciences and literature, especially drama, and literary and visual representations of 9/11 with a special emphasis on Arab-American literature and televisual fiction.

Her most recent publications include:

- "Utopía y poder en Estados Unidos: Una unión más perfecta: utopías de pluralidad en los Estados Unidos." *Utopía y poder en Europa y América*. Ed. Moisés González García and Rafael Herrera Guillén. Madrid: Tecnos Editorial, 2015 pp. 251–268
- "Reimagining the Body in Post-Singularity Techno-Utopias" *Spaces of Utopia: An Electronic Journal*, 2nd series, no. 3, 2014 pp. 70–83
- *Performing Identities and Utopias of Belonging*. Ed. Teresa Botelho/Iolanda Ramos. Newcastle: Cambridge Scholars Publishing, 2013
- "Dystopias of Belonging: Defamiliarizing Sameness in Danzy Senna's *Symptomatic*" *Performing Identities and Utopias of Belonging*. Ed. Teresa Botelho/Iolanda Ramos. Newcastle: Cambridge Scholars Publishing, 2013 pp. 52–68
- "Finding an Aesthetic of her Own: Partnering Identities in the Work of Faith Ringgold" *Women and the Arts: Dialogues in Female Creativity*. Ed. Diana V. Almeida. Bern: Peter Lang, 2013 pp. 109–124
- "Becoming Tony Alfama: The American Education of a Portuguese Kid in Charles Reis Felix' Tony: A New England Boyhood." *Gávea-Brown: A Bilingual Journal of Portuguese-American Studies* Vol. XXXIV–XXXV, 2013 pp. 131–146.

Peer review

Scholars have been invited, through an international CFP, to submit researches on theoretical and methodological aspects related to Utopia in the scientific fields of Architecture, Industrial Design, and Landscape, Arts and Humanities and present their technical and theoretical research results carried out on the Congress core theme: Utopia(s); Worlds and Frontiers of the Imagination.

All full paper proposals were subjected to double blind peer review, distributed according to each paper scientific area to senior researchers for avaluation. All accepted papers were invited for oral presentation and publication as chapters of a volume with the same title as the Conference.

Conference report

Abstracts and full papers accepted from:

Brazil, Greece, Italy, Netherlands, Poland, Portugal, Romania, Serbia, Switzerland, Sweden, Ukraine, United Kingdom, United States of America, and Turkey

Sponsors

Publication sponsored by the strategic project of CHAM, FCSH, Universidade NOVA de Lisboa, Universidade dos Açores, which is funded by the Fundação para a Ciência e a Tecnologia—UID/HIS/04666/2013.

Author index